MODERN HOME GARDENING

Life before agriculture (and gardening). Along and within the borders of Botswana, Angola, and Namibia (South West Africa) live the !Kung San, who until recently were a semi-nomadic, hunting and gathering people. Here a !Kung woman gatherer digs at the base of a *sha* vine for its turnip-flavored bulbs, an important part of the !Kung diet (Courtesy of Marjorie Shostak/Anthro Photo.)

MODERN HOME GARDENING

Clyde L. Calvin
Portland State University

Donald M. Knutson
Portland State University

John Wiley & Sons

New York • *Chichester* • *Brisbane* • *Toronto* • *Singapore*

Covery by: Phil McKenna

Library of Congress Cataloging in Publication Data:

Calvin, Clyde L., 1934-
    Modern home gardening.

    Includes indexes.
    1. Vegetable gardening.   I. Knutson, Donald M.
II. Title.
SB321.C24   1983   635   82-15978
ISBN 0-471-02486-4

Printed in the United States of America

10 9 8 7 6 5 4 3 2 1

PREFACE

Modern Home Gardening is intended for use by beginning college students, and others at the same educational level. While parts of the text are relatively detailed, we have simplified complicated terms, concepts, and processes whenever possible. The book can be read without difficulty by those having no specialized training in botany and in the agricultural sciences. Many well-chosen illustrations are used. Significantly, all the illustrations pertain directly to garden plants. Many have never before been published.

The first part of the book deals with the *principles* of gardening. In this part there is a comprehensive coverage of plants, soils, and climates—those basic elements with which the gardener must deal. The second half pertains to the *practices* of gardening—the manipulation of the basic elements. Here the growing of important vegetables, herbs, perennial food plants, and flowers is discussed in detail.

We planned the book to fill the need we see for an introductory college textbook on vegetable gardening. A majority of community colleges, colleges, and universities offer gardening courses. While a large number of popular books on gardening are available, none is suitable for an introductory college text. Most of these books are incomplete and deal with gardening in a "cookbook" approach. The scientific texts in the horticultural area present problems too. Most are either too broad in scope, too technical, or both. They deal with such topics as commercial production, marketing, storage, world crops, turf crops, landscape design, floriculture, and so on. One popular text illustrates the problem, stating, "The term *vegetable gardening*, formerly used, no longer defines commercial vegetable growing since a large part of the production of the major crops is conducted on a large scale as a specialized type of farming rather than as gardening." We feel that by focusing on the principles and practices of "backyard" vegetable production we overcome the objectionable features of both the popular and technical gardening books.

Over the years two major philosophies of gardening have emerged: organic and chemical. We try throughout the text to discuss fairly the advantages and disadvantages of each approach. Ultimately we encourage readers, on the basis of their newly acquired gardening knowledge, to form their own philosophies of gardening.

A valuable feature of *Modern Home Gardening* is the appendix material containing: (1) summary of plant classification, (2) conversion factors, (3) sources of agricultural information, (4) climatic data for selected locations within Canada and the United States, (5) a listing of commercial seed companies that specialize in mail order sales, and, (6) nutritional guide to some common garden crops. Here the gardener can find a complete list of vegetables belonging to a given plant family, how many square feet are in an acre, the nearest place to get soil tested, growing season precipitation in New York City, mail order seed companies that specialize in vegetables for northern climates, the vitamin C content of a normal serving of tomatoes (raw or cooked),

v

and much, much more. The appendixes make our book not only an excellent source of information on gardening, but also a valuable reference book on those detailed, hard-to-find aspects of gardening.

All gardening authors are faced with one especially challenging problem, that of *regionality*. Each region has its own unique gardening practices and gardening problems. Obviously, we cannot deal with a multitude of details in a text of limited size. We have, however, dealt directly with the problem through several approaches. First, the principles material presented in early chapters can be applied broadly to all gardening situations. Second, we have traveled widely throughout Canada and the United States acquainting ourselves with many facets of the gardening experience. Third, we have drawn information from many regional gardening publications to ensure applicability in all sections of the country. Finally, we have included as guest authors three outstanding horticulturists, Drs. Wesley P. Judkins, Mary Tingley Compton, and George Whiting. These three discuss their subjects authoritatively from a broad historic and geographical perspective.

Many people have contributed in various ways in the preparation of our manuscript. We wish to acknowledge particularly Dr. John Ambler, Dr. Gordon Kilgour, Dr. Joann Loehr, Dr. Richard Tocher, and Dr. John Walls for critical reviews of specific chapters. Also, Carol Anne Wilson, who read the manuscript in its entirety and made many helpful suggestions. Others have made notable contributions too. Peter Chan and Richard Null, skilled gardeners and teachers, made valuable suggestions on organization and content. Ron Sera of the British Columbia Department of Agriculture has graciously provided us with data on sources of agricultural information for Canada. Elizabeth R. Wallmann compiled and organized our appendix tables. Sharon Swanson, Mary Dozark, Karen Bennett, and Amy Rowe toiled unswervingly with the many clerical aspects of manuscript preparation. Martha Brookes and Tawny Blinn provided invaluable assistance in proofreading of galleys. Finally, we thank the many home gardeners who have shared their gardens and gardening experiences with us.

Clyde L. Calvin
Donald M. Knutson

CONTENTS

MODERN HOME GARDENING

1

INTRODUCTION

*"And the Lord God planted a garden eastward in Eden . . .
and. . . . took the man, and put him into the garden . . . to
dress it and to keep it."*

Genesis, Chapter 2

Gardening in a practical sense encompasses only the act. Gardening in a strict scientific sense encompasses not only the act, but the bases for acting as well. Gardening in an artistic sense covers not only the act and the bases for acting, but instills in the "actor" a love and appreciation—a genuine feeling—for his or her actions. We wish to pursue the subject of gardening in the broad sense. We begin by briefly examining the origins of agriculture in general, and gardening in particular.

ORIGINS OF AGRICULTURE AND GARDENING

Archeological evidence indicates that humans have roamed the earth for thousands of years.

The sparse human population appears to have been divided into extended family groups that might include up to 20 persons. People subsisted in this preagricultural phase by hunting, fishing, and collecting edible wild plants and animal products such as honey. One can visualize that a great deal of time each day was needed for food procurement and that all capable members of a group participated in this activity with the men doing the bulk of the hunting and fishing and the women, with children in tow, collecting the edible plant and animal products (see frontispiece). This early type of subsistence is technically called a **hunting and gathering society**. As we will see, it was drastically and irreversibly altered—for much of humankind—about 6000 to 10,000 years ago.

The enlightenment that brought an end to

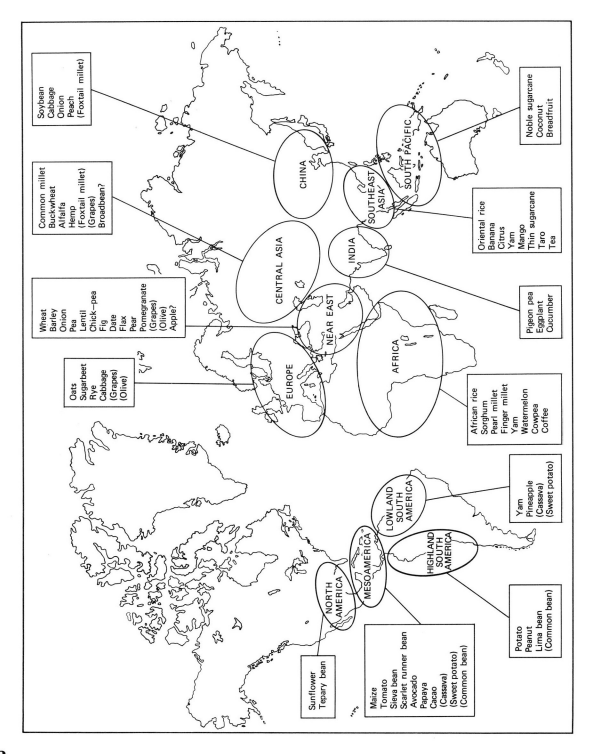

Soybean
Cabbage
Onion
Peach
(Foxtail millet)

Common millet
Buckwheat
Alfalfa
Hemp
(Foxtail millet)
(Grapes)
Broadbean?

Noble sugarcane
Coconut
Breadfruit

Oriental rice
Banana
Citrus
Yam
Mango
Thin sugarcane
Taro
Tea

Wheat
Barley
Onion
Pea
Lentil
Chick--pea
Fig
Date
Flax
Pear
Pomegranate
(Grapes)
(Olive)
Apple?

Pigeon pea
Eggplant
Cucumber

Oats
Sugarbeet
Rye
Cabbage
(Grapes)
(Olive)

African rice
Sorghum
Pearl millet
Finger millet
Yam
Watermelon
Cowpea
Coffee

CHINA

SOUTH PACIFIC

SOUTHEAST
ASIA'

CENTRAL ASIA

INDIA

NEAR EAST

EUROPE

AFRICA

Yam
Pineapple
(Cassava)
(Sweet potato)

Potato
Peanut
Lima bean
(Common bean)

LOWLAND
SOUTH
AMERICA

HIGHLAND
SOUTH
AMERICA

NORTH
AMERICA

MESOAMERICA

Sunflower
Tepary bean

Maize
Tomato
Sieva bean
Scarlet runner bean
Avocado
Papaya
Cacao
(Cassava)
(Sweet potato)
(Common bean)

this long-established societal form was the realization that plants could be manipulated for food purposes to the advantage of people. These early efforts on the part of humans to care for plants represent primitive forms of **agriculture**. The implications of the manipulation of plants for agricultural purposes were enormous. For example, a nomadic existence was no longer necessary, plant improvement through selection was made possible, the domestication of animals for food began, and the formation of agricultural communities that were perhaps 10 to 20 times as large as the nomadic groups was made possible. In these larger groups specialization of labor could occur and trades such as those of the weaver and the potter developed.

We do not wish to imply that the concept of deliberate plant cultivation was one that came suddenly to the mind of some primitive. To the contrary, the evidence that has accumulated in recent years, particularly in the past decade, tends to suggest an almost opposite view. In the words of the American agronomist, Jack Harlan,

> Agriculture is not an invention or a discovery and is not as revolutionary as we had thought; furthermore, it was adopted slowly and with reluctance. . . . The current evidence indicates that agriculture evolved through an extension and intensification of what people had already been doing a long time.

The first attempts at plant cultivation were no doubt feeble and yields were low and uncertain compared to present day standards.

Nevertheless, the agricultural revolution had started, and in the broadest sense so had *gardening*.

The development of agriculture had an additional profound effect upon humankind. Improved agricultural skills eventually made possible large gatherings of persons not directly engaged in food production—that is, cities were formed. Thus the modern **industrial society** was born.

The cultivation of plants appears to have begun at an early time in each of several areas of human activity around the world (Fig. 1-1). These areas were somewhat isolated from the others through much of their cultural development and consequently many cultivated plants were already highly modified by plant selection and cultural manipulations—that is, they had become **domesticated**—before they were transported to other parts of the world. In the case of potatoes, tomatoes, and grapefruit, for example, widespread distribution has been a recent event, occurring in post-Columbian times. Each of the areas of plant domestication possessed different native plant populations and therefore different vegetable, cereal, and fruit crops developed, as illustrated in Fig. 1-1. Brief descriptions of two of these areas of plant domestication are given in the following paragraphs.

The Near East offered unique opportunities for the domestication of crop plants, their number constituting an impressive list (Fig. 1-1). In this region three continents—Europe, Africa, and Asia—and many cultures merge

FIGURE 1-1 Areas where plants were domesticated are indicated on this map; area boundaries have been generalized. Except for wheat, where different genera or species were independently domesticated in different areas, the plant name appears in each area: onion, yams, and millets are examples. Where the same species was certainly or probably domesticated independently the name appears enclosed in parentheses in each area; among the examples are the common bean, the sweet potato, olive, and grape. Where the area of domesication is in doubt a question mark follows the plant name. (From Jack Harlan, The Plants and Animals that Nourish Man. Copyright © Sept. 1976 by *Scientific American, Inc.* All rights reserved.)

together, as shown in Fig. 1-2. Climatic conditions favor growth of a great diversity of plant species. An area known as the Fertile Crescent (darkened in Fig. 1-2) contains some of the oldest remnants of an agricultural society. Kernels of wheat have been uncovered in the ruins of Jarmo, a thriving city in 6000 B.C. Barley was also domesticated in the region. Important vegetable crops from the Near East are the onion, pea, and lentil, while important fruits would include the fig, date, pear,

and pomegranate. Last, but certainly not least, the all-American favorite—the apple—is thought to have first appeared in the hilly land around the Black and Caspian Seas (Fig. 1-2).

Cultivation and domestication of crops in the Americas probably had its beginnings in Central America and the northwestern part of South America, Mesoamerica, and Highland South America (Fig. 1-1). Principal among New World crops is maize, or corn as Amer-

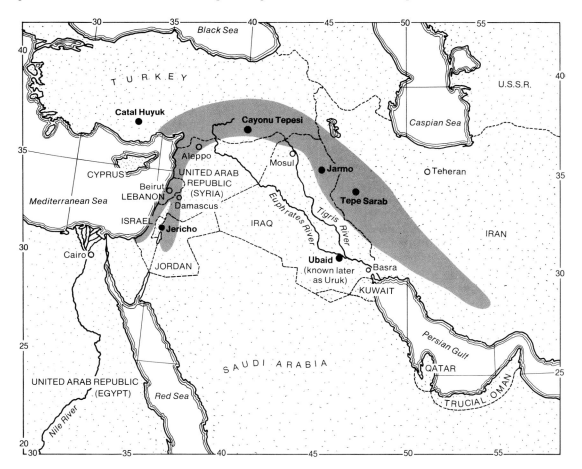

FIGURE 1-2 Near East region showing the Fertile Crescent (darkened) where certain cereals and garden crops are believed to have been first domesticated. (Reprinted with permission of Macmillan Publishing Co., Inc. from A. S. Boughey, *Man and the Environment.* Copyright © 1971 by Macmillan Publishing Co., Inc.)

icans call it, whose cultivation extends far back in prehistory. The old Anglo-Saxon word "corn" means grain of any kind, and, except in the United States, it does not refer specifically to the Indian corn, *Zea mays*. Maize apparently went through its early development in the Andes, probably in southern Peru. Later it appears that the peoples from the Andes carried their primitive maize to Central America where it hybridized with certain closely related grass plants. Products of these hybridizations later spread northward into Mexico. In caves near Tehuacan, Mexico, a series of corn ears has been unearthed (Fig. 1-3) that shows the increase in cob size from less than one inch long to more than six inches in length over a period of about 6000 years. One can almost imagine that the ears came from ancient gardens that may have graced the entrances of Tehuacanian dwellings.

Regardless of the source of the corn ears,

FIGURE 1-3 Increase in size of maize ears between 5000 B.C. and A.D. 1500 at Tehuacan, Mexico. The oldest cob (at left) is slightly less than one inch long. (From Douglas S. Byer, *The Prehistory of the Tehuacan Valley.* Copyright 1967 by University of Texas Press. Used with permission of the Robert S. Peabody Foundation for Archeology, Andover, Mass.)

the accomplishment of crop improvement, through selection, is dramatically illustrated. The Tehuacan people doubtless saved the seeds of the biggest and best corn plants to grow again and again. Many moderns are surprised to learn that crop improvement was underway so long ago. In the words of the horticulturist Dr. Edward Borchers,

> Many people consider plant breeding, the method by which crop varieties are improved, to be an intricate, complicated and perhaps mysterious process understood only by highly trained scientists. In reality, however, the basic idea is very simple and consciously or unconsciously has been practised . . . by man for thousands of years . . . Perhaps the simplest form of crop improvement is mass selection. A . . . population is merely examined plant by plant and all individuals that do not possess the desired characteristics are removed. The remaining individuals are then permitted to produce seed from which another population is grown that again is subjected to selection.

Precisely this sort of thing occurred at Tehuacan.

Some evidence suggests that deliberate cultivation in the Americas predated the development of corn. In an arid area, in the Mexican state of Tamaulipas, gourds, squashes, beans, and chili peppers have been found at levels dated as between 7000 and 5500 B.C. And in the Tehuacan area squash, avocado, chili pepper, and amaranth, as well as corn, have been unearthed that date back to about 5000 B.C. More recent garden crops in Tehuacan include the guava, pineapple, and peanut. These plants are of particular interest, as the peanut is definitely South American, and the pineapple is believed to have originated in Brazil. The presence of these plants in Mexico provides evidence that the peoples of the area established contact with South America at an early date.

Interestingly, the region now within the continental United States is the source of very few of our domesticated plants (Fig. 1-1). Perhaps you can think of reasons as to why this is so. Only one, the sunflower, ranks among the world's 30 most important crops. Another, the Jerusalem artichoke, is prized by many gardeners for its fleshy tubers. The Jerusalem artichoke is not to be confused with the true artichoke. Both are members of the sunflower family; however, the Jerusalem artichoke is more closely related to the sunflower and is often described as a tuber-bearing sunflower. While the North American area has not given us many domesticated plants, it cannot be said that it has not provided us with a multitude of important improved kinds of domesticated plants. We as gardeners owe a great debt to the modern-day plant breeders.

The technology involved in modern home gardening has, like agricultural technology in general, evolved slowly and until recently mainly through trial-and-error discoveries. The Egyptians cultivated cucumbers, grapes, olives, melons, and lettuce as early as 3000 B.C. By the year 1500 B.C. they had developed extensive formal gardens in which plants were grown for their beauty alone.

The Greeks inherited from the Egyptians the idea of formal gardens, although theirs showed much less rigidity of form than was characteristic of Egyptian gardens. Greek contributions in the realm of gardening were to create natural parks for the training and inspiration of athletes, and to develop natural study sites for educational purposes—the original academies. And it was the Greeks who handed down the first detailed and systematic studies of plant parts and plant behavior.

We find a sharp contrast in attitudes toward gardening as we follow the development of gardening from the Greeks to the Romans. Rich valleys, a favorable climate, and a people who loved to grow plants combined to make vegetable production a dominant trait among Romans long before the days of the Empire. While many aspects of the great Roman civilization were built upon ideas borrowed from the Egyptians and Greeks, their appreciation for agriculture was distinctly their own. Plant culture was an important contributor to the Roman economy.

During the 500 years of the Roman reign many improvements were made in the culture of plants, including improved knowledge of grafting, manure application, green manuring, crop rotation, dry farming, storage of fruits, irrigation, and useful garden tools, including the iron plow.

In the chaotic years after the fall of the Roman Empire knowledge of gardening and other agricultural arts largely died out in Europe. This cultural stagnation lasted for hundreds of years, but gradually influences, internal and external, began to bring a new awareness. The great Swedish botanist and physician Linnaeus (1707–1778) created new interest in plants by developing a scientific system of plant classification; a system we still use today. In the 1800s, the Austrian monk, Gregor Mendel (1822–1884) did his classic work breeding table peas. Although his results went largely unnoticed for many years, he established the field of plant genetics. In England, Charles Darwin further stimulated scientific inquiry by publishing his theory of survival of the fittest. He also extensively studied plant development. Also in England, John Lawes founded the Rothamsted Experiment Station in 1843, the oldest agricultural experiment station in the world.

Immigrants to America came not only with their knowledge of gardening but also with garden seed. Early gardening efforts in America were enhanced by techniques provided by native Indians. In 1862 Congress created the U.S. Department of Agriculture, and passed the Morrill Act, establishing the land-grant colleges to promote the agricultural and mechanical arts. The Hatch Act, passed in 1887, provided for the development of agricultural research programs in the various states. Finally, the Smith-Lever Act of 1914 created the Cooperative Extension Service to disseminate research information to the people. Through this trinity of organizations, basic information on the production of garden crops has been made available to generations of home gardeners.

ORGANIC AND CHEMICAL GARDENING DEFINED

Humans have cultivated plants for thousands of years. During this long period of time, people labored *without* the aid of power equipment, pesticides, or chemically defined fertilizers. However, the addition of organic residues to the soil was a common practice. In essence, throughout most of agricultural history the method of gardening practiced has been **organic** gardening.

There are probably as many definitions of organic gardening as there are organic gardeners. The common point in almost all of these definitions is that no synthetic fertilizers and pesticides should be used. On the other hand, organic matter and natural sources of nutrient elements essential to plant growth can be used, and pests can be controlled by naturally occurring pesticides or biological agents (handpicking insects, insect predators, crop rotation, etc.).

Although organic gardening in reality predates recorded history, Sir Albert Howard (1873–1947) is considered to have founded the modern organic gardening movement. This popular movement was initiated because of concern about the effects of the extensive use of chemical fertilizers in agriculture.

The German chemist, Justus von Liebig (1803–1873), who died the year Howard was born, is commonly considered to be the founder of chemical agriculture, and more specifically **chemical** gardening. Indeed, in the history of agriculture a period of about 90 years, from 1840 to 1930, is sometimes called the Liebig Era. Liebig is known to chemists for the discovery of chloroform, the isolation of the first amino acid, and the development of baking powder. He is known to agricul-

turists—and gardeners—particularly for his work on plant nutrition. Liebig rejected the old theory that plants "ate" humus as food. He showed that plants took carbon dioxide, water, and ammonia from the air and soil. Research by Liebig indicated that soil fertility might be established and maintained by the addition of defined mineral elements. This does not mean that he did not recognize the value of soil organic matter and humus as related to soil tilth. Nevertheless, the abuses of chemical agriculture are frequently "laid at the feet" of Liebig. Sir Albert Howard in his book *Agricultural Testament* says of Liebig,

> In his onslaught on the humus theory he was so sure of his ground that he did not call in nature to verify his conclusions. It did not occur to him that while the humus theory, as then expressed might be wrong, humus itself might be right.

In any event, excesses in the use of inorganic fertilizers accompanied by a neglect of organic matter content in soils led to a gradual deterioration of the structure of many agricultural soils. It was out of concern for this situation that the modern organic gardening movement was born.

WHY GARDEN?

Terrestrial plant life is the prerequisite for terrestrial animal life. "All flesh is grass" is the biblical phrase expressing this truth. The human race gets essentially all carbohydrates and about 75 percent of its protein directly from plants and the rest indirectly through eating other plant-eating animals. Plants make our existence possible because they are able to manufacture food materials (carbohydrates, proteins, fats) out of water, carbon dioxide, and minerals by the process of photosynthesis. It is this activity of the plant world that supports the ever-enlarging human population. As a matter of fact, the burgeoning human population is perhaps the prime reason

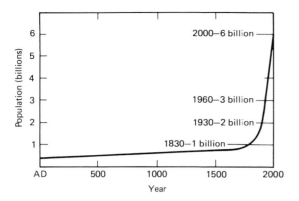

FIGURE 1-4 World population growth since the time of Christ's birth. (Reprinted with permission of Macmillan Publishing Co., Inc. from Nyle C. Brady, *The Nature and Properties of Soils.* Copyright © 1974 Macmillan Publishing Co., Inc.)

for a strong commitment to gardening. If the world's population continues to expand at the present rate it will double itself every 35 to 40 years; that is, the world population could conceivably reach 6 billion by the year 2000 and 12 billion by the year 2040 (Fig. 1-4). The problems brought about by such increases would be truly monumental.

It is evident that the earth is undergoing a population explosion that is without precedent. Never before in the history of our planet have human numbers increased so dramatically. Experts tell us that there is little new farmland to be developed, and we cannot wholly rely on food from the sea, miracle grains, or the transfer of people to nearby planets. What can we do? One positive, productive activity available to most people is gardening, whether in our backyard or in community vegetable gardens. A one-fourth-acre garden (100 × 100 feet) will produce the majority of food for a family of four. For every 30 gardens measuring 30 × 50 feet another acre of food-producing land is added to the nation's resource base. We must begin to garden on an unprecedented scale, and indeed, surveys tell us this is happening across the face of America.

In 1982, 38 million American families had gardens—a 5 million increase from 1974. Many more would garden if space were available. **Community gardening**, that is gardening by individuals on land that they do not own themselves, increased markedly from 1974 to 1982, prompting encouraging support from private and governmental agencies.

The present gardening movement is not the first our government has supported. Some people will remember when posters like those illustrated in Fig. 1-5 reading "Every Garden a Munition Plant" and "We'll have lots to eat this winter, won't we Mother?" were as common as the poster slogan "Uncle Sam Wants You."

There are other very good reasons for raising a garden. A garden of your own lets you choose varieties that are frequently tastier, have a more delicate texture, or ripen over a longer period. Many supermarket varieties have been developed for mechanical picking, in-transit ripening, and long shelf life. Plant breeders have had to develop varieties with thick skins and uniform ripening, often compromising flavor considerations of tastier varieties that could not be handled on a large-scale basis. Additionally, many tasty vegetables and herbs

FIGURE 1-5 Gardening posters from World war I (left) and World War II. (Courtesy of Karima Me-Kelburg Wallmann and the Oregon Historical Society, respectively.)

Annuals, Biennials, and Perennials

A common, and ancient, system of classifying plants on the basis of their growth characteristics is the division into annuals, biennials, and perennials (Table 2-1). **Annuals** normally complete their life cycle—from seed to seed—in a single growing season. **Biennials**, on the other hand, normally complete their life cycle over two growing seasons (Fig. 2-1). The first year, growth is entirely vegetative and a **rosette** (a cluster of spreading or radiating basal leaves) may be formed. At the beginning of the second season the biennial will form an elongated stem or seed stalk. **Perennials** are plants that grow year after year, often taking several years to mature. In temperate regions the aboveground portion of a **herbaceous** (developing only a small amount of woody tissue) perennial may die back each year, but most are woody. If woody perennials lose their leaves during a portion of the year they are said to be **deciduous**; if their leaves persist they are called **evergreens**.

The classification of plants based on their life span is somewhat arbitrary; the distinction between annuals, biennials, and perennials is not always sharp. For example, eggplant, potato, and tomato are actually subtropical perennials that grow as annuals in more temperate climates. Many biennials are grown by gardeners for their fleshy storage organs, rather than fruits or seeds and thus are treated culturally as annuals. Where the climate is mild some annuals can be planted in the fall and harvested the next spring or summer, thereby displaying a growth pattern much like that of biennials. Some perennials, such as asparagus, produce seed the first year and so fit the definition of annuals. Some normal biennials may also complete their life cycle in one season, but they then die. In the genus *Rubus* (blackberry, raspberry, and other brambles) the roots are perennial and the shoots biennial.

Hardy and Tender Crops

Plants are also placed into categories based on their temperature tolerances and temperature requirements. Accordingly, plants may be described as hardy or tender. **Hardy** plants will endure ordinary frosts without injury, whereas **tender** plants would be killed. While this classification scheme is most often used with reference to perennials that must survive low winter temperatures, vegetables are also

TABLE 2-1 Examples of Garden Crops Commonly Classified as Annuals, Biennials, and Perennials

Life span	Crops
Annual	Bean, corn, cucumber, eggplant, lettuce, muskmelon, pea, peanut, pumpkin, radish, rice, soybean, spinach, squash, sunflower, tomato, watermelon, wheat
Biennial	Beet, Brussels sprouts, cabbage, carrot, cauliflower, celery, chard, leek, onion, parsnip, rutabaga, turnip, Welsh onion
Perennial	Apple, artichoke, asparagus, cherry, chive, garlic, Jerusalem artichoke, peach, pear, potato, rhubarb, sweet potato

TABLE 2-2 Some Common Vegetables Grouped by Resistance to Spring Frosts[a]

Cool-season crops		Warm-season crops	
Hardy	Half hardy	Tender	Very tender
Broccoli	Beet	Cowpea	Cucumber
Brussels sprouts	Carrot	New Zealand	Eggplant
Cabbage	Cauliflower	spinach	Lima bean
Collards	Celery	Snap bean	Muskmelon
Garlic	Chard, Swiss	Soybean	Okra
Kale	Chinese cabbage	Sweet corn	Pepper
Kohlrabi	Endive		Pumpkin
Leek	Mustard		Squash
Lettuce	Parsnip		Sweet potato
Onion	Potato		Tomato
Pea	Rutabaga		Watermelon
Radish	Salsify		
Spinach			
Turnip			

Courtesy of U.S. Department of Agriculture.
[a]This classification is used to determine earliest safe date to plant vegetables. Hardy crops can be planted four to six weeks before the average date of the last 32°F freeze in spring, half-hardy vegetables two to four weeks before this date, tender vegetables not sooner than the last 32°F freeze; and very tender crops one to two weeks after the last average 32°F freeze. This classification is only approximate, and particular crops will vary in positioning from differing authorities.

categorized in this manner (Table 2-2). Each major category may be subdivided, as illustrated by designations such as half hardy and very tender.

Cool-Season and Warm-Season Crops

A widely used classification scheme based on the temperature requirements of vegetables describes them as **cool-season** or **warm-season** crops (Table 2-2). While this scheme is generally used to describe temperature requirements over the growing season, it is sometimes related directly to seed-germination requirements. The major growth of cool-season crops should occur during the cool parts of the year. If planted in spring, for example, they must have time to mature before temperatures become too warm. Warm-season crops, on the other hand, thrive on the high temperatures of the summer months. The cool-season vegetables originated in temperate climates, whereas the warm-season crops come primarily from subtropical and tropical regions.

Fruits Versus Vegetables

Garden plants are divided into two broad categories—fruits and vegetables. No other pair of words is so widely used by the gardener; yet no other pair of words is so poorly understood. Botanically a **fruit** is "the seed-bearing structure derived from a flower." It is the end product of the sexual process in angiosperms. A **vegetable**, in contrast, is an edible plant part other than the fruit. Familiar examples of the fruit include the apple, cherry, orange, tomato, and grape. Familiar vegetables include the carrot, radish, lettuce, cabbage, and potato. **Horticulturists** (those concerned with garden crops) define fruits and vegetables somewhat differently, related to tradition, culture, and table use. **Fruits** are the plants from which a more or less succulent fruit (botanical) or closely related structure is commonly eaten as a dessert or snack. **Vegetables** are those herbaceous plants of which some portion is eaten, either raw or cooked, during the main part of the meal. It is not difficult to find differences in the positioning of garden crops when comparing the horticultural and botanical definitions. Green beans, pod peas, eggplant, okra, squash, and pumpkin are quite respectable botanical fruits, but are universally regarded as vegetables. The **grains** of corn and other cereals (wheat, oat, barley), and the unshelled "seeds" of sunflower are also fruits in a botanical sense. In contrast, in rhubarb the stalk of the leaf is the edible portion. Botanically a vegetable, this crop is commonly regarded as a fruit because it ends up primarily in sauces or pies, eaten as desserts. Clearly, the botanical and horticultural distinction between fruits and vegetables can cause confusion. Consider the tomato. Botanically this specimen is a fruit, a berry; but horticulturally it is regarded as a vegetable. The matter was decisively settled, however, in 1893, when, in relation to import duties, the U.S. Supreme Court declared the tomato a vegetable! More will be said about fruits and vegetables in Chapter 3.

FROM SEED TO SEEDLING

Seed are of central concern to the gardener, not only for their role in plant propagation, but also as a source of human food. Seed of the grasses rice, wheat, and corn, are basic in the human diet. Others are important sources of plant oils, flavoring agents, and snacks. What ballgame would be complete without peanuts? Many definitions exist for the seed: botanically, the **seed** is a mature ovule, which in angiosperms is enclosed within the ovary or

fruit. Pea seed, for example, are mature ovules, and are enclosed within the ovary, the pea pod in this case.

Seed Structure

Seed vary greatly in size, shape, internal structure, germination requirements, and longevity; but all seed have certain features in common. Every seed has three basic parts; an embryo, a food storage tissue, and an external covering called the seed coat. These parts are illustrated in Figure 2-3 for four of the most common garden plants—bean, pea, corn, and onion. (Technically, corn "seed" are fruits in which the ovary wall is fused to the seed covering.) Our discussion of the seed and its germination will deal mainly with these four plants, as they illustrate many of the features applicable to all garden plant seed.

Bean and pea seed appear quite similar (Fig. 2-3). Each has an **embryo**, a young plant that on resumption of growth—**germination**—becomes the new seedling. Development of the new seedling requires an energy source; in the case of bean and pea the food stored in the **cotyledons**, or seed leaves. The two cotyledons occupy the bulk of the seed, and are delimited to the outside by the **seed coat**, the outer portion of the ovule that matured into the seed. The seed of corn and onion, while having the same basic parts as pea and bean, show some differences (Fig. 2-3). Being monocotyledons, each embryo possesses only one cotyledon. Furthermore, the food reserve needed on germination is not stored in the cotyledon, but rather in the endosperm tissue. The **endosperm** is the nutritive tissue that is formed in the maturing ovule. In certain cases, as with bean and pea, it is consumed as the embryo grows, and the cotyledons take on the function of food storage. In corn and onion the endosperm remains; and the cotyledons aid in removing the food material from the endosperm and channeling it into the grow-

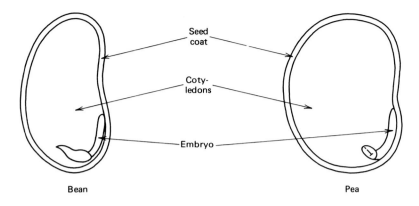

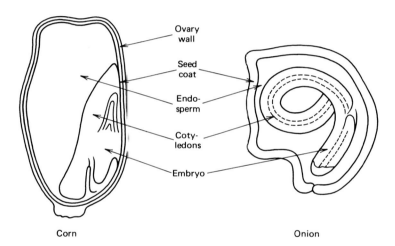

FIGURE 2-3 Seed of bean, pea, corn, and onion to illustrate parts.

ing portions—shoot and root—of the newly developing seedling.

Seed Germination

The gardener is primarily concerned with getting seed to germinate and to develop into seedlings. This sequence of events involves several possible complications: (1) stimuli to germination must reach the inside of the seed; (2) the embryo in resuming growth must escape, at least in part, from the seed; (3) the developing plant must reach the soil surface, often after growing through an inch or more of soil; and (4) the emergent seedling must be able to withstand injury by external agents. Fortunately, each plant has its own unique methods for coping with these potential hazards.

Figures 2-4 through 2-7 illustrate seed germination and seedling development of our representative plants. In all four of these plants it is a simple task to activate the embryo; one simply saturates the seed with water and places it in a suitable environment. Within a few days the developing seedling will appear. Ease of

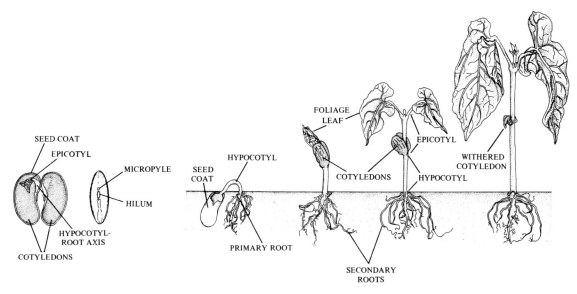

FIGURE 2-4 Stages in the germination of bean. (Reprinted with permission of Worth Publishers, Inc. from Peter H. Raven, Ray F. Evert, and Helena Curtis, *Biology of Plants.* Copyright © 1981 by Worth Publishers, Inc.)

germination is not common to all kinds of seed, however. Some will not germinate when placed in a suitable environment, unless special conditions are met. These seed are said to be **dormant**. Fortunately, dormancy, which can be advantageous in nature, is rarely encountered among vegetable crops.

The seed of our four representative plants imbibe water and begin to grow, producing first the **primary root**. From this point on the seed of the different species diverge in their developmental pattern. In bean the **hypocotyl** (the portion of an embryo or seedling situated between the cotyledons and the em-

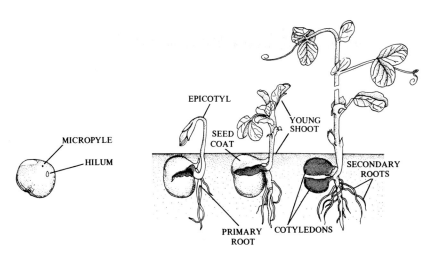

FIGURE 2-5 Stages in the germination of pea. (Reprinted with permission of Worth Publishers, Inc. from Peter H. Raven, Ray F. Evert, and Helena Curtis, *Biology of Plants.* Copyright © 1981 by Worth Publishers, Inc.)

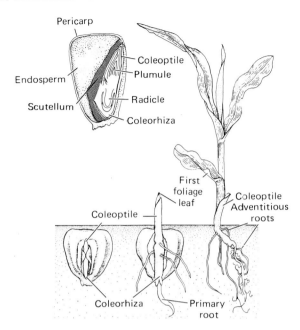

FIGURE 2-6 Stages in the germination of corn. (Reprinted with permission of Worth Publishers, Inc. from Peter H. Raven, Ray F. Evert, and Helena Curtis, *Biology of Plants*. Copyright © 1981 by Worth Publishers, Inc.)

bryonic root; Fig. 2-4) elongates, *pulling* the cotyledons and protected shoot tip through the soil. In pea (Fig. 2-5) and corn (Fig. 2-6) the cotyledons remain in place, and the seedling shoot *pushes* through the soil. In pea the developing shoot is hooked near the tip and the leaves are not developed, both facilitating the upward movement. In corn the growing shoot is protected by a specialized leaf, the **coleoptile**. It is elongation primarily in the region below the coleoptile that brings the coleoptile and its enclosed leaves to the soil surface. In onion (Fig. 2-7) the elongation of the middle portion of the cotyledon initially pushed the root, shoot bud, and lower portion of the cotyledon out of the seed. Subsequently the elongating middle portion of the cotyledon, in the form of a loop, emerges from the soil.

Germination Requirements

The successful germination of vegetable seed requires favorable environmental conditions. Among the factors essential to germination are good aeration, an adequate supply of moisture, and suitable temperature.

Because respiration rates are high in germinating seed an adequate supply of oxygen is required. While generally present in quantity, oxygen can become limiting when the soil is saturated with water. Under these conditions germination rate and percentage germination of most garden seed will decrease markedly. The seed of cucurbits are particularly sensitive to low levels of oxygen. Add-

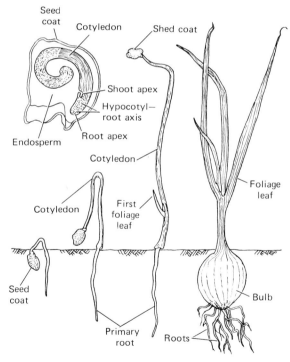

FIGURE 2-7 Stages in the germination of onion. (Reprinted with permission of Worth Publishers, Inc. from Peter H. Raven, Ray F. Evert, and Helena Curtis, *Biology of Plants*. Copyright © 1981 by Worth Publishers, Inc.)

ing sand to the garden soil or planting sensitive seed in mounded areas will improve soil aeration near seed and enhance germination.

As indicated, the relationship between soil moisture and aeration is a close one. Enough water must be present to initiate growth processes within the seed. At the same time, soil aeration should not be reduced by excess water. Vegetable seed can be divided into five groups based on soil moisture required for seed germination (Table 2-3). It is important to realize that the crops are ranked on the basis of *minimum* soil moisture required for germination; most will germinate better at somewhat higher moisture levels.

Soil temperature can strikingly influence the percentage and rate of seed germination too. Temperature requirements vary from crop to crop as shown in Table 2-4. As you might expect, seed of cool-season crops germinate

TABLE 2-3 Soil Moisture Required for Vegetable Seed Germination

Group 1. Seeds that give nearly as good germination at the permanent wilting percentage as at higher soil-moisture contents.

Cabbage	Pumpkins
Broccoli	Radish
Brussels sprouts	Sweet corn
Cauliflower	Squash
Kohlrabi	Turnip
Muskmelon	Watermelon
Mustard	

Group 2. Vegetable seeds that require a soil moisture content at least 25% above the permanent wilting percentage.

Bean, snap	Peanut
Carrot	Pepper
Cucumber	Spinach
Leeks	Tomato
Onion	

Group 3. Vegetable seeds needing a soil moisture content at least 35% above the permanent wilting percentage.

Lima bean	Pea

Group 4. Vegetable seeds needing a soil moisture content above 50% of the permanent wilting percentage.

Beet	Endive
Chinese cabbage	Lettuce

Group 5. Vegetable seeds needing a soil-moisture content close to field capacity.

Celery

Used with permission of AVI Publishing Co. From Walter E. Splittstoesser, Vegetable Growing Handbook. *Copyright 1979 by AVI Publishing Co., P.O. Box 831, Westport, Conn. 06881.*

TABLE 2-4 Soil Temperatures Required for Vegetable Seed Germination

Minimum

32°F	40°F		50°F
Endive	Beet	Parsley	Asparagus
Lettuce	Broccoli	Pea	Sweet corn
Onion	Cabbage	Radish	Tomato
Parsnip	Carrot	Swiss chard	Turnip
Spinach	Cauliflower	Celery	

60°F	65°F
Bean, lima	Eggplant
Bean, snap	Muskmelon
Cucumber	Pumpkin
Okra	Squash
Pepper	Watermelon

Optimum

70°F	75°F	80°F
Celery	Asparagus	Bean, lima
Parsnip	Endive	Carrot
Spinach	Lettuce	Cauliflower
	Pea	Onion
		Radish
		Tomato
		Turnip

85°F		95°F
Bean, snap	Parsley	Cucumber
Beet	Pepper	Muskmelon
Broccoli	Sweet corn	Okra
Cabbage	Swiss chard	Pumpkin
Eggplant		Squash
		Watermelon

Maximum

75°F	85°F	95°F		105°F
Celery	Beans, lima	Asparagus	Eggplant	Cucumber
Endive	Parsnip	Bean, snap	Onion	Muskmelon
Lettuce	Pea	Beet	Parsley	Okra
Spinach		Broccoli	Pepper	Pumpkin
		Cabbage	Radish	Squash
		Carrot	Swiss chard	Sweet corn
		Cauliflower	Tomato	Turnip
				Watermelon

Used with permission of AVI Publishing Co. From Walter E. Splittstoesser, Vegetable Growing Handbook. *Copyright 1979 by AVI Publishing Co., P.O. Box 831, Westport, Conn. 06881.*

best at temperatures 18 to 27°F (10 to 15°C) below those most favorable to warm-season crops.

In addition to the factors discussed above, light, or darkness, is sometimes essential to germination. The seed of celery, lettuce, and many flowers require light. As a consequence these should be planted very shallow to assure good germination. The seed of *Allium* and phlox, on the other hand, require darkness and must be planted deep enough to avoid light.

Seedling Establishment

As our representative seedlings grew through the soil it was important that the growing point and young leaves—known collectively as the **epicotyl**—be protected from mechanical injury. This was accomplished in bean by the hooked hypocotyl, in pea by the hooked shoot tip and undeveloped leaves, in corn through the tubular-shaped coleoptile, and in onion by a cotyledonary sheath. Continued development results in striking changes in seedling form, and as we will see, it is the action of light that initiates these changes.

The importance of light in regulating the development of seedlings is shown in Fig. 2-8. The dark-grown seedlings have longer shoots and smaller leaves and are said to be **etiolated**. Etiolation is common, for example, in young bean sprouts that one purchases in the supermarket.

Let's now look more carefully at the effects of light on the form of our seedlings. In bean, light is essential for the straightening of the hypocotyl hook, and, of course, for reducing stem elongation and encouraging leaf expansion. As the hypocotyl straightens the cotyledons and enclosed epicotyl are brought above the soil; then the cotyledons move apart and the shoot begins to elongate. In the pea plant, light initiates the straightening of the hooked shoot tip. In corn, light inhibits further elon-

(a)   (b)

FIGURE 2-8 Pea seedlings grown in light (*a*) and in darkness (*b*). Note the elongated stem, minute scale leaves, and the hooked shoot tip of the dark-grown plant. (From Peter Martin Ray, *The Living Plant.* Copyright © 1963 by Holt, Rinehart and Winston, Inc. Reprinted by permission of Holt, Rinehart and Winston, CBS College Publishing.)

gation of the mesocotyl and the coleoptile, and the developing leaves thus push through the top end of the coleoptile. In onion the action of light results in a straightening of the hooked cotyledon; concomitantly, the end of the cotyledon, still curled within the endosperm of the seed, is elevated above the soil surface.

Seedlings as Immature Plants

Established seedlings have a vertical axis terminated at either end by a growing point (Fig. 2-9). If lateral branches of stem or root are present they are also terminated by growing points. Each growing point is a localized mass or body of cells that does not mature but remains capable of further growth and division. Cells displaying such characteristics are said to be **meristematic**, and the growing points in which they are found are called **meristems**. Because these meristems occur at the tips of stems and roots they are known as **terminal** (or **apical**) **meristems**. The meristems are regions of cell and tissue initiation. It is meristems that give plants the unique capacity for more or less unlimited growth. In the words of one botanical writer, certain plants can achieve great age and "give the impression of immortality" because of continued meristem activity. In the next section we explore the development of the mature plant from a small seedling, through the activity of meristems.

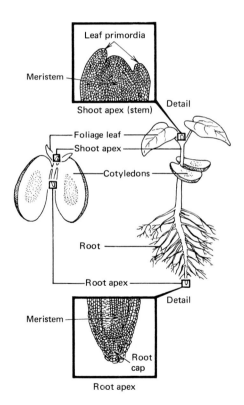

FIGURE 2-9 The positions of apical meristems in seed and seedling of bean. (Reprinted with permission of Wadsworth Publishing Company, Inc. from William A. Jensen and Frank B. Salisbury, *Botany: An Ecological Approach.* Copyright © 1972 by Wadsworth Publishing Company, Inc.)

FROM SEEDLING TO MATURE PLANT

The Root System

The first event normally observed in a germinating seed is the emergence of the primary root. This main root elongates rapidly and gives the seedling immediate and intimate contact with the soil—the source of water and minerals essential for plant growth. The root system also serves to securely anchor the plant. The developing root system of a plant, like the shoot system, has a characteristic form for each species (Fig. 2-10). In some garden plants, for example, beet, carrot, and soybean, the main root grows directly downward, deeply into the soil. This root gives rise at intervals to lateral, or branch roots, and these in turn may give rise to smaller rootlets. This type of root system is known as a **tap root system**. In other plants the main root is not so predominant and the entire system spreads more or less uniformly in the upper depths of the soil. This type of root system is termed a **fibrous root system**. The fibrous root system may develop entirely from branches of the main root, or it may come partially or predominantly from roots that arise near the ground level of the stem, as exemplified by the prop roots in corn (Fig. 2-11). When roots arise from stems as in corn they are interpreted as arising from other than the "normal" place, namely from preexisting roots, and are said to be **adventitious**. Many plants with deep tap roots form an up-

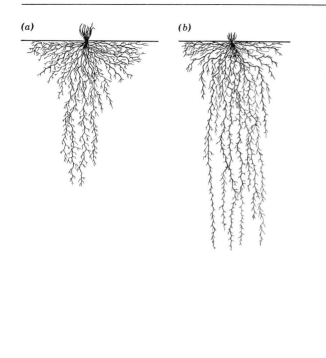

FIGURE 2-10 Different types of root systems; (a), shallow, fibrous root system; (b), fibrous root system penetrating the soil evenly from 1 to 1.5 m; (c), tap root system, in which main primary root penetrated soil 2.5 m or more. (Reprinted with permission of John Wiley & Sons from T. Elliot Weier, C. Ralph Stocking, Michael G. Barbour, and Thomas L. Rost, *Botany: An Introduction to Plant Biology.* Copyright © 1982 by John Wiley & Sons.)

per network of fibrous feeder roots, thus combining both the ability to obtain water from deeper soil levels and nutrients from the more fertile upper layers of soil.

While each crop species has a characteristic root pattern, this pattern is not absolute. A plant with a normally deep tap root system, for example, may develop a shallow, spreading system if the tap root encounters an impermeable soil layer or a high water table. Root system form can also be modified directly by man. Nursery workers commonly prune back the tap roots of developing shrubs and trees to encourage the development of a more fibrous root ball concentrated near the soil surface. This practice makes it possible to transplant even relatively large plants.

Understanding the form of the root systems of plants is important to the gardener. To illustrate, tap root systems generally extend more deeply into the soil than do fibrous root systems. Because of this plants of each kind will "feed" (take in water and minerals) at different levels within the soil, and even if grow-

ing in close proximity may not compete directly for these life-giving commodities. Variations such as this explain why weeds having a long tap root may thrive in a lawn in which insufficient water has caused the grasses, with their fibrous root systems, to turn brown. Feeding characteristics directly influence the selection of and spacing requirement for garden plants, a topic discussed in Chapter 7. A knowledge of root systems is also important in relation to watering, cultivation, fertilization, and other cultural practices, subjects dealt with in Chapter 8. The depth characteristics of the root systems of a number of garden plants are presented in Table 2-5.

The Shoot System

The shoot begins its development shortly after the initiation of root growth in the seed. The shoot commonly has a modified form as it extends through the soil; a form that makes growth through this medium possible. After emergence from the soil, however, the shoot

FIGURE 2-11 Prop roots of corn, a familiar example of adventitious roots. (Reprinted with permission of A. & L. Publications from Samuel R. Aldrich, Walter O. Scott, and Earl R. Leng, *Modern Corn Production.* Copyright © 1975 by A. & L. Publications.)

gradually assumes a form characteristic of the species. Regardless of variations in form, all shoots share certain morphological features (Fig. 2-12). The cylindrical stem forms the main axis of the shoot, and flattened leaves occur at intervals along the axis. The points along the stem where the leaves are attached are known as **nodes**, and the segments of stem between the nodes are termed **internodes**. It is important to note that the internodes become progressively longer as one moves away from the apex of the plant. This increase in length comes about through the activity of meristematic cells in the internodal regions. Note also that just above each leaf, in its axil, a lateral bud is present. The **lateral buds** are condensed shoots, just as is the terminal bud, and under suitable conditions can develop into lateral branches similar in form to the main shoot. Certain buds, both terminal and lateral, are used directly by the gardener. Most notable of these are the very large terminal buds that form the heads of cabbage and lettuce, and the smaller lateral buds of the Brussels sprouts.

Stems show great variation in both form and function. To illustrate, the stems of some vegetables, such as cabbage, endive, and radish, remain condensed (lack internodal elongation), whereas others do not. The elongate stems of corn, eggplant, and okra are upright, and those of the cucurbits are prostrate. Regardless of form, stems carry out a number of important functions. They provide support for leaves and flowers, serve as storage organs (kohlrabi), provide an organic connection between leaves and roots, and assume a major role in photosynthesis (asparagus).

TABLE 2-5 Depth of Root Systems in Selected Garden Plants When Grown in Well-Drained Soils

Shallow rooted (down to 2 feet)	Moderately deep rooted (down to 4 feet)	Deep rooted (down to 6 feet)
Brussels sprouts	Beans, snap	Artichokes
Cabbage	Beets	Asparagus
Cauliflower	Carrots	Cantaloupes
Celery	Chard	Lima beans
Lettuce	Cucumber	Parsnips
Onions	Eggplant	Pumpkins
Potatoes	Peas	Squash, winter
Radish	Peppers	Sweet potatoes
Spinach	Squash, summer	Tomatoes
Sprouting broccoli	Turnips	Watermelons
Sweet corn		

Courtesy of University of California Cooperative Extension Service.

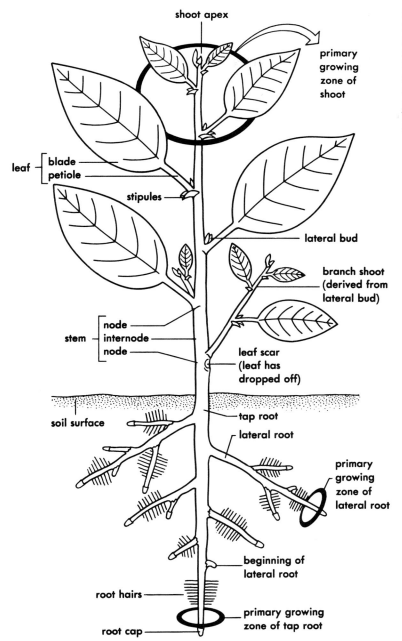

FIGURE 2-12 Diagram illustrating the external features and growing regions of shoot and root systems. (From Peter Martin Ray, *The Living Plant.* Copyright © 1963 by Holt, Rinehart and Winston, Inc. Reprinted by permission of Holt, Rinehart and Winston, CBS College Publishing.)

Leaves also show much variation in form (Fig. 2-13). In most dicotyledonous plants each leaf has a stalk, the **petiole**, and an expanded portion, the **blade**. Plants such as the garden pea have leaflike appendages known as **stipules** at the juncture of leaf and stem. Leaves may be simple or compound. In **simple** leaves the blade is not divided into distinct parts, although it may be deeply lobed. In **compound** leaves the blade is divided into a number of smaller segments known as **leaflets**. Simple leaves characterize beet, cucumber, eggplant, and lettuce, while compound leaves are found in bean, carrot, pea, and tomato. In plants with compound leaves the leaflets can be distinguished from the leaves in that the leaflets *do not* have buds in their axils. This distinction is often important as relates to altering the growth habit of a plant through pruning and pinching. In monocotyledonous plants, such as corn, the leaf consists of a blade and a sheath (Fig. 2-13). The **sheath** envelops the stem to the point of attachment of the blade. A small membranous structure, known as the **ligule**, is usually visible at the juncture of sheath and blade. In onion, a monocotyledon, the leaves are tubular.

Both leaves and leaflets may show striking modifications from the types described above. Sometimes leaves (cucumber) and leaflets (pea) are modified as clasping structures, called **tendrils** (Fig. 2-13), to aid plants of climbing habit.

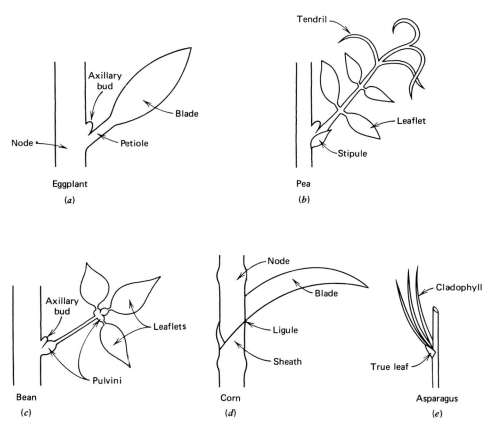

FIGURE 2-13 Leaves of some representative garden plants.

Leaves also have modifications to aid in their movements. In bean, enlarged (swollen) areas known as **pulvini** occur at the bases of leaves and leaflets (Fig. 2-13). Pulvini are found in many other plants too, including monocotyledons. The leaves of asparagus are reduced to scales and the major photosynthetic organs are leaflike stems known as **cladophylls** (Fig. 2-13). Gardeners who have not grown asparagus may have seen cladophylls in the closely related asparagus fern (not a fern but an asparagus), a common houseplant.

Leaves perform a number of functions the most important of which is photosynthesis. Lesser known but vital roles include the manufacture of certain growth regulators, the perception of external stimuli, and storage of water and food materials.

Leaves are rather transitory organs, and on older regions of stems they may be absent. While this feature goes almost unnoticed in most vegetable crops it is conspicuous in woody perennials, especially the deciduous types. In these perennials the study of naked twigs reveals much about the growth patterns of the species in question (Fig. 2-14). In walnut twigs, for example, yearly growth increments are quite distinct, because the **bud scales** that surround and protect the terminal bud during the winter season drop off, leaving definite **bud-scale scars** along the stem axis. The number of leaves produced in a given year can be determined by a count of the leaf scars present between any two adjacent sets of bud-scale scars. The fate of the lateral buds present in the leaf axils is also evident. Some of these buds assume a dormant stance, but remain as potential sources of future growth. Other buds become active and develop into lateral branches. Not all buds produce vegetative branches, however. Some produce both flowers and vegetative increments, and still others produce only flowers. A thorough understanding of the morphology of woody twigs is essential to the successful management of many perennial plants.

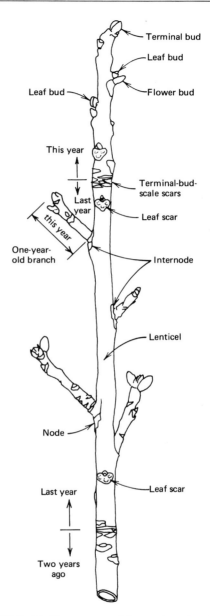

FIGURE 2-14 Three-year-old woody twig of walnut. (Reprinted with permission of John Wiley & Sons from T. Elliot Weier, C. Ralph Stocking, Michael G. Barbour, and Thomas L. Rost, *Botany: An Introduction to Plant Biology*. Copyright © 1982 by John Wiley & Sons.)

MATURE PLANT STRUCTURE

Stems and Roots

Stems and roots are cylindrical plant organs developed through the growth activity of terminal meristems. Although these organs at maturity differ somewhat in structure (Fig. 2-15) they have much in common and can be discussed together.

Epidermis

The **epidermis** is the outermost, protective covering of plant organs. On aerial plant parts the epidermis is covered by a layer known as the **cuticle** (Fig. 2-20). The cuticle layer is composed of a fatty material termed **cutin**, which functions to reduce the loss of water from plant surfaces—a feature of great importance in the drying environment of the garden. The shiny appearance often apparent on aerial plant parts is due to the presence of the cuticle.

In addition to the ordinary cells present in the epidermis, several kinds of specialized cells are found. On aerial parts hairs, each comprised of one or more cells, occur. These hairs, technically known as **trichomes**, are often abundant, as in the tomato (Fig. 2-16). While the functions of trichomes are not always clear, some are known to secrete materials. The glandular hairs in tomato, for example, se-

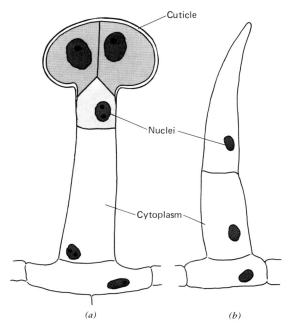

FIGURE 2-16 Glandular (*a*) and nonglandular (*b*) hairs from the stem of tomato. (Courtesy of Guerry Dean.)

crete substances that make the stem sticky to the touch. Another kind of specialized cell of the epidermis is the **guard cell**, which occurs in pairs to comprise the **stomates** (Fig. 2-17). The paired guard cells of the stomates have unequally thickened walls that allow for an opening, or pore, to form between them when water enters the guard cells. These openings make for an efficient exchange of gases between the internal tissues of the shoot and the external environment. More on this important feature of stomates later in the chapter.

In roots the epidermis also consists of ordinary cells, as well as one or more kinds of specialized cells. Unquestionably, the most important specialized epidermal cells are the root hairs (Fig. 2-18). The **root hairs**, which are trichomes on the root epidermis, originate as simple extensions of the epidermal cells. In most plants the individual root hairs are short lived; they are constantly developing near the

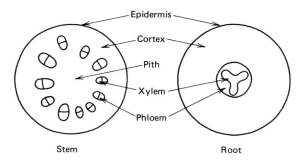

FIGURE 2-15 Cross sections of stem and root.

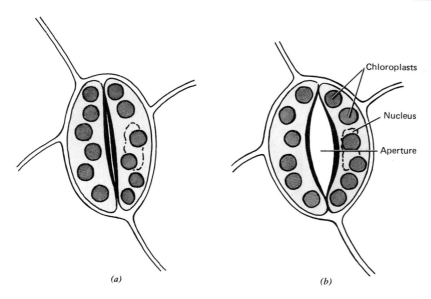

FIGURE 2-17 Closed (*a*) and open (*b*) stomates from the epidermis of a bean stem. Note the numerous chloroplasts present in the guard cells. (Courtesy of Guerry Dean.)

Chloroplasts

Nucleus

Aperture

(*a*)

(*b*)

tip of the root while being lost on older portions. The presence of enormous numbers of root hairs vastly increases the absorbing surface of roots, and thereby enhances water and mineral uptake.

The Cortex and Pith

The cortex and pith comprise the greatest tissue volume in stems and roots (Fig. 2-15). The predominant type of cell present in these two regions is a relatively unspecialized cell termed **parenchyma** (Fig. 2-19). The living parenchyma cells, which retain much of their meristematic potential, play vital roles in photosynthesis, energy release, aeration, storage, and support throughout the plant body.

The Vascular Tissues

The parenchyma tissue of plant organs is classified as **simple tissue**, because it is composed of a single cell type, the parenchyma cell. In contrast, the vascular tissues—xylem and phloem—are regarded as **complex tissues**, because, at maturity, each of these conducting tissues contains several different kinds of cells. In both tissues parenchyma cells and fibers

are found. The **fibers** are elongate, thick-walled cells especially adapted to a support function (Fig. 2-19). In the **xylem**—the water and mineral conducting tissue of the plant—two unique cell types are found, the **tracheids** and **vessel members** (Fig. 2-19). The two cell types are responsible for the conduction of water and minerals throughout the plant. It is interesting to note that both the tracheid and the vessel member are dead at maturity; that is, they are specialized to such a degree that they can carry out their primary function only after a loss of cellular contents. Obviously, these cells, being dead, lack any meristematic potential. The living parenchyma cells of the xylem and phloem can, however, become meristematically active.

In the **phloem**—the food conducting tissue of the plant—specialized cells are also found. The cell designed for conduction purposes is the **sieve-tube member**, or **sieve element** as it is sometimes called. The sieve elements always occur in association with one or more specialized parenchyma cells known as **companion cells** (Fig. 2-19). The sieve elements, unlike the conducting elements of the xylem, do not lose their cellular contents at maturity,

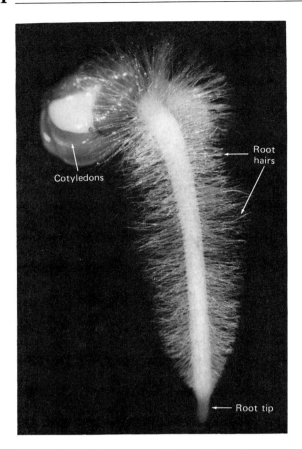
Cotyledons

Root hairs

Root tip

FIGURE 2-18 A young radish seedling with an enormous number of root hairs. (Reprinted with permission of John Wiley & Sons from T. Elliot Weier, C. Ralph Stocking, Michael G. Barbour, and Thomas L. Rost, *Botany: An Introduction to Plant Biology.* Copyright © 1982 by John Wiley & Sons.)

although they do become highly modified. Even though the sieve elements are definitely alive, they lack certain features that characterize living cells. As you might expect, these highly specialized cells, and the parenchyma cells associated with them, lack the ability to resume meristematic activity.

A major difference between stems and roots is in the arrangement of the vascular tissue within the ground tissue (Fig. 2-15). In roots the vascular tissue occurs as a solid central core, whereas in stems the vascular tissue is in the form of separate bundles arranged either in a ring (most dicotyledons) or scattered throughout the stem (most monocotyledons). Note that in a given vascular bundle of the stem the vascular tissues have a collateral arrangement along a radius, with the phloem to the outside and the xylem to the interior. In roots, where a solid core of vascular tissue is present, the xylem and phloem have an alternating arrangement.

Leaves

While roots and stems are usually elongate, cylindrical organs, leaves are typically flattened. The tissues of the leaf show a strikingly different arrangement from those of the stem (Fig. 2-20). These observed variations in structure are related in no small way to the primary function of the leaf—photosynthesis. Basically, each side of the leaf is limited by a cuticle-covered epidermis containing numerous stomates. In apples, the stomates are confined entirely to the lower epidermis, but in cabbages and corn they are about equally abundant on both the lower and the upper epidermis. The vascular tissue in these flattened organs is in the form of a ramifying network embedded within the parenchyma tissue of the leaf, the **mesophyll**. Because of the vascular network it is said that no cell in the mesophyll is more than a few cells removed from the conducting tissue. In many plants, particularly dicotyledons, two distinct kinds of parenchyma comprise the mesophyll. Near the upper surface barrel-shaped **palisade parenchyma** cells stand end to end and side by side (Fig. 2-20). Although these cells appear to be tightly packed together, each is in contact with a great deal of air space. On the lower side of the leaf occurs the **spongy parenchyma**, the other type of mesophyll tissue of the leaf. The two parenchyma types are seen to merge near the center of the leaf, at about the same level as the veins. Consequently, the veins are in direct contact with both forms of parenchyma tissues.

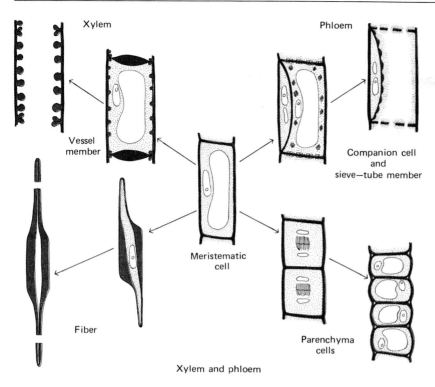

FIGURE 2-19 Diagram of some basic cell types derived from the meristematic cells. [of the meristems.] (Reprinted with permission of Worth Publishers, Inc. from Peter H. Raven, Ray F. Evert, and Helena Curtis, *Biology of Plants.* Copyright © 1981 by Worth Publishers, Inc.)

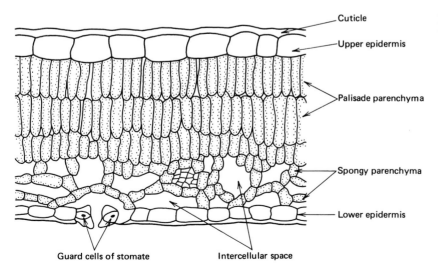

FIGURE 2-20 Cross section of an apple leaf.

One aspect of cellular activity in the mesophyll can be seen externally and is familiar to many gardeners. In some vegetables such as cabbage and spinach there are cultivars with smooth leaves and others with wrinkled or crumpled leaves (Fig. 2-21). The wrinkled forms arise through the development of the tissue between the veins, a phenomenon known as **savoying**. Cultivar names may reflect this leaf condition as with 'Savoy Perfection Drumhead' cabbage and 'Savoy Supreme' and 'Virginia Savoy' spinach.

Primary Versus Secondary Growth

To this point we have been discussing plant growth originating from the meristematic activity of terminal meristems. These meristems contribute to the initial growth in *height* of a plant and are thus termed **primary meristems**. The growth resulting from such meristems is called **primary growth**. Monocoty-

FIGURE 2-21 A savoy-leaved cultivar of cabbage. (Courtesy of Asgrow Seed Co., Kalamazoo, Michigan.)

ledonous plants such as corn and onion are derived entirely from the activity of primary meristems; that is, they show only primary growth. In many other garden plants, however, continued growth of plant organs occurs as a result of meristematic activity in two additional meristems—the vascular cambium and the cork cambium. Functionally, the **vascular cambium** gives rise to additional increments of xylem and phloem, while the **cork cambium** forms an outer protective layer, known as **cork** (Fig. 2-22). The activity of these meristems results in an increase in the *diameter* of both stems and roots. Because this increase can occur only after the primary plant body is established these meristems are termed **secondary meristems**. Logically, the growth resulting from the activity of secondary meristems is termed **secondary growth**.

Some Examples of Secondary Growth

Examples of secondary growth in garden plants are numerous. Nearly all herbaceous dicotyledons, including bean, eggplant, and tomato, show at least modest additions of secondary xylem and phloem. The same is true of many garden herbs and flowers. More conspicuous examples of secondary growth are offered by the tree fruits. These plants add new increments of vascular tissue every year. The annual increments of xylem, which together comprise the **wood** of these trees, form in rings of ever larger diameter. It is these annual rings that make possible the determination of age in felled trees.

In addition to secondary growth from a vascular cambium, many herbaceous and most woody perennials develop a cork layer through the activity of the cork cambium. The brownish skin covering a sweet potato represents the cork layer of this important, edible storage root. The skin of an Irish potato is also a cork layer. As a matter of fact, in nearly all plants that have secondary growth the epidermis of both stem and root is replaced by a cork layer.

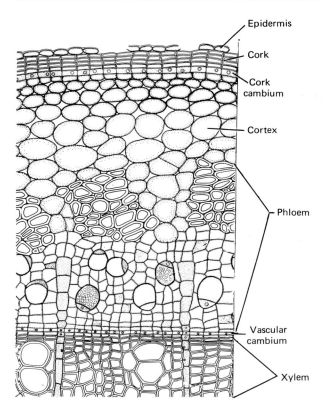

Epidermis

Cork

Cork cambium

Cortex

Phloem

Vascular cambium

Xylem

FIGURE 2-22 Cross section of a dicotyledonous stem showing secondary growth activity from cork and vascular cambia. Remnants of the original outer covering, the epidermis, are still visible. (Reprinted with permission of Holt, Rinehart and Winston, Inc., from Carl L. Wilson, Walter E. Loomis, and Taylor A. Steeves, *Botany.* Copyright © 1971 by Holt, Rinehart and Winston, Inc.)

The cork layer and the phloem tissue beneath it constitute the **bark** of a plant. Leaves, which have a more transitory nature, generally lack a cork layer, although they may develop a small amount of secondary xylem and phloem.

SOME IMPORTANT PLANT PROCESSES

Over three centuries ago the English botanist, Robert Hooke, studied the cork tissue of trees using a crude microscope. He described the cork as being made up of "many little boxes," and named these units **cells**. Later, other workers extended Hooke's observations, and in the 1830s two prominent German scientists, Matthias Schleiden and Theodor Schwann, independently published their work, which formed the basis for the cell theory.

The **cell theory** states simply that all living things are composed of cells. The theory, which has become a commonplace part of established biological thought, was quite revolutionary in the mid 1800s. Cells are the basic units of plant (and animal) life. We have seen that they occur in the meristems and that they produce derivatives that mature as xylem, phloem, and so on. The numerous plant processes we wish to discuss occur at the cellular level. Thus, it is important that we begin our discussion by examining briefly the plant cell.

The Plant Cell

Each plant cell is surrounded by a **cell wall** (Fig. 2-23). It is this wall that imparts properties of rigidity to plant tissues—and provides us humans with our dietary fiber. Adjoining plant cells are cemented together by

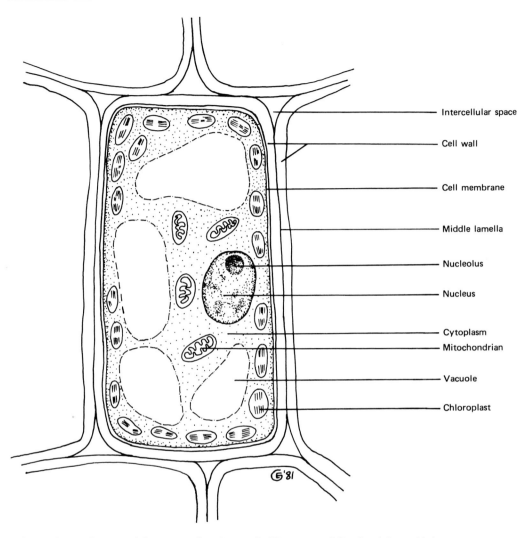

Intercellular space

Cell wall

Cell membrane

Middle lamella

Nucleolus

Nucleus

Cytoplasm
Mitochondrian

Vacuole

Chloroplast

FIGURE 2-23 Structural features of a plant cell. (Courtesy of Portland State University Biology Department and John H. Wirtz.)

a layer of pectic materials that comprises the **middle lamella**. The cell contents, known as **protoplasm**, are separated from the cell wall by an outer, limiting membrane, the **cell membrane**. Each living cell has a protoplast. As cells differentiate their cellular contents become modified. In fibers, for example, the protoplast, while still present, is regarded as metabolically inactive. In vessel members and

tracheids the living protoplasts are no longer present; only the thick, rigid walls remain.

Within the protoplasm of cells are numerous small bodies of variable size known as organelles (Fig. 2-23). The **organelles** are definite entities of the cell and include the dictyosomes, endoplasmic reticulum, mitochondria, nucleus, plastids, and vacuoles. Each organelle performs one or more specific functions within the cell, as summarized in Table 2-6. As we explore the various plant processes we will gain a greater appreciation of these cellular structures.

Photosynthesis and Respiration

Photosynthesis is the process whereby plants make themselves out of water (H_2O), carbon dioxide (CO_2), and minerals in the presence of light. The essentials of the photosynthetic process are outlined in Fig. 2-24. Additionally, the process can be described in chemical terms, as follows:

$$6CO_2 + 6H_2O \longrightarrow C_6H_{12}O_6 + 6O_2$$

| Carbon dioxide | Water | Carbo-hydrate | Oxy-gen |

Note from the above illustration that our atmosphere is the source of the CO_2 and that the H_2O and minerals are provided by the soil solution. One product of photosynthesis, the O_2, is released into the atmosphere. Oxygen is required by all higher forms of animal life, including humans.

It is virtually impossible to overemphasize the importance of photosynthesis, because it is the process that sustains all forms of life. Life as we know it could not exist without green plants. Through photosynthesis plants convert raw materials into **foods**—carbohydrates, proteins, fats—which they utilize for their own growth. Fortunately for us, they normally manufacture more food than they consume in maintenance and growth. The surplus food is stored in roots, tubers, seed, fruits, and other plant parts. These storage sources can be tapped directly by us, as when we grow a vegetable garden. They are also tapped indirectly when we eat beef, pork, or chicken; products from animals that lived and grew at the expense of plants.

The energy-storing process of photosynthesis can occur only in the presence of light. Another vital process, respiration, goes on in all living cells—plant and animal—around the clock. **Respiration** is an energy-releasing process that can be summarized in chemical terms, as follows:

$$C_6H_{12}O_6 + 6O_2 \longrightarrow 6CO_2 + 6H_2O + Energy$$

| Carbo-hydrate | Oxy-gen | Carbon dioxide | Water |

TABLE 2-6 Organelles Present in Plant Cells and Some Presumed Functions of Each

Organelle	Functions
Dictyosomes	Secretion
Endoplasmic reticulum	Protein synthesis, cellular transport
Mitochondria	Respiration
Nucleus	Cell control center, inheritance
Plastids	
Chloroplasts	Photosynthesis
Chromoplasts	Coloration, storage
Leucoplasts	Storage of food products
Vacuoles	Storage of cellular products, pigment deposition, water balance in the cells, breakdown of macromolecules

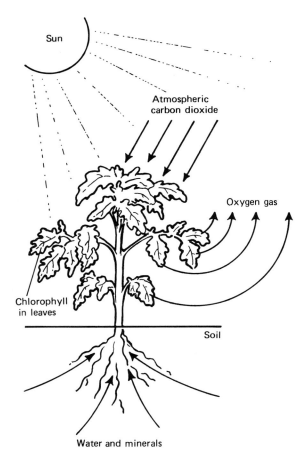

FIGURE 2-24 The essentials of photosynthesis. In the presence of sunlight, green (chlorophyll-containing) plants incorporate atmospheric carbon dioxide into organic compounds such as sugars. Oxygen in gaseous from is released into the atmosphere. Water and dissolved minerals in the soil play an important role too.

Note that the basic reaction for respiration is just the reverse of that given for photosynthesis. Note also that oxygen (O_2) is required in respiration. The dependency of animals on O_2 mentioned earlier is directly related to its use in the respiratory process. Plants, of course, also use oxygen in their respiration.

Since photosynthesis is an energy-storing process and respiration an energy-releasing one, it stands to reason that plants must produce more food in photosynthesis during the daylight hours than they consume in respiration around the clock if they are to increase in size—**grow**. The intimate relationship between the two processes is illustrated graphically in Fig. 2-25. In darkness plants respire with a net uptake of O_2 and a net release of CO_2. As the light of early morning increases, photosynthesis begins, consuming some of the CO_2 released in respiration. Soon a point is reached where the amount of CO_2 utilized in photosynthesis just equals that released in respiration. At this point, known as the **compensation point**, the rate of photosynthesis exactly equals the rate of respiration. That is, there is neither a net gain nor a net loss in food reserves. At higher light intensities, the rate of photosynthesis increases further, resulting in a net uptake of CO_2, a net release of O_2, and the accumulation of stored food. Thus, during the daylight hours when photosynthesis is proceeding at a rapid rate a plant adds "credits" to its energy account, whereas at any light level below the compensation point "debits" occur.

For luxuriant plant growth the gardener must strive to maximize energy credits while minimizing debits. If over an extended period

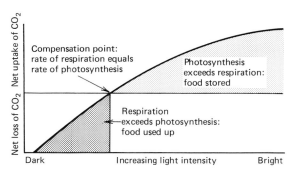

FIGURE 2-25 The relation of photosynthesis to respiration in the daily cycle of a plant. (Reprinted with permission of Saunders College from David L. Rayle and Hale L. Wedberg, *Botany: A Human Concern.* Copyright © 1980 by Saunders College, Philadelphia.)

carbohydrate synthesis does not exceed use plant food reserves will become depleted, a condition known as **carbohydrate exhaustion**. This condition may be displayed by crowded seedlings, for example. Since the presence of abundant carbohydrates is essential for the growth of healthy plants, the maintenance of a favorable balance is of great practical significance. Manipulations by the gardener—thinning, weeding, watering, fertilization, pest control—play a crucial role in preventing carbohydrate exhaustion. Understanding the factors that influence the rates of photosynthesis and respiration will also help the gardener to enhance the accumulation of food reserves by plants.

Photosynthesis can proceed only in the presence of light. Said another way, at least one reaction in the photosynthetic process is light driven. A characteristic of all such light reactions is that they proceed in the light more or less independent of temperature. The exposure of film, for example, represents a light reaction. Accordingly, the length of exposure at a given light intensity remains constant regardless of the ambient temperature. As it turns out, many additional reactions of the photosynthetic pathway are known to proceed in the absence of light. These so-called dark reactions, in contrast to the light reactions, are driven by enzymes that are quite sensitive to temperature. The entire series of reactions comprising respiration can, of course, proceed in the dark too. An important property of these temperature-dependent reactions is that over the range of temperatures at which biological reactions occur their rate is directly proportional to the temperature. In general, for each 10°C (18°F) rise in temperature the reaction rate approximately doubles.

What does all of this mean to the gardener? Simply this. Since both photosynthesis and respiration involve temperature-dependent reactions, the rate at which they proceed is somewhat under the gardener's influence. As

examples, mulches (see Chapter 8) can drastically alter soil and aboveground temperatures and thus can be utilized to enhance food production through more rapid photosynthetic rates. Garden seed can be stored at lower temperatures (see Chapter 7), where the lowered respiration rates prolong greatly their storage life. Vegetables grown indoors (see Chapter 14) can be exposed to lower nighttime temperatures resulting in a greater differential between the manufacture and the utilization of the products of photosynthesis. Additional examples of the influence of temperatures on plant growth are given elsewhere in the text.

The rate of photosynthesis depends not only on light intensity and temperature, but also on other factors including the CO_2 supply, water availability, and nutrient status of the plant. The CO_2 necessary for photosynthesis diffuses into the leaves from the external atmosphere mainly through the stomates (Fig. 2-17). The amount of CO_2 in the air is quite low (0.03 percent or 300 parts per million) and can become limiting under certain conditions, as on a sunny day when the air is calm. Experiments have shown that photosynthetic rates can be enhanced by raising the level of CO_2 in the air. While it is not feasible to elevate CO_2 levels in the garden, many commercial growers do install CO_2 generators in their greenhouses, increasing CO_2 levels by as much as five times the normal level. Under these conditions photosynthetic rates will increase, provided the process is not limited by some other factor.

When water is in short supply in the soil garden plants may become flaccid, or limp, due to excessive loss of water, a phenomenon known as **wilting** (Fig. 8-9). As the plant wilts the stomatal pores become greatly constricted or closed. In this condition the exchange of gases and the water milieu needed for photosynthesis are in short supply. As a result the process of converting light energy into chemical energy slows or stops altogether.

About 16 nutrient elements are needed by the plant for normal growth, as discussed in Chapter 5. Nearly all of these are supplied by the medium in which the plant grows. If any of these nutrients are in short supply the rate of photosynthesis will be limited, even if all of the other requirements are present in abundant amounts. Consider the element magnesium, for example. Magnesium is needed by plants in only small amounts, yet its presence is absolutely essential because the element is a vital part of the chlorophyll molecule. When magnesium is limiting, chlorophyll synthesis is lessened and photosynthesis is reduced.

Movement of Materials in Plants

Numerous materials move throughout the plant on a regular basis. Manifestations of some of these movement phenomena are plainly visible to the gardener, and an understanding of other transport patterns can enhance our gardening efforts. Although an almost endless number of materials are being transported we will center our attention on the following: (1) water and dissolved minerals, (2) food materials, mainly sucrose, and (3) certain growth regulators.

Transport of Water and Minerals

An abundant supply of water and minerals is essential to normal plant functions. In the usual course of events water and minerals from the soil are taken in by roots and transported via the xylem tissue to aerial parts of the plant. This is well enough, but what drives this process? In certain cases, when the soil is rich in water and the atmospheric humidity is high, water and minerals may actually be pushed into aerial parts by pressures originating in the roots. You have probably observed this phenomenon because of specialized openings, known as **hydathodes** at the margins and tips of leaves which allow excess water to be expelled. When conditions conducive to root

pressure occur, dewlike droplets of water containing mineral salts can often be seen at the margins of leaves (Fig. 2-26). In essence, this **water of guttation** is being forced from the leaf by the pressure originating in the roots. The phenomenon of guttation is most commonly observed in the early hours of the day. Water of guttation can be distinguished from dew by its pattern of distribution and by the larger size of droplets.

Gardeners may encounter certain problems with water of guttation, because of its mineral content. In the heat of the day the water droplets evaporate, but the mineral salts remain as encrustations on leaves. These salts can build up to toxic levels, killing portions of the leaf. **Necrosis**(death) of tissue near the tips or margins of leaves—sites where guttation fluid accumulates—is a manifestation of this problem. Generally the problem is not severe with plants grown outdoors, but is often encountered with plants grown indoors in pots. The problem can be prevented by periodic washing of susceptible areas.

Most of the time the water moving in plants is not pushed upward by pressures originating in the root system, but rather pulled upward by tensions originating in the shoot sys-

FIGURE 2-26 Water of guttation on leaflets of strawberry. (Reprinted with permission of Macmillan Publishing Co., Inc., from Victor A. Greulach, *Plant Function and Structure.* Copyright © 1973 by Victor A. Greulach.)

tem. But what is the source of the tensions that pull water upward? A mechanical apparatus will aid our understanding of this question (Fig. 2-27). First, a porous clay pot is filled with water and attached to the end of a long glass tube also filled with water. Then, the water-filled tube is placed with its lower end below the surface of a volume of water. Soon water begins to evaporate from the pores in the pot and is replaced by water pulled up from below in a continuous column through the narrow glass tube. In other words, the evaporation of water creates the driving force for pulling water upward. The ability of water molecules to cling together, a property termed **cohesion**, allows pull over long distances. Similarly, it has been determined that the loss of water in vapor form from plant surfaces, known as **transpiration**, creates a negative pressure (tension) in the piping (xylem) of the plant. Furthermore, the tensions created have been shown to be great enough to lift water to the tops of even the tallest trees.

While the transpirational loss of water is essential to the operation of the system just described, it also means that much of the water taken up by plants is merely transpired into the atmosphere. As a matter of fact, it has been shown repeatedly that plants transpire several hundred units of water for each unit of dry matter produced (Fig. 8-17). This great water loss is related in no small way to the presence of stomates on aerial parts. The stomates are normally open during the daylight hours but greatly constricted or closed at night. It is, of course, important that the stomates remain open during the daylight hours for the efficient exchange of gases necessary for rapid photosynthesis. Thus, open stomates, while essential for efficient gas exchange, result in the loss of much water vapor from the plant. For this reason transpiration is sometimes called the "necessary evil" of the plant kingdom. But even evils have their advantages! For one, transpirational water losses create the tensions needed for water uptake. Then too, the conversion of water into vapor form within the leaf and its accumulation around the leaf act as coolants to leaf surfaces during the heat of the day. Putting this biological dilemma aside, the important thing for the gardener to recognize is that plants must be supplied with much more water than is necessary for dry matter production alone, because large amounts are transpired.

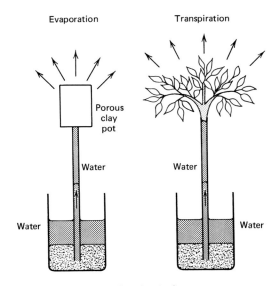

FIGURE 2-27 A simple physical apparatus to demonstrate the upward pull of water created by evaporation from the surface of a porous clay pot. A similar mechanism, driven by transpiration, operates in plants. (Reprinted with permission of Worth Publishers, Inc., from Peter H. Raven, Ray F. Evert, and Helena Curtis, *Biology of Plants.* Copyright © 1981 by Worth Publishers, Inc.)

Conduction of Food Materials

The carbohydrates manufactured in photosynthesis are moved throughout the plant in the phloem tissue, specifically in the sieve elements. While this movement is visualized to be predominantly downward to roots and storage organs, elaborated foods, termed **assimilates**, have been shown to move in other directions as well (Fig. 2-28). In general, young,

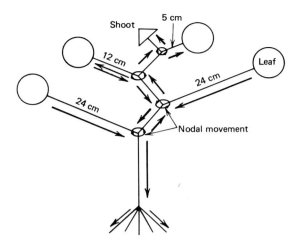

FIGURE 2-28 Movement of food materials throughout a 15-day-old squash plant. The arrows indicate direction, and their thickness suggests the quantities of material being moved. (Reprinted with permission of John Wiley & Sons from E. Epstein, *Mineral Nutrition of Plants.* Copyright © 1972 by John Wiley & Sons.)

rapidly growing leaves require an input of assimilates, but as they mature they become exporters of food materials. Similarly, developing fruits, regardless of their location on the shoot, draw on plant assimilates, but as they mature these demands slow or stop altogether.

What is the mechanism of transport of assimilates in the phloem tissue? While this question has not been completely answered, great strides have been made towards understanding phloem movement. Established facts relating to translocation in the phloem include the following: (1) assimilates, mainly sucrose (table sugar is almost pure sucrose), move in the sieve elements; (2) the material within sieve elements is under a positive pressure; (3) the rate of movement can approach 100 centimeters (a little over a yard; see Appendix B for metric-English conversion factors) per hour; and (4) the major direction of flow is from sites of manufacture, known as **sources**, to sites of utilization or storage, known as **sinks**. Any concept of movement must take into account these four factors.

The most widely accepted concept of phloem movement is the **pressure-flow** mechanism illustrated in Fig. 2-29. According to this concept the sugars manufactured in the leaf enter sieve tubes in quantity. This causes water, which always moves towards sites of its relative lower concentration, to enter the sieve tube, increasing turgor pressure. At the other end of the line, at sinks, sugar is removed from sieve tubes to be utilized in respiration, converted to other compounds, or stored as starch. The removal of sugar causes water to leave the sieve tube, and turgor pressure falls. As a result of the pressure differential at the two ends of the line—source and sink—the sugar solution flows through the sieve tube toward the sink. In Fig. 2-29 the root system is designated as a sink, but, as we have indicated, developing shoots and fruits can also be sinks.

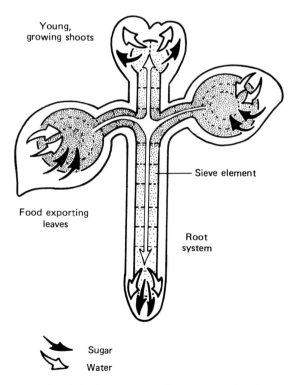

FIGURE 2-29 The pressure-flow mechanism for the transport of sugar in the phloem. (Courtesy of Guerry Dean.)

The source-sinks relationships that exist throughout the plant in relation to the movement of assimilates need to be understood by the gardener. Examples will serve to illustrate this point. In some garden vegetables, such as the summer squashes, the fruits are harvested while immature. Timely harvest of the developing fruits is important, because if left on the plant they continue to act as sinks, thereby drawing on assimilates that would otherwise be channeled to new sinks (other developing fruits). Thus, failure to harvest properly results in a decreased yield of quality fruits. In daffodil and tulip, flowering occurs early in the season and an extended period of vegetative growth follows. The showy flowers can be harvested for indoor use or left on the plant. In the latter case the flowering stalks, which act as sinks, can be removed as the flowers fade to prevent the needless development of fruits and seeds. The numerous leaves, which by now act as sources, should be left on the plant, however, because the assimilates they manufacture are essential to a renewal of food reserves in the bulb, a sink. The following spring the reserves of the bulb are mobilized and utilized in the development of flowers and leaves again. In both of these examples the gardener is, in reality, manipulating source-sink relationships. In many other situations, including the establishment of perennial food plants, the production of giant pumpkins or other fruits, and the eradication of certain weeds through starvation, a clear understanding of the movement of assimilates is required.

Movement of Growth Regulators

The **growth regulators**, are chemicals produced by plants, which, in low concentrations, regulate physiological processes. Generally these substances move within the plant from a site of production to a site of action, as illustrated by the following example. Long ago the eminent biologist, Charles Darwin, and his son, Francis, studied the phenomenon of curvature of plants toward light. They found that the natural curvature of seedlings could be prevented if the tip of the seedling was covered with a lightproof collar, whereas if the collar was placed below the tip normal curvature took place. Correctly, the Darwins concluded that, in response to light, an "influence" that causes bending is transmitted from the seedling tip to the area below, where bending occurs. The stimulus that causes bending was eventually isolated and identified. The substance has a complicated technical name, but even before its chemical identity was known the substance was named **auxin**. Auxin conforms nicely to our definition of a growth regulator, because it is transported in low concentrations from a site of production to a site of action and there induces a response.

The downward movement of auxin in the stem is believed to occur mainly in the cells of the cortex and pith, rather than in the xylem or phloem. While the actual mechanism of movement remains obscure the transport process is thought to involve an interaction between auxin and the outer membranes of plant cells.

Since the early discovery of auxin, several additional growth-regulating substances have been identified. Certain of these substances and some responses they are said to elicit are given in Table 2-7. While their mechanisms of movement within the plant are of relatively little interest to the gardener the responses they elicit are not. Indeed, many standard horticultural practices aimed at regulating plant growth and development are based on our understanding of the behavior of growth regulators. To illustrate, pinching back of the shoot tips of plants to encourage branching is based on the knowledge that the high levels of auxin production at the tips inhibit the development of lateral buds. The root-promoting compounds, now sold at most garden centers, became available only after the realization that auxin stimulates root initiation and growth. The storage life of many fruits was enhanced

TABLE 2-7 Some Plant Growth Regulators and Responses They Are Reported to Elicit

Growth regulator	Some reported actions
Auxin	Promotes cell elongation
	Controls **phototropism** (movement in response to light)
	Inhibits the growth of lateral buds
	Stimulates root initiation and growth
	Abscission correlated with diminished production
	Influences differentiation of vascular tissue
	Promotes growth of the cambium in woody plants
	Promotes fruit growth
Gibberellins	Promotes stem elongation
	Stimulates cell elongation and division
	Overcomes dwarfing in certain plants
	Overcomes dormancy in certain seeds
	Promotes maintenance of juvenile characteristics
	Promotes bolting (and subsequent flowering) in certain plants
	Stimulates pollen germination and growth
	Promotes fruit growth
Ethylene	Promotes leaf abscission
	Causes blanching of chlorophyll
	Promotes lateral expansion of cells, as opposed to elongation
	Hastens leaf abscission
	Promotes downward curvature of petiole
	Hastens fruit ripening
Florigen (?)	Induces flowering

only after it became known that the ethylene gas produced by developing fruits hastens ripening. Additional examples of the roles of growth regulators in garden plants are given elsewhere in the text.

Questions for Discussion

1. Define the following terms:
   (a) Taxa
   (b) Perennial
   (c) Vegetable
   (d) Etiolated
   (e) Adventitious
   (f) Cuticle
   (g) Compensation point
   (h) Hydathode

2. Common garden plants fall into either of two broad taxonomic groupings. Identify the two categories and give examples of plants in each.

3. Describe the system of classification developed by the Swedish physician, Linnaeus, and use it to identify a specific vegetable.

4. Give three reasons why the use of common names is not a good way to identify garden plants.

5. Describe the life cycle characteristics of biennial plants and name three vegetables that fall into this category.

6. Compare and contrast warm-season versus cool-season vegetables.

7. Sketch a seed of corn, bean, or pea and identify its major parts.

8. Identify morphological adaptations that allow germinated seed to successfully grow through the soil to the surface.

9. Make a diagram of a vegetable such as eggplant and label its numerous parts.

10. Distinguish between the growth activities of primary and secondary meristems.

11. List three organelles common in plant cells and state their functions.

12. What are the major raw materials needed for photosynthesis? How does the plant procure each of these materials?

13. Give reasons why the gardener should understand source-sink relationships in plants.

14. What are the criteria for classifying a plant chemical as a "growth regulator"?

Selected References

Edmond, J. B., T. L. Senn, F. S. Andrews, and R. G. Halfacre. 1975. *Fundamentals of Horticulture*. McGraw-Hill, New York.

Esau, K. 1965. *Plant Anatomy*. John Wiley, New York.

Esau, K. 1977. *Anatomy of Seed Plants*. John Wiley, New York.

Halfacre, R. G. and J. A. Barden. 1979. *Horticulture*. McGraw-Hill, New York.

Janick, J. 1972. *Horticultural Science*. Freeman, San Francisco.

Janick, J., R. W. Schery, F. W. Woods, and V. W. Ruttan. 1974. *Plant Science*. Freeman, San Francisco.

Leopold, A. C. 1975. *Plant Growth and Development*. McGraw-Hill, New York.

Porter, C. L. 1967. *Taxonomy of Flowering Plants*. Freeman, San Francisco.

Ray, P. M. 1972. *The Living Plant*. Holt, Rinehart and Winston, New York.

Rayle, D. and L. Wedberg. 1980. *Botany: A Human Concern*. Saunders College, Philadelphia.

Splittstoesser, W. E. 1979. *Vegetable Growing Handbook*. AVI Publishing Co., Westport, Conn.

Steward, F. C. 1964. *Plants at Work*. Addison-Wesley, Reading, Mass.

Weier, T. E., C. R. Stocking, M. G. Barbour, and T. L. Rost. 1982. *Botany*. John Wiley, New York.

Wilson, C. L. and W. E. Loomis. 1971. *Botany*. Holt, Rinehart and Winston, New York.

LEARNING OBJECTIVES

By the time you have finished this chapter you should be able to:

1   List several advantages associated with propagating plants by asexual means.

2   Name and describe three natural vegetative propagules formed by plants.

3   Define the term "crown" as it applies to garden crops.

4   Explain the process of layering.

5   Identify two or three food plants in which grafting plays an important role in their culture.

6   Outline the sexual life cycle in corn or other annuals.

7   Describe basic flower structure.

8   Discuss methods of pollination common among vegetables, including appropriate examples.

9   List three mechanisms that encourage crossing in plants.

10   Explain the ripening process in tomato or other fruits.

11   Identify three morphological features that create variation in the structure of fruits.

3

HOW PLANTS MULTIPLY

"Like a great poet nature produces the greatest results with the simplest means. There are simply a sun, flowers, water, and love."

Heine

Plants have evolved two methods of reproducing themselves—sexual and asexual. **Sexual reproduction** occurs within flowers and involves the union of sex cells and the subsequent development of embryos within seeds. Plants arising by the sexual process express characteristics of both parents, but are not identical to either. **Asexual reproduction** is any reproductive process that does not involve the union of gametes. This form of reproduction results in offspring that are genetically identical to the parent plant. Sexual reproduction is discussed later in the chapter.

THE IMPORTANCE OF ASEXUAL REPRODUCTION

Asexual reproduction normally involves the vegetative parts of the plant (stems, roots, leaves), and for this reason is often called vegetative reproduction or vegetative propagation. Plants arising by asexual means do not differ genetically from the parent plant. The term **clone** is used to designate all of the descendants of a single plant, produced vegetatively. All of the individual plants of a given cultivar of apple or pear, such as 'Delicious' or 'Bartlett' comprise a clone, for example. So do all of the Irish potato plants derived from the growth of vegetative buds from a single tuber.

The vegetative propagation of plants is of immense importance. It has been estimated that about half of our cultivated plants are reproduced vegetatively, even though most produce viable seed. A partial list of those familiar to the gardener is given in Table 3-1. There are many reasons for propagating plants vegetatively, the most important of which are the following.

TABLE 3-1 Methods of Propagating Some of Our Important Cultivated Crops

By vegetative methods	By seed
Artichoke	Beet
Asparagus	Brassicas (broccoli, Brussels sprouts, cabbage, cauliflower, collards, kale, kohlrabi)
Banana	Carrot
Citrus fruits (orange, etc.)	Celery
Garlic	Cereals (corn, oat, rice, wheat)
Horseradish	Cucurbits (cucumber, gourds, melons, pumpkin, squash)
Jerusalem artichoke	Eggplant
Nut crops (English walnut, filbert, pecan)	Legumes (bean, pea, peanut, soybean)
Pineapple	Lettuce
Pome fruits (apple, pear)	Onion
Potato (sweet, Irish)	Parsnip
Rhubarb	Pepper
Shallots	Tomato
Small fruits (blackberry, grape, strawberry)	Turnip
Stone fruits (apricot, cherry, peach, plum)	

the desired type and perpetuation of hybrid vigor.

2. Some plant cultivars lack sexual reproduction entirely and must be propagated vegetatively. Such is the case with the cultivated banana, the pineapple, the navel orange, and certain grapes.

3. Often a fully developed and healthier plant can be obtained in much less time through vegetative reproduction. Seeds of some species have a long dormancy period or their seedlings grow slowly, whereas vegetative portions, for example stem cuttings, become established rapidly. Furthermore, shoots of favorable plant types can be grafted to roots of closely related plants to give better adaptation to specific soil conditions, disease resistance, or dwarfing effects.

Two broad categories of asexual reproduction are recognized. The first makes use of *natural* plant structures such as bulbs, tubers, and rhizomes. The second encompasses situations where *artificial* techniques such as grafting, budding, and certain forms of layering are used to multiply plants. We discuss examples from each of these areas as they relate to modern home gardening.

1. It makes possible the retention of desirable features of plants that do not "breed true" from seeds. Many prized plant cultivars are **hybrids** (offspring of parent plants differing in one or more characters). Plants grown from seed produced by hybrids will differ greatly from the parents and from each other. If seed from a 'Delicious' apple are planted, for example, it is quite unlikely that the resulting trees would have all the desirable qualities of the parent tree. Only asexual propagation assures retention of

VEGETATIVE PROPAGATION USING NATURAL STRUCTURES

Natural methods of vegetative propagation occur by means of specialized stems and roots. We deal first with stem modifications and then those involving roots.

Vegetative Propagation by Stems

Runners

A **runner** is a specialized stem that develops from the axil of a leaf near the **crown** of a plant (Fig. 3-3), grows horizontally along the ground, and forms new plants at intervals. The garden plant noted for reproduction by runners is the strawberry, *Fragaria* (Fig. 3-1). In strawberries the runners produce a new plant at every second node. The daughter plants also produce runners and these runners, in turn, produce new plants at alternate nodes. The new plants take root but remain attached to the mother plant. The connecting stems die in the late fall and winter thus separating the individual plants, which can be transplanted to new locations when they have become well rooted. Runner formation in the strawberry has been found to be related to both length of day and temperature. Runners are produced in long days of 12 or more hours and at high midsummer temperatures.

Several weed pests owe their troublesome nature to rapid propagation by means of runners. The tropical grass, Bermuda grass, is a good example. The grass is used in the southern United States as a forage for livestock, and in lawn mixes. In this same region it is often referred to as "wiregrass" or "devilgrass" because it invades gardens and cultivated fields with its vigorous runners. As we will see, many troublesome weeds reproduce extensively by vegetative means. The control of weed pests is discussed in Chapters 8 and 9.

Rhizomes

Many important garden plants reproduce asexually by means of another specialized stem structure in which the main axis grows at or just below the ground surface. This modified stem, known as a **rhizome**, is characteristic of several economically important plants including bamboo, banana, iris, Kentucky bluegrass, and sugarcane. Some of our most serious weed pests, including Canada thistle, johnsongrass, and quackgrass also spread by means of rhizomes. Rhizomes may be short, thick, and fleshy as in canna (Fig. 3-2) and iris, or relatively long, thin, and nonfleshy as in Kentucky bluegrass. Regardless of form, rhizomes are stems that possess definite nodes and internodes. Each node has an axillary bud subtended by a scale leaf. Upright, aboveground shoots are produced either terminally from the rhizome tip or from branches developing from the lateral buds. Adventitious roots may also develop along the rhizome axis. These

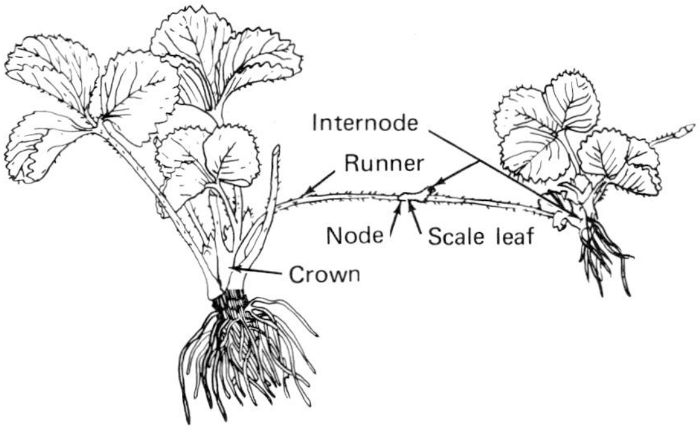

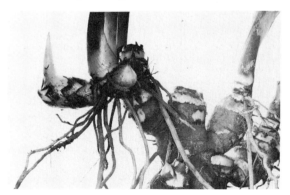

FIGURE 3-1 Runner of strawberry; roots and shoots are produced at every other node. (Reprinted with permission of John Wiley & Sons from T. Elliott Weier, C. Ralph Stocking, Michael G. Barbour, and Thomas L. Rost, *Botany: An Introduction to Plant Science.* Copyright © 1982 John Wiley & Sons.)

FIGURE 3-2 The short, fleshy rhizomes of canna produce shoots and roots at every third node. (Reprinted with permission of John Wiley & Sons from T. Elliott Weier, C. Ralph Stocking, Michael G. Barbour, and Thomas L. Rost, *Botany: An Introduction to Plant Science.* Copyright © 1982 John Wiley & Sons.)

perennial, underground stems are most commonly associated with monocotyledons, but some dicotyledons, for example, lowbush blueberry, have underground stems regarded as rhizomes. If the tubers of the Irish potato and the Jerusalem artichoke are interpreted as arising at the end of a rhizome, then these two plants can also be added to the list of rhizomatous dicotyledons.

Tubers

A **tuber** is defined as a much-enlarged, short, fleshy underground stem. These storage structures have been variously interpreted as arising at the **distal** (farthest from the place of origin or attachment; see Fig. 3-3) end of a runner or rhizome. The Irish potato, *Solanum tuberosum* (Fig. 3-4*a*), and the edible portion of the Jerusalem artichoke (Fig. 3-4*b*) are both tubers. In a mature potato a scar is visible where the tuber was broken from the elongate, underground stem. The tuber, being a true stem, has definite nodes and inter-

nodes, lateral buds, and a terminal bud. The "**eyes**" present in a spiral arrangement around the tuber represent the nodes. Each eye consists of one or more small buds subtended by a leaf scar. The eye farthest away from the point of attachment to the stem is the terminal bud. Characteristically in stems, the terminal bud inhibits the development of the lateral buds below it, a phenomenon termed **apical dominance**. The terminal bud of a potato shows the same apical dominance as the terminal bud of other stems.

If a whole tuber is planted the terminal bud usually inhibits the other buds, whereas if smaller pieces, known as **seed pieces**, are planted each will give rise to a new potato plant. In the fall when potato tubers are dug the buds are dormant. This dormancy, which lasts from six to eight weeks, must, of course, be broken before sprouting will take place. Supermarket potatoes are usually treated chemically to prevent sprouting and will not grow if planted.

Tuber formation begins with the inhibition of stem elongation and the enlargement and division of cells in the region of the stem tip. Reduction in day length, reduced temperatures (especially at night), and a lowered mineral content stimulate tuberization in potato. In general, conditions that favor rapid growth of the aboveground portion of the plant do not promote tuber formation.

Corms and Bulbs

Both corms and bulbs are the enlarged bases of aerial stems, and both are produced by certain herbaceous monocotyledons. The two differ in basic structure, however.

A **corm** is the swollen base of a stem axis enclosed by dry, scalelike leaves. It is a solid stem structure with definite nodes and internodes. No garden vegetables are propagated by means of corms, but most gardeners are familiar with the corms of gladiolus (Fig. 3-5*a*) and crocus. When a large, mature corm

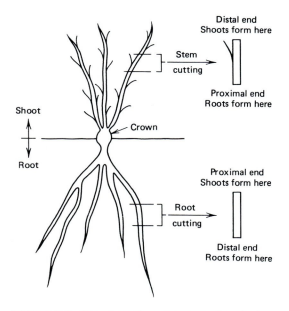

FIGURE 3-3 Shoot, root, and crown of a plant. Where shoots and roots originate on cuttings from stems and roots is also shown.

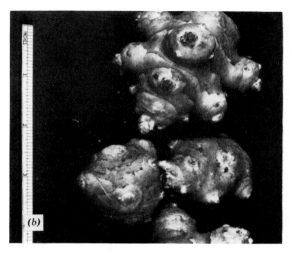

FIGURE 3-4 Edible tubers of the Irish potato (*a*) and Jerusalem artichoke (*b*). (*a*, reprinted with permission of John Wiley & Sons from T. Elliott Weier, C. Ralph Stocking, Michael G. Barbour, and Thomas L. Rost, *Botany: An Introduction to Plant Science.* Copyright © 1982 John Wiley & Sons. *b*, courtesy of U.S. Department Agriculture.)

is planted roots develop from its base. Soon leaves appear above ground and before long a flowering shoot is produced. The food supply in the corm is used up in flower production, and as the season advances the shoot forms a new corm above the old. It is important that the gardener not remove the vegetative foliage after flowering, because the leaves continue to manufacture food materials that are stored in the new corm. Miniature corms, termed **cormels**, may develop between the old and new corm. If planted these will grow vegetatively and produce new, larger corms. These require one to two years to reach flowering size.

The onion is a good example of a **bulb**, a short, usually vertical stem axis with many fleshy scale leaves filled with stored food. Toward the center of the bulb the scale leaves are more leaflike, and in the center there is either a vegetative meristem or an unexpanded flowering shoot. The leaves of the bulb form meristems (buds) in their axils, some of which produce miniature bulbs known as **bulblets**. When grown to full size these bulblets are known as **offsets**.

There are two basic types of bulbs—tunicate and nontunicate. **Tunicate** bulbs are characterized by dry outer scales called a **tunic**. The tunic provides protection from drying and mechanical injury. In onions, those with a red-colored tunic have a disease resistance not found in white tunicate types. **Nontunicate** bulbs, such as those of lily, have no dry covering and consequently require more care in handling to reduce bruising (Fig. 3-5*b*). Scaly bulbs generally have roots that persist through two growing seasons.

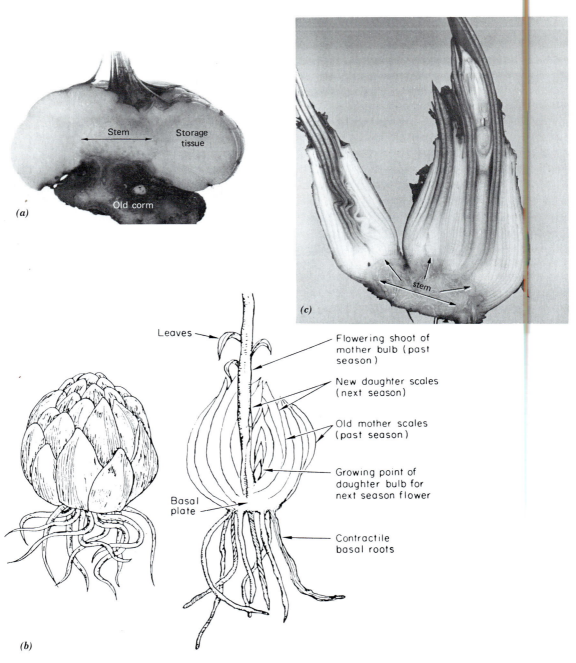

FIGURE 3-5 Corms and bulbs. The stem of gladiolus (*a*) illustrating storage tissue in corm of the current year and decayed corm of preceding year. The nontunicate bulbs of the lily (*b*) have no dry covering and thus require more care in handling. A longitudinal section through the stem of daffodil (*c*) showing two bulbs united by a short stem. (*a* and *c* reprinted with permission of John Wiley & Sons from T. Elliott Weier, C. Ralph Stocking, Michael G. Barbour, and Thomas L. Rost, *Botany: An Introduction to Plant Science.* Copyright © 1982 John Wiley & Sons. *b* same as for 3-6 on p. 55.)

Several garden vegetables and flowers multiply by means of bulbs. The vegetables include these important members of the lily family—onion, garlic, shallot, and chive. The shallot develops several small bulblets, or **cloves**, held together at the base, and garlic forms a group of small bulbs or cloves enclosed in a membranelike skin. Bulbous flowers include the tulip, hyacinth, lilies, daffodils (Fig. 3-5c), and ornamental alliums. The manipulation of bulbs will be discussed in the *practices* section of the book.

Vegetative Propagation by Roots

Tuberous Roots

Several species of herbaceous perennials produce thickened, somewhat fleshy roots that contain stored food. These storage structures, termed **tuberous roots**, are best known in the sweet potato, *Ipomoea batatas* (no relation to the Irish potato) and the dahlia (Fig. 3-6). Tuberous roots, like other roots, lack nodes and internodes. The areas on a sweet potato that look superficially like "eyes" are places where branch roots were attached. When examining a sweet potato note that the eyelike areas are in more or less vertical rows characteristic of branch roots. In contrast, the eyes of the Irish potato have a spiral arrangement around the tuber, just as do leaves on the aboveground portion of the stem.

In propagation by tuberous roots it is important in many species, including dahlia, that a portion of the crown be present, because shoot buds occur only on this end. Similarly, fibrous roots are produced only at the distal end. The sweet potato, on the other hand, forms adventitious shoots and roots along the thickened root axis. Note that in the sweet potato, shoots arise at the proximal end of the tuberous root or root segment. If an Irish potato is planted as a single unit, shoots develop only at its distal end. Can you explain why this is so?

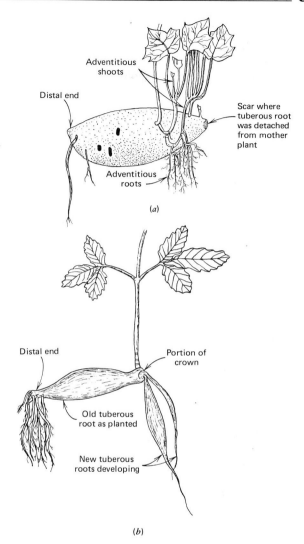

FIGURE 3-6 Tuberous roots. Sweet potato (*a*), showing adventitious shoots and roots. Dahlia (*b*). The old root pieces will disintegrate in the production of the new plant; the newly developing tuberous roots can be used for propagation next season. (Reprinted with permission of Prentice-Hall, Inc. from *Plant Propagation: Principles and Practices* by Hudson T. Hartmann and Dale E. Kester. Copyright 1975 by Prentice-Hall Inc.)

Other Forms of Vegetative Propagation

While many garden plants are propagated by vegetative structures that can be clearly designated as runners, rhizomes, tubers, and so on, several others increase vegetatively from nonspecific organs.

In herbaceous perennials, for example, the aboveground portion may die back each fall, with new shoots arising the following spring from the crown region. The common garden perennial, rhubarb, is multiplied by a crown portion that is variously interpreted as a rhizome or a root. Asparagus possesses both fleshy roots and rhizomes, and both are used in propagation. Garden flowers such as aster and chrysanthemum can be multiplied by **division** of crowns too, and in the strawberry and daylily (Fig. 3-7) lateral shoots result in a number of distinct crowns that can be divided.

Many plants give rise to new shoots from the vicinity of the crown or some distance out from the crown. These new shoots, which arise from below ground, are known as **suckers**.

FIGURE 3-7 Propagation by means of crown division in daylily. [*Hemerocallis*]. (Reprinted with permission of Prentice-Hall, Inc. from Hudson T. Hartmann and Dale E. Kester, *Plant Propagation: Principles and Practices*. Copyright 1975 by Prentice-Hall Inc.)

Most commonly suckers arise from adventitious buds that form on roots, as in the red raspberry (Fig. 3-8a). After suckers are well rooted they can be separated from the parent plant and established in a new location. Transplanting is usually done during the dormant season.

ARTIFICIAL VEGETATIVE PROPAGATION METHODS

Gardeners sometimes propagate plants vegetatively using methods that do not occur in nature, or by extending natural methods to other plant types. The most common techniques employed are layering, cuttings, and grafting.

Vegetative propagation by layering and cuttings hinges upon the natural ability of plants to *regenerate* missing parts. For example, a root must initiate a new shoot, a shoot must initiate new roots, and a leaf must form both shoots and roots. In **layering**, the plant part remains *attached* to the parent plant, whereas **cuttings** are plant parts *detached* from the parent plant. Propagation by layering offers somewhat better chances for success because the plant part in question can receive nutrition from the parent plant until it forms the needed portions. The timing and techniques used in layering are not as critical either.

Layering

In layering we are generally trying to induce the formation of adventitious roots on a stem while it is attached to the parent plant, after which the newly rooted stem can be detached and will continue growth on its own root system. Layering occurs naturally as when strawberry runners touch the ground and take root. Layering occurs both naturally and is induced in **brambles** (red and black raspberries, blackberries, loganberries, and other members of the genus *Rubus*) when drooping stem tips contact the soil and take root (Fig. 3-8b).

Several methods of layering are illustrated in Fig. 3-9. In layering many plants it is necessary to form a wound along the stem. This causes an interruption in the downward movement of food materials and growth regulators and thereby stimulates root formation. Common garden plants propagated by layering include the brambles, currants, gooseberries, filberts, some citrus fruits, grapes, and many woody ornamentals.

Cuttings

Stems, roots, and leaves are all used as cuttings, but stems are probably the most popular. Stem cuttings are sometimes called **slips**, especially if they are nonwoody. The general procedure in making cuttings is to remove from a plant a stem section 3 to 12 inches or even more in length, bearing several nodes with lateral buds. When such stems are placed in sand (a rooting medium), roots grow from the basal end and the uppermost buds develop into shoots (Fig. 3-10a). Plant hormones, such as indolebutyric acid (IBA), added

to the cut end will often promote rooting (Fig. 3-10b).

Successful rooting of cuttings hinges on several environmental factors, including humidity, temperature, light, and the rooting medium used. In general the humidity should be kept high to prevent desiccation before rooting and to reduce transpiration. A temperature of about 75°F in the rooting medium promotes rooting. On the other hand, the aerial portion can be kept cooler to reduce transpiration and respiration. A lower light intensity favors more rapid rooting but reduces food manufacture through photosynthesis. In

FIGURE 3-8 Vegetative propagation (a) Suckers arising as adventitious shoots from the roots of a red raspberry plant. (b) Tip layer in boysenberry. The rooted tip can be severed from the parent plant and transplanted to a new location. Since new tip layers are tender and dry out easily, replanting should be done soon after digging. (Reprinted with permission of Prentice-Hall, Inc., from Hudson T. Hartmann and Dale E. Kester, *Plant Propagation: Principles and Practices.* Copyright 1975 by Prentice-Hall Inc.)

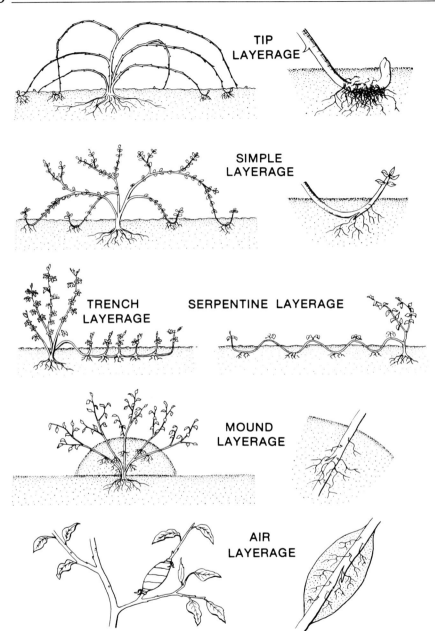

TIP
LAYERAGE

SIMPLE
LAYERAGE

TRENCH
LAYERAGE

SERPENTINE LAYERAGE

MOUND
LAYERAGE

AIR
LAYERAGE

FIGURE 3-9 Common methods of layering. (Reprinted with permission of Macmillan Publishing Co., Inc., from Ervin L. Denisen, *Principles of Horticulture*. Copyright 1979 by Macmillan Publishing Co., Inc.)

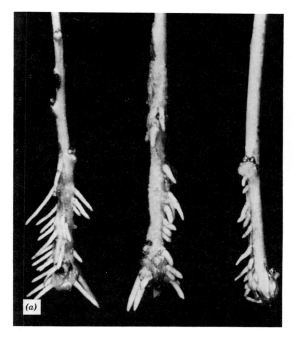

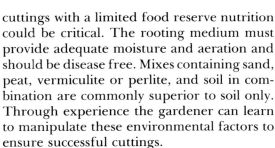

FIGURE 3-10 Vegetative propagation by cuttings. (a) Formation of adventitious roots on stem cuttings of plum. (b) Tomato leaf cuttings treated with a plant hormone and placed in sand for 12 days. Left to right: no treatment, plant hormone (in talc) at 1000 ppm, at 3300 ppm, and at 8000 ppm. (Reprinted with permission of Prentice-Hall, Inc. from Hudson T. Hartmann and Dale E. Kester, *Plant Propagation: Principles and Practices.* Copyright 1975 by Prentice-Hall, Inc.)

cuttings with a limited food reserve nutrition could be critical. The rooting medium must provide adequate moisture and aeration and should be disease free. Mixes containing sand, peat, vermiculite or perlite, and soil in combination are commonly superior to soil only. Through experience the gardener can learn to manipulate these environmental factors to ensure successful cuttings.

Few food plants are propagated by cuttings (grape and pineapple are exceptions) but a vast number of indoor plants and ornamentals, including African violet, begonia, cacti, camellia, coleus, geranium, holly, juniper, rhododendron, rose, viburnum, willow, and yew are commonly propagated by this method.

Grafting

Grafting is an ancient art. The Chinese apparently knew of this art as early as 1000 B.C., and it was a common practice throughout the Roman Empire. **Grafting** is defined as the art of joining parts of plants together in such a manner that they will unite and continue their growth as one plant. The graft segment that is to become the upper part of the new union is termed the **scion**, and that which is to become the lower or root portion is termed the **rootstock**, or simply the **stock**.

In grafting, the plants to be joined must have a close natural relationship. Generally grafting within a species results in successful (or **compatible**) unions, but grafts made between species of the same genus or between species of closely related genera vary in degree of compatibility.

The vascular cambia of the graft partners must also be properly aligned, and they must be held firmly together. This is necessary because cells in the cambial region of the stock and scion can multiply (Fig. 3-11a), thereby forming an interlocking mass of parenchyma tissue known as **callus**. Subsequently a new cambium develops in the callus mass and joins the plant parts by a structural and physiological connection (Fig. 3-11b). Temperatures from about 80 to 85°F are most conducive to

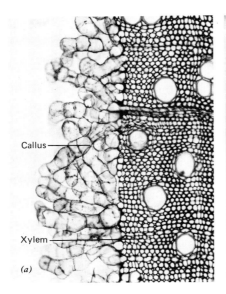

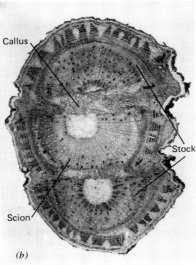

FIGURE 3-11 Callus formation and grafting in hibiscus. (*a*) Callus production from incompletely differentiated xylem, ×120. (*b*) Cross section of graft union showing callus tissue and vascular tissue produced by cambium of callus, ×10. (Reprinted by permission of John Wiley & Sons from Katherine Esau, *Plant Anatomy.* Copyright 1965 John Wiley & Sons.)

callus formation, a prerequisite to successful union. Finally, the scion, severed from the parent plant, is vulnerable to drying out. To prevent this the graft union should be thoroughly waxed. Prompt waxing of the graft union also helps prevent infection by pathogenic organisms.

There are many different kinds of grafts, the more popular being the cleft, whip-and-tongue, and the side graft (Fig. 3-12). The cleft graft technique allows one to graft a small scion (one-quarter to one-half inch diameter works best) onto a branch of large diameter (one to four inches commonly). The whip-and-tongue graft technique lends itself to the union of graft partners of a more or less uniform diameter. Specimens with a diameter of one-quarter to one-half inch are considered ideal, but the technique works with pieces up to one inch in diameter. The side graft is used extensively, especially on certain evergreens. The Japanese employ this grafting technique on bonsai plants. Commonly in side grafting the stock piece has a diameter of about one inch, and the scion a lesser diameter.

A successful variation of standard grafting is to use a single bud for the scion, a proce-

dure known as **budding**. As in regular grafting, the scion can take several shapes at the region of attachment to the stock.

Grafting offers several advantages. It can promote growth and productivity of the scion through a superior root system, or it can restrict scion growth, resulting in a smaller plant—a process known as **dwarfing**. By grafting a desired cultivar onto a disease-resistant rootstock one can combat soil-borne disease organisms. Or a new, more useful cultivar can be grafted onto an established stock, thereby making use of an already extensive root system. Grafting can be used to enhance pollination, and to repair damaged parts of trees. Of interest to those with only a small garden space is the fact that several cultivars of a fruit species can be grown on a single tree of that species.

Generally grafting is not used in the production of vegetables. An exception occurs in Europe where greenhouse cucumbers and tomatoes are grafted onto disease-resistant rootstocks. On the other hand, many tree fruits, nuts, and ornamentals are cultivated commonly as grafted specimens. Almond, apple, apricot, camellia, cherry, citrus fruits, grape,

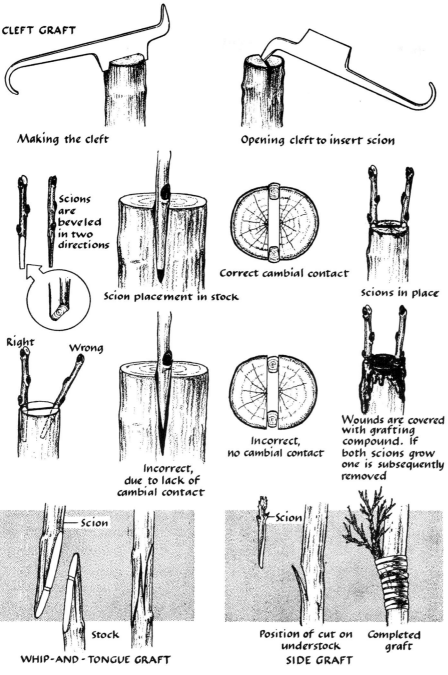

CLEFT GRAFT

Making the cleft

Opening cleft to insert scion

Scions are beveled in two directions

Scion placement in stock

Correct cambial contact

Scions in place

Right   Wrong

Incorrect, due to lack of cambial contact

Incorrect, no cambial contact

Wounds are covered with grafting compound. If both scions grow one is subsequently removed

Scion

Stock

WHIP-AND-TONGUE GRAFT

Scion

Position of cut on understock

SIDE GRAFT

Completed graft

FIGURE 3-12 Grafting. (From Jules Janick, *Horticultural Science*, 2nd ed., W. H. Freeman and Company. Copyright © 1972.)

hibiscus, holly, lilac, peach, pear, plum, rose, and walnut are examples. Note that this list does not include any monocotyledons. Most members of this group lack a discrete vascular cambium and therefore cannot be grafted.

Our somewhat brief discussion of vegetative propagation only outlines its many aspects. Readers who wish to pursue the topic further are directed to the references at the end of the chapter, and especially to the text *Plant Propagation, Principles and Practices* by Hudson T. Hartmann and Dale E. Kester.

SEXUAL REPRODUCTION AND THE FLOWERING PLANT LIFE CYCLE

In garden plants sexual reproduction occurs within flowers, and results in the formation of fruits that contain seed. Sexual reproduction involves the formation of male and female gametes, or sex cells, and their fusion to form a **zygote** (first cell of a new individual). Growth of the zygote results in an embryo contained within a seed. These seed, when planted, grow into new plants that display characteristics of both parents.

The life cycle of a typical garden annual, corn, is shown in Fig. 3-13. On a mature corn plant the female flowers are represented by the ear shoots ("silks") and the male flowers by the tassels. Within the ear shoots, female gametes, termed **egg cells**, are produced. Each egg cell contains only one-half (n) as many **chromosomes** (heredity carriers) as the large parent cell ($2n$). At the same time that egg cells are being formed in ear shoots, male gametes, or **sperm**, are being produced in the male flowers of the tassel. The sperm also contain only one-half the number of chromosomes as the parent cells from which they were derived. The male gametes develop in pairs inside of specialized structures termed **pollen grains**, or simply **pollen**.

For the life cycle of a corn plant to be completed the male gametes must be brought into proximity of egg cells. This allows the gametes to fuse together, forming zygotes that develop into embryos within the seed. The chain of events begins when pollen from the tassels is transmitted by wind to the silks—the act of **pollination**—and there it germinates. The germinated grains form long tubes as they grow down the silks (styles) to their base where the egg cells are located. When a pollen tube comes into contact with an egg cell one of the sperms fuses with the egg cell—the act of **fertilization**—and a zygote is formed. The other sperm fuses with an adjacent cell initiating the endosperm, or storage tissue. Because both of the sperms fuse with cells contained within the base of silks the fertilization act in corn (and other angiosperms) is said to be one of **double fertilization**. Soon after fertilization the endosperm starts its development and a short time later development of the zygote into an embryo begins.

The fusion of egg and sperm yield a zygote that over time develops into an embryo within a seed. As indicated, during the formation of gametes the chromosome number of cells was halved. It is important to emphasize that with the fusion of gametes to form the zygote, the larger chromosome number of cells is restored so that the embryo that gives rise to a new corn plant of the next generation has cells with a chromosome number equal to that of the parent plants of the previous generation. The nature of the hereditary characters carried by these chromosomes depends, of course, on the characteristics of the parent plants. A key aspect of sexual reproduction is that each parent contributes equally (one-half of the chromosomes) to the offspring and so it is not at all surprising that the offspring reflect the genetic contribution of each parent.

Many of the gardeners' activities relate directly or indirectly to the sexual cycle of plants. The gardener plants seeds, worries about—and sometimes aids—pollination, and harvests fruits and seeds. Indeed, success or failure in gardening can hinge upon an understanding of the sexual cycle in plants, and for

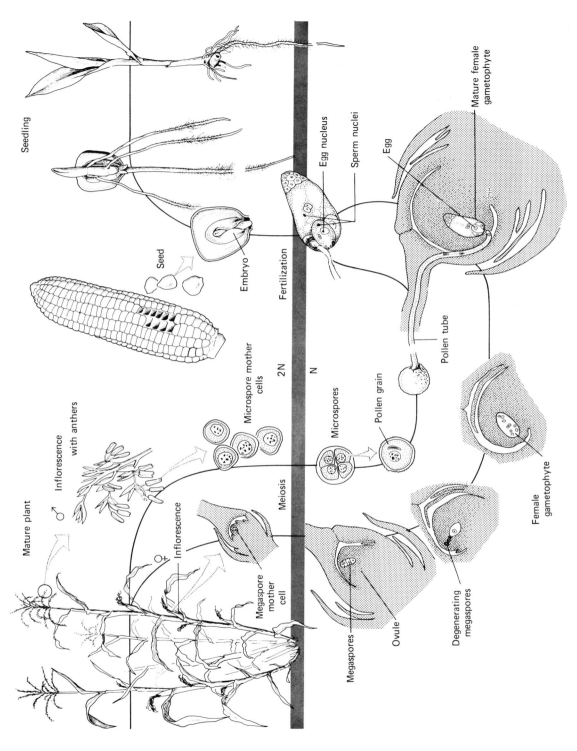

FIGURE 3-13 Life cycle of corn. (Reprinted with permission of Wadsworth Publishing Company, Inc. from William A. Jensen and Frank B. Salisbury, *Botany: An Ecological Approach.* Copyright © 1972 by Wadsworth Publishing Company, Inc.)

Seedling

Egg nucleus

Sperm nuclei

Egg

Mature female gametophyte

Seed

Embryo

Fertilization

Pollen tube

Microspore mother cells

2N

N

Pollen grain

Female gametophyte

Inflorescence with anthers

Microspores

Meiosis

Mature plant

♂

Inflorescence

♀ Inflorescence

Megaspore mother cell

Megaspores

Ovule

Degenerating megaspores

this reason we now look at the flower, flowering, and fruit set and maturity.

The Flower

Nearly 200 years ago Johann Wolfgang von Goethe, the great German poet and naturalist, described the flower as a shortened stem bearing modified leaves. Numerous other interpretations of the flower have been offered since, but no other has stood the test of time. Consequently, most botanists continue to adopt the theory formulated by Goethe.

The shortened stem, to which the modified leaves of the flower are attached, is known as the **receptacle** (Fig. 3-14). The specialized leaves of the flower occur in four distinct whorls, identified as sepals, petals, stamens, and pistils. The lowermost whorl, the **sepals**, are usually green and leaflike. The **petals**, found just above the sepals, are usually larger and brightly colored. The **stamens** generally are not leaflike. Each stamen terminates in an anther in which pollen is formed. The **pistil** or pistils of a flower are at the tip of the receptacle. Each pistil consists of a stigma, style, and ovary. It is within the ovary that the ovules which produce egg cells are found.

The flower of Fig. 3-14 has four kinds of specialized leaves—sepals, petals, stamens, and pistils. This represents the totality of floral appendages, and any flower with all four is termed **complete**. If any of the four is lacking the flower is **incomplete**. Of the four specialized leaves only two—the stamens and pistils—are directly involved in reproduction. These are called the **essential** parts of a flower. Flowers are **perfect** if they contain both of the essential floral organs, **imperfect** or **unisexual** if they contain only one. Some cucurbits form imperfect (male flowers) and perfect flowers on the same plant (Table 3-2).

The flowers of corn (silks and tassels) are unisexual (Fig. 3-13). When both staminate and pistillate flowers occur on one plant, as they do in corn and several other vegetables (Table 3-2), the plant is said to be **monoecious**. When unisexual flowers are found on different plants, as in asparagus and spinach, the plant species is said to be **dioecious**.

Flowers may occur either singly—that is, they are **solitary**—or in natural groups, known as **inflorescences**. Several different inflorescence types are found among common garden plants (Fig. 3-15). Flowers and inflorescences are not always easily distinguished. In plants such as the sunflower, for example, the flowers within an inflorescence are compacted and of two different forms resulting in an

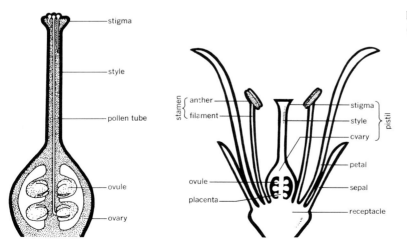

FIGURE 3-14 Parts of a flower. (Courtesy of U.S.D.A.)

inflorescence that superficially resembles a single flower.

With respect to the formation of inflorescences the shoot growth of plants can be classified as indeterminate or determinate. In **indeterminate** growth plants form flower clusters laterally along stems and the shoot tips remain vegetative and continue to grow until stopped by senescence or unfavorable growing conditions. In **determinate** plants, on the other hand, the shoot tips change from vegetative to reproductive; that is, they form flower parts rather than foliage leaves. With the formation of flowers, shoot elongation stops.

As one might expect, these differing patterns of growth can have a striking influence on plant stature. In the tomato, for example, both the determinate (also known as self-topping or self-pruning) and indeterminate forms occur. According to a leading authority on the tomato, Dr. Charles M. Rick of the University of California, tomato plants with determinate growth characteristics grow in an orderly, compact manner, in contrast to the sprawling, unlimited growth of indeterminate plants (Fig. 3-16). In Dr. Rick's own words:

> The branches of "determinate" plants terminate their growth at approximately the same distance from the center of the plant, and the plant flowers more abundantly than the normal (indeterminate) type. As a result the grower has plants in a form that facilitates harvesting, particularly by machines, and the fruiting is concentrated in a shorter season.

Somewhat similar comments could be made about pole (indeterminate) and bush (determinate) beans.

Not all contrasting patterns of growth are related to the determinate *versus* indeterminate habit, however. In pea and cucumber, for example, both full size and compact forms occur. These differing statures are caused by genetically controlled differences in the length of fully extended internodes, rather than the termination of vegetative growth at shoot tips.

Flowering

As indicated, as a plant matures some of its meristems undergo a transition from the vegetative phase to the reproductive phase. One event that puzzled scientists for decades was the nature of the stimulus that brings about the initial transformation from an apex producing leaves to one producing floral appendages. Little progress was reported in solving this puzzle until the early 1920s when two U. S. Department of Agriculture scientists, W. W. Garner and H. A. Allard, reported that in a certain tobacco cultivar **photoperiod** (length of day) was important; that is, short days were required to trigger flowering.

It soon became apparent that a great many plants responded, and some failed to respond, to variations in day length, and among these plants three distinct patterns emerged, as follows:

Short-day plants—initiate flowers only when the day length is below about 12 hours.

Long-day plants—initiate flowers only when the day length exceeds 12 hours.

Day-neutral plants—can initiate flowers independent of day length.

Subsequent research has shown that the situation with regards to flowering is more complex than suggested above. For example, it turns out that plants respond to the length of the *dark* period rather than the light period. That is, short-day plants require an uninterrupted dark treatment *longer* than some critical length in order to flower, whereas long-day plants require a dark period *shorter* than some critical length for flowering. This concept is illustrated in Fig. 3-17. Note that if the night length is interrupted, even by a light period of less than one minute in duration, the short-day plant will not flower. Conversely, the long-day plant will flower because its dark period was shorter than the critical length.

TABLE 3-2 The Reproductive Characteristics of Some Common Vegetables

Family	Crop	Genus and species	Flowering type[a]	Selfed or crossed[b]	By[c]	Life cycle[d]	Variety[e] type	Remarks
Chenopodiaceae	Beet	*Beta vulgaris*	Perfect	C	W	B	OP	All forms cross, including sugar beet
(beet)	Swiss Chard	*Beta vulgaris*	Perfect	C	W	B	OP	Crosses with other beets
	Spinach	*Spinacia oleracea*	Dioecious	C	W	A	OP or H	Also contains perfect-flowered plants
Asteraceae	Lettuce	*Lactuca sativa*	Perfect	S		A	PL	Seed may carry lettuce mosaic
(sunflower)	Globe artichoke	*Cynara scolymus*	Perfect	C	I	P	C	Poor types from seed
	Jerusalem artichoke	*Helianthus tuberosus*	Perfect	C	I	P	C	Seedlings variable
Convolvulaceae (morning glory)	Sweet potato	*Ipomoea batatas*	Perfect	C	I	P	C	Propagate by rooted shoots from tubers—often do not flower
Brassicaceae (mustard)	Cabbage	*Brassica oleracea*	Perfect	C	I	B	OP or H	All vegetables in this group cross readily; are often self-incompatible
	Cauliflower	*Brassica oleracea*	Perfect	C	I	B	OP or H	
	Broccoli	*Brassica oleracea*	Perfect	C	I	A	OP or H	
	Brussels sprouts	*Brassica oleracea*	Perfect	C	I	B	OP or H	
	Kohlrabi	*Brassica oleracea*	Perfect	C	I	B	OP	
	Kale	*Brassica oleracea*	Perfect	C	I	B	OP	
	Collards	*Brassica oleracea*	Perfect	C	I	B	OP	
	Turnip	*Brassica rapa*	Perfect	C	I	B	OP or H	
	Rutabaga	*Brassica napobrassica*	Perfect	C	I	B	OP	
	Chinese cabbage	*Brassica pekinensis*	Perfect	C	I	A	OP or H	
	Radish	*Raphanus sativus*	Perfect	C	I	A	OP	Crosses with wild forms
Cucurbitaceae	Cucumber	*Cucumis sativus*	Monoecious	C	I	A	OP or H	Does not cross with muskmelon
(gourd)	Muskmelon	*Cucumis melo*	Andromo-noecious or monoecious	C	I	A	OP or H	Includes netted (cantaloupe), honeydew, casaba, and mango melon
	Squash	*Cucurbita maxima*	Monoecious	C	I	A	OP or H	

Family	Crop	Scientific name	Flower type[a]	[b]	Pollinated by[c]	[d]	[e]	Comments
	Pumpkins	*Cucurbita pepo*	Monoecious	C	I	A	OP or H	
		Cucurbita moschata	Monoecious	C	I	A	OP or H	
	Watermelon	*Citrullus vulgaris*	Monoecious	C	I	A	OP or H	Seedless varieties are sterile triploids
Poaceae (grass)	Sweet corn	*Zea mays*	Monoecious	C	W	A	OP or H	All corn types (pop, flint, dent, flour, sweet) cross
Fabaceae	Bean	*Phaseolus vulgaris*	Perfect	S		A	PL	Seed can carry serious diseases
(pea)	Lima	*Phaseolus lunatus*	Perfect	S		A	PL	
	Pea	*Pisum sativum*	Perfect	S		A	PL	
	Cowpea	*Vigna sinensis*	Perfect	S		A	PL	Includes blackeye and crowder peas, blackeye bean
	Soybean	*Glycine max*	Perfect	S		A	PL	
Liliaceae	Asparagus	*Asparagus officinalis*	Dioecious	C	I	P	C	Propagate by seed, not by division
(lily)	Onion	*Allium cepa*	Perfect	C	I	B	OP or H	Includes shallots (usually propagated by division)
	Garlic	*Allium sativum*	Perfect			P	C	Never produce seed
	Welsh onion	*Allium fistulosum*	Perfect	C	I	B	OP	Does not cross with *Allium cepa*
	Leek	*Allium ampeloprasum*	Perfect	C	I	B	OP	
Solanaceae	Tomato	*Lycopersicon esculentum*	Perfect	S		A	PL or H	
(nightshade)	Pepper	*Capsicum annuum*	Perfect	CS	I	A	PL or H	
	Eggplant	*Solanum melongena*	Perfect	CS	I	A	PL or H	
	Potato	*Solanum tuberosum*	Perfect	S		P	C	Rarely produces seed; propagate by tubers
Apiaceae	Carrot	*Daucus carota*	Perfect	C	I	B	OP or H	Crosses readily with wild carrots
(parsley)	Celery	*Apium graveolens*	Perfect	C	I	B	OP	
	Celeriac	*Apium graveolens*	Perfect	C	I	B	OP	
	Parsley	*Petroselinum hortense*	Perfect	C	I	B	OP	

Courtesy of James Baggett, Oregon State University.

[a] Perfect—male and female in the same flower; dioecious—male and female flowers on separate plants; monoecious—separate male and female flowers on the same plant; andromonoecious—male and perfect flowers on the same plant.

[b] S—self pollinated; C—cross pollinated; CS—both crossed and selfed.

[c] Pollinated by: I—insects; W—wind.

[d] A—annual; B—biennial; P—perennial.

[e] PL—pureline; C—clone; OP—open pollinated; H—F$_1$ hybrid.

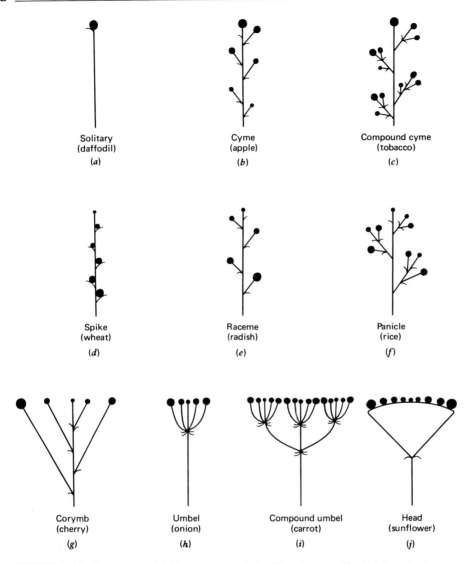

FIGURE 3-15 Flowers and inflorescences: (a) solitary flower; (b,c) determinate inflorescence; and (d–j) indeterminate inflorescences. Small dots represent immature flowers, progressively larger dots more mature flowers and developing fruits.

It has also been shown that it is the leaves, not the vegetative apex, that are the receptors of the stimulus to flower. It would seem logical that the message to form flowers is transmitted from leaves to buds by a plant hormone. As a matter of fact, the proposed hormone has been given a name—**florigen**. Unfortunately, florigen remains only a name; it has never been found. There is, however, compelling evidence that florigen exists. If a short-day plant is grown under long days with only one leaf subjected to a short photoperiod the

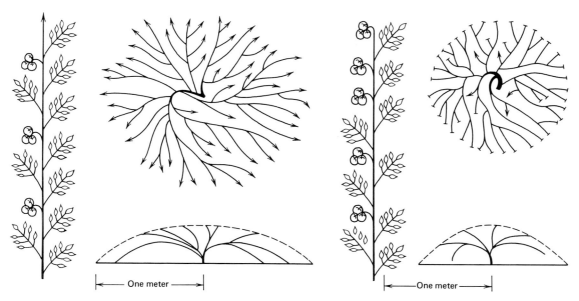

FIGURE 3-16 Growth characteristics of the normal indeterminate tomato (left) compared to that of a determinate plant. The stems differ; instead of the normal three-leaf nodes and single inflorescence per segment of main stem, the determinate stem develops progressively fewer leaf nodes per segment and terminates with an inflorescence. The mature plants, shown in top and side views, have different forms. The normal plant spreads indefinitely; the determinate plant covers less area. It is limited to a symmetrical circular shape by the termination of its branches at a fixed distance from the center. (From Charles M. Rick, "The Tomato." Copyright © 1978 by *Scientific American, Inc.* All rights reserved.)

plant will flower. However, if the leaf is removed immediately after the short-day treatment the plant will not flower. It must be left on the plant for a few hours. It has also been shown that the hypothetical hormone can cross a graft union.

Fruit Set and Maturity

For convenience we divide our discussion of fruit set and maturity into three phases: (1) pollination, (2) fertilization, and (3) postfertilization development.

Pollination

Earlier in the chapter we defined pollination as the act of transferring pollen from the anther of the stamen to the stigma of the pistil. Pollen may be transferred to the stigma by any one of several methods, including (1) direct contact, (2) gravity, (3) wind, and (4) insects. Direct contact occurs in many perfect flowers where the pollen-bearing anthers touch against the stigma of the same flower. Gravity can transfer pollen from one flower to another and within a single flower as, for example, when the anthers extend beyond the stigma in an upright flower or when the stigma extends beyond the anthers in a pendant flower.

Wind and insects are unquestionably the most common agents for transfer of pollen (Fig. 3-18). Normally, wind-pollinated flowers are small and inconspicuous and have no odor or color to attract insects. They produce copious quantities of lightweight pollen and their

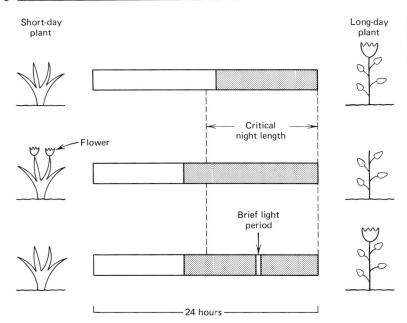

Short-day plant

Long-day plant

Flower

Critical night length

Brief light period

24 hours

stigmas are constructed so as to catch pollen as it floats by. They may be featherlike as in many grasses, or they may exude a sticky substance. Insect-pollinated flowers, on the other hand, are commonly large, have showy parts, and produce **nectar** (a sugary fluid that pollinators use as food) in specialized glands. Their pollen is usually heavy and sometimes sticky. Insect-pollinated flowers are so designed that when the organism, commonly a bee, visits the flower for nectar its body, which carries the pollen, comes into contact with the stigma. In addition to bees, moths, beetles, hummingbirds, and even bats can pollinate flowers. A partial list of wind- and insect-pollinated vegetables is given in Table 3-2. In addition to the crops listed in the table the important tree fruits, including apple, cherry, and pear, and most ornamental flowers are also pollinated by insects.

The transfer of pollen to any flower of the same plant is known as **self-pollination**, while the transfer of pollen to a flower of a different plant is known as **cross-pollination**. Both pol-

lination mechanisms occur in familiar vegetables, although within a given family a single method predominates (Table 3-2). In general, self-pollination maintains the genetic uniformity of a species, whereas cross-pollination introduces genetic variability. Many plants both self- and cross-pollinate naturally. For example, although wheat is normally self-pollinated, natural cross-pollination occurs in 1 to 4 percent of the flowers, according to one study.

Many plants have mechanisms to insure self-pollination. Generally self-pollinated plants have perfect flowers, but they also have structural adaptations that prevent or greatly reduce crossing. For example, flowers may remain closed during the flowering process. In beans and peas the essential parts are tightly enclosed in a tube comprised of two specialized petals, thus assuring a high degree of selfing. Or, in many tomato cultivars the anthers are joined into a tube that surrounds and protects the stigma of the pistil. And in lettuce the developing stigma grows through a tube of pollen-bearing stamens. Thus, even

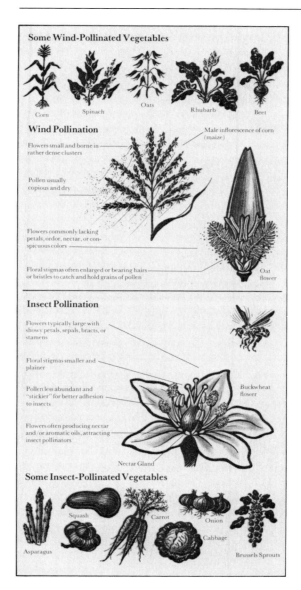

FIGURE 3-18 Adaptations to pollination exhibited by wind- and insect-pollinated flowers. (Courtesy of James Baggett, Oregon State University.)

though the flowers are open, selfing is virtually assured in these crops.

Cross-pollination, characteristic of many garden plants, is also encouraged or assured by a number of features. The most obvious of these is the development of the monoecious and dioecious habit. The monoecious plant, corn, is normally cross-pollinated. This is enhanced by the vertical separation of male and female flowers and by the fact that in a given corn plant the pollen may be shed before the silks are receptive. Many plants with perfect flowers are also cross-pollinated. This is often made possible by mechanisms such as differential maturity of male and female organs, and of course adaptations to assure wind or insect pollination.

One of the most interesting ways in which crossing is assured is the mechanism of **pollen incompatibility**, or **self-sterility**, in which pollen germination and growth is inhibited on the plant that produced it. Although a single sunflower head has innumerable perfect flowers in a tightly packed cluster it cannot produce viable seeds from selfing; it must be crossed. Another example of self-incompatibility is provided by Brussels sprouts. The stigma of this flower is comprised of numerous protuberances called **papillae** (Fig. 3-19a), each of which is covered by a waxy layer. Incompatible pollen (Fig. 3-19b) does not stick to and penetrate the wax layer and germination does not occur. Compatible pollen (Fig. 3-19c, d) does stick and the outer wall of the pollen grain penetrates the wax layer. Subsequently the papillae collapse, the pollen grain germinates, the pollen tube grows toward the ovary, and fertilization results.

The phenomenon of pollen incompatibility is also common among the tree fruits, where the problem is complicated by the fact that all of the trees of a given cultivar are members of a clone, and thus identical genetically. To overcome incompatibility in these crops it is important to have present another cultivar to act as a pollinizer (Table 3-3). Even in those

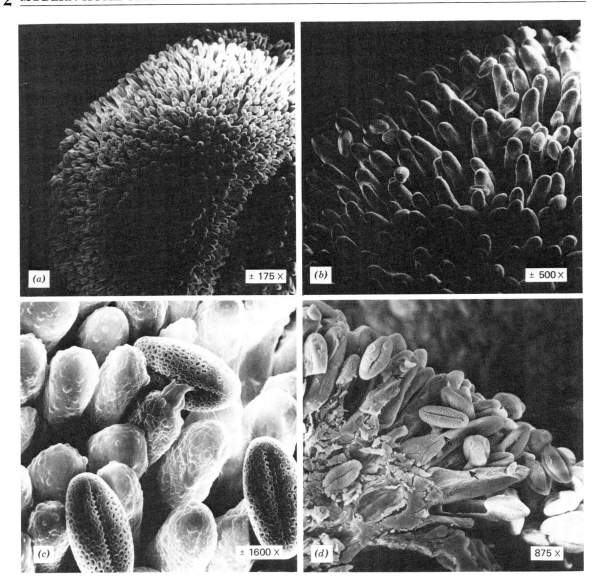

FIGURE 3-19 Pollen incompatibility and pollination in Brussels sprouts. [*Brassica oleracea.*] (*a*) Stigmatic surface showing papillae, each of which is covered by a waxy layer. (*b*) Incompatible pollen does not stick to and penetrate waxy layer and germination does not occur. (*c,d*) Compatible pollen does stick and the outer wall of the pollen grain penetrates the wax layer. Subsequently the papillae collapse, the pollen grain germinates (at arrow in *d*), and the pollen tube grows toward the ovary where fertilization occurs. (Courtesy of H. P. Roggen, Wageningen, Nederland.)

TABLE 3-3 Some Common Tree Fruits and Their Pollination Requirements

Fruit	Cultivar	Needs pollinizer	Best pollinizer
Apples	Gravenstein	Yes	Golden Delicious, Yellow Transparent
	King	Yes	Golden Delicious, McIntosh
	Northern Spy	Yes	McIntosh
	Red Delicious	Yes	Golden Delicious
	Rome Beauty	No	
	Tydeman's Red	Yes	Lodi, Empire
	Winesap	No	
	Golden Delicious	No	
	Yellow Transparent	No	
Pears	Anjou	Yes	Bartlett
	Bartlett	Yes	Anjou, Fall Butter
	Bosc	Yes	Comice
	Comice	Yes	Bosc
Peaches	Elberta	No	
	Imp. Early Elberta	No	
	J. H. Hale	Yes	Early Elberta
	Pacific Gold	No	
	Red Haven	No	
	Rochester	No	
Sweet cherries	Bing	Yes	Van, Sam, Corum
	Corum	Yes	Royal Anne
	Lambert	Yes	Van, Sam, Corum
	Lamida	Yes	Van
	Rainier	Yes	Sam or Van
	Royal Anne	Yes	Corum
	Van	Yes	Bing, Lambert, Sam
Apricots	Perfection	Yes	Tilton
	Tilton	No	
	Wenatchee	No	

cultivars where an outside pollinizer is not a definite requirement, as in certain apple cultivars, yields are generally higher if a pollinizer is present. The gardener with abundant space can utilize separate trees as pollinizers, whereas those with limited space can graft on branches of suitable pollinizers thereby eliminating the need for additional trees. More on this in Chapter 12.

The subject of pollination as relates to the topic of saving garden seed for future planting is a topic discussed in Chapter 7.

The above discussion suggests that the only function of pollination is to provide male gametes for the fertilization of egg cells. Actually, pollination serves another less publicized but equally important role—it bring about an inhibition of flower abscission. This inhibition results in what is termed **fruit set** (really a *potential* for fruit to develop). This phenomenon is clearly illustrated by experimental situations where dead pollen is applied to a stigma, resulting in fruit set but obviously not fruit maturation. The latter normally occurs only after fertilization.

Fertilization

Once in contact with the stigma the pollen grains germinate forming pollen tubes. The elongating pollen tubes grow down the style to the ovary at its base (Fig. 3-14). Once in the ovary the tubes grow into the ovules. Generally the pollen tube enters the ovule through a specialized opening, the **micropyle,** that occurs at the base of the ovule (Fig. 3-14). Once inside the ovule the sperms are released and fertilization occurs.

The events described above set the stage for normal fruit development, that is, the development of fruit containing viable seed. In certain cases, however, fruit set occurs without seed production. This phenomenon, known as **parthenocarpy,** can occur: (1) without pollination, (2) with pollination but not fertilization, (3) when both pollination and fertilization occur but the developing embryos abort.

The development of fruit without pollination occurs in banana, pineapple, the 'Washington' navel orange, and some kinds of figs and grapes. It also occurs occasionally in cucumber, pumpkin, pepper, and tomato. Fruit development with pollination but without fertilization can occur in tomato when the temperatures are cool. Below about 60°F pollen tubes do not grow and fertilization does not occur. Under these conditions the gardener can apply plant hormones to induce fruit development. Parthenocarpic fruits with aborted embryos occur in the seedless watermelons. Although the embryos have aborted, seedlike objects consisting of small seed coats are present.

Postfertilization Development

Fertilization initiates striking changes in the rate of growth of the ovary and the ovules contained therein. In lettuce, for example, fertilization occurs only hours after pollination and shortly thereafter the **pericarp** (ovary wall) begins to increase in size, and within five or six days reaches full size (Fig. 3-20). The endosperm of the seed also starts its growth relatively early, and shortly thereafter the embryo beings to enlarge. In lettuce the endosperm is completely absorbed by the developing embryo. Such seed are said to be **exalbuminous.** In other plants, including onion and corn (Fig. 2-3) a portion of the endosperm remains. Seed of this type are said to be **albuminous.** In exalbuminous seed the stored food reserves are in the embryo itself, particularly in the cotyledons.

The stimulus for fruit growth comes from the developing seed within the fruit. In strawberries the enlarged, fleshy receptacle is the edible part of the fruit (Fig. 3-21a). If seed are removed from two sides of a developing fruit the final result is a disc-shaped fruit (Fig. 3-21b), whereas removal of all seed stops fruit growth entirely (Fig. 3-21c). Apparently the stimulus provided by the seed is in the form of auxin, because if auxin is applied to young fruits from which all seed have been removed the resultant fruit resemble normal strawberries (Fig. 3-21d). There is a direct relationship between fruit size and seed number in many plants. Fruit size is also influenced by the conditions under which a plant is growing. A robust plant receiving ample sunlight, water, and the right temperature will produce more and larger fruit than one growing under less than ideal conditions. Competition between developing fruits is also important in determining fruit size. In general, the larger the number of fruit developing the smaller the size of individual fruits. The total weight of fruit

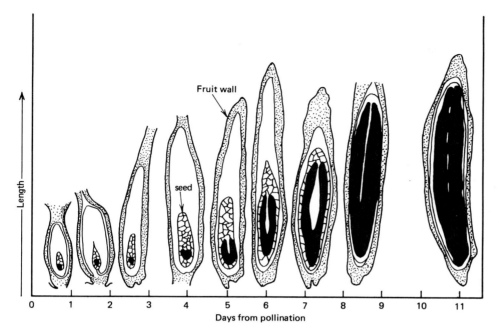

FIGURE 3-20 Growth of the fruit and seed of lettuce. (Courtesy of the University of California Division of Agricultural Sciences.)

FIGURE 3-21 Strawberry fruits: (a) normal; (b) fruit resulting from removal of seeds on two sides of the receptacle; (c) fruit in which all seeds were removed; and (d) fruit in which all seeds were removed but auxin was applied to the receptacle. (Courtesy of J. P. Nitsch.)

produced will remain about constant, however.

As fruits reach the end of their growth period they normally undergo a number of changes that collectively is referred to as **ripening**. The more common of these changes are the following: (1) change in color, reflecting change in pigment composition, (2) softening of the fruit flesh, and (3) conversion of starches to sugars, or vice versa, as in corn and tomato. Other ripening changes include changes in acidity, vitamin content, and soluble solids, particularly sugars.

One final aspect of ripening is worthy of mention. In the 1930s it was discovered that a great many ripening fruits undergo spectacular changes in respiration rates. These changes include a lowered rate in the full size (mature) fruit, followed by a large increase during ripening, and a subsequent return to a rate about equal that before the rapid rise. The phase of ripening characterized by the somewhat abrupt increase in respiration is known as the **climacteric**. Although the gardener cannot *see* this phenomenon directly, its occurrence often coincides with other visible changes, for example, the color changes mentioned above. The practical importance of this information relates to the fact that for many fruits the time of optimal eating quality coincides with the climacteric peak or the period just after the peak. Picking hastens the climacteric rise in many fruits, and certain fruits, avocado for example, will start the climacteric only after being picked. On the other hand, ripening citrus fruits normally seem to lack a climacteric. After the climacteric, fruits start their senescent decline, and their eating qualities decline accordingly.

Certain storage conditions delay or prevent the climacteric. Every gardener knows that the storage life of fruits is longer in the refrigerator. An atmosphere low in oxygen and high in carbon dioxide, such as created by fruits in polyethylene bags, can greatly im-

prove storage. In recent years it has been found that ethylene is produced by the ripening fruits themselves. The finding that ethylene promotes ripening has found widespread practical application. By storing fruits under conditions whereby the ethylene is constantly removed from the atmosphere the ripening activities, including the climacteric, can be greatly delayed. In this regard, did you ever wonder why your bananas ripen so fast when you bring them home from the supermarket?

KINDS OF FRUITS

Botanists define fruits as mature ovaries and other flower parts that may be associated with them. Differences in fruit type result from the manner in which the ovary wall, or **pericarp**, matures (Fig. 3-22). Accordingly, fruits may be fleshy or dry, and dry fruits may **dehisce** (split open) at maturity or remain closed. In fruits of apple portions of the sepals, petals,

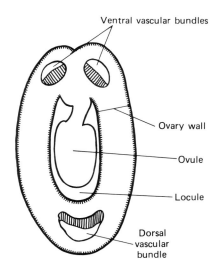

FIGURE 3-22 A bean ovary shortly after fertilization showing an ovule (developing seed) and the ovary (fruit) wall. (Courtesy of Guerry Dean.)

and stamens fuse to form a floral tube that makes up the outer part of the apple. All of the fruits just described have one feature in common; they develop from a *single* ovary and are termed **simple** fruits (Fig. 3-23).

In some plants the fruits consist of several mature ovaries attached to a common receptacle. These **aggregate** fruits are exemplified by strawberry and blackberry. In other plants the individual fruits develop from the ovaries of several flowers. These **compound** fruits are illustrated by the pineapple and fig (Fig. 3-23).

SENESCENCE

Every gardener is aware that once corn plants have flowered and set fruit they begin a slow decline toward death. No effort on the part of the gardener can keep them alive. Simi-

larly, each fall the tops of asparagus, dahlia, and rhubarb plants die back though the crown and root system continue their perennial existence. In many fruit trees all of the leaves die each autumn, and in evergreens there is a continual loss of older leaves. The four situations just described are familiar examples of **senescence**—the degradative processes that result in death of the plant or some portion of it. Actually two patterns of senescence emerge (Fig. 3-24). The first, as with corn, is **complete**, the others are **partial**. Although senescence introduces an apparent negative element into the plant life cycle, all is not lost; the continuity of the life chain is assured by the seed formed within fruits, or by some asexual propagule.

This concludes our basic discussion of the plant and its life cycle. We now turn our attention to a medium that makes the flowering plant life-style so successful—the living soil.

Pea
(legume)

Carrot
(schizocorp)

Tomato
(berry)

Apple
(pome)

Peach
(drupu)

Simple fruits

Strawberry
Aggregate fruit

Pineapple
Compound fruit

FIGURE 3-23 Several types of fruits found among garden plants.

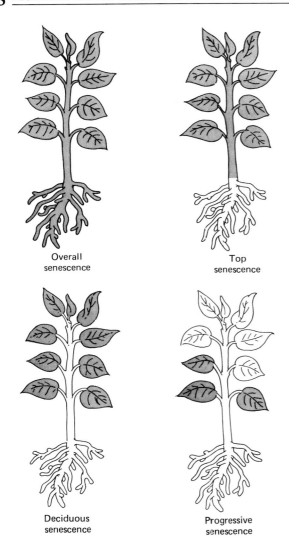

Overall
senescence

Top
senescence

Deciduous
senescence

Progressive
senescence

FIGURE 3-24 Patterns of plant senescence. The dead portions of the plant are shaded. (Modified from A. C. Leopold, 1961, by Guerry Dean.)

Questions for Discussion

1. Define the following terms:
   (a) Clone
   (b) Rhizome
   (c) Budding
   (d) Double fertilization
   (e) Dioecious
   (f) Short-day plant
   (g) Pollination
   (h) Parthenocarpy

2. Name three garden crops that are propagated primarily or entirely by asexual means, and for each crop state why this form of propagation is employed.

3. Describe the vegetative propagules of the Irish and sweet potatoes and explain the planting procedures and growth characteristics for each.

4. Outline procedures for propagation by layering.

5. List the criteria that enhance success in starting new plants from cuttings.

6. Why is it important in grafting that the vascular cambia of stock and scion be closely aligned?

7. Outline the sexual cycle of a common garden annual such as pea or bean. Be sure to indicate places where chromosome numbers are halved or doubled.

8. What is the relationship between a flower and a fruit?

9. What factors control flowering in garden crops?

10. Distinguish clearly between self- and cross-pollination.

11. Describe the various patterns of senescence common to a broad range of garden crops.

Selected References

Baggett, J. 1978. *Saving your own vegetable seed: A pollination primer. Horticulture*, 56:18–25.

Cronquist, A. 1982. *Introductory Botany.* Harper & Row, New York.

Edmond, J. B., T. L. Senn, F. S. Andrews, and R. G. Halfacre. 1975. *Fundamentals of Horticulture.* McGraw-Hill, New York.

Esau, K. 1965. *Plant Anatomy.* John Wiley, New York.

Hartmann, H. T. and D. E. Kester. 1975. *Plant Propagation, Principles and Practices.* Prentice-Hall, Englewood Cliffs, N. J.

Janick, J. 1972. *Horticultural Science.* Freeman, San Francisco.

Leopold, A. C. and P. E. Kriedemann. 1975. *Plant Growth and Development.* McGraw-Hill, New York.

MacGillivray, J. H. 1953. *Vegetable Production.* Blackiston, New York.

Raven, P. H., R. F. Evert, and H. Curtis. 1976. *Biology of Plants.* Worth, New York.

Rick, C. M. 1978. *The tomato. Scientific American,* 239: 76–87.

Weier, T. E., C. R. Stocking, and M. G. Barbour. 1974. *Botany, An Introduction to Plant Biology.* John Wiley, New York.

Wright, R. C. M. 1975. *The Complete Handbook of Plant Propagation.* Macmillan, New York.

LEARNING OBJECTIVES

By the time you have finished this chapter you should be able to:

1 Define "soil" from a gardener's point of view.

2 Characterize the major horizons making up a soil profile.

3 Identify the major factors influencing soil formation.

4 Distinguish between residual and depositional soils.

5 Describe some of the useful types of information that one could expect to find in soil survey reports.

6 Discuss soil texture, including (a) particle size groups, (b) the textural classes and how they are determined, and (c) the importance of clay in soils.

7 Indicate the importance of maintaining good soil structure.

8 Name three practices gardeners can employ to maintain the organic matter level of their soils.

9 Explain the concept of carbon/nitrogen ratios.

10 Discuss the intimate relationship that exists between soil water and soil air.

11 Illustrate by use of diagrams what is meant by the phrase "available water."

12 Identify at least two kinds of problem soils and tell how the gardener can deal with each.

4

INTERPRETING GARDEN SOILS

"It is expedient then, as I was saying, to study each kind of soil to determine for what it is, and for what it is not, suitable. . ."

Varro (first century B.C.)

Gardeners know that soil, which forms the growing medium for their plants, determines, in large part, the quality of the garden. Not all soils have the same ability to produce crops, but all soils will yield to the gardener's skills. Crops vary in their soil needs; a good soil for acid-loving blueberries would be a poor soil for most garden crops. With knowledge of soil properties, and by careful selection of vegetables, gardeners can produce food from almost any soil. Soil-modifying practices such as liming, composting, or installation of drain tile can markedly improve a soil and thereby increase the variety of vegetables, herbs, fruits, and flowers that can be grown.

Let us review the fundamental aspects of soils to understand better the scientific bases for many garden practices that are carried out either to maintain or to improve the quality of our soil. Initially, we need a working definition of what it is that comprises soil.

A DEFINITION OF SOIL

Soil scientists refer to the entire accumulation of mineral material over bedrock as the **regolith**. The regolith may be shallow or hundreds of feet deep. The **soil** is the upper part of the regolith (Fig. 4-1), and may be defined as the unconsolidated cover of the earth, made up of mineral and organic components, water and air, and capable of supporting plant growth. This definition seems appropriate for gardeners and other agriculturists because it includes the four major com-

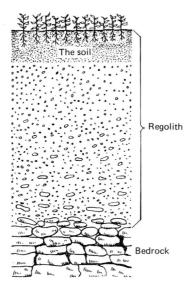

FIGURE 4-1 The regolith, its soil, and the underlying bedrock. If the regolith is thin it may be converted entirely into soil; then the soil rests directly on bedrock. (Reprinted with permission of Macmillan Publishing Co., Inc., from T. L. Lyon, H. O. Buckman, and N. C. Brady, *The Nature and Properties of Soils.* Copyright © 1957 Macmillan Publishing Co., Inc.)

ponents of soils. It also states the most important function of the soil—to grow plants. As a medium for plant growth the soil performs four functions:

1. Anchors plant roots.
2. Supplies water to plants.
3. Provides air to plant roots.
4. Yields the minerals needed by plants.

The soil is distinguished from the unconsolidated material beneath by its relatively high content of organic matter, including numerous plant roots and soil organisms. In addition, there is usually evidence of more intense weathering and the presence of a number of more or less distinct horizontal layers, one upon the other. Study of these layers tells us a great deal about soils.

THE SOIL PROFILE

When exposed soil is seen along road cuts, river banks, or similar vertical exposures, a number of horizontal layers, often of different color, are evident. This entire vertical section of soil is called a **soil profile**, and the individual layers are called **horizons**. The development of horizons is brought about largely by the action of water, which leaches material from the soil surface into the deeper soil. This results in the formation of three main horizons or layers, termed the "A," "B," and "C" horizons (Fig. 4-2), with the A being uppermost. Individual horizons are not always distinct. For one thing, some soils—such as those on flood plains or disturbed areas—may be too recent to have developed distinctive layering. And even if horizons are present, they may be all but obscured by a gradual transition from one layer to the next. Or, each main horizon may contain a number of sublayers. In addition to the horizons referred to above, one other horizon can be recognized. Above the mineral material lies an accumulation of organic matter derived from the remains of dead plants and animals. This layer, which is of variable thickness, is designated the "O" horizon.

Each soil horizon is characterized by definite features. The O horizon is the organic layer and is comprised or recognizable (above) and nonrecognizable (below) plant and animal remains. The lowermost portion of this organic layer is sometimes termed **leafmold**.

The A horizon is the uppermost mineral horizon, and varies in thickness from two or three inches to a foot or more. This horizon has high levels of organic matter in varying states of decomposition. The dark color common to the A horizon is due to this high organic matter content. The layer is teeming with living organisms, including the roots of plants, animals, and microorganisms. Worms, insects, fungi, and bacteria are present in vast

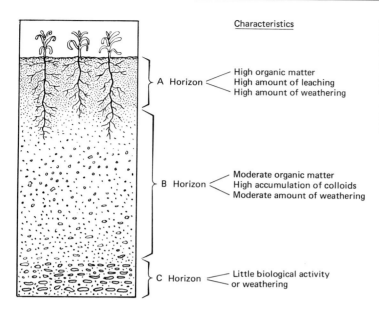

Characteristics

A Horizon — High organic matter
High amount of leaching
High amount of weathering

B Horizon — Moderate organic matter
High accumulation of colloids
Moderate amount of weathering

C Horizon — Little biological activity
or weathering

FIGURE 4-2 Soil profile, the major horizons, and their characteristics. (Reprinted with permission of Macmillan Publishing Co., Inc., modified from T. L. Lyon, H. O. Buckman, and N. C. Brady, *The Nature and Properties of Soils.* Copyright © 1957 Macmillan Publishing Co., Inc.)

numbers, and by their activity greatly improve the condition of the soil. Typically the A horizon has lost some of its clay, iron and aluminum, and organic colloids by the leaching action of water. For this reason it is designated as the "zone of maximum leaching." The A horizon is of greatest importance to the gardener since this is the seedbed and the region where the most root growth occurs.

We have said that the A horizon is characterized by a loss of colloidal materials through leaching. These leachates collect at a lower level in the soil where they can have a profound influence on soil structure. This "zone of maximum accumulation" is designated the B horizon. The B horizon usually contains less organic matter, and is often considerably deeper—up to three or four feet total—than the A horizon. It can vary in color from whitish or gray to red or yellow. These color differences usually result from variations in the amount and form of iron present, but they may be related to other factors as well. The B horizon is sometimes called the subsoil. In actuality, the **subsoil** represents that part of the soil below the **plow layer**, the soil ordinarily altered by tillage. Ideally the plow layer is confined to the A horizon, but if the A horizon is shallow it may include a portion of the B horizon. In some areas of the continent severe **erosion** (wearing away of the land surface) has resulted in removal of part or all of the vital A horizon.

The C horizon contains unconsolidated material that is only slightly weathered. It is below the zone of major biological activity, and since it receives few leachates, shows little structural detail. The C horizon is of little direct concern to the gardener, but it is important to remember that in time, its upper portion may become a part of the more active B horizon. Some of the characteristics of the major horizons are summarized in Fig. 4-2.

SOIL FORMATION

In simplified terms soil formation can be visualized to proceed somewhat as follows. A specific loose or weathered rock material, known as the **parent material**, provides the physical substance from which our soil will form. For our purposes this material can be designated as a C horizon (Fig. 4-3a). Over a period of time the uppermost layers of this parent material will be colonized by both plants and animals. As these organisms die their remains are broken down by decay, and are gradually incorporated into the upper material. This remaining more or less stable organic material is known as **humus**, and the process leading to its formation **humidification**. It is the humus that over a period of many decades imparts a dark color to this upper several inches of parent material. Concomitant with the humus buildup mechanical and chemical weathering of the parent material results in the formation of very small inorganic particles called **clay**. Both humus and clay play vital roles in soil fertility as we will see in Chapter 5. Technically, the dark-colored, clay-containing layer can be designated as an A horizon. Young soils commonly have A and C horizons, but not B horizons (Fig. 4-3b).

Over prolonged periods of time additional changes can occur in our soil profile if it is in a stable land position—that is, neither being lowered by erosion nor being raised by deposition of materials. The downward movement of water carries with it colloidal materials. These fine particles accumulate at some depth and there result in the formation of large soil aggregates having a blocky or prismatic structure. This layer is the B horizon (Fig. 4-3c). Generally, the formation of large soil aggregates is beneficial because it promotes favorable features such as drainage and aeration. In some cases, however, the accumulation of clay colloids in the B horizon is so great as to result in the creation of a layer, known as **hardpan** or **claypan**, that acts as a barrier to water penetration and root growth. This and other problem soils will be discussed later in the chapter.

The Factors of Soil Formation

Several natural factors interact to create soil. These factors include the following:

- Parent material.
- Climate.
- Topography.
- Organisms.
- Time.

Expressed verbally, a given soil is derived from a specific parent material as modified by climate, topography, and organisms over a period of time. One profound formative influence is missing in our list of formative factors—the effects of mankind. The effects of humans

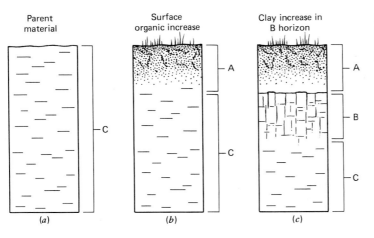

Parent material   Surface organic increase   Clay increase in B horizon

(a)   (b)   (c)

FIGURE 4-3 Soil formation. (Courtesy of the University of California, Agricultural Experiment Station.)

are relatively short term as compared to geologic time and they may be of either a positive or a negative nature.

Parent Material

The nature of the parent material, particularly its texture and structure, and its chemical and mineral composition have a strong influence on the properties of a given soil. A number of basic materials can serve as the stuff from which soils are made; consequently the influence of parent material on soil characteristics can be quite variable. Igneous rocks, sedimentary rocks, volcanic ash, sand dunes, and swamp vegetation all act as parent material for specific soils. Some soils, called **residual soils**, form directly above their weathered parent material. These soils usually occur on mountain slopes and in the foothills. Residual soils are abundant, for example, in the Appalachians, the Ozarks, and in the southern Great Plains.

Another group of soils, called **depositional soils**, form from material that has been transported and deposited many miles from the original parent rock. Depositional soils may be transferred by gravity, water, ice, or wind. Many soils have been deposited by rivers and streams in the form of flood plains, terraces, deltas, and alluvial fans. These soils are usually deep and fertile. The Mississippi River flood plain is the largest single alluvial deposit in the United States. Some material transported by water ends up in the bottom of lakes that eventually fill. The Red River Valley of Manitoba, Minnesota, and North Dakota was once the bed of ancient glacial Lake Agassiz and is now famous for its deep, fertile soils and high crop yields. Materials transported by glacial action are mixtures of soil and rock and occur in irregular deposits called **moraines**. Large areas of Canada and the northern United States are coverd by glacial deposits. These fertile deposits usually have a favorable topography for agricultural activities. Soils derived from materials transported by wind are represented by volcanic ash, sand dunes, and loess. Soils of volcanic ash occur over areas of Montana, Idaho, Kansas, and Nebraska. They are light and porous and of variable agricultural value. Sand dunes, such as the sandhills of western Nebraska, provide poor gardening media unless modified. This is not true of **loess**. These upland deposits are silty in character and cover wide expanses of the central United States. They also occur in parts of eastern Oregon and Washington and in Idaho. In general these soils are highly fertile and excellent for gardening.

In summary, parent materials may be of residual or depositional origin. Generally depositional soils are more productive than residual soils, and soils that are a mixture of residual and transported material are more productive than those of residual material alone. Try to describe the parent material of the soils in your area.

Climate

Of all the factors involved in soil formation, climate—particularly temperature and precipitation—is regarded as exerting the dominant influence. The relationship of temperature and precipitation to the organic matter content of soils in the central United States illustrates the strong influence climatic factors play on soil properties (Fig. 4-4). Moving from Canada to the Gulf Coast the organic matter content of comparable soils is seen to decrease. This is related to the faster decomposition of the organic matter in warmer climates (for every 10°C rise in temperature the rate of chemical reactions doubles). On the other hand, in moving from Colorado to Indiana one finds a gradual increase in the organic matter content of soils. This increase seems to be related to the more complete breakdown of organic matter in the areas of lower rainfall to the west, but also to the sparser vegetation of the western region.

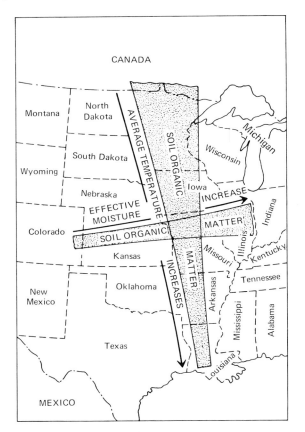

FIGURE 4-4 The relation of temperature and precipitation to the organic matter contents of Great Plains soils. In general, organic matter decreases with higher temperatures but increases with greater moisture. (Reprinted with permission of Macmillan Publishing Co., Inc. from Nyle C. Brady, *The Nature and Properties of Soils.* Copyright © 1974 Macmillan Publishing Co., Inc.)

The influences of climate on soil formation as described for the Midwest occur on a broader scale over the whole of North America. Thus, the gradual organic matter changes illustrated in Fig. 4-4 continue to the north and to the south. Northward through Canada the cooler temperatures, abundant moisture, and acid soil conditions favor the preservation of organic residues. These residues often accumulate in a partially decomposed organic layer known as **peat**. In northern Canada and Alaska there are extensive deposits of peat derived primarily from sedges and mosses. These deposits, known as **muskegs**, are often underlain by material that is perpetually frozen. South of the border, in Mexico, the increasingly tropical conditions favor rapid decomposition of organic residues. Many of these highly weathered soils become brick red in color as a consequence of the abundant oxidized iron present.

West and east through temperate America similar trends are evident. In large areas of the West the sparse rainfall supports little natural vegetation, but the soils support abundant crops if irrigation is employed. Eastward the greater precipitation supports more abundant natural vegetation and, in combination with moderate temperatures, results in dark soils high in organic matter (Fig. 4-5). The black soils of the corn belt probably provide the richest and most productive garden soils in the world. Beyond the corn belt lies a vast area of primarily residual soils that developed under forested conditions. In this area the average annual precipitation increases from west to east and temperatures from north to

FIGURE 4-5 A sample of very dark, humus-rich soil taken from McLain County, Illinois. (Courtesy of Samuel R. Aldrich.)

south. Thus, climatic forces are most intense in the southeast as relates to soil formation. These woodland soils are generally less fertile than those of the corn belt but as they lie in a region of ample rainfall and favorable temperatures crop production is quite good if they are fertilized properly.

Topography

The surface configuration of the earth, its **topography**, affects soil formation in several ways. Most pronounced are those that modify climatic forces. As examples, a southern slope is warmer and dryer than a northern slope. A depression is supplied with more moisture and nutrients than adjacent hillsides and quite likely supports different natural vegetation and different organic matter production. Excessive slope encourages erosion, which may prevent the formation of a deep soil. Or, in a depression where water stands much of the year, soil formation is slowed because climatic and organismic actions are moderated.

Organisms

Soil organisms play important roles in soil development. Small organisms, most invisible to the human eye, are abundant in all soils (Table 4-1). These organisms contribute about 1/1000 of the weight of the soil they occupy, but their contribution is far more important proportionately. They are responsible for the breakdown of organic materials with the release of minerals that become available for plant growth and for the secretion of mucilaginous substances that promote soil aggregation. One group, the **mycorrhizal fungi**, act as "living bridges," promoting the transfer of scarce mineral nutrients from infertile soils into the roots of plants. Somewhat larger organisms such as insects and worms, and even moles and gophers, are plentiful in the soil too (Fig. 4-6). Earthworms make channels in the earth by passing soil through their digestive tracts and depositing it in the form of

TABLE 4-1 The Average Weight (in lb/acre) of Organisms in the Upper Foot of Soil

Organism	Low	High
Bacteria	500	1000
Fungi	1500	2000
Actinomycetes	800	1500
Protozoa	200	400
Algae	200	300
Nematodes	25	50
Other worms and insects	800	1000
	4025	6250

Taken from O. N. Allen, Experiments in Soil Bacteriology, 1957. By permission of Burgess Publishing Co., Minneapolis, Minnesota.

castings. In a single year, their combined activities may produce from 10 to 200 tons of castings per acre. These castings are noted for their favorable effect on soil productivity.

Higher plants are the predominant soil organism, however. Their roots and fallen aboveground parts are the primary source of organic matter returned to enrich the earth. The addition of high-energy compounds, such as cellulose, starch, sugars, fats, and proteins,

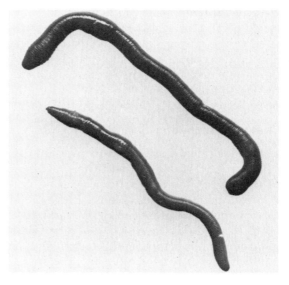

FIGURE 4-6 Earthworms. (Courtesy of U.S.D.A.)

supports the mass of smaller soil organisms. Living roots evolve carbon dioxide which, in the form of carbonic acid, increases the rate at which soil minerals dissolve. Elongating roots can penetrate small crevices in rocks and later split the rocks through forces exerted in growth. The channels remaining after these roots die serve as pathways for the movement of water.

Time

Soil formation takes place over long time intervals. A centimeter of new soil can be formed at rates estimated from 5 to 1000 years or more. These estimates come from information pertaining to known past events. For example, soils occur in Alaska on glacial deposits about 1000 years old. And in Russia, soils 4 to 16 inches thick have developed on limestone slabs abandoned in the 1700s. Many conditions hasten or retard the rate of soil development. Warm, humid climates, forest vegetation, flat topography, and a permeable subsoil help to speed the process; whereas cold, dry climates, grass vegetation, irregular topography, and impermeable subsoils retard soil development. Said another way, the time it takes a soil, or a horizon, to develop is related to the parent material, climate, vegetation, and so on; a given soil is a product of the interaction among the various factors affecting soil formation.

Human Influence

It is clear that humans play an important role in soil genesis. Most human activities have been of a harmful nature. Excessive grazing, unwise cultivation, and destruction of forests and other habitats have created deserts and dustbowls throughout the world. Under prudent management, however, human activity can be a positive force for soil betterment. Contour farming, the rotation of crops, and the veg-etating of steep slopes and gullies all have positive effects on the soil. Gardeners all over the world are improving their soil by using compost, green manure crops, and other soil-building practices. Through the individual efforts of these concerned gardeners thousands of acres of poor soils are being improved and world food production is being increased. As a matter of fact, large-scale agriculture is taking a long and careful look at many of the small-scale practices of gardeners that have proven to be so successful.

SOIL CLASSIFICATION

The field of soil classification is so voluminous and technical as to be beyond the scope of an introductory text on gardening. Nevertheless, some knowledge of classification is of importance to the gardener, and with this thought in mind we have "distilled" some of the key points in soil classification as they relate to the home gardener.

The classification scheme for soils, like that used for plants, consists of categories of increasing specificity, as is illustrated in Table 4-2. *Cucumis sativus*, cucumber, is a specific plant that can be distinguished from all other members (approximately 30) of the genus *Cucumis* by a specific set of characteristics. The same is true of Miami silt loam, which can be distinguished from all other members of the Miami series. Furthermore, both the genus *Cucumis* and the Miami series have certain characteristics in common with other closely related genera or series that allow their grouping into common families, and so on up the line. Our comparison differs in one important respect, however: The rules for naming plants are part of an *international code* and apply uniformly the world over, whereas the soil classification scheme presented is relatively new and while used in the United States has not found universal application.

TABLE 4-2 Comparison of the Classification of a Common Garden Plant, Cucumber (*Cucumis sativus*), and a Soil, Miami Silt Loam

Plant classification	Soil classification
Division—Tracheophyta	Order—Alfisol
Class—Angiospermae	Suborder—Udalf
Subclass—Dicotyledoneae	Great Group—Hapludalf
Order—Loasales	Subgroup—Typic Hapludalf
Family—Cucurbitaceae	Family—Fine loamy, mixed, mesic
Genus—*Cucumis*	Series—Miami
Specific epithet—*sativus*	Textural class—silt loam

Reprinted with permission of Macmillan Publishing Co., Inc. From Nyle C. Brady, The Nature and Properties of Soils. *Copyright 1974 by Macmillan Publishing Co., Inc.*

Of the soil categories presented in Table 4-2, two—the soil series and textural class—are of particular value to the home gardener. The Soil Conservation Service, in cooperation with the various states, has surveyed extensive land areas and prepared reports of the survey results. These reports, published on a county, or counties, basis (Fig. 4-7) can be utilized by the home gardener. The reports contain maps to help you locate your soil, and once located provide a wealth of details on profile characteristics, soil use, proper management, and even productivity ratings. A sample description is given in Fig. 4-8. Note that the specific **soil type** is identified by the series name followed by the textural class of the soil. Often further descriptions, such as "gently sloping" or "strongly eroded" follow the soil type name. These terms identify soil **phases**; subdivisions based on some important deviation such as coarse surface texture, slope, erosion, or stoniness.

Many Canadian soils have also been surveyed and the results published. Fortunately, soils at the series and texture level are described in about the same manner in both Canada and the United States and thus the material presented above applies directly to both situations.

FIGURE 4-7 Soil survey reports.

Chehalis Series

The Chehalis series consists of deep, well-drained soils that formed in mixed, recent alluvium on alluvial bottom lands. Slopes are 0 to 3 percent. In most areas these soils are subject to overflow about once in 3 to 5 years.

Where these soils are not cultivated, the vegetation is Douglas-fir, ponderosa pine, ash, bigleaf maple, and oak and an understory of vines and shrubs. Elevation ranges from 190 to 300 feet. Average annual precipitation is 40 to 45 inches, average annual air temperature is 52° to 54° F., and the average frost-free season is 165 to 210 days.

In a representative profile the surface layer is dark-brown and very dark brown silty clay loam about 11 inches thick. The subsoil is very dark brown and dark-brown silty clay loam about 34 inches thick. The substratum is dark yellowish-brown silty clay loam that extends to a depth of 60 inches or more.

Chehalis soils are used for cereal grain, pasture, hay, vegetable and specialty crops, wildlife habitat, and recreation.

Chehalis silty clay loam (Ch).—This soil occupies large areas on alluvial bottom lands along the major streams and rivers. Slopes are 0 to 3 percent.

Representative profile, 2 miles northeast of Oliver Butte, in the NW¼SE¼ sec. 11, T. 14 S., R. 5 W.:

> Ap1—0 to 6 inches, dark-brown (10YR 3/3) light silty clay loam, dark brown (10YR 4/3) dry; moderate, very fine and fine, granular structure and weak, medium, subangular blocky structure; hard, friable, slightly sticky and slightly plastic; many very fine roots; many very fine interstitial pores; common worm casts; very dark grayish-brown (10YR 3/2) and very dark brown (10YR 2/2) coatings; neutral; clear, smooth boundary. 5 to 9 inches thick.
>
> Ap2—6 to 11 inches, very dark brown (10YR 2/3; 10YR 2/2, uncrushed) silty clay loam, dark brown (10YR 4/3) dry; moderate, fine granular structure and weak, medium, subangular blocky structure; hard, friable, slightly sticky and slightly plastic; common very fine

> roots; many very fine interstitial and tubular pores; neutral; gradual, smooth boundary. 0 to 7 inches thick.
>
> B21—11 to 20 inches, very dark brown (10YR 2/3; 10YR 2/2, uncrushed) silty clay loam, dark brown (10YR 4/3) dry; strong, fine granular structure and very fine, subangular blocky structure; hard, friable, sticky and plastic; common very fine roots; many very fine and few fine tubular pores; neutral; gradual, smooth boundary. 6 to 10 inches thick.
>
> B22—20 to 35 inches, very dark brown (10YR 2/3; 10YR 2/2, uncrushed) silty clay loam, brown (10YR 5/3) dry; moderate, very fine and fine, subangular blocky structure; hard, friable, sticky and plastic; common very fine roots; many very fine and fine tubular pores; neutral; clear, smooth boundary. 12 to 16 inches thick.
>
> B3—35 to 45 inches, dark-brown (10YR 3/3) silty clay loam, brown (10YR 5/3) dry; moderate, fine and medium, subangular blocky structure; hard, friable, sticky and plastic; many very fine tubular pores; few very fine roots; neutral; clear, smooth boundary. 6 to 12 inches thick.
>
> C—45 to 60 inches, dark yellowish-brown (10YR 4/4) silty clay loam, yellowish brown (10YR 5/4) dry; few, medium, faint, very dark grayish-brown (10YR 3/2) mottles; massive; hard, friable, sticky and plastic; many very fine pores; no roots; neutral.

The A and B horizons have a moist value and chroma of 2 or 3. Some faint mottling occurs below a depth of 40 inches in some places. Depth to underlying strata of sand and gravel ranges from 5 feet to many feet.

Included with this soil in mapping were about 10 percent Cloquato or Newberg soils and 5 percent McBee and Wapato soils.

Runoff is slow on this Chehalis soil, and the hazard of erosion is slight to moderate. Rooting depth is 60 inches or more. Permeability is moderate. Available water capacity is 11 to 13 inches. Workability is good.

This soil is used for cereal grain, hay, pasture, and orchards. It is used for pole and bush beans, sweet corn, mint, berries, and other row crops when irrigated. Capability unit IIw–2; wildlife group 1.

FIGURE 4-8 A soil-type description. (Courtesy of U.S.D.A.)

THE MAJOR COMPONENTS OF SOILS

The four major components of mineral soils are (1) mineral material, (2) organic matter, (3) air, and (4) water. Mineral material occupies the greatest volume as is illustrated by the silt loam soil shown in Fig. 4-9. The organic fraction, on the other hand, occupies the smallest volume, usually averaging from 3 to 5 percent by weight of the topsoil. These components, which represent the **soil solids**, comprise about 50 percent of the volume of a given soil. The solid components together determine the native fertility of soils. Between the solid particles lies a great deal of **pore space**, space that can be occupied by either air or water. When a gardener waters his or her garden he or she increases the amount of water within this space system, whereas use of water by plants and losses through drainage and evaporation increase the amount of air. Obviously, the volume occupied by these two components can show extreme variation. Indeed, maintaining a favorable balance between the two determines in large part the suitability of the soil for plant growth.

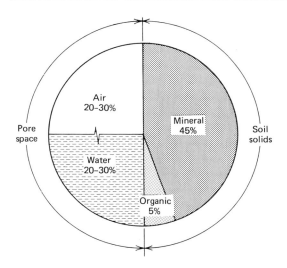

FIGURE 4-9 Volume composition of a silt loam surface soil in good condition for plant growth. Mineral material and organic matter make up the soil solids; water and air fill the pore space. (Reprinted with permission of Macmillan Publishing Co., Inc. from Nyle C. Brady, *The Nature and Properties of Soils.* Copyright © 1974 Macmillan Publishing Co., Inc.)

Mineral Material

The mineral fraction of soils consists of inorganic particles of different sizes and types. To simplify the description of the mineral portion of the soil these particles are classified according to size, without regard to their chemical composition, color, or weight. Three size groups are recognized, as follows:

Soil Separate	Size Range
Sand	0.05–2.0 mm
Silt	0.002–0.05 mm
Clay	less than 0.002 mm

Each size group is called a **soil separate**. All soils are mixtures of sand, silt, and clay, but in each soil their proportions vary. The pro-

portion of each separate is determined in the laboratory by use of a series of sieves, but the gardener can make a rough approximation by simply rubbing moist soil between the thumb and fingers, using the following guidelines:

Kind of Soil	General Feeling
Sandy	Scratchy or gritty
Silty	Smooth and slippery, but not sticky
Clayey	Slippery and sticky

Soil Texture

The proportions of the different particle sizes in a soil—its **texture**—is used in its classification. If a soil has moderate amounts of sand, silt, and clay it is called a **loam**. If one particle is present in somewhat greater abundance its name is added to the description; thus there are sandy loams, silt loams, and clay loams. If a given particle is present in an even higher proportion, the soil may bear the textural name of that separate alone. These relationships are shown in the following diagram:

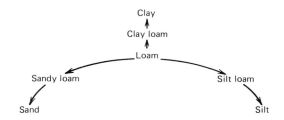

In the laboratory the soil scientist assigns descriptions more precisely by use of a **soil textural triangle** (Fig. 4-10). In all, 12 textural classes are recognized. Textural class designations when joined with the soil series name identify a specific soil type.

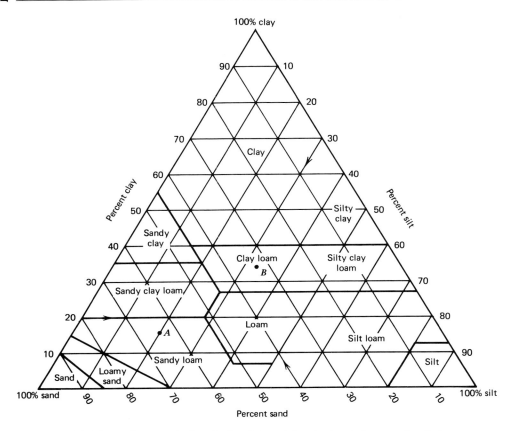

FIGURE 4-10 The soil textural triangle for determining textural classes of soil. A soil at point A has 15 percent clay, 20 percent silt, and 65 percent sand; thus it is a sandy clay. The soil at B is a clay loam. (Courtesy of U.S.D.A.)

Soil texture is of special importance to the gardener because soil particle size affects water percolation, water retention, aeration, nutrient supply, and soil strength. In general sandy soils have faster water percolation and better aeration than finer-textured soils but their ability to hold water and supply nutrients is lower. They also have less strength and therefore are easier to cultivate. Loam soils, as we have seen, combine all of the particle sizes in intermediate proportions and because of this have favorable properties of water percolation, water retention, aeration, and so on. Loam soils are versatile soils capable of growing a wide variety of vegetable crops. As any one size class becomes more predominant in a soil, the less versatile that soil becomes for growing crops, and the more exacting are its management requirements for good plant growth.

But how can particle size have such profound influences on soil properties? As it turns out, most available nutrients and water are adsorbed on the *surfaces* of soil particles, and as particle size decreases total surface area of the soil volume increases proportionately. The relationship between size and surface area is illustrated in Fig. 4-11. Clay usually has thousands of times more surface area than silt and about a million times more surface area than

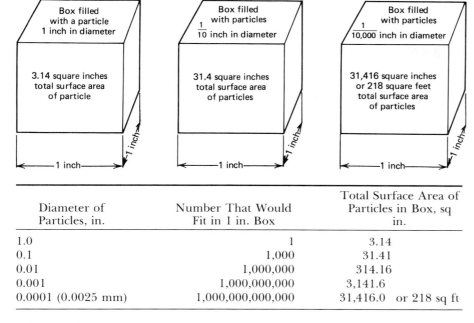

Diameter of Particles, in.	Number That Would Fit in 1 in. Box	Total Surface Area of Particles in Box, sq in.
1.0	1	3.14
0.1	1,000	31.41
0.01	1,000,000	314.16
0.001	1,000,000,000	3,141.6
0.0001 (0.0025 mm)	1,000,000,000,000	31,416.0 or 218 sq ft

FIGURE 4-11 The relation between particle size and surface area as illustrated by the number of round particles that will fit into a square box with sides one-inch.

coarse sand. It has been estimated that a handful of loam soil has from 5 to 10 acres of surface area.

You may have concluded, and correctly so, that clay is the most important of the soil separates. Because of its vastly greater surface area it has a far greater ability to enter into chemical and physical reactions in the soil. The sand and silt are merely broken-down rock fragments and when compared to the clay are essentially inert, although they do play an important skeletal role. The minute clay particles, on the other hand, have an electrical charge and can therefore adsorb and retain nutrients for use by plants. The soil textural triangle (Fig. 4-10) in essence takes into consideration the vital role of clay colloids. Soils with over 40 percent clay are called clay soils, whereas they must contain over 85 percent sand to be regarded as sandy soils.

Soil Structure

We have seen that texture is of great importance in determining many characteristics of soils. But examination of your garden soil will show that it is not composed of very small particles alone, but of *aggregates* of small particles (Fig. 4-12). The term soil **structure** is used to describe these aggregates of soil particles. Several structural types are found in mineral soils, with different types commonly characterizing the different horizons. Large prismatic and columnar aggregates, for example, are usually found in subsoils. Their formation is promoted by the accumulation of colloids from above. Granular and crumb type structures are common to surface soils. Their formation is encouraged by high levels of organic matter. Of the various types of soil aggregation, the granular and the crumb are

FIGURE 4-12 A well-granulated soil (left) and a soil with poor granulation (right). The soil at right is puddled.

of direct concern, because they are the only ones influenced by the day-to-day activities of the gardener.

Soil structure, like texture, influences many soil properties, including water movement, aeration, density, and porosity. Although the totality of factors promoting good soil structure are obscure, it seems certain that the presence of clay and humus colloids plays a major role. The importance to gardeners of regular organic matter additions to the soil thus becomes obvious. Addition of lime also encourages aggregate development because the calcium present causes colloidal particles to clump, a process called **flocculation**. While this sort of clumping is desirable it is not permanent. Stabilization of these clumps can be assured, however, through the agency of organic matter.

Many factors can also reduce the degree of soil aggregation. Sodium, for example, causes colloidal particles to disperse, thus creating undesirable structure. Soils in this condition are said to be **puddled** (Fig. 4-12). Puddling is promoted by other factors also, as when clayey soils are tilled when they are too wet. In fact, tillage of all types has important effects on soil structure. In general, its short-term effects are to improve structure through the incorporation of organic matter and breaking of clods. Long-term effects are usually detrimental because the constant stirring of soil causes a rapid oxidation of soil organic material. Additionally, tillage implements tend to break down soil aggregates and heavy implements cause compaction. Excessive soil preparation has been shown to reduce crop yields in some cases.

Because of the possible adverse effects of tillage on soil structure, and for other reasons, systems of **minimum tillage**, or none at all, are being developed. Minimum tillage practices are most successful with soils of good physical condition for agricultural activities; that is, soils of good **tilth**. The heavy mulches that often accompany reduced tillage programs may retard soil warming and harbor slugs and other pests. Nevertheless, this energy-saving approach presents the gardener with important alternatives for future consideration.

In summary, an important task of the gardener is to maintain and encourage a desirable soil structure through prudent management of the soil. The gardener must strive to maintain or create a well-granulated topsoil that will allow ease of tillage and good plant growth. The physical changes imposed on the soil through tillage, drainage, liming, manuring, and addition of compost all work to improve soil structure. Thus, while the gardener can do little to alter soil texture, he or she can profoundly influence the structure of the topsoil—for better or for worse.

Organic Matter

Of the major components of mineral soils organic matter is present in the smallest amounts—about 3 to 5 percent by weight of most topsoils (Fig. 4-9). Yet, despite its small quantity, it has the most profound effects on soil properties and thus on plant growth.

The Importance of Soil Organic Matter

Among the important influences of organic matter we have cited its beneficial effect on soil structure and on soil tilth. We have also alluded to the abundance of organisms in the soil and the favorable impact of these organisms. Organic matter is the main source of energy for the soil microorganisms, without which many important soil reactions would not occur. Two additional roles of organic matter are as a source of nutrients and water-holding capacity. Organic matter contains from 5 to 60 percent of the total phosphorus, 10 to 80 percent of the total sulfur, and virtually all—about 95 percent—of the total nitrogen. It also contains smaller quantities of other nutrients. Plant growth depends on a ready supply of these essential nutrients. Organic matter also influences the amount of water a soil can hold and the proportion of this water available to plants. This function is dramatically illustrated in Fig. 4-13 where identical amounts of water are added to two soils, one low and one high in organic matter. In the soil high in organic matter, the water pene-

trated a lesser volume of soil, revealing its greater water-holding capacity. Thus emerges the most remarkable feature of organic matter: through its promotion of soil aggregation organic material enhances soil aeration while at the same time improving the ability of soils to store water.

Two forms of soil organic matter are recognized: (1) newly added tissue and its partially decomposed derivative and (2) the humus. The main source of newly added organic residues is plant tissue, which is about 75 percent water and 25 percent dry matter. The dry matter consists of carbohydrates, lignins, proteins, and lesser material (Fig. 4-14). These organics vary in their resistance to decomposition. Lignins are the last to decompose, although they eventually supply much of the total energy for organisms. The general rates of decomposition are as follows:

Compounds	Rate of Decomposition
Sugars, starches, proteins	Rapid
Cellulose, hemicellulose	
Lignins, fats, waxes, resins	Slow

Complete decomposition of organic material returns to nature all of the elements present in the tissue originally. Of these elements carbon, oxygen, and hydrogen predominate, but many other elements, including nitrogen, are present in lesser quantities (Fig. 4-14).

The Decomposition of Organic Residues

Briefly, organic decomposition proceeds somewhat as follows: initially, the gardener adds plant debris to the soil. Within hours soil microorganisms invade the tissue and begin the enzymatic breakdown of compounds. This breakdown provides energy for the soil organisms to multiply to very high numbers. Some energy is also dissipated as heat (this is

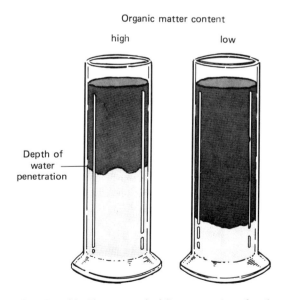

Organic matter content

high   low

Depth of water penetration

FIGURE 4-13 The water-holding capacity of soils is enhanced by the presence of organic matter. If equal volumes of water are added to two soils differing only in organic matter content, water permeates a lesser volume of the soil high in organic matter, showing its greater water-holding capacity. (Courtesy of Guerry Dean.)

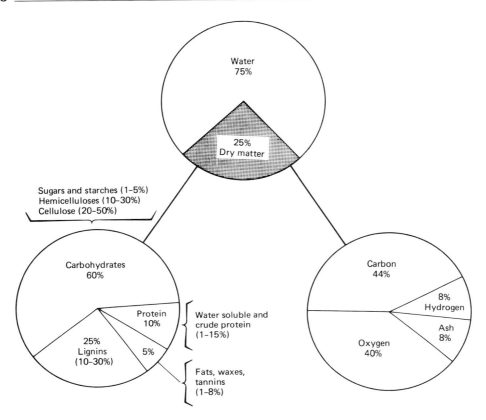

Water
75%

25%
Dry matter

Sugars and starches (1-5%)
Hemicelluloses (10-30%)
Cellulose (20-50%)

Carbohydrates
60%

Protein
10%

Water soluble and
crude protein
(1-15%)

25%
Lignins
(10-30%)

5%

Fats, waxes,
tannins
(1-8%)

Carbon
44%

8%
Hydrogen

Ash
8%

Oxygen
40%

FIGURE 4-14 The approximate composition of plant residues added to soils. All of the inorganic elements are represented in the ash constituent. (Reprinted with permission of Macmillan Publishing Co., Inc. from Nyle C. Brady, *The Nature and Properties of Soils.* Copyright © 1974 Macmillan Publishing Co., Inc.)

the reason your compost heap gets quite hot). Water and carbon dioxide are also produced. The nitrogen-containing proteins are broken into amino acids and then into ammonium compounds. These in turn are changed into nitrates, the nitrogen form taken up by plants, and also to ammonia gas that is lost to the atmosphere. As decomposition continues organic residues become scarce and the number of organisms is reduced. As a result many compounds that were a part of these organisms are released into the soil solution and become available for another cycle of plant growth.

Eventually the organic debris is reduced entirely to the dark, colloidal material known as humus. Additional humus is derived from compounds made by the microorganisms themselves. Finally, however, even the humus is degraded to the basic elements. Thus, decomposition can be viewed as a two-stage process:

Organic matter $\xrightarrow{\text{Rapid}}$ $CO_2 + H_2O$ + humus

Humus $\xrightarrow{\text{Slow}}$ $CO_2 + H_2O$

The constant recycling of organic compounds reveals their rather transitory role as soil constituents, a role that can only be maintained through the constant addition of residues by the concerned gardener.

Studies have shown that the organic matter content of soils changes slowly. In Saskatchewan, for example, 70 years of intensive cultivation resulted in a loss of approximately 50 percent of the original organic matter. The extent of organic matter reduction depends on the kinds of crops grown, however. In one Midwest study of plots cropped in different ways for 32 years it was found that **crop rotations**, especially if they included legumes, helped to maintain soil organic matter (Table 4-3). This is a main reason why crop rotations have become such an important part of large-scale agriculture practice, but the same technique can be employed by gardeners.

Maintaining Organic Matter In Soils

What are the sources of organic residues available to the gardener? The most obvious is the remains of currently growing crops. Their residual root systems, which may have comprised up to 50 percent of the dry weight of the intact plant, make the most important contribution, but unused aboveground parts may be turned into the soil too. Alternatively, unused aerial parts, along with a variety of plant and animal remains, may be used to make compost and this added to the soil. The subject of composting is discussed in Chapter 7.

Another important method of organic matter addition is through **green-manuring**—the turning under of crops while immature. Being immature, green-manure crops decompose relatively rapidly. A variety of plants serve as green-manure crops (Table 4-4), but the legumes and cereals are perhaps most notable. These crops may be grown singly or in combination. Combinations, such as oats and vetch, work well in some cases as a larger amount of green material is produced and one plant may physically support the other. Green-manure crops may be planted at any time of year (Table 4-4), and their use integrated with vegetable production. Remember, however, that it is important that you allow time for the decomposition of your crop before planting another. In warm weather and under moist conditions almost complete decomposition takes place within about six weeks, but rates vary depending on the succulence of the material turned under.

Farm manure, another important source of organic material, consists mainly of undigested plant parts. A combined total of about 25 pounds of nitrogen, phosphorus, and potassium are present in each ton of fresh, undried manure (Table 4-5). This may seem like a minuscule amount when compared to commerical fertilizers, but this value gains perspective when it is realized that farm manures

TABLE 4-3 Organic Matter and Nitrogen Contents of Unfertilized Soils at Wooster, Ohio, After Being Cropped for 32 Years

Cropping	Organic matter (tons/acre)	Nitrogen (lb/acre)
Original crop land	17.5	2176
Continuous corn	6.4	840
Continuous oats	11.4	1425
Continuous wheat	11.0	1315
Corn, oats, wheat, clover, timothy	13.4	1546
Corn, wheat, clover[a]	14.8	1780

Reprinted with permission of Macmillan Publishing Co., Inc. From Nyle C. Brady, The Nature and Properties of Soils. *Copyright 1974 by Macmillan Publishing Co., Inc.*
[a]Continued for 29 years instead of 32.

TABLE 4-4 Some Commonly Used Green-manure Crops

Green Manures	Areas for which best adapted					When to sow	When to turn under
Legumes	N.E. and N.C. States	Southern and S.E. States	Gulf Coast and Florida	Northwestern States	Southwestern States		
Soybeans	●	●	●	●	●	Spring or summer	Summer or fall
Crimson clover	●	●	●	●	●	Fall	Spring
Cowpea		●	●		●	Late spring or early summer	Summer or fall
Crotalaria		●	●		●	Spring or summer	Summer or fall
Indigo, hairy		●	●		●	Spring or early summer	Summer or fall
Lespedeza		●				Early spring	Summer or fall
Field pea	●	●	●	●	●	Early spring	Summer
Sweet clover	●	●	●	●	●	Spring	Summer
Vetch (hairy or common)	●	●	●	●	●	Spring or fall	Fall or spring
Nonlegumes							
Barley	●			●	●	Spring or fall	Summer or spring
Buckwheat	●			●		Late spring and summer	Summer or fall
Millet	●					Late spring or summer	Summer or fall
Oats	●	●	●	●	●	Spring or fall	Summer, spring, or fall
Rye, spring	●	●				Spring	Summer
Rye, winter	●	●				Fall	Spring
Sudan grass	●	●	●	●	●	Late spring or summer	Summer or fall
Wheat, winter	●			●		Fall	Spring

Courtesy of U.S. Department of Agriculture.

TABLE 4-5 The Average Nutrient Content of Several Farm Manures

Animal	Manure (lb/ton)		
	N	P_2O_5	K_2O
Dairy cattle	10.0	2.7	7.5
Feeder cattle	11.9	4.7	7.1
Poultry	29.9	14.3	7.0
Swine	12.9	7.1	10.9
Sheep	23.0	7.0	21.7
Horse	14.9	4.5	13.2

Reprinted with permission of Macmillan Publishing Co., Inc. From Nyle C. Brady, The Nature and Properties of Soils. *Copyright 1974 by Macmillan Publishing Co., Inc.*

are applied at rates of from 10 to 20 tons per acre. Manures vary in their nutrient content. Chicken manure is noted for its high nitrogen content, while sheep manure is high in potassium. Nearly all manures are deficient in phosphorus (Table 4-5), a fact that must be considered by gardeners in fertilization programs.

Carbon/Nitrogen Ratios

No discussion of soil organic matter would be complete without considering, at least briefly, a concept of great practical value to gardeners. This concept deals with the relationship, or ratio, between the amount of carbon in the soil and the amount of nitrogen, and is commonly referred to as the **carbon/nitrogen ratio**, or simply the **C/N ratio**. Carbon atoms join together to form the basic framework of *all* organic molecules, and nitrogen is a constituent of the amino acids that link up to make proteins. The richer a compound is in protein the *lower* will be its C/N ratio. As a rule young, succulent tissues with a high proportion of living cells are high in protein and older more mature plant parts are low in protein. These older parts contain cells high in cellulose, starch, and lignin, for example—all substances with little or no nitrogen. Some approximate C/N ratios are as follows:

Material	C/N Ratio
Soil microorganisms	7:1
Topsoil	11:1
Succulent plant parts	20:1
Legumes and farm manure	25:1
Strawy residues	80:1
Sawdust	300:1

It should be stressed that C/N values for any material are quite consistent; topsoils near 11:1 and soil microorganisms near 7:1. It should also be stressed that if a young seedling is to grow—add new living tissue—it will require large amounts of nitrogen for protein synthesis. The needed nitrogen comes from the soil.

The practical value of understanding C/N ratios can be illustrated by an example. If we have a soil with a C/N value of 11:1, and we incorporate into this soil a strawy residue with a C/N ratio of 80:1, what happens? We saw earlier that when residues are turned under they are attacked by microorganisms (C/N = 7:1) whose numbers greatly multiply, and only as decomposition slows do their numbers wane. But where does the nitrogen come from for this untold millions of new 7:1 bodies? It comes primarily from the soil, for little is present in our strawy residue. In fact, when strawy residues, or farm manure, are added to garden soil the nitrogen demands of the microbes of decomposition may be so great that little nitrogen is available for our crop plants (Fig. 4-15). How many gardeners have added fresh farm manure to their gardens, promptly planted a crop, and witnessed the growth of yellow, spindly (nitrogen-deficient) seedlings? Countless thousands. When residues with a high C/N ratio are turned under in an effort to maintain the organic matter level of our soils we must do one of two things; either wait until the natural cycle of decomposition has run its course before planting crops, or add a high-nitrogen fertilizer so plenty of nitrogen will be available. The primary value of composting is to create favorable C/N ratios. While doing this we retain many organic residues that would otherwise be discarded.

Soil Air

Between soil particles there are always many spaces of varying sizes, which collectively make up the pore space (Fig. 4-9). The pore space comprises a fairly constant percentage of the total soil volume, usually between 40 and 60 percent. This space is filled with water or air in varying proportions, depending on the moisture content of the soil (Fig. 4-16). Sandy soils have less total volume occupied by pore space than do more finely textured soils, but the size of the individual pores is much larger. These large pores, called **macropores**, allow for the faster movement of air and water through the soil, a process termed **percolation**. Thus, in a sandy soil percolation may be quite rapid even though the total pore space is small, simply because of the greater pore size. These relationships are summarized in Table 4-6.

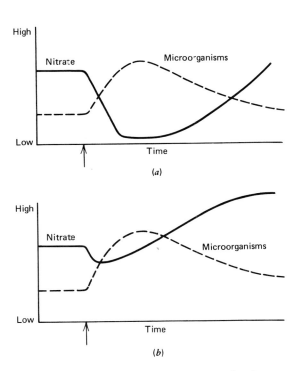

FIGURE 4-15 Changes in the amount of soil nitrate (solid line) and the activity of decomposer microorganisms (dashed line) when a nitrogen-poor residue such as straw (a) or nitrogen-rich material such as green-manure crops (b) is added to the soil (at the time indicated by the arrows). (From Maarten J. Chrispeels and David Sadava, *Plants, Food and People.* Copyright 1977 W. H. Freeman and Company.)

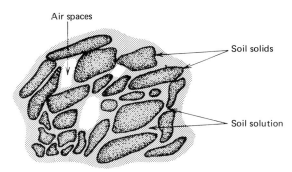

FIGURE 4-16 Soil solids and pore space. The smaller pore spaces are occupied by water but some of the larger spaces contain air. (Courtesy of Esther McEvoy.)

TABLE 4-6 The Relationship of Soil Texture to Pore Space, Size of Pores, and Rate of Percolation

Soil texture	Total pore space	Size of pores	Rate of percolation
Sandy	Low	Large	Fast
Fine textured	High	Small	Slow

Soil air is not merely an extension of atmospheric air. Indeed, the soil air does not form a continuum within the soil, but rather occupies a maze of spaces of varying sizes. Moreover, the air composition differs from space to space as well as between soil and atmosphere. The relative humidity of soil air approaches 100 percent, a condition favoring the growth of soil organisms. Soil air commonly has more carbon dioxide—up to several hundred times as much—and less oxygen than atmospheric air. In general there is an inverse relationship between the content of these two gases, with the oxygen decreasing as the carbon dioxide content increases.

The concentrations of both oxygen and carbon dioxide are closely related to biological activity within the soil; activity by microorganisms, larger animals, and by the roots of plants. As indicated earlier, oxygen is absolutely essential for the energy-yielding process of respiration, and living organisms—plant and animal—respire "around the clock." A deficiency of oxygen to roots, even for a relatively short period, can severely reduce plant size and crop yields (Fig. 4-17). Deficiency symptoms may manifest themselves in unusual ways. It is known, for example, that deficiency of oxygen can slow nutrient and water uptake by plants. Consequently, plants may show nutrient-deficiency symptoms (a subject of Chapter 5) on soils well supplied with the needed plant nutrients.

Soil oxygen will be high if good soil structure is maintained. Soils with large aggregates have an abundance of macropores. Following a rain, or irrigation, these pores are soon freed of water, thus allowing gases to move into the soil. A favorable balance of organic matter in the soil is the best way of assuring desirable structure. A second approach is through cultivation. In some very heavy soils it has been shown that light cultivation enhances soil aeration, while at the same time controlling weeds. Finally, high levels of oxygen can be maintained through manipulation of soil water. Proper drainage is important and can be facilitated by leveling, ditching, and the installation of tile. In the application of water the gardener must also be aware of the delicate balance between water and air in the pore space of the soil.

Soil Water

Water is the solvent in which soil nutrients are dissolved. This mixture of water and nutrients—the **soil solution**—is the vehicle whereby nutrients are moved throughout the plant. The many chemical reactions that take place in the soil occur in water. Water also plays an important role in regulating soil temperature, and as mentioned earlier, bears an intimate relationship to soil air.

Water—A Unique Molecule

The diverse activities of water are related to the unique structure of the water molecule. Each molecule consists of two atoms of hy-

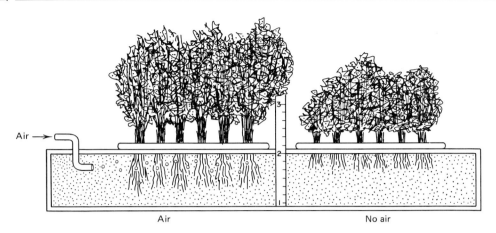

FIGURE 4-17 Effects of aeration on asparagus plants grown in a nutrient solution containing all the essential elements. (Courtesy of The University of California, Agricultural Experiment Station.)

drogen and one atom of oxygen. These are not arranged symmetrically, however, but in a "V" arrangement, with the two hydrogen atoms separated from each other by an angle of only 105 degrees (Fig. 4-18a). Because of this configuration the water molecule has a definite polarity; one side of the molecule has a positive charge and the other a negative charge (Fig. 4-18b). The positive side of the water molecule is attracted to negative surfaces, such as those offered by clay colloids. In addition, water molecules are attracted to each other—a property termed **cohesion**— and form large groupings. Soil nutrients, which themselves are charged particles, can also join with water molecules. These relationships are shown in Figure 4-18c.

To understand the behavior of water in soils we need to clarify one other point. Water molecules, in addition to having an attraction for each other, have an attraction for solid surfaces, a property termed **adhesion**. Thus, water is held on the surface of each soil particle and in the pore space between them. Also, water near the surface of a soil particle is held more tightly. This is extremely important because, as we have seen, soils vary greatly in the amount of surface they offer. Clay has much more surface than silt or sand and consequently holds a given amount of water much more tightly than the coarser textured soils. In a sand and clay of the same moisture content, the layer of water covering each sand grain would be much thicker than that covering individual particles of clay. The amount of water that a soil can release to plant roots depends on how tightly the water is held by the soil, not on the total amount of water present. In our example the loosely held water of the sand would be more readily available to plants.

The Kinds of Water in Soils

Visualize a volume of soil that has just been deluged by water. All of the pores, large and small, are filled with water; the soil is completely **saturated** (Fig. 4-19a). This saturated

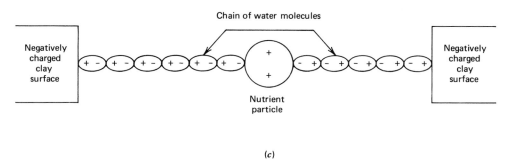

FIGURE 4-18 Soil water. The composition of a water molecule: (a) its polarity (b) and a schematic representation of water molecules in the soil (c).

condition will persist only until water begins to drain from the soil due to gravitational forces. The water in the macropores will then be replaced by air. When this weakly held gravitational water has been removed the soil is said to be at **field capacity** (Fig. 4-19b). Plants can, of course, use gravitational water if they can capture it before it drains away.

Plants growing on our soil will absorb water, and water will be lost from the soil surface through evaporation. Finally, however, a point will be reached where the water is held so

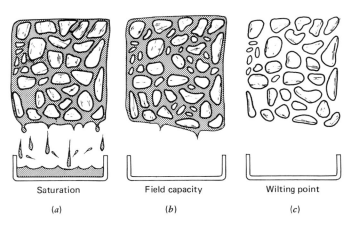

Saturation

(a)

Field capacity

(b)

Wilting point

(c)

FIGURE 4-19 A soil at saturation (a), at field capacity (b), and at the wilting coefficient (c). (Reprinted with permission of Macmillan Publishing Co., Inc. from Nyle C. Brady, *The Nature and Properties of Soils.* Copyright © 1974 Macmillan Publishing Co., Inc.)

tightly by the soil particles that plants cannot obtain it fast enough to replace water lost by transpiration. As a consequence, the plants will wilt. The soil moisture content at this stage is called the **wilting coefficient** (Fig. 4-19c). In other words, the water available to plants is that present in the soil between field capacity and the wilting coefficient. This water is called the **available water**. Soil water in the available range moves from wetter to dryer locations in the soil more or less independent of gravity. This movement, which can be up, down, or sideways, occurs through very fine pores, termed **micropores**. The diameter of micropores approximates that of hairs and as a consequence this water is also called **capillary water**, from the Latin word *capilla* meaning hair.

It is important to stress that soil at the wilting coefficient still has a considerable amount of moisture present; in fact, it still appears somewhat moist.

A Summary of Soil Moisture Characteristics

Several points emerge from our discussion of soil water. For one, drainage water is of little value to plants. This water is moved downward by gravitational forces to the **water table**, the level beneath which the soil is saturated. The depth of the water table varies for different regions, but in general tends to be near the surface in areas of high rainfall and to be fairly deep in arid regions. In soils with poor natural drainage, drainage operations, such as installation of subsurface tile, can facilitate the removal of drainage water, improve soil aeration, and effectively lower the water table.

Much water is available to plants between field capacity and the wilting coefficient, but that near the wilting coefficient is held with greater tenacity by the soil particles. The

moisture condition of the soil at which irrigation should start is sometimes called the **irrigation point**. While this point varies, depending on soil texture, type of crop, and other factors, it generally falls about midway between field capacity and the wilting coefficient. Aspects of irrigation are discussed in Chapter 8.

Finally, it bears repeating that soil moisture characteristics are intimately related to soil texture. The finer-textured soils, because of their smaller-sized particles, have a greater ability to store added soil moisture, although clay soils commonly rank somewhat below well-granulated silt or clay loams (Fig. 4-20). Said another way, the finer textured soils at field capacity have more water available for plant growth than do coarse-textured soils. Because of their greater available moisture storage the finer textured soils require less frequent irrigation. Note also that the amount of water remaining in the soil at the wilting coefficient

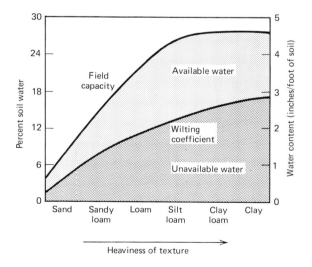

FIGURE 4-20 The general relation between soil moisture and soil texture. (Reprinted with permission of Macmillan Publishing Co., Inc. from Nyle C. Brady, *The Nature and Properties of Soils.* Copyright © 1974 Macmillan Publishing Co., Inc.)

also increases in the more finely textured soils. Thus, while they have more water available for plants their increased surface allows the retention of greater quantities of water.

SOME PROBLEM SOILS

Thus far we have painted a rather simplified picture of soils, implying that if one understands a few basic principles that soil management will present no problems. Such, however, is not the case. In conversations with an experienced Texas gardener we asked what she regarded as the number one gardening problem in her area. Her reply was short and precise: caliche! Gardeners from Florida and elsewhere have given similar replies to this question. Caliche is just one of the several forms of hardpan that can occur in soils. Other soils, the saline soils, are high in neutral soluble salts, and still others, the sodic soils, contain unfavorable levels of sodium. All of these soil conditions—hardpan, high soluble salts, excess sodium—present problems to the gardener.

Hardpan Layers

Hardpan refers to a hardened soil layer, generally in the lower A or in the B horizon, caused by the cementing of soil particles by colloidal materials. In some soils accumulation of clay colloids in the B horizon is so great that it results in the formation of a claypan that acts as a barrier to water movement and root growth (Fig. 4-21). In other soils a cemented layer forms from calcium or magnesium salts that have come out of solution. This layer, known as **caliche**, may be near the surface or deep in the profile. In some cases the caliche is at the surface because the topsoil down to the hardened layer has been lost by

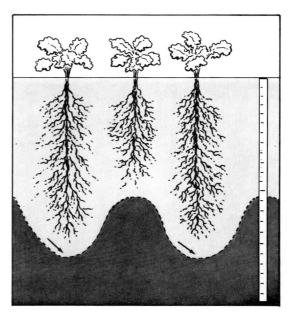

FIGURE 4-21 Hardpans can arise from either naturally or from human actions; they are essentially impervious to water, air, and growing roots. In this example, the hardpan was loosened mechanically to a depth of 16 inches (at arrows) by use of a sub-soiler in early fall, then seeded with a cover crop (dwarf rape). The more extensive root growth made possible by subsoiling is clearly shown by the dotted line marking the depth to which roots have penetrated. (Adapted from Brooklyn Botanic Garden *Handbook on Soils* by Guerry Dean.)

erosion. Not all hardpans are a result of natural processes. Continued plowing of a soil at a certain depth, for example, can result in the formation of a **plowpan** through which roots cannot penetrate.

What can be done about hardpan layers? First, one must determine their extent and depth. Soil survey reports will indicate if they are common to your area. The presence of hardpan under your garden plot can often by detected by probing with a steel rod. Hardpan can be dealt with in several ways. **Subsoiling**—tillage below the depth reached by the ordinary plow—can sometimes break and shatter a hardpan (Fig. 4-21). Large-scale agriculture employs subsoilers of several types for this operation, but the gardener can achieve similar results using a hammer and chisel or similar equipment. In some cases pan layers have been broken up by deep plowing and the mixing in of a soil conditioner such as straw. In other situations the best approach is simply to take full advantage of the soil above the pan layer. This soil layer is often shallow, however, and to get a seedbed of adequate depth one must either add additional topsoil or employ a raised bed system (Figs. 7–12, 13). Large areas in Florida are now being cropped by forming raised beds directly over caliche.

Saline and Sodic Soils

Saline (high soluble salts) and **sodic** (high sodium content) soils are widespread in western Canada and the United States. The provinces of Alberta and Saskatchewan alone each contain over five million acres of soils of this sort. These problem soils develop when leaching has not occurred to a significant degree, and when surface evaporation is excessive—conditions common in arid climates. The process of surface accumulation of salts is similar to that occurring when an open pan of salt water boils on a stove; soon the water in the pan boils away, but the salts remain. In the soil the upward movement of water brings salts to the surface. When the water evaporates the salts remain and eventually accumulate in amounts toxic to plants. Saline soils are sometimes called **white alkali** soils because their salt accumulations are whitish in color. Sodic soils, on the other hand, are referred to as **black alkali** soils because their excess sodium causes a dispersion of the dark-colored humus carried upward to the soil surface. In general, saline soils are less detrimental to plants than are sodic soils, and their salts can be tolerated in higher amounts.

Saline and sodic conditions discourage or inhibit plant growth in several ways. First, whenever the soil solution has a higher concentration of salts than do plant cells, then water will be drawn out of the cells and into the external solution. Second, increasing salinity is accompanied by increased soil moisture and this creates poor soil aeration. Third, certain salts form compounds which have a strong corrosive action on plant tissues. Finally, salts, such as sodium, cause a dispersion of colloids and thereby create a puddled condition in soils (Fig. 4-12).

What can be done about these problem soils? Garden crops vary in their tolerance to salty soils (Table 4-7). Deep-rooted crops, such as sugar beet, have an advantage since a portion of their root systems may be at a depth where soluble salts are minimal. As expected, seedlings are more sensitive to salts than are established plants, and sometimes slight differences in planting and watering technique can mean the difference between success and failure in arid-climate gardening (Fig. 4-22). Once seedlings are established, actions that retard evaporation, such as mulching, are important as they slow the upward movement of salts. Frequent irrigations and maintenance of high levels of soil fertility also favor growth of established seedlings. If soluble salts are too high in concentration for successful gardening, drainage tile can be installed, and the salts leached from the soil by heavy and repeated additions of water. Some sodic soils can be reclaimed by large additions of gypsum, with subsequent leaching by water recommended.

Questions for Discussion

1. Define the following terms:
   (a) Leafmold
   (b) Residual soil
   (c) Mycorrhizal fungi
   (d) Flocculation
   (e) Percolation
   (f) Capillary water
   (g) Subsoiling

TABLE 4-7 Tolerance of Certain Vegetable and Fruit Crops to Salty Soils

Tolerance of vegetable crops to salty soil			
High salt tolerance	Medium salt tolerance		Low salt tolerance
Garden beet	Tomato	Potato	Radish
Kale	Broccoli	Carrot	Celery
Asparagus	Cabbage	Onion	Green bean
Spinach	Bell pepper	Pea	
Sugar beet	Cauliflower	Squash	
	Lettuce	Cucumber	
	Sweet corn	Sunflower	

Tolerance of fruit crops to salty soil			
High salt tolerance	Medium salt tolerance		Low salt tolerance
Date palm	Fig	Pear	Almond
	Olive	Apple	Apricot
	Grape	Orange	Peach
	Cantaloupe	Grapefruit	Strawberry
		Prune	Lemon
		Plum	Avocado

Courtesy of U.S. Department of Agriculture.

2. Outline the major horizons seen in a profile of a mature soil and rank these as to organic matter content, degree of leaching, and amount of biological activity.

3. What criteria would you employ to determine if the soil of your garden, or other specified area, is residual or depositional?

4. Identify the major factors involved in soil formation and indicate which one is regarded as exerting the dominant influence.

5. Describe approaches you would take to obtain technical data on the soils of your area.

6. List the major components of a typical mineral soil and comment on their relative percentages.

7. Distinguish clearly between "soil texture" and "soil structure." Over which of these does the gardener have the least control?

8. Why is organic matter considered such a vital component of soils?

9. Explain in a technical manner why the addition of an organic material, such as barnyard manure, to the soil might actually depress crop yields.

10. What exactly is meant by the phrase "available water"?

11. The text states the "ideal" level of soil moisture, as relates to crop growth, is approximately midway between field capacity and the wilting coefficient. Assuming this to be true, why do not gardeners water moisture-depleted soils to exactly this point?

12. Say that you live in the arid Southwest where the salt content of the soils is high. What measures could you take to enhance your gardening efforts under these conditions?

Selected References

Brady, N. 1974. *The Nature and Properties of Soils.* Macmillan, New York.

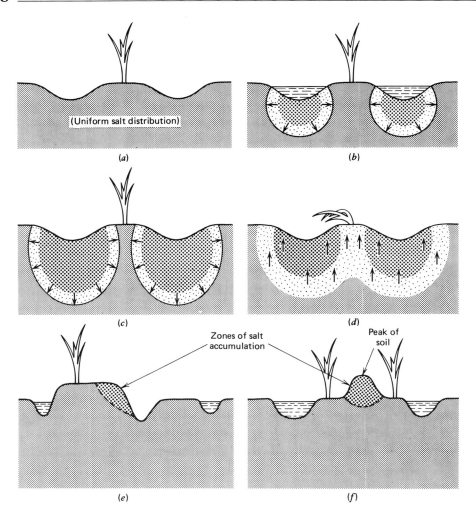

FIGURE 4-22 Seedling placement and salt accumulation with irrigation. Shown are improper placement (a–d), and two placements (e,f) that favor differential salt accumulation and seedling survival. (a–d) Reprinted with permission of Macmillan Publishing Co., Inc. from Nyle C. Brady, *The Nature and Properties of Soils.* Copyright © 1974 Macmillan Publishing Co., Inc.)

Daubenmire, R. 1974. *Plants and Environment. A Textbook of Plant Autecology.* John Wiley, New York.

Foth, H. and L. Turk. 1972. *Fundamentals of Soil Science.* John Wiley, New York.

Hunt, C. 1972. *Geology of Soils. Their Evolution, Classification, and Uses.* Freeman, San Francisco.

Kellogg, C. 1952. *Our Garden Soils.* Macmillan, New York.

Knuti, L., M. Korpi, and J. Hide. 1970. *Profitable Soil Management.* Prentice-Hall, Englewood Cliffs, N. J.

Kohnke, H. 1953. *Soil Science Simplified.* Published by Author.

United States Department of Agriculture. 1951. *Soil Survey Manual. Handbook 18*. Government Printing Office, Washington, D.C.

———. 1975. *Soil Taxonomy. Handbook 436*. Government Printing Office, Washington, D.C.

University of Saskatchewan. 1975. *Guide to Farm Practices in Saskatchewan*. Extension Division, Saskatoon, Saskatchewan.

Wildman, W. 1976. *What on Earth Is Soil?* Leaflet 2637. University of California, Davis.

Wilsie, C. 1962. *Crop Adaption and Distribution*. Freeman, San Francisco.

LEARNING OBJECTIVES

By the time you have finished this chapter you should be able to:

1 Identify the essential nutrient elements needed for normal plant growth.

2 Distinguish between "macronutrients" and "micronutrients," and give examples of each.

3 Describe some specific functions of mineral elements in the plant.

4 State the "principle of limiting factors" and tell how this principle relates to our gardening efforts.

5 Discuss some of the properties of ions.

6 Explain what is meant by the phrase "cation exchange capacity."

7 Define what is meant by the terms "acid" and "base."

8 Describe how soil pH affects the availability of mineral nutrients.

9 Discuss the nitrogen cycle as it occurs in nature.

10 Describe the fertilizer guarantee.

11 State the law of diminishing returns and tell how it relates to fertilizer applications and crop yields.

12 Explain some of the factors that need to be considered in determining optimal levels of soil fertility.

5

SOIL FERTILITY AND PLANT GROWTH

"This continual cycle of nutrients out of the soil into plants and back to the soil is perhaps the most important fact of soil dynamics. If the cycle is broken by the removal of nutrients through the harvesting of crops, the soil may deteriorate rapidly. On the other hand, farmers and gardeners can build up the soil to a much higher level of nutrients than its natural condition by adding chemical fertilizers or organic matter from over the hedge."

Charles E. Kellogg

An important principle in vegetable production is that plant growth rates should be kept at a maximum. Vegetable plants seldom recover fully if growth is retarded at any stage of development, and retarded growth usually means reduced yields. The soil plays a vital role in this scheme, because it serves as a medium for plant growth. Plants require abundant amounts of water, air, and mineral nutrients for good growth, and with few exceptions these needs are obtained from the soil by the root system of the plant. In addition, the root system must securely anchor the plant.

To fulfill its diverse roles the root system must spread extensively within the soil. That this is the case is shown by the detailed analysis of the root systems of winter rye plants by Dittmer. Data on the root system of a single plant, 20 inches high and with a clump of 80 shoots, is given in Table 5-1. As enormous as these figures are, they only begin to tell the story. Near the tip of each of these millions of roots vast numbers of root hairs are formed (Fig. 2-18). It was estimated that the rye plant mentioned above had more than 14 billion root hairs with a total area of more than 4000 square feet. These root hairs establish an intimate contact with the soil particles and the soil solution (Fig. 5-1). Through these hairs, and other surface cells of the root tip pass the water, dissolved oxygen, and mineral nutrients essential for growth and reproduction.

The root system of a plant is not only ex-

TABLE 5-1 Characteristics of the Root System of a Single Rye Plant

Kind of root	Number	Total length in feet
Main roots	143	214
Branches of main roots (secondaries)	35,600	17,800
Branches of secondaries (tertiaries)	2,300,000	574,000
Branches of tertiaries (quarternaries)	11,500,000	1,450,000
Total	14 million	2 million feet or 380 miles

tensive; it also ramifies a given volume of soil very thoroughly. In the rye plant the root system had a combined surface area in excess of 6500 square feet, more than 130 times the total area of the shoots. Yet the volume of soil penetrated and exploited by the entire root system comprised only about 2 cubic feet. These figures leave no doubt as to the ability of the root system to form a dynamic interface with the soil.

In this chapter we wish to concentrate on the nutrient elements needed by plants, their functions within the plant, the forms in which

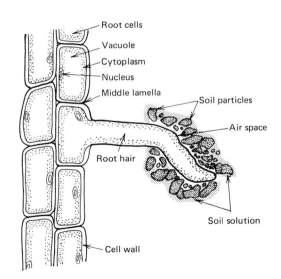

FIGURE 5-1 A root hair and adjacent cells. The root hairs, which are outgrowths of epidermal cells, form an intimate contact with the soil. (Courtesy of the Hawaiian Sugar Planter's Association, Aiea, Hawaii.)

nutrients are absorbed by plants, the importance of clay, the humus colloids, and soil reaction (pH) as they relate to nutrients, sources—natural and artificial—of mineral nutrients, and, finally, nutrient deficiency problems.

THE ESSENTIAL ELEMENTS

Approximately 90 elements occur in a stable form on earth. Of these, just two—silicon and aluminum—comprise about 86 percent of the inorganic material of the A horizon of a typical mineral soil. The others, including those essential for plant growth, are found in the remaining 14 percent of material. To date, plant scientists have shown that only 17 elements are necessary for the normal growth of plants. These 17 are termed the **essential elements.** It should be stressed that some quantity of each of these elements is *absolutely* essential at some stage in the life cycle of a large number of plant species.

Macronutrients and Micronutrients

The essential elements come from different sources and are used by plants in varying amounts (Table 5-2). Three elements—carbon, hydrogen, and oxygen—are obtained mainly from water and air. The remaining 14 elements come from the soil. Of these, 6 are used by plants in relatively large quantities and are called **macronutrients.** The other 8

TABLE 5-2 The Essential Nutrient Elements, Their Sources, and the Relative Amounts Needed by Plants

Essential elements used in relatively large amounts		Essential elements used in relatively small amounts	
Mostly from air and water	From soil solids	From soil solids	
Carbon	Nitrogen	Iron	Copper
Hydrogen	Phosphorus	Manganese	Zinc
Oxygen	Potassium	Boron	Chlorine
	Calcium	Molybdenum	Cobalt
	Magnesium		
	Sulfur		

Reprinted with permission of Macmillan Publishing Co., Inc. From Nyle C. Brady. The Nature and Properties of Soils. *Copyright 1974 by Macmillan Publishing Co., Inc.*

are needed in much smaller amounts and are thus termed **micronutrients** (also called **trace** or **minor** elements).

The Macronutrients

The macronutrients are nitrogen, phosphorus, potassium, calcium, magnesium, and sulfur. Nitrogen, phosphorus, and potassium—the big three—are the elements most commonly added to the soil as farm manure or commercial fertilizer. Thus, they are termed the **primary** or **fertilizer elements,** and any fertilizer that contains all three is said to be a **"complete"** fertilizer. Calcium, magnesium, and sulfur are termed the **secondary elements.** They are also added to the soil frequently; calcium and magnesium in the form of ground limestone and sulfur as part of commercial fertilizers such as superphosphate and sulfate of ammonia. These elements are, of course, present in farm manure too.

The Micronutrients

The micronutrients—iron, manganese, boron, molybdenum, copper, zinc, chlorine, and cobalt—are used by plants in very small amounts. They are present in farm manure and occur as impurities in many commercial fertilizers. Soils generally have an ample supply of these elements, but on soils cropped for many years, and in specific regions, one or more of the micronutrients may become deficient. The micronutrient cobalt has not been shown to be essential for a broad range of higher plants, but it is essential for the nitrogen-fixing microorganisms, including those present in legumes. The importance of the nitrogen fixation process in the overall scheme of gardening, however, warrants the inclusion of cobalt in our list (Table 5-2).

Selective Uptake of Mineral Elements

You might expect that in the course of evolution higher plants would have made use of the elements most abundant in nature, and to some extent this is so. Three abundant elements—carbon, oxygen, and hydrogen—comprise about 92 percent of the dry weight of plants (Fig. 4-14). Other elements, such as sodium, are little used by plants, however. It seems to be required by only a few desert and salt-marsh species. Silicon and aluminum are also present in large quantities in soils, yet only a few plants require silicon and none requires aluminum. On the other hand, all plants require small quantities of copper and molybdenum, elements present in the soil in minuscule amounts (Table 5-3).

Plants also take up many elements that play no known role in metabolism. Barium, gallium, gold, and mercury are found in plants, but serve no known purpose. Plants also take up silicon and aluminum. Horsetails (the genus *Equisetum*) incorporate silicon into their cell walls making the plant quite unpalatable to most animals, but useful (at least in the past) for scouring pots and pans. Aluminum is found in traces in most plants. Flowers of hydrangea are reported to turn from pink to blue if plants are grown in soils high in soluble aluminum.

Plants may accumulate elements in amounts that are toxic, either to themselves or to animals. Lithium, for example, can cause **necrosis** (localized death of tissues) of leaf margins in avocado, celery, corn, and citrus. Even essential elements such as boron and copper are toxic except in very minute concentrations. Selenium accumulates in certain plant species of western rangelands, and if eaten by grazing animals causes a disease known as "alkali disease" or "blind staggers." Humans can also suffer from selenium poisoning. We are also affected by high levels of arsenic and lead, a subject discussed in Chapter 15. Finally, certain elements taken up by plants are essential only to animals. A prime example is iodine, a deficiency of which causes **goiter** (an enlargement of the thyroid gland) in humans. Addition of iodine to table salt has virtually eliminated this disease.

FUNCTIONS OF MINERALS IN PLANTS

Every essential element taken up by plants serves one or more specific roles within the plant body. Carbon, hydrogen, and oxygen, as mentioned earlier, are the basic elements of all organic molecules. Proteins contain, in addition, nitrogen and sulfur. Several of the trace elements are essential structural parts of enzymes or serve an indirect role as activators or regulators of enzyme actions. Phosphorus

TABLE 5-3 The Percentage of Certain Essential Elements in Representative Agricultural Soils

Essential element	Percentage in representative agricultural soils	Amount (lb/acre)
Iron	3.5	70,000
Potassium	1.5	30,000
Calcium	0.5	10,000
Magnesium	0.4	8000
Nitrogen	0.1	2000
Phosphorus	0.06	1000
Sulfur	0.05	1000
Manganese	0.05	1000
Chlorine	0.01	200
Boron	0.002	40
Zinc	0.001	20
Copper	0.0005	5
Molybdenum	0.0001	2

Modified from G. W. Leeper, Introduction to Soil Science. *Melbourne University Press: New York, Cambridge University Press, 1948.*

is an essential component of cell membranes and is also involved in energy transformations within the cell. Calcium combines with pectic materials to form the middle lamella that cements plant cells together. Boron somehow regulates calcium utilization, but other than that its precise functions are unknown. Magnesium plays a central role in the structure of the chlorophyll molecule. A summary of some of the functions of the mineral elements required by plants is presented in Table 5-4.

The Principle of Limiting Factors

As you might expect, the demands for a given nutrient fluctuate during the plant cycle. Heavy nitrogen demands are associated with vegetative growth, and abundant phosphorus is a necessity in the reproductive phase (Fig. 5-2). Other elements, show selective use in the plant cycle too. The gardener needs to be aware that nutrient needs do change throughout the growing season; to some extent her or his fertilization program can be planned accordingly. It is also important to realize that no matter how minuscule the quantity of a given element required by a plant, that element must be present in adequate supply at the time it is needed or growth will be slowed.

This raises a very important question: What are the yield consequences if an element, or elements, are present in less than optimum amounts? As it turns out, the element that is *least* optimum will determine the level of crop production. This concept, which is known as the **principle of limiting factors,** states in general that *the level of crop production can be no greater than that allowed by the most limiting of the essential plant growth factors.* To illustrate this principle, the growth factors are represented by the staves of a barrel (Fig. 5-3). The level of the water in our barrel is determined by the height of the shortest stave (in this case representing nitrogen). If we raise the height of the nitrogen stave (by adding nitrogen to the soil) then the level of water (crop pro-

duction) will rise until some other factor becomes limiting, in this case potassium (Fig. 5-3b). Note also that our example applies to more than the nutrient elements. If the height of the light stave is halved, then light will become the limiting factor. Then, no matter how much nitrogen or potassium is added to the soil, crop production will not increase; growth is limited by the least optimum factor. Crop yields can also suffer from too much of a given factor, which results either in a toxic reaction or in excessive growth.

FORMS IN WHICH NUTRIENTS ARE ABSORBED BY PLANTS

Thus far we have spoken of the essential elements without reference to the forms in which they occur in the soil or are absorbed. Here we must introduce a few concepts of chemistry, starting with the element itself.

Elements, Atoms, and Ions

Elements are substances that cannot be decomposed into simpler substances by ordinary chemical means. The sugar glucose ($C_6H_{12}O_6$) is not an element because our bodies can break it down into carbon dioxide and water with the release of energy. The carbon dioxide (CO_2) can be broken down still further to give carbon and oxygen. The carbon and oxygen cannot be broken down, however, and thus fit our definition of elements.

The elements exist in the form of discrete units called **atoms.** Each atom of an element consists of a nucleus and electrons (Fig. 5-4). The **nucleus** is the central core of an atom and carries a positive electrical charge. Surrounding the nucleus occur negatively charged particles known as **electrons.** Atoms as a whole possess no charge, because the number of positive charges in the nucleus always equals the number of negative charges carried by electrons. In the soil solution, however, atoms

TABLE 5-4 A Summary of Mineral Elements Required by Plants

Element (and chemical symbol)	Form in which absorbed	Approximate concentration in whole plant (as % of dry weight)	Some functions
Macronutrients			
Nitrogen (N)	NO_3^- or NH_4^+	1–3%	Amino acids, proteins, nucleotides, nucleic acids, chlorophyll, and coenzymes.
Potassium (K)	K^+	0.3–6%	Enzymes, amino acids, and protein synthesis. Activator of many enzymes. Opening and closing of stomata.
Calcium (Ca)	Ca^{2+}	0.1–3.5%	Calcium of cell walls. Enzyme cofactor. Cell permeability.
Phosphorus (P)	$H_2PO_4^-$ or HPO_4^{2-}	0.05–1.0%	Formation of "high-energy" phosphate compounds (ATP and ADP). Nucleic acids. Phosphorylation of sugars. Several essential coenzymes. Phospholipids.
Magnesium (Mg)	Mg^{2+}	0.05–0.7%	Part of the chlorophyll molecule. Activator of many enzymes.
Sulfur (S)	SO_4^{2-}	0.05–1.5%	Some amino acids and proteins. Coenzyme A.
Micronutrients			
Iron (Fe)	Fe^{2+}, Fe^{3+}	10–1500 parts per million (ppm)	Chlorophyll synthesis, cytochromes, and ferredoxin.
Chlorine (Cl)	CL^-	100–10,000 ppm	Osmosis and ionic balance; probably essential in photosynthesis in the reactions in which oxygen is produced.
Copper (Cu)	Cu^{2+}	2–75 ppm	Activator of some enzymes.
Manganese (Mn)	Mn^{2+}	5–1500 ppm	Activator of some enzymes.
Zinc (Zn)	Zn^{2+}	3–150 ppm	Activator of many enzymes.
Molybdenum (Mo)	MoO_4^{2-}	0.1–5.0 ppm	Nitrogen metabolism.
Boron (B)	BO^{3-} or $B_4O_7^{2-}$ (borate or tetraborate)	2–75 ppm	Influences Ca^{2+} utilization. Functions unknown.
Elements Essential to Some Plants or Organisms			
Cobalt (Co)	Co^{2+}	Trace	Required by nitrogen-fixing microorganisms.

Reprinted with permission of Worth Publishers, Inc. From Peter H. Raven, Ray F. Evert, and Helena Curtis, Biology of Plants. *Copyright 1981 by Worth Publishers, Inc.*

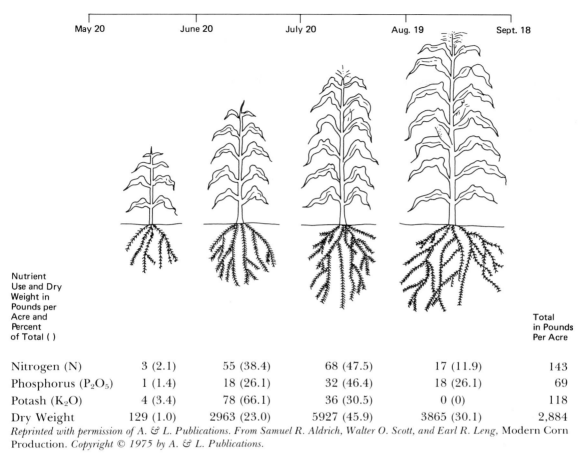

Nutrient Use and Dry Weight in Pounds per Acre and Percent of Total ( )	May 20	June 20	July 20	Aug. 19	Sept. 18	Total in Pounds Per Acre
Nitrogen (N)		3 (2.1)	55 (38.4)	68 (47.5)	17 (11.9)	143
Phosphorus (P_2O_5)		1 (1.4)	18 (26.1)	32 (46.4)	18 (26.1)	69
Potash (K_2O)		4 (3.4)	78 (66.1)	36 (30.5)	0 (0)	118
Dry Weight		129 (1.0)	2963 (23.0)	5927 (45.9)	3865 (30.1)	2,884

Reprinted with permission of A. & L. Publications. From Samuel R. Aldrich, Walter O. Scott, and Earl R. Leng, Modern Corn Production. *Copyright © 1975 by A. & L. Publications.*

FIGURE 5-2 Nutrient needs change throughout the growing season. (Courtesy of the Potash and Phosphate Institute, Atlanta, Ga.)

may either gain or lose electrons. It is important to note that when this occurs they are no longer electrically neutral, but rather become electrically charged atoms, known as **ions.** It is these ionic forms that occur in the soil solution and are taken up by plants (Table 5-4).

Chemical Notation

Chemists deal with the elements on a day-to-day basis, and have symbols for each element (Table 5-4). These symbols may represent abbreviations of the English name for the element; for example, "N" for nitrogen and "O" for oxygen. Others represent abbreviations of the Latin; for example, "K" for potassium whose Latin name is *kalium*. When designating ions, electrical charges are indicated. Some examples are H^+, K^+, Ca^{2+}, Mg^{2+}, Cl^-, NO_3^- and SO_4^{2-}. Chemists call ions that possess a positive charge **cations,** and those that possess a negative charge **anions.** Note that most of the positively charged ions consist of only one element, whereas in many of the negatively charged ions the crucial element is joined to

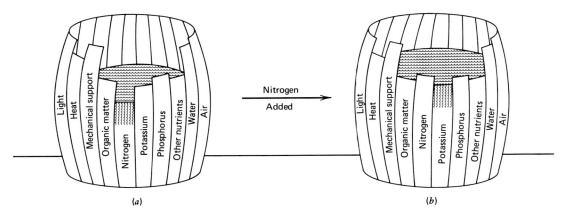

FIGURE 5-3 The principle of limiting factors. The level of water in the barrel represents the level of crop production. The level is controlled by the factor (stave) present in the least amount. (Reprinted with permission of Macmillan Publishing Co., Inc. from Nyle C. Brady, *The Nature and Properties of Soils.* Copyright © 1974 Macmillan Publishing Co., Inc.)

other elements and they together form the structure of the total ion. The nitrate ion, NO_3^-, for example, consists of one atom of nitrogen and three atoms of oxygen, and carries a single negative charge. Note also that some elements are absorbed in only one ionic form, while others are available to plants in two forms (Table 5-4). Regardless of these variations, it is important to remember that ions are electrically charged atoms, and, as we will see, their behavior in the soil solution depends in large part on the nature of this electrical charge.

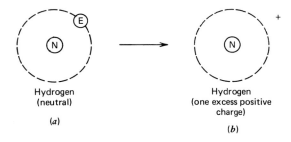

Hydrogen
(neutral)

(a)

Hydrogen
(one excess positive
charge)

(b)

FIGURE 5-4 An atom of hydrogen (a), showing the nucleus (N) and surrounding electrons (E). The ionic form is also shown (b).

Organic Versus Inorganic Nutrient Sources

Thus far in our discussion of the essential nutrients and the forms in which they are taken up by plants we have made no distinction between nutrients from inorganic sources, such as weathered minerals, and nutrients from organic sources, such as decomposed organic residues. We have made no distinction because there is none to be made! In the words of Warren Schoonover of the University of California,

> To a plant physiologist, the term organic has no meaning when applied to green plants. Plants grow and build organic matter by using ions of the essential elements. Whatever their source, the plant absorbs these ions in their inorganic forms. The plant does not distinguish between ammonia produced from fermentation in a manure pile and ammonia produced in a chemical factory or distilled from coal in coke manufacturing.

These words apply equally to the gardener. This does not mean, however, that organic

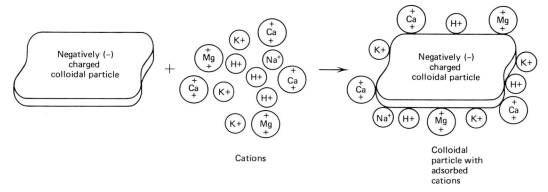

FIGURE 5-5 Negatively charged colloidal particles attract and hold the positively charged cations present in soil solutions.

fertilizers are of lesser value. Their ability to improve soil structure, provide food for soil organisms, and help maintain an ecological balance in nature is well known. They are of great value, but for reasons other than a uniqueness in their nutrient contribution.

CLAY AND HUMUS—THE DYNAMIC DUO

Clay and humus colloids are extremely small, negatively charged particles. They are the chemically active ingredients of the soil solids, and together can be regarded as the "dynamic duo" of the soil system. But what makes these ingredients so important? For one thing, they offer a large negatively charged surface area to which cations are attracted and held (Fig. 5-5). Adsorbed cations cannot be easily leached from the root zone and therefore represent a storehouse of nutrients available to growing plants.

Cation Exchange

The negatively charged soil colloids do not hold the different cations with equal strength. In general, when the cations are present in equal quantities their relative strengths of adsorption are in the following order:

Element	Strength of Adsorption
Aluminum (Al^{3+})	Strong
Hydrogen (H^+)	
Calcium (Ca^{2+})	
Magnesium (Mg^{2+})	
Potassium (K^+)	
Sodium (Na^+)	Weak

If one cation, say calcium, is present in greater abundance it will tend to exchange for ions, such as those of hydrogen, present in lesser abundance. This mass action exchange reaction can be represented as follows:

$$\boxed{\begin{array}{l}\text{Negatively}\\\text{charged}\\\text{colloid}\end{array}}\begin{array}{l}-\text{H}\\-\text{H}\end{array} + Ca^{2+} \quad \rightleftharpoons \quad \boxed{\begin{array}{l}\text{Negatively}\\\text{charged}\\\text{colloid}\end{array}}-Ca + 2H^+$$

The arrows in the middle are shown as pointing in both directions because the reaction can proceed either way depending on the relative concentration of the two ions. The phenomenon just described, termed **cation exchange,** goes on in the soil constantly. Any time we add fertilizer, any time nutrients are released from the soil solids, or any time nutrients are absorbed by roots, cations are exchanged. Our example using calcium represents in simplified form what happens when we add lime to our soil.

We have seen that the cations are held by clay and humus colloids, but what about the negatively charged anions? To some extent they are attracted to positively charged colloidal materials, but in general they are held less tightly than are cations. Phosphate, sulfate, and nitrate are the most important of the anions in plant nutrition. Of the three, phosphate is held most tightly by the soil solids. The others are essentially free and are able to move with the soil water. This freedom of movement has important consequences, because the ions can readily be leached from the root zone. One reason nitrogen is so commonly added to garden soils is because supplies are constantly leached away. Proper water management can reduce these losses somewhat.

Cation Exchange Capacity

Cation exchange capacity (CEC) is defined as the sum total of exchangeable cations that a soil can adsorb. Soil scientists describe this capacity in technical units,[1] but our point is well illustrated using stated cation exchange values simply as arbitrary units. Soil colloids vary in their ability to hold cations, as illustrated by the following data:

Adsorbing Material	Average CEC
Humus	200
Clay	
Montmorillonite	100
Illite	30
Kaolinite	10

Montmorillonite clay has approximately 10 times the cation holding ability of kaolinite. But compare the clays to humus! These organic colloids have a much higher CEC than do the mineral colloids. Clearly, adding organic residues to garden soils is the best way to maintain their ability to hold nutrient ions.

The CEC values for a soil include both organic and inorganic colloids. Since soils vary greatly in clay content and composition as well as in amount of humus it is not surprising to find that soils vary widely in their CEC (Table 5-5). The coarse textured soils have a much lower CEC than do the finer textured soils; they simply have less clay colloids to hold cations. This is the main reason why the larger particled sandy soils require fertilizer additions much more frequently than do clayey soils. Variations within a textural class relate both to the kinds of clay colloids and the amount of humus present.

CEC is important to us because research has repeatedly shown that the ability to hold cations for exchange is the single best index of potential soil fertility. Exchangeable cations are generally available to higher plants. By cation exchange, hydrogen ions from roots replace nutrient ions from the soil colloids. The less tightly held nutrient ions are forced into the soil solution where they can be taken up by plants, or, in some cases, leached away. When gardeners have their soils tested, determination of CEC is one of the many tests available. Every gardener should have some idea as to the ability of his or her soil to hold cations. Information on soil-testing services is presented in Appendix C.

[1]Expressed in milliequivalents per 100 grams of oven-dried soil or other adsorbing material. The term milliequivalent may be defined as one milligram of hydrogen or the amount of any other ion that will combine with or displace it.

TABLE 5-5 Cation Exchange Capacity of Surface Soils from Various Parts of the United States

Soil type	Exchange[a] capacity	Soil type	Exchange capacity
Sand		Silt loam	
Sassafras (NJ)	2.0	Fayette (MN)	12.6
Plainfield (WI)	3.5	Dawes (NJ)	18.4
		Miami (WI)	23.2
Sandy loam		Clay and clay loam	
Sassafras (NJ)	2.7	Cecil clay loam (AL)	4.0
Coltneck (NJ)	9.9	Gleason clay loam (CA)	31.5
Colma (CA)	17.1	Sweeney clay (CA)	57.5
Loam			
Sassafras (NJ)	7.5		
Collington (NJ)	15.9		

Reprinted with permission of Macmillan Publishing Co., Inc. From Nyle C. Brady, The Nature and Properties of Soils. *Copyright 1974 by Macmillan Publishing Co., Inc.*
[a]Milliequivalents per 100 grams of dry soil.

SOIL REACTION (OR pH)

Acids and Bases

Two classes of ionic compounds—acids and bases—have such a profound influence on soil properties that they merit our special attention. A multitude of definitions exist for acids and bases, but for our purposes it can be said that a solution—such as the soil solution—is **acid** when it contains a preponderance of hydrogen ions (H^+) over hydroxyl ions (OH^-). Conversely, a solution is **basic,** or **alkaline,** when it contains more hydroxyl ions than hydrogen ions. When the two ionic forms are present in equal numbers a solution is said to be **neutral.**

To illustrate, pure water is comprised of three components; water molecules, hydrogen ions, and hydroxyl ions. The number of molecules in a volume of water is very large, and the numbers of ions is very small. In pure water the number of hydrogen ions exactly equals the number of hydroxyl ions and thus, according to our definition, is neutral (Fig. 5-6). If something is now added to our container of pure water that alters the relative

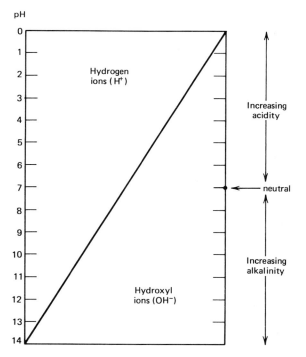

FIGURE 5-6 In an acid solution hydrogen ions predominate; in a neutral solution the numbers of hydrogen and hydroxyl ions are identical; and in an alkaline solution hydroxyl ions predominate.

concentrations of the two ions the resulting solution will become acid or alkaline (Fig. 5-6). The *degree* of acidity or alkalinity can, of course, vary. To express these variations in the acidity or alkalinity of solutions chemists have devised a scale known as the **pH scale.** (In the expression pH, the p stands for "power" and the H stands for the hydrogen ion.) The pH scale goes from 0 to 14 with a pH of 7 being neutral (Fig. 5-7). Solutions with a pH of less than 7 are acidic, while those with a pH greater than 7 are basic. Going in either direction from neutrality increases the acidity or alkalinity of a solution.

All of us deal on a daily basis with both acids and bases (Fig. 5-7). Our drinking water is seldom if ever neutral, and the sour taste of our citrus juices is due to their highly acid contents. Common household products such as ammonia and lye are quite basic in reaction. More to the point, as gardeners we deal with soil solutions that vary greatly in pH. In arid regions where calcium and other salts are abundant in the soil the pH commonly ranges from 7 to 9, whereas in humid regions, where leaching has played a more important role, the soils are generally somewhat acid. These pH differences are of great importance to the gardener, because they have a direct bearing on soil fertility.

Soil pH and Nutrient Availability

Soil pH directly influences the availability of plant nutrients, as shown in Fig. 5-8. Above pH 5 the availability of iron, manganese, zinc, copper, and cobalt show a marked decrease. The availability of molybdenum, potassium, calcium, and magnesium, on the other hand, increase above pH 5. Many of the nutrient elements have a rather broad range of maximum availability, but one, phosphorus, shows maximum availability over only a narrow pH range (Fig. 5-8). This presents a serious problem, because phosphorus is required in large amounts by plants, and even at its optimum

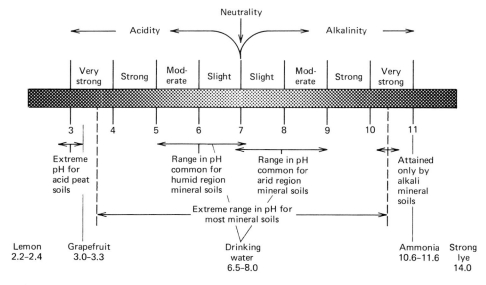

FIGURE 5-7 A portion of the pH scale, and the pH's of some soils and common household products. (Reprinted with permission of Macmillan Publishing Co., Inc. from Nyle C. Brady, *The Nature and Properties of Soils.* Copyright © 1974 Macmillan Publishing Co., Inc.)

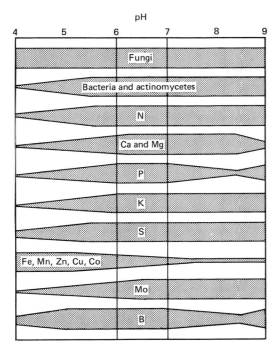

FIGURE 5-8 The relationships between the pH of mineral soils and the availability of plant nutrients. Overall optimal availability of the mineral nutrients occurs between pH 6.0 and 7.0. (Reprinted with permission of Macmillan Publishing Co., Inc. from Nyle C. Brady, *The Nature and Properties of Soils.* Copyright © 1974 Macmillan Publishing Co., Inc.)

pH—around 6.5—is still only sparingly available. One soil scientist has estimated, for example, that the supply of available phosphorus in the soil turns over 1000 times per day. That is, at any one time only 1/1000 of the daily supply is in a form useful to plants. This shows clearly the critical nature of the relationship between soil pH and nutrient availability. The key position of phosphorus in the nutrient availability scheme has prompted many soil scientists to recommend that the pH of mineral soils be maintained in the range of maximum availability of phosphorus. This range not only helps assure an adequate supply of phosphorus but also gives optimal availability of the other mineral nutrients.

The pH Requirements of Garden Crops

Every garden crop has its own optimum pH range, which normally extends over at least one pH unit (Table 5-6). Virtually all vegetables thrive at a pH range of 6.0 to 6.5—the range of optimal nutrient availability—but by careful selection of crops gardening is possible on mineral soils ranging from below pH 5 to pH 8. Sometimes soil pH can be adjusted to correct for disease problems. Thus, while the optimum pH range for the Irish potato is from 5.0 to 6.5, they are sometimes grown on soils of low pH (4.8–5.4) to reduce the damage from scab. Similarly, cabbage may be grown in somewhat alkaline soils to minimize damage from club root. The optimum pH range for crops grown on organic soils is lower than for those grown on mineral soils. This relates to the fact that optimum nutrient availability in organic soils occurs at somewhat lower pH than in mineral soils.

We have seen that each garden crop has its own optimum pH range. In many regions, however, pH values are below this optimum and the growing of many vegetables is restricted or impossible. The reasons why acid soils are injurious to plants are not fully known, but deficiency of calcium and magnesium, toxicity from aluminum and manganese, inactivation of phosphates, and reduction in microorganism activity are believed to be contributing factors. Problems stemming from each of these factors are known. For example, beet and celery are particularly sensitive to aluminum toxicity, and beans and members of the cabbage family are especially sensitive to excess manganese. The beneficial soil bacteria, including those involved in nitrogen fixation, and certain fungallike organisms, show greatly reduced activity in highly acid soils (Fig. 5-8). Gardeners living in regions where soil acidity is a problem can limit their gardening to the more acid-tolerant crops. This is not essential, however, because most soil

TABLE 5-6 Optimum pH Range for Certain Vegetable Crops

Crop	pH
	5.0 5.5 6.0 6.5 7.0 7.5 8.0
1. Asparagus Beet Cabbage Muskmelon	
2. Peas Spinach Summer squash Celery Chives Endive Horseradish	
3. Lettuce Onion Radish Cauliflower Sweet corn	
4. Pumpkin Tomato Snap beans Lima beans Carrot Cucumber	
5. Parsnip Pepper Rutabaga Hubbard squash Eggplant	
6. Watermelon	
7. Irish potato	

5.0   5.5   6.0   6.5   7.0   7.5   8.0

←——————— Acid ——————◆—— Alkaline ——→

Neutral

From George W. Ware and J. P. McCollum, Producing Vegetable Crops, *3rd ed., 1980. Used with permission of The Interstate Printers and Publishers, Inc., Danville, IL.*

acidity problems can be corrected easily by the addition to the soil of liming materials, a topic discussed in Chapter 7.

SOURCES OF THE MINERAL NUTRIENTS

In a broad sense there are two sources of the mineral nutrients obtained by plants from the soil—natural sources and artificial sources in the form of additions by humans.

Natural Sources of Plant Nutrients

We saw earlier that the surface soil is composed mainly of chemicals that are not needed by plants or are needed in very small quantities. Oxides of silica and aluminum comprise up to 88 percent of topsoil, and iron commonly adds another 3 to 4 percent. Thus, some 92 percent of topsoil accounts for only one micronutrient, iron; the other 8 percent of soil must contain the remainder. Of the six macronutrients obtained from the soil, only potassium is present in a percentage greater than one (Table 5-3). Even though the amount of potassium in soils is high, ranging from 20 to 40 tons per acre in most soils, plants commonly respond to potassium fertilization. The reason for this is simple. The major potassium-containing compounds—mica and feldspar—are very resistant to weathering. As a result, over 90 percent of all soil potassium is unavailable for immediate plant use. There is, of course, a slow weathering of these minerals with a release of potassium. The slow release serves to conserve the soils' supply of this important nutrient.

The Nitrogen Cycle in Nature

Nitrogen is a unique plant nutrient. Plants require more nitrogen than any other soil nutrient (Table 5-4). Ironically the supply of nitrogen in the soil is small and easily lost through leaching and in other ways. In addition, unlike the other mineral nutrients, nitrogen does not occur as a mineral in any of the parent rocks from which soils are derived. Soil nitrogen is derived largely from special microorganisms that can take nitrogen gas from the atmosphere and incorporate it into their bodies. When these organisms die and decompose, their nitrogen is released in a form useful to plants.

Biological nitrogen fixation. The ocean of air surrounding the earth is 79 percent nitrogen. However, as it exists in the atmosphere, nitrogen is an inert gas and cannot be utilized by plants. To be available this nitrogen must be "fixed" in usable form. The process of converting nitrogen into a usable form is called **biological nitrogen fixation** and includes the conversion of atmospheric nitrogen (N_2) into ammonia (NH_3). Nitrogen fixation is carried out by two groups of organisms, the symbiotic bacteria and certain free-living organisms.

In symbiotic nitrogen fixation bacteria of the genus *Rhizobium* invade roots and induce the formation of root **nodules.** These nodules become quite large and can be seen easily without magnification (Fig. 5-9). Active legume nodules have a distinct pinkish color owing to the presence of a hemoglobinlike substance. A pale color indicates low nitrogen-fixing activity. The relationship between the infected plant and the bacterium is beneficial to both; the plants supply the microorganism with food and the microorganisms supply the plant with fixed nitrogen. Such a mutually beneficial relationship is termed **symbiosis.** Not all plants have evolved such a convenient relationship. Among garden plants only the legumes—peas, beans, and peanuts—fix ni-

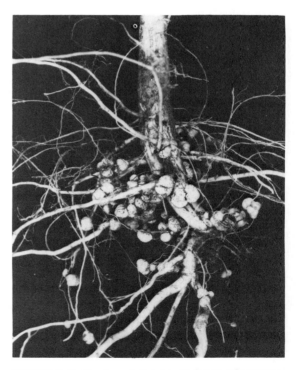

FIGURE 5-9 Root nodules on soybean. (Courtesy of U.S.D.A.)

trogen. The phenomenon is known to occur in at least six additional plant families, however. Recent evidence indicates, for example, that certain members of the cactus family are capable of nitrogen fixation.

Not all legumes fix nitrogen with equal facility. Some determined rates for important crops are as follows:

Crop	lb N/acre/yr
Alfalfa	251
Soybean	105
Field bean	58
Field pea	48

Alfalfa appears to be without equal in its ability to fix atmospheric nitrogen. The amount of nitrogen fixed depends on several factors, however, including soil nutrient content, aeration, moisture, and pH, as well as the amount of readily available nitrogen present in the soil (high levels discourage nodulation). Because of their ability to fix nitrogen it is commonly assumed that legumes enrich the soil nitrogen supply. This is not always so. Beans and peas, for example, support low levels of fixation, and since much of the plant may be removed in harvest soil nitrogen levels are lowered. In general, however, the demands legumes make on soil nitrogen are much less than those made by nonlegumes. For this reason legumes should be regarded as nitrogen *conservers* rather than nitrogen increasers.

Atmospheric nitrogen is also fixed by free-living organisms. Foremost among these are certain blue-green algae, bacteria, and fungi. The blue-green algae are important in agriculture. They are widely distributed in surface soils of all kinds and tolerate a wide range of temperatures and oxygen levels. Among the nonsymbiotic bacteria *Azotobacter* and *Clostridium* are best known. *Azotobacter* is an **aerobic** (oxygen requiring) form, *Clostridium* **anaerobic.** This fact is regarded as significant since even the best soils have areas of low oxygen supply. As with the symbiotic bacteria, their ability to fix nitrogen seems to be highest when the available nitrogen supply of the soil is low. Finally, recent research has produced evidence that certain microorganisms living very near the roots of crop plants such as corn and rice are able to fix nitrogen. These organisms may derive their energy from organic substances exuded by the plant root. The extent and significance of this reported type of nitrogen fixation has yet to be determined.

Nitrification. We have seen that the conversion of atmospheric nitrogen to ammonia can proceed through two distinct groups of organisms. In addition, there are microorganisms in the soil, known as **decomposers,** that break down the proteins in organic matter with the subsequent release of ammonia. While ammonia, in the form of the ammonium ion (NH_4^+), can be readily used by plants, much of the nitrogen is commonly absorbed in nitrate form (NO_3^-). The reason for this is that ammonia is rapidly converted to nitrate in the soil by a process of **nitrification.** The organisms involved in this conversion are a group of bacteria collectively called the **nitrobacteria.** The conversion requires oxygen (Fig. 5-10) and is favored in warm, well-aerated soils with modest acidity. Conversely, cold, wet soils of strong acidity provide an unfavorable environment for nitrification. You might conclude from the above discussion, and correctly so, that plants that grow well under acid conditions (potato) or in cold, saturated soils (rice) may absorb much of their nitrogen as the ammonium ion. As a matter of fact, some scientists have suggested that substances, known as **nitrogen stabilizers,** be added to the soil to retard or inhibit nitrification. This suggestion stems from the fact that nitrates are easily lost by leaching and through a process of denitrification, whereas the ammonium ion is retained by soil colloids. To date, however, nitrogen stabilizers have been used to only a limited extent. Although an inexpensive sub-

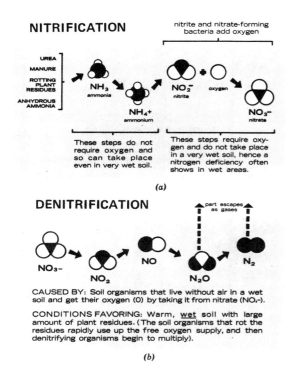

NITRIFICATION

nitrite and nitrate-forming
bacteria add oxygen

UREA
MANURE
ROTTING
PLANT
RESIDUES
ANHYDROUS
AMMONIA

NH_3
ammonia

NH_4^+
ammonium

NO_2^-
nitrite

oxygen

NO_3^-
nitrate

These steps do not require oxygen and so can take place even in very wet soil.

These steps require oxygen and do not take place in a very wet soil, hence a nitrogen deficiency often shows in wet areas.

(a)

DENITRIFICATION

part escapes
as gases

NO_3^-

NO_2

NO

N_2O

N_2

CAUSED BY: Soil organisms that live without air in a wet soil and get their oxygen (O) by taking it from nitrate (NO_x-).

CONDITIONS FAVORING: Warm, <u>wet</u> soil with large amount of plant residues. (The soil organisms that rot the residues rapidly use up the free oxygen supply, and then denitrifying organisms begin to multiply).

(b)

FIGURE 5-10 Nitrification and denitrification; two important aspects of the nitrogen cycle. (Reprinted with permission of A. & L. Publications from Samuel R. Aldrich, Walter O. Scott, and Earl R. Leng, *Modern Corn Production.* Copyright © 1975 by A. & L. Publications.)

stance (nitropyrin) that retards bacterial conversion has been discovered, the effects of this or other retardants on the ecological balance of the soil are unknown.

Denitrification. **Denitrification** is the process whereby certain anaerobic soil bacteria convert nitrate nitrogen back into gaseous forms, either as nitrous oxide (N_2O) or molecular nitrogen (N_2). This gaseous nitrogen escapes into the atmosphere and is thereby lost from the soil nitrogen pool. Denitrification is the reverse of nitrification in that oxygen is removed rather than added (Fig. 5-10). Many crop scientists feel that denitri-

fication is the main way in which nitrogen is lost from poorly drained soils, and losses are regarded as significant even on soils with better drainage. On the other hand, the process is regarded as important because leached nitrates would eventually end up in the ocean and thus be lost from use by land plants, to say nothing of their effects on ocean ecology. The gardener can slow denitrification by improving the drainage of wet soils, by adding nitrogen to the soil as near as possible to the time it is needed, and by adding the ammonium form rather than the nitrate form of nitrogen.

Fertilizer Additions by Humans

Commercial fertilizers come in many different ways. They may be from organic or inorganic sources. They may be gaseous, liquid, or solid, although gardeners use predominantly solid forms. They may contain only one nutrient element, in which case they are called **fertilizer materials.** Or, they may contain two or three nutrients, in which case they are termed **mixed fertilizers.** We said earlier that any fertilizer that contains all three of the fertilizer elements (N, P, K) is referred to as a complete fertilizer. Fertilizers also come in regular and premium grades. The premium grades contain some special features, such as secondary elements and/or micronutrients, a nutrient in several forms, or a more uniform size material. Finally, fertilizers come in different granular forms aimed at controlling the speed of release of a nutrient or nutrients. Despite these many variations, all commercial fertilizers have one very important feature in common—their fertilizer guarantee.

The Fertilizer Guarantee

The package label of every commercial fertilizer is usually required by law to provide the following information:

1. Net weight of fertilizer contents.
2. Name, brand, or trademark.
3. Name and address of manufacturer.
4. Guaranteed chemical composition.
5. Potential acidity.

The first three items are self-explanatory, but the last two require explanation. Every fertilizer label will show a series of three figures such as 10-20-10, 10-6-4, or 33-0-0 (Fig. 5-11). These figures represent the amount of nitrogen, phosphorus, and potassium—*always* in this order—in a fertilizer, and constitute the **fertilizer guarantee.** Unfortunately, these figures do not represent in a uniform manner the amounts of each nutrient present. Thus while a 10-20-10 fertilizer has a total of 10 percent of *elemental nitrogen* (N), it has 20 percent available phosphorus pentoxide (P_2O_5), and 10 percent water-soluble potassium oxide (K_2O), commonly called **potash.** The amounts of elemental phosphorus and potassium in a fertilizer can be calculated, however, using the following formulas:

$$\% \; P = \% \; P_2O_5 \times 0.44$$
$$\% \; K = \% \; K_2O \times 0.83$$

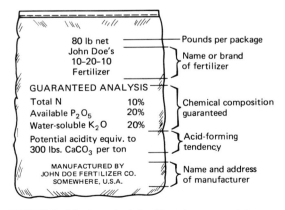

FIGURE 5-11 A typical fertilizer bag containing the information commonly required by law. (Reprinted with permission of Macmillan Publishing Co., Inc. from Nyle C. Brady, *The Nature and Properties of Soils.* Copyright © 1974 Macmillan Publishing Co., Inc.)

Some fertilizer labels already show the percent of P and K on an elemental basis as well as in the oxide forms P_2O_5 and K_2O. It is planned that eventually all manufacturers will provide both sets of information.

The fertilizer percentages given above represent ratios. In a 10-20-10 fertilizer the ratio of the three nutrients is 1:2:1. A 50-pound bag of 10-20-10 would have the same amount of nutrients as a 100-pound bag of 5-10-5. Both contain 5 pounds of available nitrogen, 10 pounds of phosphorus (as P_2O_5), and 5 pounds of potassium (as K_2O).

In a bag of 10-20-10 fertilizer, 10 percent of its bulk supplies the nitrogen, 20 percent its phosphorus, and another 10 percent its potassium. This means that 60 percent of its bulk is filler. In the 5-10-5 fertilizer 80 percent of its bulk is filler. While the filler generally has little or no fertilizer value, it does serve to dilute the fertilizer ingredients making them much easier to handle. In diluted form rates of application can be controlled more accurately, and the possible caustic action of the fertilizer itself is overcome. The organic fertilizers usually have much lower analysis figures, because their nutrients are in less concentrated forms. The nitrogen content of fresh cow manure (0.6) is only about one-fiftieth that of ammonium nitrate fertilizer (33.0).

The Acidity Equivalent of Fertilizers

In addition to stating the percentages of the fertilizer elements, the label may reveal the presence of secondary elements and/or micronutrients. Quantities of one or more of these nutrients may be present even if not stated, but do not rely on their presence unless they are guaranteed by the label. Finally, the label will state the potential **acidity equivalent** of the fertilizer. The value given is the quantity of calcium carbonate required to offset the acidity caused by one ton of the fertilizer. Most often the acidity is associated with the nitrogen carrier. Some examples, listed in order of most acid forming to least acid forming—per pound of nitrogen—are as follows:

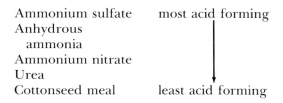

Ammonium sulfate — most acid forming

Anhydrous ammonia

Ammonium nitrate

Urea

Cottonseed meal — least acid forming

Inasmuch as most fertilizers are added to acid soils, and since soil acidity is of great importance in soils as relates to nutrient availability (Fig. 5-8), the consideration of the acidity potential of a fertilizer becomes an important criterion in fertilizer selection.

Important Fertilizer Sources

There are many sources of the fertilizer elements available to the gardener, both organic (Table 5-7) and inorganic (Table 5-8). Choice for any given situation will depend on such

TABLE 5-7 Some Organic Fertilizers (Organic Materials and Natural Deposits) and the Nitrogen, Phosphorus (P$_2$O$_5$), and Potash (K$_2$O) Content of Each Material

Material	% N	% P$_2$O$_5$	% K$_2$O
Animal manures			
Cow (fresh)	0.6	0.15	0.45
Horse (fresh)	0.7	0.25	0.55
Poultry (75% water)	1.5	1	0.5
Poultry (15% water)	6	4	3
Sheep (fresh)	0.5	0.3	0.5
Swine	0.3	0.3	0.3
Cow (dehydrated)	1	1	1
Sheep (dehydrated)	1	2	1
Apple pomace	0.2	0.2	0.15
Blood meal[a]	15	1.3	0.70
Bone meal (raw)[b]	4	10	—
Bone meal (steamed)[b]	2	20	—
Coffee grounds (dried)	2	0.4	0.7
Cottonseed meal[a]	6	2	1
Fish emulsion[a]	5	2	2
Fish meal[a]	10	6	2
Granite meal[c]	—	—	3–5
Greensand[c]	—	1.5	6
Guano[a]	16	10	2
Kalnit[c]	—	—	12–16
Milorganite	5	2–5	2
Rock phosphate (ground)[b]	—	20–30	—
Seaweed[c]	1.5	0.6	5
Sodium nitrate[a]	16	—	—
Soybean meal[a]	7	1.5	2
Sludge (sewage)[a]	6	4	—
Tankage	6	8	—
Wood ash[c]	—	1	6

Courtesy of U.S. Department of Agriculture.
[a]Rich nitrogen source.
[b]Rich phosphorus source.
[c]Rich potassium source.

TABLE 5-8 Some Important Inorganic (or Synthetic) Nitrogen, Phosphorus, and Potassium Carriers[a]

Nitrogen carriers	% N
Ammonium phosphate	12
Sodium nitrate	16
Ammonium sulfate	21
Ammonium chloride	23
Ammonium nitrate	33
Urea	46
Phosphorus carriers	**% P_2O_5**
Basic slag	15
Single superphosphate	20
Triple superphosphate	46
Ammonium phosphate	48
Mono calcium phosphate	50
Potassium carriers	**% K_2O**
Manure salts	20–30
Potassium nitrate	44
Potassium sulfate	48–50
Potassium chloride	48–60

[a]Many commercial fertilizers are premixed (blended) and contain proportions of two or all three of the fertilizer elements.

factors as preference, availability, price, quickness of response desired, and the kinds of plants to be fertilized. For example, say your growing corn plants are exhibiting signs of a nitrogen deficiency which you would like to correct without undue delay. In this case you might choose to add a high-analysis, water-soluble fertilizer material such as ammonium nitrate (33-0-0). In this fertilizer material the nutrients dissolve almost instantly in a well-watered soil, and can be taken up by plants within a matter of hours. Had you chosen instead to fertilize with well-rotted barnyard manure the recovery response would have been slower. This is but one of the many problems encountered in fertilization practices. Several additional aspects of fertilization will be discussed in Chapter 8 under the topic of fertilization practices.

The Law of Diminishing Returns

Gardeners frequently ask how much fertilizer to apply. There is no one answer to this question, as fertilization rates depend on many factors, such as the past fertilization and cropping history of your garden plot, kinds of crops to be grown, and desired yields. One important principle that sheds light on the subject is the law of **diminishing returns** (Fig. 5-12). This law states that beyond a certain point each added increment of fertilizer produces a proportionately smaller increase in yield. Eventually a point is reached where the cost of the additional fertilizer exceeds the return from yields. Our illustration shows a nitrogen curve, but the principle applies to any nutrient that is present in amounts that limit growth.

NUTRIENT DEFICIENCY PROBLEMS

What Constitutes a Nutrient Deficiency

To illustrate plant response to various fertility levels of nutrients. we have constructed a nutrient level scale (Fig. 5-13). In the midregion of the scale, plants will respond to increases in fertility with greater yields, although they do not show obvious deficiency symptoms. In other words, they have a need for the nutrient in question, but the need is hidden. This condition has been termed **hidden hunger.** If we continue to add fertilizer to satisfy this hidden need a point will be reached where no further yield response is evident, although the plant may continue its nutrient absorption. This is termed **luxury consumption.** Finally, if the nutrient continues to be added the plant may even show a reduction in yield due to a toxic reaction to the nutrient (see also Fig. 5-12). Towards the other end of the scale we see that if the fertility level continues to decline through the hidden hunger range a point is

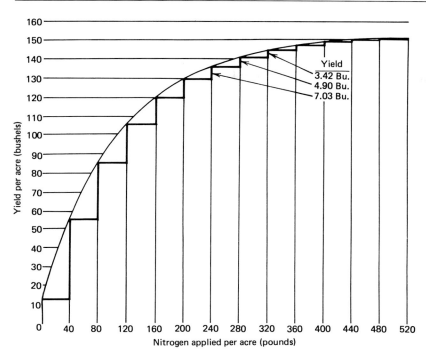

FIGURE 5-12 The yield response of irrigated corn in Washington State to nitrogen fertilization. Note that response diminished with increasing quantities of fertilizer. (Courtesy of U.S.D.A.)

reached at which the plant begins to show *visible* signs of a nutrient shortage; that is, it displays **nutrient deficiency symptoms.**

Nutrient Deficiency Symptoms

Several sorts of symptoms characterize nutrient deficiencies. Among the more common are the following:

1. **Chlorosis;** development of a pale green or yellow color.

2. **Necrosis;** localized death of tissues such as buds, leaf tips, or older leaves.

3. **Anthocyanin formation;** a red vacuolar pigment.

4. **Stunted growth.**

5. **Slender** or **woody stems.**

6. **Poor reproductive development.**

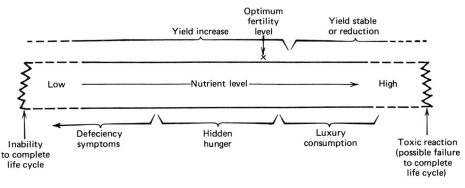

FIGURE 5-13 The relation of fertility level to plant performance.

These symptoms can be mild or severe, but in many plants severe deficiencies of certain nutrients produce rather characteristic responses, as illustrated by boron deficiency of beet (Fig. 5-14). The constancy of response to specific nutrient deficiencies has made possible the construction of diagnostic keys (Table 5-9).

Note from the nutrient deficiency key that deficiencies may show up initially either in younger or older plant parts. With nitrogen, for example, early symptoms are uniform chlorosis of older leaves. Younger leaves may show little or no chlorosis because the nitrogen of older leaves is mobilized for transport to these young developing leaves. If the deficiency is severe, however, all leaves become chlorotic. The symptoms of a calcium deficiency are quite different. Calcium has an important role in the plant as a constituent of the middle lamella that cements plant cells together. As you might expect, this element is quite immobile once utilized and therefore cannot be reactivated for movement to growing areas. As a result, calcium deficiency symptoms occur first in the buds and in younger leaves. Blossom-end rot of tomato is caused by a deficiency of calcium, and is aggravated by drought stress to plants. Similarly, the internal browning of Brussels sprouts is thought to be related to insufficient calcium or a calcium imbalance.

Correcting Nutrient Deficiencies

Macronutrient Deficiencies

The first elements to become deficient in garden soils are generally those used by plants in greatest amounts (Table 5-3), that is, the fertilizer elements. Virtually all garden soils will profit from their addition. The secondary and micronutrients differ somewhat from the "big three" in that they are likely to be deficient in specific regions or in specific soils, rather than being generally required. Calcium and magnesium deficiencies can be corrected by liming, and it is not necessary to add them unless the soil has an unfavorable pH level. Sulfur additions are seldom needed at

FIGURE 5-14 Boron deficiency in beet. The beet on the left received all essential elements, but the other three were deficient in boron. Large, black, necrotic areas develop in roots of boron-deficient plants. (Reprinted with permission of Macmillan Publishing Co., Inc. from Victor A. Greulach, *Plant Function and Structure.* Copyright © 1973 by Victor A. Greulach.)

TABLE 5-9 A Key to Plant-Nutrient Deficiency Symptoms

Symptoms	Deficient element
A. Older or lower leaves of plant mostly affected; effects localized or generalized	
B. Effects mostly generalized over whole plant; more or less drying or firing of lower leaves; plant light or dark green	
C. Plant light green; lower leaves yellow, drying to light-brown color; stalks short and slender if element is deficient in later stages of growth	Nitrogen
CC. Plant dark green, often developing red and purple colors; lower leaves sometimes yellow, drying to greenish brown or black color; stalks short and slender if element is deficient in later stages of growth	Phosphorus
BB. Effects mostly localized; mottling or chlorosis with or without spots of dead tissue on lower leaves; little or no drying up of lower leaves	
D. Mottled or chlorotic leaves, typically may redden, as with cotton; sometimes with dead spots; tips and margins turned or cupped upward; stalks slender	Magnesium
DD. Mottled or chlorotic leaves with large or small spots of dead tissue	
E. Spots of dead tissue small, usually at tips and between veins, more marked at margins of leaves; stalks slender	Potassium
EE. Spots generalized, rapidly enlarging, generally involving areas between veins and eventually involving secondary and even primary veins; leaves thick; stalks with shortened internodes	Zinc
AA. Newer or bud leaves affected; symptoms localized	
F. Terminal bud dies, following appearance of distortions at tips or bases of young leaves	
G. Young leaves of terminal bud at first typically hooked, finally dying back at tips and margins, so that later growth is characterized by a cut-out appearance at these points; stalk finally dies at terminal bud	Calcium
GG. Young leaves of terminal bud becoming light green at bases, with final breakdown here; in later growth, leaves become twisted; stalk finally dies back at terminal bud	Boron
FF. Terminal bud commonly remains alive; wilting or chlorosis of younger or bud leaves with or without spots of dead tissue; veins light or dark green	
H. Young leaves permanently wilted (wither-tip effect) without spotting or marked chlorosis; twig or stalk just below tip and seedhead often unable to stand erect in later stages when shortage is acute	Copper

TABLE 5-9 (*Continued*)

Symptoms	Deficient element
HH. Young leaves not wilted; chlorosis present with or without spots of dead tissue scattered over the leaf	
I. Spots of dead tissue scattered over the leaf; smallest veins tend to remain green, producing a checkered or reticulating effect	Manganese
II. Dead spots not commonly present; chlorosis may or may not involve veins, making them light or dark green in color	
J. Young leaves with veins and tissue between veins light green in color	Sulfur
JJ. Young leaves chlorotic, principal veins typically green; stalks short and slender	Iron

Courtesy of the Potash and Phosphate Institute, Atlanta, Ga.

present, at least near industrial centers, because of the abundance of sulfur dioxide, an atmospheric pollutant. Plants can absorb this gas directly from the atmosphere, or from that added to the soil by precipitation. Efforts to clean up air pollution will no doubt reduce the "value" of this sulfur source in the future. As mentioned, sulfur is also a constituent of many commercial fertilizers and, of course, farm manures. To date, deficiencies of sulfur have been most common in the Southeast, the Great Plains, and along the Pacific Coast. Vegetables vary in their sulfur needs. Cabbage, onions, turnips, and soybeans are reported to have high sulfur requirements, and consequently would be among the first crops to show deficiency symptoms.

Micronutrient Deficiencies

Micronutrient deficiencies are reported from several states and from Canada. Most serious deficiencies occur on strongly weathered soils, coarse-textured soils, alkaline soils, or organic soils of higher pH. As you might expect, crops vary in their need for specific micronutrients too. Some garden crops reported to have high requirements for specific micronutrients are presented in Table 5-10. Note that chlorine is not included on our list. In spite of the fact that it is used in larger quantities than most micronutrients, chlorine has not been shown to be limiting to plant growth under field conditions. A main reason for this is that an adequate supply of chlorine is virtually assured

TABLE 5-10 Garden Crops Reported to Have High Requirements for Specific Micronutrients

Element	Crops with high requirement
Boron	Apples, beets, cauliflower, celery, sugar beets
Copper	Onions, fruit trees, citrus trees, small grains
Iron	Blueberries, cranberries, grapes, peaches, nut trees
Manganese	Beans, soybeans, onions, potatoes
Molybdenum	Cauliflower, broccoli, celery
Zinc	Corn, soybeans, beans, fruit trees, citrus trees

by the large amount of sodium chloride dust present in the atmosphere. It has been estimated that each acre receives 10 to 20 pounds of dissolved sodium chloride annually by precipitation. Many commercial fertilizers, limestone, and farm manures carry the micronutrients as impurities. This incidental component should largely prevent the occurrence of micronutrient deficiencies. Generally, except in areas of proven deficiencies, the micronutrients need not be a major concern of the gardener.

Questions for Discussion

1. Define the following terms:
   (a) Macronutrient
   (b) Complete fertilizer
   (c) Ion
   (d) Symbiosis
   (e) Nitrification
   (f) Luxury consumption

2. Seventeen elements are known to be necessary for the normal growth of garden plants. List at least 10, and for each one indicate its major source in nature.

3. In most cases adding nitrogen to garden soils enhances crop yields. In specific situations, however, the addition of nitrogen does *not* result in greater yields. Explain.

4. Nutrient ions in the soil occur in two forms, based on their electrical charge. Identify these two forms and describe their "behavioral differences" in the soil.

5. Why do the authors refer to clay colloids and humus as the dynamic duo?

6. Why is an understanding of cation exchange capacity important to the home gardener?

7. Indicate measures that can be taken to correct for excess soil acidity or basicity.

8. Comment on changes in the availability of soil nutrients as the pH of the soil solution changes.

9. Outline the nitrogen cycle as it occurs in nature and discuss ways in which the home gardener alters or manipulates the cycle to her or his benefit.

10. What are the major factors to be considered when purchasing a commercial fertilizer?

11. What do we mean when we say that crop plants have "hidden hunger"?

12. Identify any micronutrients that may limit crop growth in your area. How would you correct for these deficiencies?

Selected References

Agrios, G. 1969. *Plant Pathology*. Academic Press, New York.

Aldrich, S., W. Scott, and E. Leng. 1975. *Modern Corn Production*. A & L Publishers, Champaign, Ill.

Black, C. 1968. *Soil-Plant Relationships*. John Wiley, New York.

Brady, N. 1974. *The Nature and Properties of Soils*. Macmillan, New York.

Devlin, R. 1975. *Plant Physiology*. D. Van Nostrand Co., New York.

Donahue R., J. Shickluna, and L. Robertson. 1971. *Soils: An Introduction to Soils and Plant Growth*. Prentice-Hall, Englewood Cliffs, N. J.

Gauch, H. 1972. *Inorganic Plant Nutrition*. Dowden, Hutchinson, and Ross, Inc., Stroudsburg, Pa.

Greulach, V. 1973. *Plant Structure and Function*. Macmillan, New York.

Hewitt, J. and T. Smith. 1975. *Plant Mineral Nutrition*. John Wiley, New York.

Kellogg, C. 1950. Soil. *Scientific American*, 182:30–39.

Kohnke, H. 1953. *Soil Science Simplified*. Published by Author.

Lunt, H. (Ed.) 1956. *Handbook on Soils*. Brooklyn Botanic Garden, Brooklyn, N. Y.

Pratt, C. 1965. Chemical fertilizers. *Scientific American*, 212:62–72.

United States Department of Agriculture. 1957. *Soil. Yearbook of Agriculture*, Government Printing Office, Washington, D. C.

LEARNING OBJECTIVES

By the time you have finished this chapter you should be able to:

1 Distinguish clearly between "macroclimate," "microclimate," and "local climate."

2 Describe the major characteristics of the macroclimate of your region.

3 Discuss the distribution of solar radiation over the earth's surface.

4 Discuss in general terms atmospheric circulation in the Northern Hemisphere.

5 Describe three different kinds of "fronts" and explain how they affect weather.

6 Explain ways in which climate is modified by nearby bodies of water.

7 Name the major elements of weather.

8 Comment on how gardeners can utilize plant hardiness zone information, and identify precautions that should be taken in interpreting hardiness zone data.

9 Define what is meant by the phrase "frost-free period."

10 Identify the factors that determine day length, and describe how variations in day length influence plant growth.

11 Tell why for a given area the distribution of precipitation is generally a more important consideration than the total amount of precipitation.

12 Explain why the authors regard the garden as a solar energy retrieval system.

6

CLIMATE AND WEATHER

"There are probably more ways of getting around climatic handicaps in the case of vegetables than there are with any other crops; yet it is also true that climate is still the boss in vegetable production."

Victor R. Boswell and Henry A. Jones

Climate has a profound influence on plants and soils, and indeed, crop production is climate controlled. Nevertheless, as Boswell and Jones point out, even if climate is still the "boss," there are numerous ways of getting around climatic handicaps. In this chapter we explore the idea of climate. The many cultural manipulations that allow the gardener to circumvent climatic barriers are discussed in the practices section of the book.

CONCEPTS AND DEFINITIONS

Climate governs the successful production of garden crops, but what is climate? **Climate** is defined as the characteristic weather conditions of an area, averaged over an extended

period (several decades) of time. Since the average weather conditions make up climate, then to understand climate we must also understand weather. **Weather** may be defined as the general atmospheric conditions at a given place at a particular time. The atmospheric conditions referred to in our definition of weather include the effects of temperature, light, and moisture.

In discussions of climate it is usually necessary to designate the space, area, or region to which the conditions apply. Climatologists have developed words and phrases to fit this need. At one extreme, the term **macroclimate** refers to the climate of a relatively large part of the earth's surface. We will devote only limited space to the discussion of these broad climatic zones. At the other extreme, the term

microclimate refers to the climatic conditions in the immediate vicinity of an object or organism. We can apply this term to the average conditions around a plant, its root or shoot systems, a leaf, or a smaller segment. Another expression, **local climate,** is employed to designate an area (actually a volume of space) intermediate in size between that indicated by macroclimate and microclimate. Local climate could refer to the average conditions within a garden patch, a field of soybeans, or in the environs of Lethbridge, Alberta, Canada.

To the gardener local climate and microclimate are of immediate importance. The productivity of the garden is influenced by local climate. However, each and every plant contributing to that productivity responds specifically to the microclimate of its immediate vicinity. The nature of microclimatic variations were strikingly illustrated by two Purdue University scientists, James Newman and Byron Blair. These workers measured temperatures at the interface between soil and air, and at different altitudes and soil depths in an Indiana cornfield. They found that fluctuations were greatest at the soil surface—often 40°F or more (Fig. 6-1). Soil temperatures were shown to fluctuate less widely than air temperatures. At a depth of about two feet the soil temperature remained constant. These data illustrate the well-documented fact that climatic variations near the ground are great. Most weather instruments are placed in unobstructed positions (and sheltered from sun, wind, and precipitation) 5 to 7 feet above the ground to escape these extremes.

CLIMATES OF NORTH AMERICA

The following description of the agriculturally important macroclimates of Canada and the United States are presented to help the reader better understand the broad climatic regions, to illustrate the intimate relationship

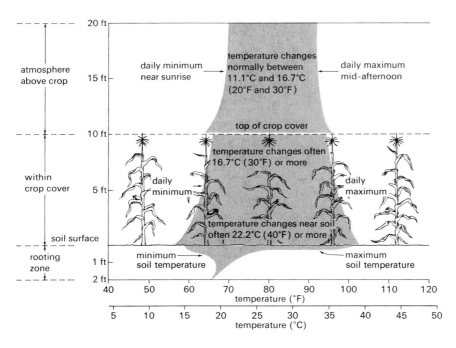

FIGURE 6-1
Temperatures around crop plants are subject to very regular variations, which decrease with both altitude and soil depth. (Courtesy of J. E. Newman and B. O. Blair, "What Is the Energy Balance in Your Corn Field," *Crops and Soils*, June-July 1964, p. 11, published by the American Society of Agronomy.)

between vegetable production and climate, and to introduce basic climate and weather terminology.

There are many ways of classifying climatic regions. One older system categorized climates on the basis of temperature as tropical, temperate, and polar. Another system classified climatic regions on the basis of average annual rainfall as arid, semiarid, subhumid, and humid. Modern systems make use of both temperature and moisture data, and some systems correlate these data with vegetation patterns. Volumes have been written on the subjects of climate classification. We will describe only relevant parts of one simplified system.

The map in Fig. 6-2 shows the climate regions for much of North America. These climates fall into four broad categories, as follows: tropical rainy, dry, humid mesothermal, and humid microthermal. Within each of these broad categories, subdivisions into more uniform climatic zones are made.

Tropical Rainy Climates

Tropical rainy climates occur in an irregular belt extending almost 30° north of the equator. This region is characterized by constantly high temperatures, absence of a winter, and an annual rainfall usually exceeding 30 inches. Within this broad climatic region two climatic types are recognized. One type—the **tropical rainy**—is constantly wet and gives rise to a lush and varied vegetation. The other type, known as **savanna,** has wet and dry seasons. Grass is the dominant vegetation, growing abundantly during wet periods and going dormant in dry periods. The main horticultural crops of the tropical region grow continuously and have no dormancy requirements. Included in this group would be banana, cacao, coconut, mango, papaya, pineapple, and sugar cane. In the United States a tropical climate occurs only at the southern tip of Florida (Fig. 6-2) and in Hawaii.

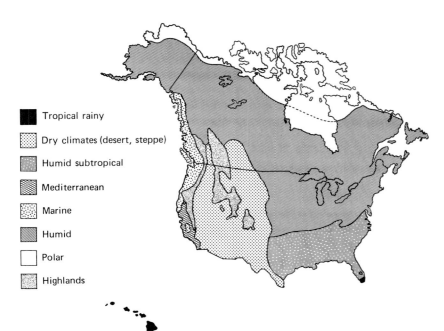

- ■ Tropical rainy
- ▦ Dry climates (desert, steppe)
- ▨ Humid subtropical
- ▨ Mediterranean
- ▨ Marine
- ▨ Humid
- ☐ Polar
- ▨ Highlands

FIGURE 6-2 Climates of North America. (Reprinted by permission of McGraw-Hill Book Company. Modified by Guerry Dean, from Glenn T. Trewartha, *An Introduction to Climate,* 3rd ed. Copyright 1954 by McGraw-Hill Book Company.)

Dry Climates

Extending northward from the tropics, across a portion of the United States and into Canada, are the **dry** climates (Fig. 6-2). These are characterized by sparse and irregular rainfall, low humidity, high light intensity, and frequent winds. Dry climates are subdivided on the basis of moisture and temperature into **desert** (arid) and **steppe** (semiarid). The latter is a climate transitional between desert and humid types. Both climatic types have soils high in nutrients. Because of the limited rainfall crop production in steppe climates requires tillage methods that conserve moisture, crops adapted to dry farming regions, or irrigation. Wheat is a crop that is well adapted and widely grown under dryland conditions. Irrigation, of course, makes possible the growth of a broad range of horticultural crops. In the southwestern United States, melons, figs, and citrus fruits are grown as are a number of cool-season vegetables.

Humid Mesothermal (Mild Winter) Climates

A major portion of North America is characterized by climates that show distinct seasonal contrasts in temperature. One, the **humid mesothermal,** has winters that are mild and short. The other, the **humid microthermal,** has winters that are harsh and long. The boundary between the two is arbitrarily established on the basis of an average coldest month temperature of 32°F.

The humid mesothermal (mild winter) climates are found at lower latitudes and along the western side of the continent (Fig. 6-2). Within this category three distinct climates are identified: humid subtropical, Mediterranean, and marine. All of these climatic regions are excellent for the production of horticultural crops.

Humid Subtropical Climate

The **humid subtropical** region is represented in the United States by the region we call the "cotton belt." In this area rainfall is abundant (30 to 80 inches), and distributed over the entire year. The humid subtropical area is well suited to vegetable production, owing to a long growing season, high summer temperatures, and abundant rainfall. Sweet potatoes, dried beans, peppers, and melons are among the crops that do well. Two crops of many vegetables can be grown owing to the extended growing season.

Mediterranean Climate

A **Mediterranean** climate occurs over only a small portion of the United States, being confined to central and coastal southern California (Fig. 6-2). This climatic type is characterized by a modest amount of precipitation (15 to 25 inches), and this comes mainly during the winter season. Los Angeles, for example, receives nearly 80 percent of its precipitation in the period December to March. Summers are warm (coastal) to hot (inland) and the winters are mild. Bright, sunny weather prevails. The growing season lasts nearly the whole year, but during the three winter months frosts do occur. Despite its limited size the Mediterranean climatic region is of immense horticultural importance, providing some 80 percent of the country's fresh fruit and vegetables. Apricot, avocado, citrus, dates, figs, grapes, and olives are produced in abundance. Many cool-season vegetables are grown throughout the winter. The home gardener can grow every vegetable imaginable in an irrigated garden.

Marine Climate

The coastal region from northern California to Alaska has a **marine** climate (Fig. 6-2). Perhaps the most notable feature of this climatic region is the cool summers. Seattle, for example, has an average daily high of 73° in July. Night cooling is not as great as in Mediterranean climates. The average daily minimum in July for Seattle is 55°. Winters are generally mild and the growing season is long. The effectiveness of the lengthy season is reduced somewhat by the large number of cloudy days. Precipitation may reach as high as 150 inches annually. Some marine climates of the world have adequate rainfall at all seasons but many show a distinct summer minimum. Vancouver, Canada, has only 2.9 inches of total rain in July and August while 16.6 inches accrue in December and January (Fig. 6-3). Although marine climates are nearly ideal for human comfort they are usually too cool for certain horticultural crops, such as melons.

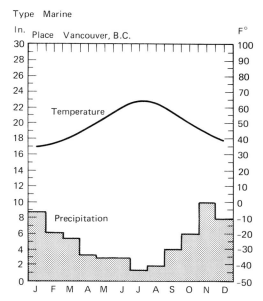

FIGURE 6-3 A marine climate station in western North America. Here the precipitation (60 inches) shows a pronounced maximum in winter. Note also that the mean temperature range for the year is narrow. (Reprinted by permission of McGraw-Hill Book Company. Modified from Glenn T. Trewartha, *An Introduction to Climate*, 3rd ed. Copyright 1954 by McGraw-Hill Book Company.)

On the other hand, cool-season vegetables such as peas, lettuce, onions, beans, and the crucifers do exceptionally well. Hardy tree fruits such as apple and pear are also well suited to this area, as are the brambles and strawberries.

The Humid Microthermal (Severe Winter) Climates

The humid microthermal climates occupy a vast portion of the North American continent (Fig. 6-2). The microthermal climates, like the mesothermal, are divided into three climatic types, as follows: **humid continental, warm summer; humid continental, cool summer;** and **subarctic.** Unlike the mesothermal climates, however, the severe-winter climates differ from each other only in degree, and that mainly in one element—temperature. Because of their similarities they can be described as a unit.

The climatic forces of strongest influence in the microthermal climates originate over land. For this reason these climates are described as **continental** in nature. (Climates in which the main influences originate over water are described as **oceanic.**) As mentioned above, the most distinctive climatic feature is the temperature. Rigorous winters occur throughout the region, but a heavy cover of snow makes temperatures at ground level less extreme. A genuine, but sometimes short, summer combines with the harsh winters to make sharp seasonal contrasts. As expected, annual temperature ranges are large. The frost-free season within the region ranges from less than 75 days in the north (Inuvik, Northwest Territories, has a frost-free period of 52 days) to more than 200 days in the south. Interestingly, the temperature gradients from north to south are much stronger in winter than summer. For example, between Winnipeg and St. Louis the January differential is 35° whereas the July differential is only 13°. This means that northerners can go south to escape winter cold much more easily than southerners can go north to avoid summer heat. The lesser summer gradient also has important implications for the gardener.

The amount of precipitation—and its distribution—are unquestionably the most important assets of the continental climates. Over much of the area summer is the season of maximum precipitation (Fig. 6-4). One has only to drive across the Corn Belt during the summer to attest to the agricultural productivity of the region, a productivity related in no small part to the high summer rainfall. For the gardener a fine garden is generally possible without irrigation. A wide range of vegetable crops can be grown. In the southern warm-summer region the various melons and tomatoes do well. As one moves north the cooler summers more typically support the cool-season vegetables and hardy fruits. In the far north the growing season is short and the various root crops and crucifers are favored.

Other Climatic Regions

Two additional broad climatic zones are shown in Fig. 6-2. The **polar** climates are found at high latitudes (and at high altitudes) and are characterized by the average warmest monthly temperature being 50°F or less. In North America the 50°F warmest month line extends well north of the Arctic Circle over Alaska and a portion of the adjacent Northwest Territories, but falls south of the 66½° parallel over the eastern portion of the Continent. Inuvik, Northwest Territories, which lies about 2° north of the Arctic Circle, has a mean July temperature of 55°F and is therefore subarctic. In the area around Inuvik it is possible to grow fine vegetables over permafrost. In regions where the mean warmest month temperature is much below 55°F outdoor vegetable growing is difficult or impossible. The polar climates, of course, fall into this category. The **highlands** do not represent a distinct climatic type, but refer to those mountainous regions where there are an endless variety of local

climates. Many of these local climates support excellent gardens.

The climatic differences described above are due *directly* to variations in latitude and altitude, ocean currents, distances from large bodies of water, and the direction and intensity of winds. These factors, however, are secondary. A source of energy is needed to drive the winds and the ocean currents. This *ultimate* source of energy comes from the sun. To understand climatic patterns we must have an appreciation for the nature of solar radiation and its distribution over the earth's surface.

SOLAR RADIATION AND ITS CONSEQUENCES

Hot bodies such as the sun and objects heated by the sun give off rays that travel in straight lines. This phenomenon is termed **radiation,** and represents an organized flow of energy through space. The sun emits tremendous amounts of radiant energy in all directions and at all times. It has been calculated that only 1/2,000,000,000 part of this solar output is intercepted by the earth. Of this minute fraction, only a portion reaches the earth's surface. For example, damaging ultraviolet radiation is absorbed by the ozone layer. Water vapor—the most important atmospheric gas from a climatological standpoint—absorbs an estimated 14 percent of incoming radiation. Microscopic dust particles tend to scatter incoming radiation, and, consequently, are partially responsible for the vivid sunrise and sunset colors. Much of the radiation that enters the earth's atmosphere, however, reaches the earth's surface where it impinges upon solid and liquid surfaces.

The solar radiation striking the earth's surface follows one of two pathways. Immediately, a part of this radiation is reflected back into space. Reflected radiation is, for all practical purposes, the same as radiation never received. Things upon our planet vary in their

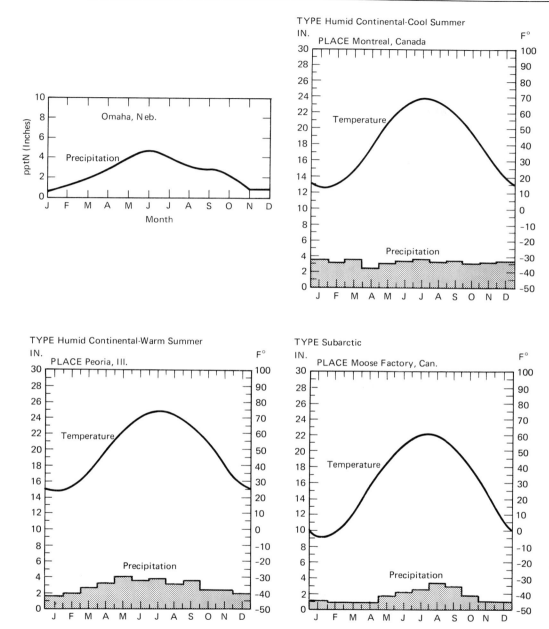

FIGURE 6-4 Precipitation patterns in four continental climates. Note that three of them have a pronounced summer maximum. (Reprinted by permission of McGraw-Hill Book Company. Modified from Glenn T. Trewartha, *An Introduction to Climate*, 3rd ed. Copyright 1954 by McGraw-Hill Book Company.)

reflectivity, or **albedo,** with some examples being as follows:

Surface	Albedo (%)
Calm ocean	2
Fields, meadows, tilled soil	15–30
Desert	30–35
Clouds	50
Fresh snow	80–85

The snow covering a garden patch melts slowly because of its high albedo. Once the snow is gone, however, the soil warms relatively quickly due to its lower albedo.

Radiation not reflected from the earth's surface is absorbed and converted into heat. This heat energy warms the soil and the air above it. Of major importance, the heated earth becomes itself a radiating body. Because the earth radiates at a drastically lower temperature than does the sun it emits a larger proportion of longer wavelength radiation. Much of this radiation is in the form of infrared, or heat, rays that have profound effects on our climate.

The Greenhouse Effect

The water vapor in our atmosphere absorbs only a small percentage (about 14 percent) of incoming short-wave radiation, but about 85 percent of the earth's outgoing long-wave radiation. As a result of this differential absorption by water vapor, temperatures near the surface of the earth are considerably higher and less variable than they otherwise would be. The phenomenon just described is known as the **greenhouse effect** of the earth's atmosphere, and is illustrated in Fig. 6-5. The panes of glass in a greenhouse or the windows in a car display the same absorptive properties as does the water vapor of the atmosphere. As a result both the greenhouse and the car can become quite hot on a sunny day. Conversely,

in desert and steppe climates the dry air and clear sky permit a greater escape of energy. As a consequence a more rapid night cooling occurs than in more humid areas.

The Distribution of Solar Radiation

Different parts of the earth's surface receive different amount of solar radiation. Near the equator, for example, the sun's rays strike more or less vertically whereas toward the poles the rays strike at progressively more oblique angles (Fig. 6-6). The oblique solar ray is spread out over a larger surface and thus delivers less energy per unit area. The energy level is further reduced, because the oblique ray passes through a thicker layer of atmosphere.

Atmospheric Circulation—The Movement of Heated Air

The differential distribution of solar energy over the earth's surface creates a pattern of atmospheric circulation. To begin, high radiation input at the equatorial region produces a mass of warm air. This warmed air, which is less dense than cool air, ascends in much the same manner as a hot-air balloon rises. The ascending equatorial air in turn creates a region of lower surface pressure into which cooler air from the north flows. The rising equatorial air cools and begins to fall, at the same time flowing poleward over the body of surface air moving toward the equator. Thus, a system of circulation is established.

The simplified circulation system just described is complicated by the rotation of the earth. As a consequence of this rotational movement three atmospheric circulation cells are formed (Fig. 6-7). As shown, this three-cell system accounts for the trade winds, prevailing westerlies, and polar fronts.

Air Circulation and Precipitation

An understanding of the physical characteristics of air also explains precipitation. As warm

① = Ultraviolet radiation absorbed by ozone layer.

② = Visible light strikes earth, warming it.

③ = The earth reradiates long (infrared) rays to the atmosphere

④ A = Some infrared absorbed by water vapor

④ B = Some is also reflected back to earth

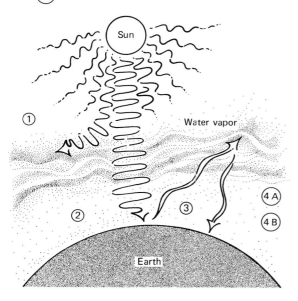

FIGURE 6-5 The "greenhouse effect" produced by earth's atmosphere. (Courtesy of Guerry Dean.)

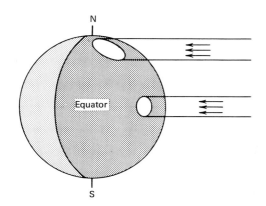

FIGURE 6-6 The effect of latitude on the dispersion of solar energy over the earth's surface during the equinox. (Courtesy of U.S.D.A.)

air rises it cools. Eventually it cools below its saturation or **dew point** (the point at which it can hold no more water vapor) and the moisture condenses to form clouds. If the moisture content is high, precipitation can result. In equatorial regions the ascension and cooling reaches a peak shortly after the sun has passed overhead. This results in afternoon thundershowers that support lush tropical rainforests.

At about 30° air sinks downward toward the earth surface. As it sinks it warms and can therefore hold more water. This results in low rainfall areas over the horse latitudes. It is for this reason that these latitudes are associated with the earth's major deserts. Finally, atmospheric cooling at the polar fronts at about 60° generates another high rainfall region across central Canada (Fig. 6-7).

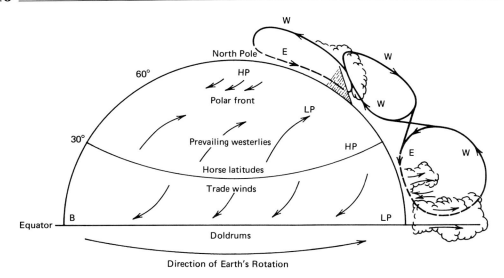

FIGURE 6-7 The pattern of atmospheric circulation for the Northern Hemisphere. Heavy precipitation bands occur at the equator and the polar front. Surface winds are distorted from straight north-south movement by the earth's rotation and declining circumference toward the pole. (Courtesy of U.S.D.A.)

How Barriers of Terrain Affect Air Circulation

Barriers of terrain also act to induce precipitation as is illustrated diagrammatically in Fig. 6-8. Air forced upward by a mountain range cools, becomes saturated, and finally produces rain or snow. The effect of mountain barriers is seen prominently in North America, where several major mountain ranges are present. The Coast Ranges of the far west, the Coast Mountains of western British Columbia and their continuation within the United States—the Cascade Range, Sierra Nevadas, Rockies, and Appalachians being most prominent. Each of these creates its area of high and low precipitation. The same is true in the Hawaiian chain where one side of an island is

FIGURE 6-8 Air forced upward by a mountain range cools, becomes saturated, and produces precipitation. (Courtesy of Guerry Dean.)

lush and green, the other dry and sparsely vegetated. In Hawaii, however, the winds are the easterly trades, whereas in the middle latitudes of the continent they are the prevailing westerlies (Fig. 6-7).

But what about the vast areas of the North American continent that are neither equatorial (obviously) nor mountainous, and yet receive ample precipitation (for example, the Corn Belt)? As it turns out exactly the same principle is operative; water vapor carried in the air is "wrung out" as warm air rises and cools below its dew point. The rise can be as abrupt and violent as when air is forced up a mountain slope. The rise is not brought about by a high intensity of solar radiation or by a mountain range, however, but rather by another air mass.

A peculiarity of air masses is that they do not mix freely with each other unless their temperature and moisture content are similar. Couple this with the fact that warm air is less dense than cold air and you will begin to see the direct source of much of our weather. When air masses with different properties are brought together they remain distinct with more or less sloping contact surfaces, termed **fronts,** between them (Fig. 6-9). A front is not an abrupt boundary, but rather a zone of some

width (3 to 50 miles commonly) within which changes in the weather elements are much more rapid than within the air masses themselves. It is along fronts that most of our weather changes originate.

Cold Fronts, Warm Fronts, and Stationary Fronts

There are several kinds of fronts (Fig. 6-10). A **cold front** occurs when an advancing mass of cold dry air pushes under a mass of warm moist air and a contact zone (front) is formed. A **warm front** is a front along which warm air replaces the colder air at the surface. When

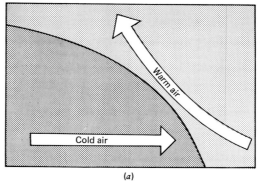

(a)

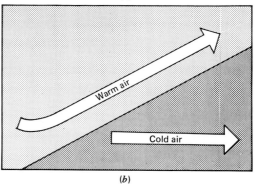

(b)

FIGURE 6-10 Cold (a) and warm (b) fronts. Note the more oblique leading edge of the cold front. (Reprinted by permission of McGraw-Hill Book Company. Modified from Glenn T. Trewartha, *An Introduction to Climate*, 3rd ed. Copyright 1954 by McGraw-Hill Book Company.)

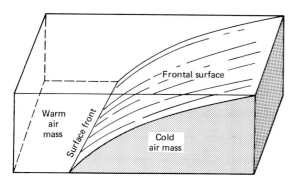

FIGURE 6-9 Three-dimensional representation of an atmospheric front. (Reprinted by permission of McGraw-Hill Book Company from Glenn T. Trewartha, *An Introduction to Climate*, 3rd ed. Copyright 1954 by McGraw-Hill Book Company.)

neither mass prevails the front shows only limited movement and is called a **stationary front.** If it does move it can become either a cold or a warm front, depending on the mass exerting the dominant influence.

Commonly a warm front forms over the trailing edge of a cold air mass. Consequently the slope of the warm front is rather gentle as compared to the steep cold front (Fig. 6-10). Generally, the abrupt change of temperature produced by the steep edge of a cold front results in more violent storm activity than that associated with a warm front. The warm air wedged aloft produces clouds, thunderstorms, and accompanying winds. By the same reasoning the effects of a passing warm front are usually more widespread and longer lasting.

High- and Low-pressure Systems

The movement of fronts is associated with low-pressure systems; high-pressure areas act as barriers. As it turns out there are semipermanent high- and low-pressure regions, and these fluctuate seasonally. In general, winter high-pressure systems dominate the continents and low-pressure systems dominate the oceans. The reverse is true in summer. The seasonal shift of highs is thus correlated with a seasonal shift in frontal activity. For example, the Pacific High shields western Washington and Oregon from rain during much of the summer but shifts south during the winter and these areas are drenched with rain (Fig. 6-3). The semipermanent high over the Midwest during the winter shifts in late March, allowing the influx of warm moist air from the Gulf of Mexico. The precipitation that follows makes the Corn Belt a superb agricultural region.

Oceanic Circulation—The Movement of Heated Water

Atmospheric and oceanic circulation have much in common. Both water and air gather heat near the equator and distribute it towards the poles. Water is the more important of the two fluids in this regard, as it covers nearly 75 percent of the earth's surface. Additionally, the oceans and other bodies of water absorb and store nearly all of the solar energy reaching them, whereas on land only a small percentage is stored. Not only do oceans store more energy, they heat and cool more slowly than do land surfaces. Thus, in summer they remain cooler than the air above them, and in winter they are warmer. When ocean influences are carried overland by air masses they act to make land temperatures less extreme, and in effect to delay the seasons.

Several oceanic currents flow along the North American Continent, and each has a profound influence on coastal climates. Approaching the west coast the warm North Pacific Drift is deflected northward and southward. The northward flowing branch, known as the **Alaska Current,** allows extension of the mild marine climatic zone to about 60° latitude, much higher than would be expected on the basis of latitude alone. The mainstream, which becomes mixed with colder water, flows southward along the west coast as the cool **California Current.** The Mediterranean climate of coastal southern California owes much to this current. The San Francisco Bay region has cool summers and mild winters. Inland, the winters are still quite mild but the summers are distinctly hotter due to the diminished oceanic influence. In Sacramento, for example, summer daily highs of 100°F or more are not uncommon. Along the Atlantic Coast the warm **Gulf Stream** is the dominant current, softening winters as far north as Cape Hatteras, North Carolina. Northward from the Cape, the cold waters of the **Labrador Current** flow south along the coastline forming the so-called "cold wall." This current contributes to the severity of New England winters, but fortunately has a moderating influence on summer heat.

While oceans are the major bodies of water with respect to climate modification they are not the only ones. Unquestionably, Puget Sound in the Northwest, the California Delta system, the Gulf of Mexico, the Gulf of St. Lawrence, and Hudson Bay have important effect on adjacent lands. Inland the Great Lakes, for example, have a profound influence on contiguous land areas. Lake Erie and Lake Huron moderate the climate of the southern tip of Ontario to the extent that the area is important in tobacco production. To the summer traveler the area around Hamilton, Ontario, resembles very much a tobacco region in the Carolinas. Similarly, Lake Michigan has a moderating influence upon the climate, turning western Michigan into an important horticultural area. And residents of Chicago and Milwaukee recognize a distinct marine influence even though they are far from any ocean.

THE ELEMENTS OF WEATHER

We have discussed some of the major factors that regulate climate in any region. We have seen that climate is influenced by latitude and altitude, mountain barriers, air masses and their movement; persistent low- and high-pressure cells, ocean currents, and so on. To the gardener a gradual understanding of these climatic *controls* will enhance her or his appreciation of weather and climate. On a day-to-day basis, however, the gardener deals not with these controls but with the direct **elements** of weather—temperature, light, and moisture—as they affect growing plants. How many gardeners have awakened with one of the following thoughts? Will it rain today or will I need to water my garden? Will it be hot enough to grow corn? To wilt cucumbers? Did it freeze last night? Will the day be cloudy or bright? The answer is that most gardeners think daily about these aspects of weather, and how they influence plant growth and de-

velopment. Let us now look at these environmental factors as they relate to gardening.

Temperature and Plant Growth

Plant Hardiness Zones

For many garden plants the most important factor in adaptation is the ability to survive low winter temperatures. The United States Department of Agriculture (USDA) has published a useful map of the plant hardiness zones for much of North America (Fig. 6-11). The continent is divided into 10 zones on the basis of approximate range of average annual minimum temperatures, with zone 10 having the highest temperature range. In Table 6-1 are listed some plants and the coldest zone in which they will normally survive and grow.

Several precautions are in order in interpreting plant hardiness zones. First, a plant may grow in a zone where it is not particularly useful. Pecan will grow in zone 5 but will generally not fruit unless in a warmer zone. The northern limit of hardiness of peach is zone 6. This does not mean that it is well suited to zones 6-10, as its southern limit is determined by chilling requirements. As a matter of fact, many fruits have, in addition to a northern limit, a cold requirement that limits their southern distribution (Table 6-2). Second, zone limits are based on temperatures averaged over many years. No single year is ever quite average and so a plant, particularly if young, may not survive in a zone where it is recommended. Or, it may survive for several years in a zone with more harsh winters. Third, the zones are broad and do not take into consideration the many "climatic islands" that may occur within a zone. Furthermore, the temperature of contiguous zones become increasingly similar near their common boundary.

The Frost-free Season

With most vegetables the gardener can forget about low winter temperatures. Instead, con-

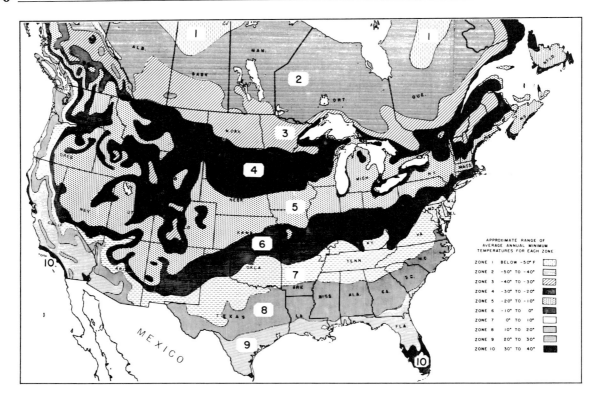

FIGURE 6-11 Zones of plant hardiness in the United States and Canada. (Courtesy of U.S.D.A.)

cern is for the interval between the last 32°F frost of the spring (Fig. 6-12) and the first 32°F frost of the fall (Fig. 6-13)—the **frost-free period** (see also Appendix D). This interval must be long enough for the vegetable in question to successfully complete its growth.

As with the hardiness zone temperatures, we are dealing with averages computed over a long period of time. If one were to plant on the mean date for the last spring frost in a given locality (Fig. 6-12) there is a 50-50 chance of escaping any future spring freeze. Planting

TABLE 6-1 Cold Hardiness Ratings for Some Perennial Food Plants

Scientific name	Common name	Zone
Prunus nigra	Canadian plum	2
Prunus americana	Wild plum	3
Vaccinium corymbosum	Highbush blueberry	4
Carya pecan	Pecan	5
Diospyros kaki	Oriental persimmon	6
Prunus dulcis	Almond	7
Cynara scolymus	Globe artichoke	8
Vaccinium ashei	Rabbiteye blueberry	9
Musa ensete	Banana	10

TABLE 6-2 Classification of Common Fruit and Nut Crops by Temperature Requirements[a]

		Temperate	
Tropical	Subtropical	Mild winter	Severe winter
Coconut			
Banana			
Cacao			
Mango			
Pineapple			
Papaya			
	Coffee		
	Date		
	Fig		
	Avocado		
	Citrus		
	Olive		
	Pomegranate		
		Almond	
		Blackberry	
		Grape (European)	
		Persimmon (Japanese)	
		Quince	
		Peach	
		Cherry	
		Apricot (blossoms tender)	
		Strawberry ⎱ (very hardy under snow)	
		Blueberry ⎰	
		Raspberry	
		Cranberry	
			Pear
			Plum
			Grape (American)
			Currant
			Apple
Low-temperature sensitive	Slightly frost tolerant	Tender	Winter hardy
Noncold requiring		Cold requiring	

From Jules Janick, Horticultural Science. *3rd ed., W. H. Freeman and Company. Copyright © 1979.*
[a]Variation in tolerance depends to a large extent on species, variety, plant part, and stage of growth.

later reduces the chance of encountering a freeze, and planting earlier enhances the chance. The probabilities related to these two planting alternatives are given in Table 6-3. You may choose your own odds.

Vegetables, like the perennials discussed above, vary in their tolerance to cold and heat.

Table 2-2 groups the common vegetables primarily on the resistance of young plants to spring frosts, and gives approximate planting times in relation to the mean frost-free date. The hardy crops, for example, can be planted up to six weeks before the frost-free date and most, if not all, actually grow better in cool

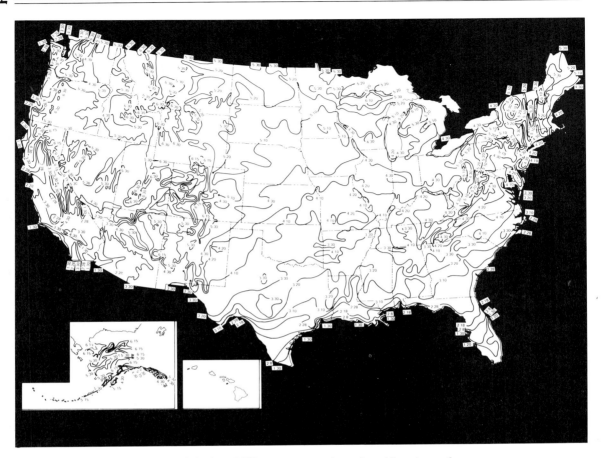

FIGURE 6-12 Average dates of the last 32°F temperature in spring. (Courtesy of U.S.D.A.)

weather. The tender crops, on the other hand, not only lack cold hardiness, they actually require hot weather for best growth. More will be said about these vegetable groups in Chapter 7 in relation to determining planting dates.

The Concept of Heat Units

Attempts have been made to express in mathematical terms the heat required by crop plants. One method makes use of heat units. **Heat units** are temperature-time values (degree-day) computed in relation to temperature above a certain minimum. For example, 40°F has been

established as the minimum (base) temperature necessary for the growth of peas. On a day when the mean temperature is 54 there would accumulate $54 - 40 = 14$ degree-days of heat units. A day with an average temperature of 70°F would provide 30 heat units, and a day with an average temperature of 37°F would provide zero degree-days of heat units.

What are the values of the heat unit concept? For one, the harvest date of certain crops can be determined by an accounting of accumulated heat units. In Wisconsin it has been found that 1150 to 1250 heat units are re-

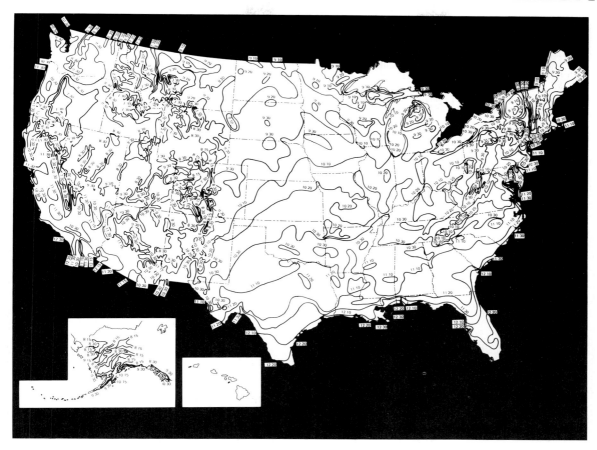

FIGURE 6-13 Average dates of the first 32°F temperature in fall. (Courtesy of U.S.D.A.)

quired to bring the 'Alaska' pea to maturity. Thus the grower or processor can predict maturity dates and thereby ensure an orderly harvest with the most efficient use of people and machines. Closer to home, we have probably all applied—or failed to apply—the heat unit concept. For example, most of us have made successive plantings of corn, say at 10-day intervals, visualizing that we would harvest crops at these same 10-day intervals. To our surprise, however, these interval plantings matured at nearly the same time. The heat unit concept helps us to understand this apparent anomaly. For sweet corn, each de-

gree above a daily mean temperature of 50°F is a heat unit. In the early part of the season, say between our first and second interval planting, few heat units accumulated. Later when temperatures were high they accumulated rapidly. Since maturity depends on the accumulated heat both mature at nearly the same time.

The situation is not quite as simple as outlined above. The later a crop is planted the less the number of heat units required to bring it to maturity. Furthermore, soil temperatures, night-time temperatures, and the distribution of accumulated heat units also have

TABLE 6-3 The Probability of Having a Freeze X Days Before or After the Mean Date of the Last Spring Freeze

Planting date (mean date)	Probability of freeze (%)
18 days before	95
14 days before	90
9 days before	80
6 days before	70
3 days before	60
0 days before	50
3 days after	40
6 days after	30
9 days after	20
14 days after	10
18 days after	5

a bearing on maturity dates. Nevertheless, the concept is useful and serves as a guide to crop planting and harvest forecasting. The heat unit concept is widely used in commercial production, particularly as it relates to perishable crops such as bean, pea, spinach, and sweet corn.

The Influence of Latitude and Altitude

Temperature is directly related to the amount of solar radiation received, and this in turn depends on **latitude** (distance from the equator) and **altitude** (distance above sea level). Each degree of latitude is about 69 miles. In the warm season there is a decrease in temperature in the neighborhood of 1°F for each degree in latitude going north through the center of the continent. Temperature also decreases with increasing elevation, averaging 3.6°F for each 1000 feet.

The changes in temperature with latitude and altitude affect the timing of biological events such as seed planting. This is expressed in a concept known as **Hopkins' bioclimatic law.** The "law" states in part that biotic events vary four days for each one degree of latitude and 400 feet of altitude. We witness this delay in the northward harvest of wheat through the Great Plains, and in the later flowering of a plant species at higher elevations. In general, gardening efforts conform to this pattern, although gardeners have devised many ingenious ways to avoid the climatic limitations imposed by temperature.

The short growing season at higher latitudes and elevations poses unique problems to gardeners. Those experienced in gardening with growing season limitations offer the following suggestions:

1. Use transplants whenever possible. Large, vigorous specimens are preferred.

2. Make use of cool weather crops (Table 2-2) and plant at the *earliest* possible date.

3. Choose rapid-maturing cultivars. Some cabbages mature in 85 days, others in 55 days—a whole month's difference. Similar variations can be found for almost any vegetable.

4. Use season-lengthening climate modifications whenever possible (see Chapter 8).

Temperature Requirements of Garden Plants

Each garden plant has its own unique temperature requirements; its minimum, optimum, and maximum for growth. Moreover, these **cardinal temperatures** vary with the stage of plant development. In general, flower buds are more sensitive to low temperatures than vegetative buds, seedlings more sensitive than established plants, new growth more sensitive than old, and roots more sensitive than stems.

Few garden plants grow at temperatures below 40°; most develop best between 60° and 90°F. Generally plants require a lower night than day temperature. As you might expect the optimum night temperature depends on the species and its stage of development. Many plant processes—including photosynthesis and respiration—show a direct relationship to temperature. This relationship is expressed by the **Van't Hoff-Arrhenius law,** which, in modified form, states that within the normal range of biological reactions there is a doubling of activity for each rise in temperature of 10°C (18°F). The reverse would hold true, of course, for a decrease in temperature. You will recall that the growth of a plant results from the accumulation of products of photosynthesis during the day. At night respiration depletes the accumulated food reserves. When the nights are too warm, respiration losses are excessive and overall growth of the plant is less (Fig. 6-14).

Many vegetable crops do better in cool weather and many fruit crops, large and small, have chilling requirements (Table 6-2). For fruit trees a cold treatment is necessary to break winter dormancy. Studies have shown that the cold requirement varies from species to species and within species (Table 6-4). Different peach cultivars, for example, have cold requirements that vary from 350 to 1200 hours of temperature below 45°F. The hours of cold appear to be cumulative in somewhat the same manner as the heat units described earlier. Cultivars with a low chill requirement can be grown further south. If the cold requirement is not met the tree will leaf out poorly or not at all in the spring.

The Effects of Unfavorable Temperatures on Plant Growth

Unfavorable temperatures elicit a number of plant responses. Some biennials, such as cabbage, beet, spinach, and onion will produce a flower stalk the first year if temperatures are too high. High temperatures also increase transpiration rates and may cause plants to become limp and flabby, a condition known as **wilting.** Wilting helps to reduce water loss because of the partial or complete closure of the stomates. This reduces the amount of carbon dioxide available for photosynthesis, however; thus the rate of food manufacture slows accordingly.

Wilting is not an uncommon phenomenon. Tomato plants wilt nearly every day in the middle of the summer, but regain turgidity at night when the air becomes cooler and more humid—their wilted condition was temporary (**temporary wilting**). If the unfavorable water balance should be maintained for a prolonged period of time, however, the plant will reach a stage of wilting from which it will not recover, even though placed in a saturated atmosphere, unless water is added to the soil. The plant has reached the **permanent wilting**

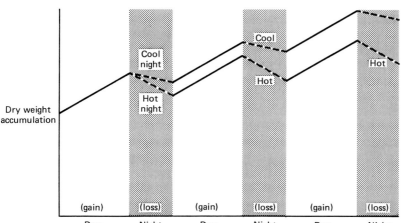

FIGURE 6-14 Growth represents a balance between the accumulation of food reserves from photosynthesis during the day and their loss by respiration at night. (Reprinted with permission of A. & L. Publications from Samuel R. Aldrich, Walter O. Scott, and Earl R. Leng, *Modern Corn Production.* Copyright © 1975 by A. & L. Publications.)

What Happens When Nights Are Too Hot?

TABLE 6-4 The Chilling Requirement of Certain Horticultureal Plants

Crops with Short Chilling Requirements

Apricot; fig; Japanese plum; peach var. Babcock; persimmon; quince; vinifera grape; muscadine grape; currant; flowering peach; forsythia; wisteria

Crops with Moderately Long Chilling Requirements

Peach (most varieties, e.g., Golden Jubilee, Belhaven, Hiley); pear var. Kieffer, Garber, LaConte; plum (most varieties); sweet cherry; walnut (northern varieties); gooseberry

Crops with Long Chilling Requirements

Apple var. Wealthy, Rhode Island Greening, Winter Banana; pear var. Bartlett, Bosc, Tyson; sour cherry; peach var. Mayflower, Salway, Dixie Red

Crops with Very Long Chilling Requirements

Apple var. McIntosh, Rome Beauty, Wagner, Northern Spy

Reprinted by permission of the McGraw-Hill Book Company from J. B. Edmond, T. L. Senn, F. S. Andrews, and R. G. Halfacre, Fundamentals of Horticulture. *4th ed. Copyright © 1975 by the McGraw-Hill Book Company.*

point. Other forms of high-temperature injury are also related to dessication. Young transplants often "burn off" at the soil line if unusually warm, dry weather follows transplanting. Direct heat injury can also occur, as when the cambium of a tree is killed by high temperature. This form of injury, known as **sunscald,** is common on species of trees with thin bark. Removal of protective shading through pruning can result in sunscald.

Low temperatures also produce undesirable plant responses. Frost-tender plants die when temperatures go much below the freezing point. Many vegetables (Table 2-2) and flowers are in this group. Nearly all crop plants stop growing at below 40°F, and tropical plants such as banana can incur chill injury at temperatures less than 50°F. In many perennials the belowground portion is frost hardy but the shoots and flowers are not. A late spring frost may kill the blossoms and young leaves of fruit trees, but the branches will leaf out again. Many garden flowers and some vegetables die back to the ground in fall but grow back from the crown the following spring.

Light and Plant Growth

Light affects the development of green plants through their entire life cycle. In Chapters 2 and 3 we discussed the roles of light in chlorophyll synthesis, vegetative growth through photosynthesis, the flowering process, and in fruit development. In each of these situations light of a certain quality (wavelength) was required for a specific process. Now we wish to turn our attention to light, not as it affects specific plant processes, but, as it relates to plant adaptation and distribution. Two aspects of light emerge as most important in this regard—its intensity and its duration.

Light Intensity

Light intensity refers to the concentration of light waves, and is frequently measured by comparison with a **standard candle.** The amount of light received at one foot from a standard candle is termed a **footcandle.**[1] Light

[1] A footcandle equals 10.76 luxes. Measurement in lux units is now accepted as the standard international unit of light intensity.

from different sources varies greatly in its intensity.

Direct sunlight for example, is 500,000 times brighter than moonlight. Conditions of the atmosphere, such as clouds or smog, can greatly reduce the intensity of light. Natural cloudiness is more prevalent in certain parts of the country (Fig. 6-15) and this has important consequences for vegetable production. For those who garden indoors or on an apartment balcony light intensity can also be critical and supplemental lighting may be required.

Light intensity requirements for the many **photoreactions** (light-induced reactions) also vary. Floral induction may require just one exposure to light of low intensity. Many crop plants probably achieve maximum photosynthesis at about one-fourth to one-half of full sunlight. Some, however, have unusually low or unusually high requirements, allowing their classification as **shade plants** and **sun plants.** In general, vegetables that are grown for their leaves or storage roots do pretty well with limited light. On the other hand, plants grown for their fruits, like tomatoes and peppers, need many hours of bright sun each day. Light considerations are thus important in planning the garden layout; a topic discussed in the next chapter.

Photoperiod

The gardener is interested in the progression of the seasons and the changing photoperiods, primarily because many responses of the plants she or he deals with are keyed to photoperiod (Fig. 6-16). The induction of dormancy and leaf abscission in colder climates are short-day responses, and in some trees the termination of dormancy in the spring is a long-day response. The formation of tubers by potato, as mentioned earlier, is also promoted by short days. The stimulus to flower and reproduce sexually is no doubt the most familiar photoperiodic response. Recall that long-day plants flower when the photoperiod *exceeds* a certain critical value (usually 12 to 14 hours) and a short-day plant flowers when the day length is *less than* a critical value. In the interim, both grow vegetatively. In nature, long-day plants flower when the days are lengthening (spring to early summer) and

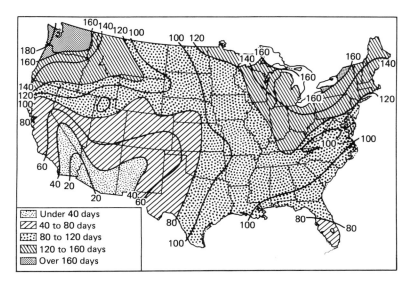

Under 40 days
40 to 80 days
80 to 120 days
120 to 160 days
Over 160 days

FIGURE 6-15 Average annual number of cloudy days. Note that the Pacific Northwest is the cloudiest part of the country. Fortunately, most of this cloudiness is in the cooler seasons. (Reprinted by permission of McGraw-Hill Book Company from Glenn T. Trewartha, *An Introduction to Climate*, 3rd ed. Copyright 1954 by McGraw-Hill Book Company.)

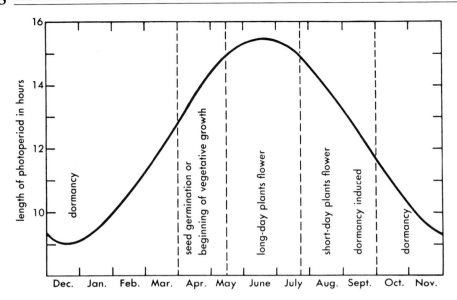

FIGURE 6-16 Yearly variation in length of the day, showing its relation to photoperiodic responses of plants in nature. Day-length data are given for 43° north latitude. The growing season indicated is approximately what prevails in northeastern United States. The positions of boundaries separating different types of behavior are approximate. (From Peter Martin Ray, *The Living Plant*. Copyright © 1963 by Holt, Rinehart and Winston, Inc. Reprinted by permission of Holt, Rinehart, and Winston, CBS College Publishing.)

short-day plants flower as the days are short-ening (midsummer to fall). In some plants, including the cultivated sunflower, flowering is not related to day length.

Inasmuch as the flowering of many plants is tied to the photoperiod and photoperiod is tied to latitude, one would expect that the distribution of plant species would be in some part determined by their photoperiodic response. Such is the case. For example, in the tropics where the days are of about equal length throughout the year, long-day plants cannot reproduce sexually. They can be grown vegetatively but they will not flower or fruit.

The Interaction of Light Intensity and Duration

One final aspect of light deserves mention. Lighting at any latitude, as it relates to vital plant processes such as photosynthesis, in-

volves both the intensity and duration of light. At the time of the spring equinox the sun is directly overhead at the equator and days are of about the same duration everywhere. The intensity of light, however, diminishes in a gradual manner toward the poles, due to the effects of the atmosphere. Thus while day length is the same everywhere the total amount of light received is not. Later, at the summer solstice, the sun is overhead at a latitude of 23½° N. Light intensity now diminishes in the direction of both the equator and the pole. This reduction in intensity is compensated for, at latitudes above 23½° N, by the longer days. The result is that even though the intensity is reduced from a maximum at higher latitudes the rate of plant growth may be great owing to the very long days. The gigantic cabbages and potatoes produced in the Tanana and Matanuska Valleys of Alaska are manifestations of this phenomenon.

Moisture and Plant Growth

Of the three elements of weather—temperature, light, and moisture—the last is the most likely to limit plant growth. Many a gardener who has gone on vacation without taking precautions to assure adequate water for growing plants can attest to this fact!

The term **moisture,** as used in relation to climate, refers to both precipitation and atmospheric humidity. Rainfall is the most common form of precipitation from a gardening standpoint and, as we have seen, is tied directly to atmospheric circulation patterns. In our discussion we consider three important aspects of precipitation—the amount, the distribution, and its availability for use by plants.

Amounts of Precipitation

On a worldwide basis the average annual precipitation varies in amount from 0.02 inches in one Chilean location to more than 900 inches in India. In the continental United States measurements vary from 1.45 inches in Death Valley, California, to 130 inches at Quinault, Washington. In Hawaii the annual rainfall varies from 2 to 450 inches. Obviously, crop production is not a mainstay in areas of precipitation extremes but instead occurs in the vast areas of moderate average annual precipitation. In that regard, the great agricultural area, the Corn Belt, receives about 40 inches of precipitation annually. Canada, another major producer and exporter of food, has about 75 percent of its farmland in the three Prairie Provinces. In this vast region the annual precipitation is between 13.5 and 20 inches.

The total precipitation in an area may fluctuate widely from year to year. For example, in one 10-year period the annual precipitation in North Platte, Nebraska, in the Great Plains, varied from 10 to more than 40 inches (the average is 18 inches). And for Glendora, California, a suburb of Los Angeles, records show that over the last 80 years annual rainfall has varied from less than 10 inches to more than 50 inches (the average is 22 inches). Thus, long-term averages (see Appendix D) provide only a realistic expectation based on past performance, and as such can be quite misleading.

Distribution of Precipitation

The seasonal distribution of precipitation is more critical than the yearly amount. Garden plants need rain during the growing season. Data on growing season precipitation is presented in Fig. 6-17 and Appendix D.

Rather specific precipitation patterns characterize the broad climatic zones discussed earlier in the chapter. For example, contrasting patterns are not uncommon when comparing a continental climate to a mediterranean or marine climate (Fig. 6-18). Due to the very favorable distribution of precipitation in central Illinois and Indiana a gardener may not need to water the garden even once during an entire summer. On the other hand, a gardener in western Washington or Oregon may need to water on a weekly basis, even though the annual precipitation in that region is twice that of the Midwest location.

Availability of Precipitation to Plants

A favorable seasonal distribution of precipitation does not tell the whole story either. Only that precipitation available to plants is **effective precipitation.** Due to evaporation, rain showers of less than ½ inch are of little value in supplying soil moisture for plant growth unless they immediately precede or follow other showers. Similarly, if the rainfall is too heavy much moisture will be lost as **runoff.** Temperature plays an important role in precipitation effectiveness. In cooler northern areas 20 inches of rainfall will be about as effective in crop growth as 30 inches in southern areas. Higher temperatures not only increase the rate of evaporation from the soil, they also increase the rate of transpiration by

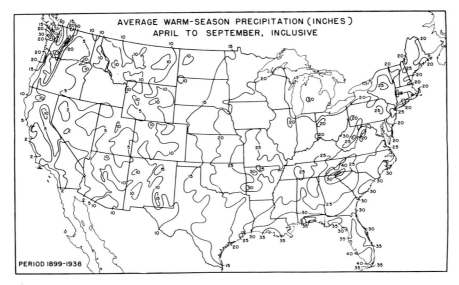

FIGURE 6-17
Average growing-season precipitation for the continental United States. (Reprinted with permission of Macmillan Publishing Co., Inc. from John H. Martin and Warren H. Leonard, *Principles of Field Crop Production*, 2nd ed. Copyright © 1967, Macmillan Publishing Co., Inc.)

plants. Combined water losses by **evapotranspiration** are also related to wind velocity, relative humidity, light intensity, and plant species. In general, the above comments also apply to irrigation water, a subject of Chapter 8.

GARDEN CROP ECOLOGY

Ecology is defined as the study of the relationships between organisms and their environment. **Crop ecology** restricts considerations to the relationships between crop plants and their environment. A crop ecologist must evaluate climatic, edaphic (soil), and biotic factors as these relate to the growth of crop plants. In reality every gardener is a practicing crop ecologist. The success of his or her efforts depend in large part on an ability to understand and deal with the environmental variables.

During the winter a gardener may look at a relatively bare garden plot, knowing that soon the area will be teeming with young vegetable plants. These plants capture solar energy and convert it into a usable form. In essence, a garden is a small-scale solar energy retrieval system. The success of the retrieval units—the individual garden plants—depends on their vigor. Plant vigor in turn is related to soil properties, climatic factors, and competition by other organisms. The more favorable these features are, the greater the success of the overall system.

The first section of the book gave you insights into the main environmental parameters affecting garden plants. Theoretical insights, however, do not a garden make. Basic information about plants, soils, and weather need to be translated into garden practices. The remaining chapters of this text are occupied with specific activities of plant production in your garden.

Questions for Discussion

1. Define the following terms:
   (a) Weather
   (b) Local climate
   (c) Dew point
   (d) Heat unit
   (e) Permanent wilting point
   (f) Evapotranspiration

2. The continent is divided into several broad climatic regions. Identify and characterize the region in which you live and garden.

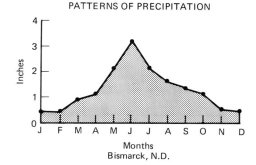

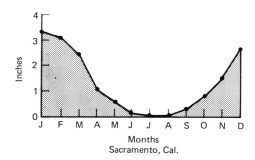

FIGURE 6-18 Contrasting patterns of precipitation distribution. Bismarck, North Dakota, has a typical continental dry-winter, wet-summer pattern, whereas Sacramento, California, has a wet-winter, dry-summer pattern. (From Carroll P. Wilsie, *Crop Adaptation and Distribution* 1962. Used with permission of Maria Wilsie Campbell. Los Altos, California.)

3. Which climatic region has the most favorable distribution of precipitation from the standpoint of gardening?

4. Describe the phenomenon known as the "greenhouse effect." How does this phenomenon relate to our gardening efforts?

5. Describe atmospheric circulation in the Northern Hemisphere.

6. Name at least two different kinds of fronts and indicate how these fronts influence weather.

7. Identify the oceanic currents influencing the North American continent and describe the nature of these influences.

8. What is the value of hardiness zone data to gardeners? What precautions should be used in interpreting hardiness zone data?

9. The mean date of the last 32°F freeze of the spring in Indianapolis, Indiana is April 15. Does this mean that frost-sensitive vegetables can be seeded or transplanted to the garden on that date? Explain.

10. Explain why successive crops of corn, planted at 15-day intervals, may reach maturity at about the same time.

11. Discuss some of the effects of unfavorable temperatures on garden plants.

12. Describe the precipitation in your area in terms of amount, distribution, and availability for use by crop plants.

Selected References

Byers, H. R. 1959. *General Meteorology*. McGraw-Hill, New York.

Calder, N. 1976. *The Weather Machine*. Viking Press, New York.

Gates, D. M. 1972. *Man and His Environment: Climate*. Harper & Row, New York.

Geiger, R. 1959. *The Climate Near the Ground*. Harvard University Press, Cambridge, Mass.

Gubbels, G. H. 1963. *Gardening in the Yukon*. Canada Department of Agriculture, Publication 1192. Queen's Printer, Ottawa.

Harris, R. E. 1970. *Gardening on Permafrost*. Canada Department of Agriculture, Publication 1408. Queen's Printer, Ottawa.

Hughes, H. D. 1972. *Crop Production*. Macmillan, New York.

Mitchell, I. 1976. High-altitude gardens call for an early harvest. *Organic Gardening and Farming*, 23:116–120.

Ponte, L. 1976. *The Cooling*. Prentice-Hall, Englewood Cliffs, N.J.

Rosby, C. G. 1941. The scientific basis of modern meteorology. In *Climate and Man*, Yearbook of Agriculture, 1941. U.S. Government Printing Office, Washington, D. C.

Twewartha, G. T. 1954. *An Introduction to Climate*. McGraw-Hill, New York.

Wilsie, C. P. 1962. *Crop Adaptation and Distribution*. Freeman, San Francisco.

LEARNING OBJECTIVES

By the time you have finished this chapter you should be able to:

1 State the major factors to be considered in the selection of a garden site.

2 Discuss the criteria to be considered in deciding what vegetables should be planted in the garden.

3 Distinguish between open-pollinated and hybrid vegetables and give examples of each.

4 Describe how to determine when the soil should be tilled, and explain tillage procedures.

5 Explain the composting process, and tell of its value in home gardening.

6 Discuss the pros and cons of using raised beds in the garden layout.

7 Name at least four planting and cropping patterns.

8 Identify those planting and cropping patterns that make for maximum utilization of space.

9 Describe an easy and effective way to store garden seed at home.

10 State the values of using transplants in the home garden.

11 Tell how to select transplants properly at the garden store or supermarket.

12 Explain the process of "hardening off" and tell of its importance.

13 Discuss the steps involved in successful transplanting.

7

PLANNING, PREPARING, AND PLANTING THE GARDEN

" . . . . at this stage of armchair gardening . . . we have it on paper, in all perfection, for the garden of the mind's eye knows no weed or pest or disease; it flourishes in soil effortlessly rich and loamy, under skies never darkened by clouds because it rains only on Mondays and Thursdays from 1 to 6 A.M."

Ruth Matson

Each garden is a unique event. Where one gardener plants vegetables, another prefers fruit trees or a beautiful display of flowers. Since each garden is different and plants all require specific conditions, the development of a successful garden demands thoughtful planning as well as the ability to properly manage growing plants. Several important aspects of garden development are dealt with in this chapter, starting with planning, then preparing the garden site, and finally planting the garden. Let us explore these three P's of gardening.

CHOOSING THE GARDEN SITE

We often choose our house and our neighborhood, but rarely our garden. Usually the garden must be developed from space surrounding the buildings. Too often this is infertile subsoil left after house construction, exhibiting poor drainage, or excessive soil compaction. But we must accept the limitations of our garden space—whether of poor soil, too much shade, root competition from a neighbor's apple tree, or a north-sloping garden space. It remains for the gardener to be aware of these limitations and to modify the soil, the weather, or the choice of plants so as to best overcome the shortcomings of the garden site.

Of cardinal importance in selecting a garden site is sunlight—the more the better. If possible, choose a site that provides at least six hours of direct sunlight each day (Fig. 7-1). Second, try to garden in a well-drained area. Ideally, the garden site should be nearly

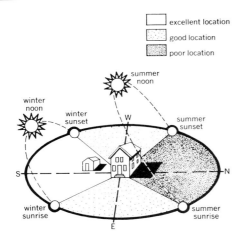

excellent location
good location
poor location

summer noon
winter noon
winter sunset
summer sunset
W
S — N
winter sunrise
summer sunrise
E

FIGURE 7-1 Choose a garden site that provides maximum exposure to direct sunlight. (Courtesy of U.S.D.A.)

level or have a slight southern exposure. A garden on a slope that allows free movement of air is more likely to escape late-spring and early-autumn frost damage, but the greater the degree of slope the greater the likelihood of erosion. In windy regions the garden should be placed so that it is protected from the wind by a hedge, fence, or other windbreak; but it is important that the windbreak not create shade or send competing roots into the garden area. In dry areas locate where the plants can be irrigated easily. Also, consider the proximity of the garden to the living area. A garden near the house is easier to care for, because you can do garden work any time you have a few spare minutes. Finally, the garden should, as much as possible, blend into the overall design of the landscape. It is not necessary that the garden be in a squared-off area of its own, although the large-scale gardens described in this chapter are of this type. The use of vegetables in the overall landscaping scheme is discussed in Chapter 14.

Garden Size

How big a garden is depends on the availability of space and water, the variety and amount of vegetables wanted, and the amount of time available to work the garden. Generally, the beginning gardener should choose an area no larger than 300 square feet. An excellent plan for a garden of this size is offered by the National Garden Bureau (Fig. 7-2). If garden space is limited choose crops that use space most effectively. Vegetables in this category include:

Beans	Eggplant	Radish
Beets	Lettuce	Spinach
Broccoli	Onions	Swiss chard
Cabbage	Peas	Tomatoes
Carrots	Pepper	Turnips

If space considerations are not a major factor then use space-consuming vegetables such as corn, cucumbers, melons, and squash. If a very large garden area is available it may be possible to set aside space to be planted to crops that help maintain soil fertility, such as the clovers.

THE GARDEN PLAN

A garden plan is the best tool for helping the gardener to effectively utilize the garden area, while dealing with the limitations of the garden space. It should be a written plan, preferably on graph paper. As a first step, determine the length of row for each vegetable. This determination can be difficult, as future yields must be anticipated. Table 7-1 serves as a guide to yields and also the quantities needed for eating fresh, canning, and freezing. The amount of space between rows must be shown in the plan. Minimum suggested spacings between rows are presented in Table 10-3. Actual spacings will vary, however, depending on garden layout, type of planting pattern used, and cultivation method employed. In garden layouts like that shown in Fig. 7-2, for example, rows can be closer together, because the individual beds can be worked from either side. Also, rows that are

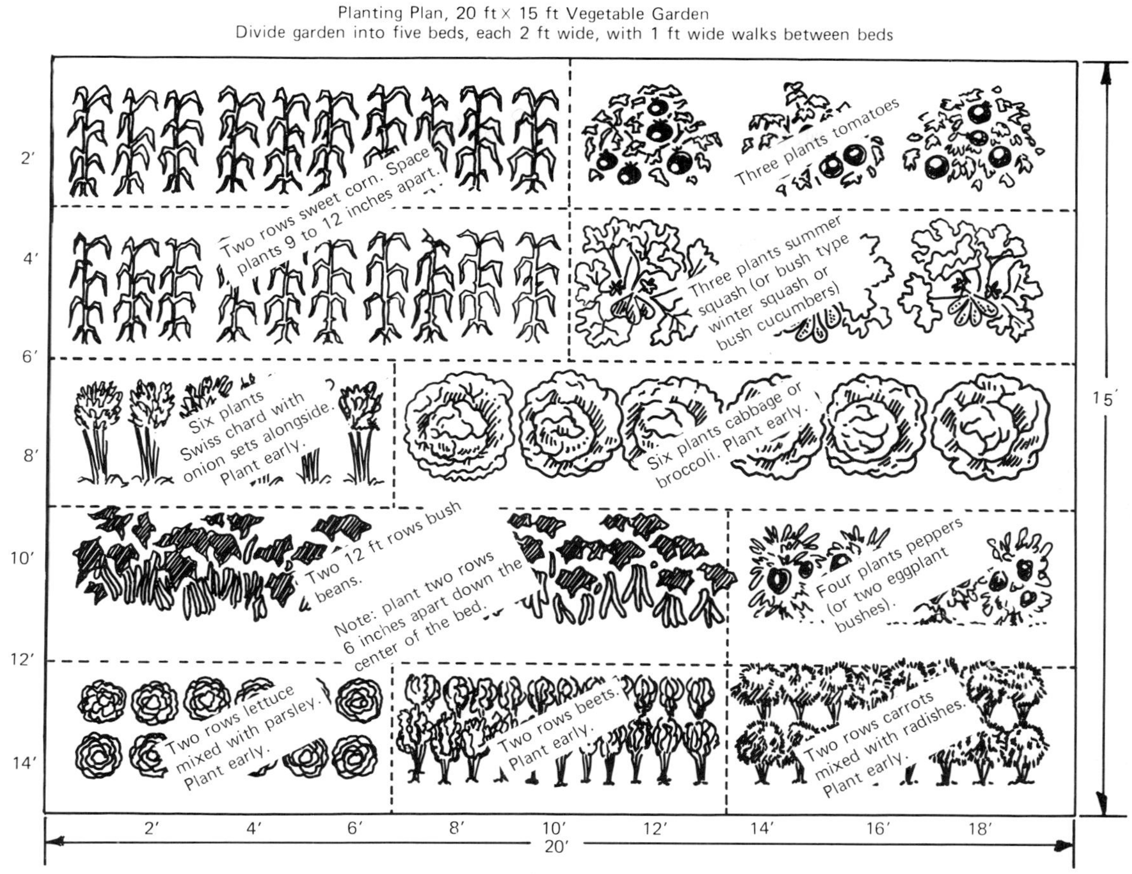

Planting Plan, 20 ft × 15 ft Vegetable Garden
Divide garden into five beds, each 2 ft wide, with 1 ft wide walks between beds

FIGURE 7-2 A garden of 300 square feet is adequate for most beginning gardeners. (Courtesy of the National Garden Bureau, Sycamore, Illinois. Redrawn by Guerry Dean.)

hand cultivated can be closer together than those that are machine cultivated. Once these aspects of the garden plan are established, the amount of seed and the number of transplants needed can be determined (Table 10-3).

Row orientation is also important. In general, when crops of more or less uniform height are grown there seems to be little difference—in terms of exposure to the sun—between rows running a general east-and-west direction as opposed to a north-and-south direction. When crops of variable height are grown, however, the tall-growing ones are best planted on the north side (or along one side) of the garden to reduce shading effects. Under these conditions east-and-west oriented rows are most convenient. The possibility of soil erosion is often a more important factor in determining row orientation. If the slope of the garden exceeds 2 percent, especially on more sandy soils, rows should be placed at right angles to the slope to minimize soil movement and at the same time enhance water infiltration.

Crop rotation is an important part of long-term gardening. Crops repeatedly planted in the same place deplete the soil of those nutrients required by a particular crop. Fur-

TABLE 7-1 Yields of Home-Grown Vegetables

Vegetable	Approx. yield/50 ft of row (lb)	Fresh veg. needed for family of 3 (lb)	Approx. amounts of fresh veg. needed for 1 quart (lb)	
			Canned	Frozen
Asparagus	28	15	4	2–3
Beans, lima	12	6	4–5	4–5
Beans, snap	30	22	11/2–2	11/2–2
Beets	35	30	21/2–3	21/2–3
Broccoli	27	20	—	2
Cabbage	75	25	—	—
Carrots	45	50	21/2–3	21/2–3
Chard	30	15	—	—
Corn, sweet	5 (doz)	10 (doz)	4–5	4–5
Cucumbers	45	12	—	—
Eggplant	50	15	—	—
Lettuce, head	30	10	—	—
Lettuce, leaf	30	10	—	—
Muskmelon	40 (melons)	15 (melons)	—	—
Onions	50	50	—	—
Parsnips	50	25	—	—
Peas (pods)	25	15	4–5	4–5
Peppers	23	10	—	—
Potatoes	60	60	—	—
Pumpkins	125	15	—	—
Radishes	40 (doz)	10 (doz)	—	—
Rhubarb	45	10	1	11/2
Spinach	22	8	2–3	2–3
Squash, summer	60	10	21/2–3	2–3
Squash, winter	80	20	2	3
Sweet potatoes	45	50	21/2–3	2–3
Tomatoes	100	30	3	—
Turnips	40	15	—	—
Watermelon	20 (melons)	10 (melons)	—	—

Courtesy of University of Illinois Cooperative Extension Service.

thermore, harmful organisms often build up to damaging levels unless crops are planted in different places each year. Let two or three years pass before replanting a given vegetable (or any of its close relatives) in the same place. By keeping your garden plans from past years you will have an excellent record of rotation practices.

In addition to utilizing and retaining garden plans, the wise gardener keeps a permanent record of planting dates, blossom and harvest dates, yields of different crops, and the susceptability of favorite cultivars to pests and diseases. Records of fertilizer applications—when, what kind, how much—will save money and prevent improper applications to the garden soil. The age-old adage "experience is the best teacher" applies especially well in gardening. Keeping good records of successes and failures, of new ideas, and of the do's and don'ts gained over the years will prevent these hard-earned experiences from fading.

Crops should be grouped within the garden. To one side of the garden, or in a separate location, plant perennial crops—including asparagus, Jerusalem artichoke, horseradish, rhubarb, strawberry, other berries, and perennial herbs—so that the normal seasonal garden activities do not disturb them.

The annual vegetables should be grouped into three, four, or even five groups within the garden according to their planting times, as is illustrated in the garden plans of Fig. 7-3 and 7-4. The first planting of the spring should include the **hardy,** cool-season crops; those that can be planted as soon as the ground can be worked. Included in this group are cabbage, onion, pea, spinach, and others (Table 2-2, column 1). Frequently these crops can be put in earlier if the part of the garden where they are to occur is worked in the fall. Because the plants in this group can withstand freezing temperatures they can be grown while spring frosts are still likely.

The second planting group should include the less hardy, or **half-hardy,** cool-season crops; those that can be planted as early as 2 to 4 weeks before the last 32°F freeze of the spring. Included in this group are beet, carrot, chard, and others (Table 2-2, column 2). The plants in this group do best if planted after the soil has warmed up, but they can withstand some freezing without injury.

Planting	Row No. and width	◄─────────── 30 feet ───────────►
1st	1–12″	Early peas (Snap beans late)
	2–12″	Second early peas (Lettuce and kohlrabi late)
	3–12″	Spinach (Spinach late)
	4–12″	Leaf lettuce (Spinach late) Turnips (Spinach late) Kohlrabi (Spinach late)
	5–12″	Onion sets (Radishes late)
	6–12″	Onion seed planted with radishes (Turnips late)
2nd	7–24″	Early cabbage plants
	8–24″	Carrots planted with radishes
	9–18″	New Zealand spinach Beets planted with radishes
	10–30″	Tomato seed
	11–24″	Snap beans
3rd	12–24″	Tomato plants
	13–24″	Snap beans
4th	14–18″	Lima beans
	15–24″	Summer squash or peppers Cucumbers or eggplant
	18″	(Border strip)

25 feet

Crops in parentheses can be planted in the indicated rows after the early crops are harvested.

FIGURE 7-3 An intermediate-size garden of 750 square feet that will allow for a greater variety of crops and larger yields (greater width between rows may be desired). (Courtesy of University of Illinois Cooperative Extension Service.)

FIRST PLANTING GROUP	Early Spring Planting	Onion Sets, Onion Seed, Radishes, Lettuce, Peas, Spinach	17'
SECOND PLANTING GROUP	Mid-Spring Planting	Early Potatoes, Carrots, Beets, Parsnips, Chard, New Zealand Spinach, Transplants of Early Cabbage, Broccoli, Cauliflower	12'
THIRD PLANTING GROUP	Late Spring Planting	Tomatoes, Snap and Bush Beans, Cucumbers, Summer Squash, Winter Squash Succession of Sweet Corn—each variety blocked out in several short rows to insure good pollination	30'
FOURTH PLANTING GROUP	Early Summer Planting	Main Crop of Potatoes Late Cabbage, Broccoli, Brussels Sprouts Late Planting of Snap Beans	33'
FIFTH PLANTING GROUP	Mid-Summer Planting	Fall Crops of Carrots, Beets, Turnips, Rutabagas, Kohlrabi, Lettuce, Radishes, Spinach, Chard	20'

FIGURE 7-4 A large garden (100 by 112 feet) that allows for a greater assortment of vegetables for fresh use throughout the season, for storage, and for preservation. (Courtesy of Washington State University Cooperative Extension Service.)

A third planting group should include the **tender,** warm-season crops; those that should be planted any time after the last frost of the spring. Included in this group are snap bean, sweet corn, and others (Table 2-2, column 3). These plants grow best during warm weather and are easily injured by frost. It is useless to plant these crops earlier, because, even if they escape injury from frost, they grow little before good growing weather warms up the soil.

A fourth planting group should include the **very tender,** warm-season crops; these that should be planted no earlier than one week after the frost-free date. Included in this group are cucumber, eggplant, lima bean, tomato, and others (Table 2-2, column 4). These plants are real "heat lovers" and require hot weather for good growth.

Another planting group that gardeners may wish to consider is a midsummer planting of cool-season crops. Many of these crops do poorly if planted in the heat of the summer, but tolerate well the cooler days of late summer and early fall. Vegetables included in this group are beet, carrot, cabbage, lettuce, and others (Table 7-2); the same vegetables planted in early spring. The addition of this planting in the garden schedule can greatly extend the gardening season, prolonging the period during which fresh vegetables are available. Oftentimes the quality of vegetables from late plantings is superior. Root crops, such as carrot, turnip, and rutabaga, for fall and winter use are best grown from these late summer plantings. Fall gardening is, of course, dependent on location. In northern areas, where

TABLE 7-2 Days to Maturity and Freeze Tolerance of Some Vegetables Suitable for Fall Gardening

Vegetable	Average days from planting to harvest	Freeze tolerance
Beans, snap	50–60	None
Beets	55–65	Good
Broccoli	70–80	Moderate
Brussels sprouts	80–90	Moderate
Cabbage	70–80	Moderate
Carrot	70–80	Good
Cauliflower	65–75	Moderate
Chinese cabbage	75–80	Moderate
Collards	65–75	Moderate
Cress, upland or garden	10–20	Moderate
Endive	90	Moderate
Kale	55–65	Good
Kohlrabi	55–65	Moderate
Lettuce	35–50	Moderate
Mustard greens	30–40	Moderate
Onions, green	25–35 (from sets or plants)	Good
Radish	25–35	Good
Spinach	40–50	Moderate
Swiss chard	50–60	Moderate
Turnip	50–60	Good

Courtesy of Flower and Garden Magazine. *Mid-America Publishing Corp., Kansas City, Missouri.*

summers are cool and short, fall gardening is limited, since the crops used can be grown all summer. In locations progressively southward fall gardening becomes more practical, until, in hardiness zone 7 and beyond (Fig. 6-11) fall gardening extends well into the winter season.

Other aspects such as soil variations should be considered in mapping garden strategy. Light-textured, well-drained soils warm more quickly than do heavy-textured, poorly drained soils and thus can be worked and planted earlier. If your garden contains only a limited area of well-drained soil it should be reserved for early plantings. In this regard, it is im-

portant to remember that although we all want to begin gardening as early in the spring as is possible, attempts to till heavy soils while they are wet invariably destroys the structure, or "puddles" the soil. Plant characteristics also need to be taken into account. Corn is better pollinated if planted in a block of short rows rather than a few long rows. Similarly, if a gardener plans to save seed from a favorite, hard-to-get cultivar it will require isolation from other cultivars of the same species if crossing occurs between them. For example, when both hot and sweet peppers of the same species are grown in close proximity crossing can occur. If seed from the sweet pepper is saved and later planted, offspring resulting from crosses will produce pungent fruits. What a surprise this could be! The skills related to saving and storing garden seed are discussed in later sections.

WHAT CROPS TO PLANT

In deciding what crops to plant you need to consider such aspects as nutritional value, the crops that grow best in your area, the amount of space you have available, and family preferences.

The USDA has prepared a table dividing vegetables into nutritional groups (Table 7-3). In order to assure that your family has a nutritionally varied diet they recommend that you try to plant two or three vegetables from each of these groups.

Another useful grouping to consider in deciding what to plant is dividing garden crops on the basis of part harvested, for example, leafy crops, root crops, and fruit crops. Each of these groups has unique characteristics, and representatives of all three groups should be included in the well-balanced garden. The leafy crops are all cool season in nature, and as such their natural growing period is in the cool parts of the year. The leafy crops are also more tolerant of shade than are the root and fruit crops. The root crops, with the exception of sweet potato, are cool-season crops too. Members of this group are noted for their high production capacity per unit of area. All root crops do well in loose, well-worked soils that will not inhibit growth of their edible storage organs. Nearly all of the fruit crops are warm season in nature, and as such require a warm soil for germination and long days with high temperatures for fruit formation and ripening.

In addition to the leaf, root, and fruit crops, the balanced garden will contain selected herbs, perennials, and flowers. Remember, however, that not everyone is interested in or can grow a balanced garden. In northern areas growth of many of the fruit crops is impossible. And even where a wide variety of crops can be grown successfully a given gardener may prefer a garden of only herbs, or perhaps just perennials such as strawberries, raspberries, and asparagus. The infinite versatility of gardening can be tailored to every possible preference.

Regardless of the kinds of crops chosen by the gardener it is well to bear in mind that each crop has its own soil, climate, and space requirements. Potato, for example, thrives in a cool, acid soil, whereas asparagus demands a well-drained loamy soil nearly neutral in reaction. Swiss chard and New Zealand spinach can provide greens through the summer heat whereas spinach may bolt and go to seed. The melons do best if allowed to spread out on the warm ground. Beans and cucumbers can be trained to grow on a fence or trellis.

SELECTING THE BEST CULTIVARS

Once the kinds of crops to be grown in the garden is determined, the gardener can attend to varietal selection. Striking variations may occur among the cultivars of a crop. Snap beans, for example, may be of pole, half-runner, or bush habit, and each of these major types has cultivars with either white or colored seed coats. Cabbage cultivars include green- and red-leafed types, smooth- and savoy-leafed types, open pollinated and hybrid types, and cabbage yellows-resistant and nonresistant types. Tomatoes come in indeterminate and determinate types, and both have cultivars differing in fruit color and size, and in resistance to disease organisms. For all of these crops, there are early and late cultivars, cultivars developed for specific climates and soils, and so on. For the commercial grower and the home gardener alike, success or failure hinges on the proper choice of cultivars.

To illustrate, nearly all of the bean cultivars

TABLE 7-3 A Grouping of Common Garden Vegetables Based on Certain Nutritional Characteristics

Group I
High in vitamins A and C
  (plant 3 or more from
  this group)
	Spinach
	Collards
	Kale
	Turnip greens
	Mustard greens
	Cantaloupes
	Broccoli

Group II
High in vitamin A
  (plant 2 or more
  from this group)
	Carrots
	Sweet potatoes
	Swiss chard
	Winter squash
	Green onions

Group III
High in vitamin C (plant 3
  or more from this
  group)
	Peppers
	Brussels sprouts
	Cauliflower
	Kohlrabi
	Cabbage
	Chinese cabbage
	Asparagus
	Rutabagas
	Radishes
	Tomatoes

Group IV
Other green vegetables
  (plant 2 or more from
  this group)
	Green beans
	Celery
	Lettuce (leaf)
	Lettuce (head)
	Okra
	Peas (English)

Group V
Starchy vegetables (plant 2
  or more from this
  group)
	Lima beans
	Sweet corn (yellow)
	Onions (dry)
	Peas (field,
	  southern)
	Potatoes

Group VI
Other vegetables (plant
  from this group for
  variety in flavor, color,
  texture, etc.)
	Beets
	Cucumbers
	Eggplants
	Pumpkins
	Rhubarb
	Summer squash
	Turnips

Courtesy of U.S. Department of Agriculture.

used by the food processing industry develop a white seed coat. This characteristic is especially important in commercial canning where overmaturity in colored-seeded cultivars often discolors the liquor in the can. Tomato cultivars are distinguished as large, intermediate, and small with respect to plant size. The commercial grower utilizes mainly the small types because they are suited to mechanical harvest. The home gardener, on the other hand, generally prefers the large cultivars because they develop longer vines and set fruit over a prolonged period.

The above examples illustrate that varietal selection depends on the needs of the specific grower or gardener. A cultivar suited to the needs of one person may be entirely unfit for the needs of another. Let us look at some of the factors that can be used to evaluate cultivars for the home garden.

Evaluating Cultivars

Each year, plant breeders at agricultural institutions, both public and private, release many new cultivars of garden crops. Each of these cultivars is an assemblage of numerous traits or characteristics. While all cabbage cultivars, for example, display the overall features of cabbage, each cultivar is developed for specific features or conditions. Among the aspects most commonly emphasized in the development of new cultivars are:

- Adaptation to specific climates or soils.
- Resistance to diseases or insects.
- Growth habit or form.
- Time to maturity.
- Ripening characteristics.
- Ease of harvesting.
- Yield.
- Flavor and nutritional qualities.
- Storage and preservation characteristics.

Obviously, no single cultivar will combine all the qualities desired by a particular gardener. Through careful study of varietal character-

istics, however, the gardener can choose cultivars that most nearly fit his or her needs.

A number of varietal characteristics are of particular importance to the gardener. For many crops plant breeders have developed early, midseason, and late cultivars. The early cultivars require less heat than the others to mature. In more northern climates the use of an early cultivar can be particularly important, because the later cultivars might never receive the total amount of heat needed to mature. Also, in many areas the gardener can, by planting some early and late cultivars along with midseason ones, substantially lengthen the fresh vegetable season. Disease resistance is also an important consideration with many vegetables. The success of a tomato cultivar, for example, may depend entirely on its resistance to the fusarium or verticillium fungi, or to soil nematodes. Also, for any given area, certain cultivars are better adapted to the particular climate and soil conditions and will generally outperform less well-adapted cultivars.

For many vegetable crops hybrids, as well as the normal types, cultivars are offered. The characteristics of these contrasting breeding types must also be evaluated. The gardener is often somewhat confused by the hybrid designation. He or she may assume that there is something "almost magic" about hybrid cultivars, or that hybrids are "automatically better" than the normal breeding types, but such is not the case. To make informed evaluations a clear understanding of the reproductive characteristics of vegetable cultivars is needed.

Reproductive Characteristics of Vegetable Cultivars

Vegetables, and other garden crops, can be divided into three groups (excluding those propagated asexually; Table 3-1) based on the manner in which they are perpetuated (Fig. 7-5). One group includes the **open-pollinated** crops; those that cross pollinate freely by nature. Examples of normally open-pollinated crops include cabbage and other members of the Brassicaceae, and many members of the beet, gourd, lily, and parsley families (Table 3-2). Open-pollinated crops tend to change in characteristics over time because of natural selection or simply through chance. They are, of course, also subject to outcrossing with other cultivars and wild varieties of the same species. For these reasons seed of open-pollinated crops should generally not be saved for future planting.

Pureline cultivars, in contrast to open-pollinated types, are developed in crops that are normally self-pollinated. Purelines are the result of a selection procedure during or after several generations of natural inbreeding. Eventually, a line established from a single plant selection is increased and perpetuated as a cultivar (Fig. 7-5). Examples of pureline crops include bean, lettuce, and pea, as well as many cultivars of eggplant, pepper, and tomato (Table 3-2). Because the members of a pureline are self-pollinated the line will normally remain uniform for an indefinite number of generations. As a result, seed from such crops can be saved for later use. Even here difficulties may be encountered, however. Certain vegetable diseases—the bacterial blights of beans, for example—can be transmitted from one generation to the next through the seed. Unless the gardener is fully aware of the symptoms of the blight disease, he or she can greatly reduce the potential for a future blight disease problem by purchasing disease-free seed annually.

A **hybrid** designation means the seed comes from a direct cross of two specially selected, inbred parent strains, or lines (Fig. 7-5). Inbreeding in a plant line over several generations increases uniformity, but causes a marked decrease in both plant size and vigor. When plants from inbred lines are crossed, however, the hybrid offspring almost always show a dramatic increase in vigor, size, and yield over the original parents. Sweet corn was the first hybrid vegetable used extensively, but today many vegetables are hybridized, including broccoli, Brussels sprouts, cabbage, cucumber, onion, spinach, tomato, and others (Table 3-2). If hybrid seed is planted the succeeding crop will be highly variable, and in general inferior in quality to the hybrid parent. For this reason seed from hybrid crops should not be saved for later garden use.

It is interesting to note that hybrid cultivars can be developed from either open-pollinated or pureline breeding types (Table 3-2). When a given crop contains both hybrid and regular cultivars, the regular cultivars are frequently designated as **standard** cultivars. Thus, a popular seed catalog segregates its corn and cu-

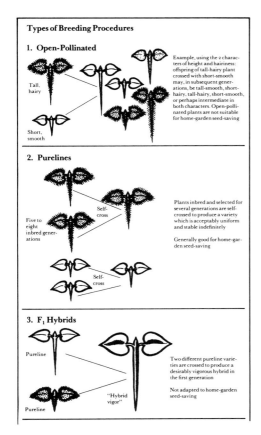

FIGURE 7-5 The types of breeding procedures characterizing vegetable cultivars. (Courtesy of James Baggett, Oregon State University.)

cumber (both crops are open-pollinated by nature) cultivars into hybrid and standard types. The same catalog also separates its tomato (a pureline breeding type) cultivars as either hybrid or standard types.[1] In some cases, the cultivar name itself indicates a hybrid origin as in 'Jade Cross Hybrid' Brussels sprouts, 'Burbank Hybrid' sweet corn, and 'Sweet Favorite Hybrid' watermelon. In other cases it is not made clear to the gardener whether he or she is dealing with a hybrid or a standard cultivar of a given crop. This important information is frequently not included on seed packets! Perhaps it is felt that the home gardener would not fully appreciate these technical differences anyway. For the increasingly sophisticated gardener, however, this information is a must, and hopefully will become an established part of seed packet labeling in the future.

Should the home gardener use standard or hybrid cultivars? There is no one answer to this question. Some crops are available only as open-pollinated or pureline cultivars, and for corn nearly all cultivars are of hybrid origin. In other crops the gardener has a wide choice of both standard and hybrid types. In cabbage, as mentioned earlier, both open-pollinated and hybrid types are available. The hybrid cultivars, as compared to the open-pollinated types, are much more uniform in plant and head type and also in maturity. They are, therefore, more suited to commercial production, but for the home gardener uniform maturity is more of a drawback than an advantage. The generally higher yields of hybrids are not as critical to the home gardener either. Then, too, the heavy expense involved in the development of hybrid seed often makes it more costly. In short, the gardener must look at the totality of features of each cultivar in question to determine which best fits his or

her needs. The problem of cultivar selection is not the gardeners alone, however. Several sources are available to aid the gardener in the decision-making process.

Sources of Information

There are several excellent sources of information on cultivars. First, and foremost, is the local Extension agent in your county or city (Appendix C). These specialists can usually provide you with the latest list of recommended cultivars, and with other information that will enhance your gardening efforts. Other sources of information include (1) seed packets and related material in the seed racks of local garden stores, (2) seed catalogs, and (3) the independent evaluation of All-America Selections.

Many seedsmen sell their products exclusively through seed packet displays at retail stores. Seed packets generally provide information on the following points:

- Cultivar name and characteristics.
- Number of seeds or seed weight per packet.
- Where and when to plant.
- How to plant.
- How to thin or transplant.
- Soil and cultural needs.
- Special characteristics, such as disease resistance.
- Days to maturity.
- Best uses (canning, freezing, etc.).

In addition, seed packets normally have a pictorial representation of the edible part of the vegetable in question (Fig. 2-2). A number of seedsmen also have available, at nominal cost, vegetable gardening guides that provide information on varietal selection and culture.

Another group of seedsmen sell their products exclusively through mail order catalogs (Appendix E). These catalogs are usually free, and contain a wealth of information on the cultivars of each crop offered. Many garden catalogs feature garden supplies as well as seed.

---

[1]Plant breeders also use the term standard to refer to tomato cultivars that develop a dark green shoulder color at the mature green stage.

Seed catalogs are an invaluable source of information on garden plants and their culture, and it is safe to say that every serious gardener should have access to one or more of them. They provide excellent reading on cold, winter nights when next year's garden is little more than a pleasant vision. Through experience you will find out which catalogs best fit your individual needs.

Newly released cultivars are given rigorous testing and evaluation both by plant breeders and by organizations such as All-American Selections (AAS). AAS is an educational, nonprofit organization established by the seedsmen in 1932 to test new seed-grown garden flower and vegetable cultivars. Tests take place at more than 50 officially designated test gardens across Canada and the United States. In test trials, experienced judges evaluate new vegetable cultivars for such factors as convenient fruit size, improved flavor, resistance to pests and adverse weather, continuing productivity, and vigor. When a gardener sees a cultivar designated as an "All-America winner," "All-America Selection," or "AAS," the gardener can be sure that it is the best in its class and adapted to a wide range of soils, climates, and cultural practices. All-America Selections also maintains more than 60 display gardens throughout the United States, Canada, and foreign countries where visitors can, at appointed times, view AAS winners "in action" (Fig. 13-11).

After deliberating on where to locate the garden, what size the garden should be, and what crops to plant we can now discuss how to prepare the garden site for planting. This topic can be divided into several categories, including (1) tools required, (2) tilling the soil, (3) maintaining and improving the garden soil, and (4) garden layout. In preparing the garden remember that considerably more effort is required to prepare a new area. A converted lawn or soil in a recent housing development will require more work than a garden plot that has been worked for several years.

TOOLS

For a small garden few tools are required. In most instances a shovel, spade, or spading fork, hoe, rake, a ball of cord or twine for laying off rows, and some wooden stakes for row ends are all that is needed. A trowel is useful in transplanting and a cultivator is useful for weeding work, but they are not essential. A wheelbarrow is also a useful tool, and an adequate length of garden hose is essential where watering is required. The tools mentioned come in many quality grades. In general it is much better to buy high-grade tools, even though they are usually more expensive.

As garden size increases the number and kinds of tools needed also increases. For gardens greater than 1000 square feet in size a wheel hoe is a useful tool, because it can be used to do much of the work formerly done with a hand hoe, and the work can be done faster. Wheel hoes come in many styles and sizes, but the single wheel type is probably easier to handle and most useful for all-purpose work. If the garden approaches or exceeds 2000 square feet in size a power tiller becomes important. A garden tiller can replace the wheel hoe for cultivation and in addition is useful in preparing all but the heaviest of soils for planting. Most garden tillers can be adjusted for either shallow or deep cultivation, and some newer models also have a width adjustment.

Tool Maintenance

An important aspect of garden tools is their care and maintenance. Shovels and hoes that are sharp allow you to work easier and faster. Tools should be stored in a garage or tool shed, because if left out in the weather they are likely to rust and wooden handles will become checked and warped. Before tools are put away after use they should be carefully cleaned. This can be accomplished by washing the tool in a stream of water, or by use of a small wooden scraper. A successful cleaning approach used by some gardeners is simply to work tools up and down in a bucket of sand after use. Using this method tools come out clean, shiny, and dry. A light coat of oil will protect tools against rust over the winter months.

TILLING GARDEN SOILS

Tillage encompasses all of the operations involved in working the soil, and includes spading or rototilling, hoeing, raking, planting, and subsequent cultivation. In this section we discuss those tillage operations used to prepare a seedbed for planting. As pointed out in Chapter 4, methods of seedbed preparation range from the traditional method of spading, hoeing, and raking, to methods of minimum tillage or no tillage at all.

Reasons for Tillage

Tillage practices are carried out for several reasons. First, tillage is used to incorporate residues into the soil. These residues can include the remains of last year's crop, mulch materials, growing plants, composts, and fertilizers. Turning these materials under helps to prepare the seedbed and improves soil tilth. A clean seedbed reduces plant diseases and is aesthetically pleasing. Second, tillage improves the physical condition of compacted soils. Through tillage, soils are loosened and aerated and thereby provide a better medium for growing roots. Third, tillage is used to prepare a suitable seedbed and medium for root growth. The ideal seedbed will be loose and granular—but not too fine—so that it gives good contact between seed and soil. Preparation of an overly fine seedbed requires the expenditure of unneeded energy, creates a soil that is likely to seal over after a rain, and in many cases actually results in reduced crop yields. Finally, tillage is used to reduce competition from weed growth. The initial tillage involves destruction of native vegetation, be it weeds or a green-manure crop. Tillage throughout the season reduces or eliminates new weed growth giving the crop a competitive advantage.

When to Till

Several factors influence when the garden soil should be worked. If you have a garden area suitable for early spring plantings rough tilling in fall is best. In dry areas, soils also are best tilled in the fall because winter precipitation is readily retained by the rough surface, with little or no runoff. It is well to remember, in this regard, that impurities in precipitation are an important source of nutrients to garden soils. Sandy soils are preferably cover cropped and tilled in the spring. The **cover crop**—a crop growth between cropping seasons to protect the soil—helps prevent the leaching of nutrients in soils with little ability to hold them, adds organic matter, and prevents erosion. If the soil has a heavy sod it should be turned well ahead of planting time to allow for decay of incorporated materials. Likewise, manure should be worked in some time before planting.

Regardless of the season chosen to till the garden soil, it should not be worked when too wet, unless followed by an extended period of freezing and thawing that will cause soil clumps to soften and crumble. A simple test to determine if your soil is too wet to work is to take a representative handful and squeeze into a ball. If the soil packs into a hard ball, or the ball does not crumble under slight pressure, the soil is too wet to work (Fig. 7-6). Soil that sticks to a spade or plow is usually too wet also. Remember, any attempt to till heavy soils while they are too wet will be disastrous. When a soil of this type is tilled its pore space is greatly reduced and it becomes practically impervious to air and water. Furthermore, as

FIGURE 7-6 The ball test is used to determine if a soil is too wet to work. If a handful of soil packs into a hard ball (*a*) it is too wet, but if the ball crumbles easily (*b*) the soil is ready to work. (Courtesy of University of California Cooperative Extension Service.)

this puddled soil dries it becomes dense and hard. On the other hand, if heavy soils are tilled when too dry, large clods form that are difficult to work into a seedbed. Timing is less critical in the tillage of more sandy soils.

How to Till

How should the garden soil be worked? If you are using a shovel or a spade the object is to loosen and aerate the soil, as well as turn under debris. This *does not* mean that each shovelful of soil must be inverted. In many cases the topsoil is only a few inches thick, and the inversion of each spadeful of soil can bring less developed soil to the surface. The best way to work the soil is to spade in rows, turning each spadeful at an angle and depositing on the opposite side of the furrow created by

this technique (Fig. 7-7). Normally, spading to a depth of 8 to 10 inches is adequate, even if a cover crop is being turned under.

The trenching, or double spading, method is used by some gardeners to till and improve heavy textured soils. This method requires breaking up the soil to a depth of 18 to 24 inches. First the surface soil is removed from a strip about 30 inches wide across the garden and piled to the side. The subsoil is then broken up and mixed with organic materials, lime, or fertilizers. After this treatment the surface soil with any possible additives is put back in its original place. While the method is laborious, heavy soils treated in this way show better drainage and aeration, moisture retention, and root penetration. Usually only a part of the garden is trenched each year until, finally, the whole area has been treated in this way.

FIGURE 7-7 Spading should be done in sequential rows with each spadeful of soil lifted and thrown to the opposite side of the furrow created by this technique. Inverting each spadeful of soil is unnecessary. (Courtesy of Brooklyn Botanic Garden, Brooklyn, New York.)

Rototilling is another popular way of preparing garden soil, and reduces the amount of physical work the gardener must do. Rototillers are usually powered by a gasoline engine, some have power-driven wheels and electric starters, but all have a set of revolving tines. The tines, or feet, rotate rapidly through the soil destroying weeds and eventually producing a loose seedbed. After the initial tillage, soil amendments can be spread and the soil tilled until the material is well distributed in the surface layer. Several precautions are necessary in rototilling. There is a tendency to till to only a shallow depth (4 to 6 inches). Care must be taken to assure deeper tillage. Also, excessive tillage tends to reduce the natural granular or crumb structure of the surface soil, leading to an unfavorable physical condition. The formation of a hardpan

layer may also be encouraged. Generally rototillers are needed only for larger garden plots; the beginner can start out with spade, hoe, and rake, then move on to the larger investment involved in a rototiller as the garden size increases.

MAINTAINING AND IMPROVING THE GARDEN SOIL

Although tillage is generally used to prepare a seedbed for planting, tillage alone is not adequate to maintain good soil tilth. As a matter of fact, a philosophy of concerned gardeners is that the soil should be passed on to the next generation in as good or better condition than when it was acquired by the gardener. To maintain and improve a soil requires basic knowledge and the ability to put this knowledge to use. Much of the needed knowledge in this area was presented in Chapters 4 and 5. We wish now to discuss putting this knowledge to work in caring for garden soils.

Dealing with Misplaced Subsoil

A major problem for many suburbanites is that their garden site ends up being the place where subsoil was deposited from excavation done for the foundation or daylight basement of the home. Subsoil presents serious problems to the gardener. Subsoil is often as rich as the overlying topsoil in mineral nutrients with the exception of nitrogen and sometimes phosphorus and potassium. It is much lower in organic matter, however, and may have a poor structure.

Two approaches are commonly used to deal with problems of misplaced subsoil. The more obvious is simply to remove trash (bricks, plasterboard, lumber, wire) from the soil, and cover with 8 to 10 inches of good topsoil. If topsoil is expensive or hard to get a lesser depth will work, particularly if the garden is laid out in raised beds. A second approach, which takes a little longer, is to work in large quantities of a well-rotted manure. Sand, ground limestone, and a complete fertilizer can also be worked in as needed. Sand in association with ample organic material will improve aeration and drainage, limestone will reduce soil acidity, and the fertilizer will supply nutrients present in short supply. The vegetable garden can be planted after these additions are made, or a green manure crop or two can be grown prior to beginning gardening. In either event, a topsoil favorable for gardening can soon be established.

Testing Soils

Accurate and regular soil testing takes much of the guesswork out of soil management. Soil testing is the best way to determine the pH or acidity of garden soils and the fertilizer requirements. Because the information provided by testing is crucial in directing plant growth, the soil test is probably the one most important step the gardener can take towards more productive gardening. Fortunately, soil testing is readily accessible to the gardener; he or she can purchase one of the many soil testing kits available commercially, send a sample of the soil to the appropriate provincial or state soil testing laboratory, or have the soil tested through a private laboratory.

Taking a Soil Sample

The most critical aspect of soil testing is taking a composite sample that represents accurately the garden soil. It is recommended that small samples be taken at from 10 to 20 locations throughout the garden, depending on garden size (Fig. 7-8*a*). Each sample should be about a tablespoon in size and include soil both from the surface and to a depth of 6 to 8 inches. An easy way to sample is simply to insert a spade or shovel into the soil, and take a representative spoonful along the exposed pro-

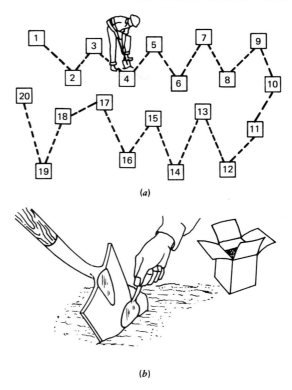

FIGURE 7-8 Individual soil samples should be taken from 10 to 20 different sites in the garden (*a*), depending on garden size. Individual samples can be taken by inserting a spade into the soil and sampling along the exposed profile (*b*).

file (Fig. 7-8*b*). Another way to sample is to take a spadeful of soil, then using a knife, remove a small, vertical sample from the soil on the spade. In either event, the samples should be put into a container and mixed thoroughly. If the soil is wet, dry the samples for a time at room temperatures before mixing. If a garden contains several strikingly different areas, these should be sampled and tested separately. For example, do not mix soil from a hilltop with that from a low, wet area, or soil from a fertilized or limed area with that from a nontreated area. Remember, the composite sample, or samples, of soil that you test or send away for testing must be representative of the thousands of pounds of soil in your garden.

Getting Your Soil Samples Tested

If you choose to send your soil to a state or provincial soil testing laboratory, they will give advice on how to sample, size of sample wanted, information you should provide with your sample, tests available, and costs involved, if any. Additionally, they may provide you with a container in which to send your sample to the laboratory. Some states require that soil samples be carefully sealed in plastic bags before mailing. Information on soil testing within your state or province can be obtained from your agricultural college (Appendix C), or from local extension agents. When your test results are returned you will generally find included recommendations on ways to correct for observed deficiencies. These recommendations are frequently given in pounds or tons of fertilizer to be applied per acre. Using Table 7-4 such recommendations can be converted to amounts needed per 50 or 100 feet of garden row or per 100 to 2000 square feet of garden area.

How Often Should Your Garden Soil Be Tested?

Generally, it is a good idea to test once a year for the first two or three years and less often thereafter. By relating changes in your soil's chemistry to your gardening practices you will come to understand what is needed to keep your soil in a highly productive condition. In short, you will develop a feeling for your soil. While soil samples can be taken at almost any time of the year, fall sampling is regarded as superior. The nutrient levels in the soil at the end of the growing season most accurately reflect natural levels. Also, if lime is required it can be added to the soil before winter, thus acting to neutralize the soil before spring planting. In spring the soil may be wet and

TABLE 7-4 Approximate Rates of Fertilizer Application per 50 or 100 Feet of Garden Row, and per 100 to 2000 Square Feet of Garden Area, Corresponding to Given Rates per Acre

Measurement	Weight of fertilizer to apply when the weight to be applied per acre is			
	100 pounds	400 pounds	800 pounds	1200 pounds
	Pounds	*Pounds*	*Pounds*	*Pounds*
Space between rows, and row length (feet):				
2 wide, 50 long	0.25	1.0	2.0	3.0
2 wide, 100 long	.50	2.0	4.0	6.0
2½ wide, 50 long	.30	1.2	2.4	3.6
2½ wide, 100 long	.60	2.4	4.8	7.2
3 wide, 50 long	.35	1.4	2.8	4.2
3 wide, 100 long	.70	2.8	5.6	8.4
Area (square feet):				
100	.25	1.0	2.0	3.0
500	1.25	5.0	10.0	15.0
1000	2.50	10.0	20.0	30.0
1500	3.75	15.0	30.0	45.0
2000	5.00	20.0	40.0	60.0

Courtesy of U.S. Department of Agriculture.

sampling is more difficult or delayed. Soil testing laboratories are commonly busier in the spring too, meaning delays in getting back test results. Regardless of how frequently or time of year at which soil samples are taken they nevertheless provide the basis for informed fertilization and liming of garden soils.

Fertilizing Garden Soils

Vegetable gardens commonly require additions of fertilizer, because the intensive plant culturing removes large quantities of soil nutrients. There is no better time to plan a fertilization program than when preparing the garden soil for planting. Unfortunately, it is hard to give general rules for adding fertilizers because soils vary widely in basic fertility, different crops have differing fertilizer demands, and the characteristics of the fertilizers themselves are variable. There are however, guidelines for making decisions about fertilization.

Factors to Consider in Deciding Fertilization Rates

Several factors should be considered in deciding on a fertilization program. First, what is the past history of the area to be used? Information on past cropping, fertilization, and crop yields all influence fertilizer decisions. If the area to be used was previously uncropped, *do not* rely on existing vegetation to determine nutrient needs, because native plant populations change to fit soil conditions. Second, what kinds of crops are to be grown? Leafy crops, like broccoli, cabbage, kale, lettuce, and spinach, need large quantities of nitrogen. The root crops, on the other hand, make larger demands on potassium than do other vegetables. Phosphorus is needed for the development of flowers and fruits. Too much fertilizer can cause root injury and tip burning of leaves, or promote excessive growth. Tomatoes, for example, if given too much nitrogen, produce luxuriant foliage, but few fruit. Third, what kinds of fertilizer are to be used? Both commercial fertilizers and animal manures will supply needed nutrients. Often a combination of the two forms gives better results than either one used alone. Choice will depend on such factors as price, availability, and preference.

Methods of Fertilizer Application

Commercial fertilizers can be applied either by broadcasting during seedbed preparation or by banding at seeding time. In **broadcasting** the fertilizer is spread uniformly over the soil surface and worked into the root zone by hoeing or rototilling. If the fertilizer is applied close to the time of seeding and mixed promptly into the soil, nitrogen losses through denitrification (Fig. 5-10) will be minimized. The broadcast method, while simple and fast, has the disadvantage of not placing all fertilizer in the optimum location for plant use. In **banding** fertilizer, distribution is more closely regulated by being deposited in rows, or bands, near the rows in which the seed, or transplants, are to be planted (Fig. 7-9). Generally the fertilizer should be placed two to three inches to the side of the seed and at the same level or slightly lower than the seed. Fertilizer placed directly under the seed is likely to injure the young seedling. Banding, while being a more time-consuming method, is regarded as a more efficient method of applying fertilizer. The banding method can also be used to sidedress growing crops, but the bands must be placed further away from the rows to avoid burning plant roots. More on this in Chapter 8.

Rates of Fertilizer Application

How much fertilizer should be applied to the garden soil? Rates will vary depending on the factors outlined above. One recommendation for preplanting fertilization for average gar-

Banding Fertilizer

FIGURE 7-9 A technique for banding fertilizer.

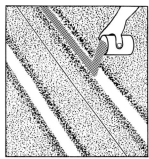

(*a*) Mark off the seed row with stakes and string; dig trenches with a hoe on both sides of the seed row, and apply fertilizer in the trenches.

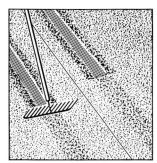

(*b*) Rake soil back over the fertilized trenches.

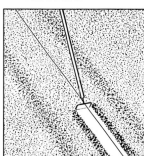

(*c*) Use a hoe to dig a seed row between the refilled trenches.

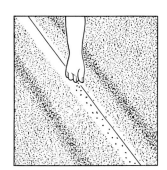

(*d*) Sow seed, and cover it. As the seedling plants begin to grow, their roots will spread into the fertilized areas.

den soils that have been cropped the previous year and given good yields is:

2 pounds of 10-10-10 per 100 square feet

If the previous crop was poor then up to 1 pound more per 100 square feet can be added at the time of garden preparation. On the other hand, if the soil has been fertilized and managed well for several years, then a lesser rate can be used. If the fertilizer available is 5-5-5 then 4 pounds would normally be applied per 100 square feet. However, if the available fertilizer has an analysis of 5-10-10 what should the rate of application be? Still 4

pounds per 100 square feet. Nitrogen is usually the critical element, the one to which fertilizer rates should be geared. Excesses of phosphorus and potassium, while creating additional expense, do not create the detrimental growth effects that can accompany excess nitrogen. Said another way, it is safe to overcompensate on phosphorus and potassium, but not on nitrogen. Regardless of fertilizers used, amounts to be applied should be carefully weighed or measured. In converting weight to measure it is safe to assume that for most commercial fertilizers 1 pound is equal to about one pint, or two measuring cups, of material.

If animal manures are to be used in pre-planting fertilization they are best applied several weeks before planting and worked well into the soil. This allows time for the manure to decompose and for nutrients to be released for plant use. If the manure to be used contains material with a high carbon/nitrogen ratio (Chapter 5), such as straw or sawdust, it may be necessary to also apply nitrogen fertilizer to avoid tying up the nitrogen already present in the soil. A number of important nitrogen carriers are given in Table 5-8. The rate of application of these supplemental nitrogen materials should range from ½ to 2 pounds per 100 square feet, depending on their nitrogen content. The rate of manure application per se will vary too, depending on if the manure is fresh or dehydrated, and on the kind of manure used. A rate of application roughly equivalent in nitrogen content to that given earlier for commercial fertilizer is as follows:

Manure	Rate
Cow (fresh)	40 pounds/100 square feet
Cow (dehydrated)	20 pounds/100 square feet

In practice, however, heavier rates of application can be used. It is not uncommon to apply cow manure at two, three, or even four times the rate given above. Poultry manure, which has about three times as much nitrogen as cow manure, should be applied more sparingly. The composition of animal manures and several other organic materials is presented in Table 5-7.

Liming Garden Soils

Most of our normal garden activities tend to lower soil pH. Cation exchange tends to increase acidity, for example, because acid-forming hydrogen ions are exchanged for the base-forming cations held by soil colloids. Acidity is also increased when acid-forming fertilizers are added to the soil. Acid rain, too, is a factor in soil acidification. Couple this with the fact that soils over a large portion of the continent are naturally acid, and you begin to see why liming is an important practice used to improve garden soils. Liming can be done at almost any time of year but is best done in the fall or early winter so that the lime will have plenty of time to react with the soil. Regardless of time of application, the added lime should be worked thoroughly into the topsoil.

Is Lime Needed?

A first step in correcting presumed soil acidity problems is to accurately determine the pH of your garden soil. This can be done either by sending a soil sample to your state or provincial soil testing laboratory (Appendix C), or by using any of a number of kits available to home gardeners. Most kits contain either pH indicator paper or an indicator dye solution along with instructions for their use. These indicators are accurate to within a half pH point. When using indicator paper make a soil paste saturated with water, immerse a paper strip, and compare with a color chart provided (Fig. 7-10). If low pH readings are obtained liming may be desired. Whether or not lime is added will depend almost entirely on the kinds of crops to be grown. Even here the situation is not always clear. For example, Irish potatoes, as mentioned earlier, may yield best at pH 6.5, but are frequently grown at pH 4.8 to 5.5 because the incidence of the soilborne disease called scab is much less below pH 5.5. Liming is not a cure-all, and not something that all gardeners should do. A soil should be limed only when its pH has been determined to be too low for good growth of desired vegetables.

Sources of Liming Materials

Liming materials are available in several forms, but limestone taken from a quarry and ground

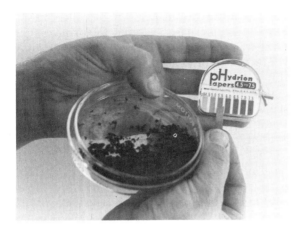

FIGURE 7-10 A strip of indicator paper touched to wet soils tells a gardener that the pH of the soil is 5.5.

to a medium consistency is the most common. The two important minerals in limestone are **calcite,** which is predominately calcium carbonate, and **dolomite,** which contains primarily magnesium carbonate. The proportions of the two minerals in limestone will vary depending on its source. Agricultural scientists analyzed 220 samples of agricultural limestone from 35 states and found the average analysis to be as follows:

Calcium carbonate	75.4%
Magnesium carbonate	17.1%
Iron	0.4%
Potassium	0.2%
Sulfur	0.1%
Manganese	330 p.p.m.
Phosphorus	210 p.p.m.
Zinc	31 p.p.m.
Boron	4 p.p.m.
Copper	2.7 p.p.m.
Molybdenum	1.1 p.p.m.

It can be seen that in addition to its main ingredients, limestone contains small quantities of other essential elements. While these additional elements are usually included, their presence is not guaranteed.

In commercial production limestone is heated, driving off carbon dioxide, and leaving calcium oxide, or **quicklime.** If water is added to quicklime, **hydrated lime** results. The neutralizing power of the forms depends entirely on the amounts of calcium and magnesium present. One hundred pounds of calcium carbonate yields 56 pounds of quicklime, and this can react with 18 pounds of water to produce 74 pounds of hydrated lime, as follows:

$$\begin{array}{c} \text{Limestone} \\ \text{100 pounds} \end{array} + \text{heat} \rightarrow \begin{array}{c} \text{quicklime} \\ \text{56 pounds} \end{array}$$

$$\begin{array}{c} \text{Quicklime} \\ \text{56 pounds} \end{array} + \text{water} \rightarrow \begin{array}{c} \text{hydrated lime} \\ \text{74 pounds} \end{array}$$

Thus, 100 pounds of pure limestone, 56 pounds of quicklime, and 74 pounds of hydrated lime all have the same neutralizing power. The use of quicklime and hydrated lime is generally not recommended for garden soils, however, because of their caustic action. Other sources of calcium include bone meal, chalk, marl, and wood ashes. In correcting soil acidity wood ashes should be applied at about double the rates recommended for limestone.

Application Rates for Liming Materials

Limestone can be ground coarse or fine. The finer the particles the more rapidly they will dissolve and react within the soil. However, finely ground limestone is unpleasant to handle and lacks lasting qualities. Generally it is recommended that limestone should be ground fine enough to sift through a 50-mesh sieve. This will assure enough fine material to produce immediate effects and sufficient coarser material for a slow release of calcium and magnesium over a period of years. Applied lime should be thoroughly worked into the topsoil. Although rates of application will vary with soil texture and pH the following rule of thumb has proven useful:

To increase the pH by one unit, for each 1000 square feet of area to be treated add:

40 pounds limestone for sandy soils

60 pounds limestone for loamy soils

80 pounds limestone for clayey soils

Remember, lime is applied to soils not as a fertilizer but as an additive to correct pH deficiencies. Applied lime raises pH and increases the percentage base saturation of soils. The influence of lime is not permanent, and may need to be added at three- to five-year intervals. While the growth of many vegetables may be enhanced by the addition of lime, the growth of others, such as potato, radish, and watermelon can easily be reduced.

In addition to correcting for soil acidity, gardeners sometimes find it necessary to deal with problems of soil alkalinity. In Chapter 4 we mentioned the treatment of sodic soils with gypsum followed by thorough leaching. In less extreme cases soils can be made less alkaline by addition of acid soil supplements such as peat moss, pine needles, and other organic residues. Many chemical fertilizers are acid in nature and as such increase soil acidity. The process of cation exchange between plant roots and soil colloids also tends to increase acidity, because acid-forming hydrogen ions are the predominant ions exchanged for the baseforming cations. This process is, of course, slow and subtle, changing the pH only gradually.

Composting

One of the major ways in which gardeners can improve their soil is through the use of compost. **Compost** consists of partially decomposed, or degraded, organic residues. The objective of composting is to convert waste organic materials into artificial manure that can be added as an amendment to the garden soil. Compost is very valuable soil additive. Its organic contents have beneficial effects on soil structure and soil tilth. Organic matter is the main source of energy for soil organisms and is an important source of mineral nutrients. The added organic material also increases the amount of water a soil can hold and the proportion of this water available to plants. Ironically, through its favorable effects on soil aggregation, added organic residues at the same time enhance soil aeration. Composting has far-reaching consequences: by this process the gardener is able to recycle—in an ecologically sound way—organic residues, many of which would otherwise be lost.

An important aspect of composting is that it consists of organic decomposition *outside* of the garden soil. You will recall from Chapter 4 that many organic residues have a high C/N ratio. If these carbon-rich materials are added directly to the soil, their decomposition by microorganisms temporarily ties up the available soil nitrogen, creating a nitrogen deficiency for garden plants. By composting, this lowering of the C/N ratio takes place within the compost pile, not in the soil. When compost is added to the soil no major transformations are needed before the nutrients contained can be taken up by plants. In short, the time that must be allowed for organic residues to decompose when added directly to the soil is eliminated, because the decomposition and concomitant lowering of the C/N ratio has occurred in the composting process.

What Kinds of Material Are Suitable for Composting?

Almost all disease-free plant residues work well. Familiar examples include leaves, lawn clippings, and garden refuse, including weeds, straw, and sawdust. The material added should be small enough to decompose rapidly. Cornstalks, large weeds, and other coarse materials should be chopped up before being added to the pile. As a rule, anything much larger than one-quarter inch in diameter will decompose more slowly than desired. Also keep in mind that materials with a high C/N ratio will decompose more slowly than those with a low ratio. Best results are obtained if carbonaceous materials like straw, sawdust, and chopped cornstalks are mixed with materials higher in nitrogen such as grass clippings and other succulent plant parts. Other obvious and important ingredients for composting are wastes from the kitchen. Useful materials in this category include lettuce trimmings, carrot tops, corn husks, and potato peelings, among other things. Animal products such as egg shells and manure can also be added, but meat scraps are to be avoided, as they attract vermin and create bad odors.

Methods of Composting

Composting is usually done in a compost pile. The pile can be simple or elaborate, fenced or unfenced, layered or unlayered. Regardless of techniques used, all composting processes have certain features in common.

The **Indore method** (named for the state in India where Sir Albert Howard perfected the process), or some modification of this method, is most commonly used to make compost. By this technique a pile is built of layers each with different contents (Fig. 7-11). The pile starts with a six-inch layer of organic residues, followed by a thin layer of commercial fertilizer or manure (to provide nitrogen and other nutrients), followed by a one-inch layer of soil (to provide beneficial microorganisms). If commercial fertilizer is used, addition of one cupful of 10-10-10 (or other material of equivalent nitrogen content) for each 10 square feet of pile surface is recommended. If manure is used, make the layer about one-half inch thick. If highly acid residues, such as pine needles, are added to the pile, or if a less acid compost is desired, ground limestone can also be added at the same rate as the commercial fertilizer. After this layering is completed the pile should be moistened, but not soaked. Repeat this layering sequence until the pile is of desired height. If the pile is not contained its sides should be sloped inward for stability. Also, if the top of the pile is concave (Fig. 7-11) it will more effectively capture rain water.

Within a few days the pile will begin to heat up due to microbial activity, often reaching a temperature of 150°F, or greater. After two to three weeks the pile should be turned to mix the contents, improve aeration, and move less decayed outer parts to the interior where they will decay more rapidly. After the pile is turned the temperature will again rise, but it will then gradually lower as the composting process reaches completion. It may be desirable to turn the pile a second time after a few additional weeks. The air introduced in mixing will accelerate decomposition by aerobic

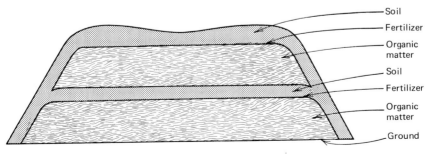

FIGURE 7-11 The compost pile should be built of layers. If the pile is not contained, the sides should taper inward. A concave top aids in capturing rainwater. (Courtesy of University of Illinois Cooperative Extension Service.)

Soil
Fertilizer
Organic matter
Soil
Fertilizer
Organic matter
Ground

organisms. Even if the pile is not turned, however, decomposition will proceed, but at a slower rate. The length of time required for thorough decomposition depends on many factors, including the nature of the organic contents, nitrogen availability, moisture content of the pile, degree of aeration, and climate conditions. In cooler climates decomposition proceeds more slowly. In any event, the contents of the pile should be ready for use within 6 to 12 months.

When compost is ready for use it will generally be dark brown in color, humuslike in appearance, and have a fragrant, earthy odor. Once ready, the compost should be used as soon as is possible. If it is to be stored for a period of time it is wise to cover the pile with black plastic or other material to prevent drying or loss of nutrients through leaching. Compost can be used virtually anywhere in the garden. For example, it can be worked into the surface soil a few weeks before planting, used as a medium for starting seeds, or used as a mulch or side dressing. Compost also serves as an ingredient in some potting mixtures (Table 14-2). Compost can be applied liberally, since there is no danger of burning due to overuse. A problem encountered by most gardeners is that they simply do not have enough compost to go around. In this case application must be selective, with a given area receiving compost every two to four years.

Three additional methods of composting merit our attention. The **14-day method,** developed at the University of California, gives ready-to-use compost in two to three weeks. By this method all of the materials to go into the compost pile are first put through a grinder or shredder. As in the Indore method, fertilizer and soil are added to the pile, but layering is not required. The shredded contents and additives are thoroughly mixed and moistened. The pile heats up rapidly and should be turned in two to three days and every three or four days thereafter. The large surface area presented by the shredded material, coupled with frequent turning to mix the contents and improve aeration, greatly accelerates the decay of organic materials.

The **anaerobic** method of composting— composting without air—is of particular interest, because it is well suited to the small-scale needs of the apartment gardener. Using this method suitable compost materials are placed in a heavy-duty garbage can liner or similar container, the excess air expelled, and the top sealed. Within hours anaerobic microorganisms begin digesting the organic residues and after a week or so the material is pretty well broken down. Composted material can be added to the soil, or stored in the plastic bags for later use. This method, like the aerobic methods mentioned above, benefits from the addition of a little soil to the organic residues, but it is not necessary to add water and fertilizers or to stir or turn the contents in order to incorporate oxygen and mix. Advantages of this method center around its adaptability to apartment-type situations. Possible disadvantages relate to the noxious odors produced in composting. For this reason it is recommended that the process be carried out in a well-aerated place, away from the center of living activity. Along this line, bags of material taken outdoors to a sunny area, not only removes a source of potential bad odors, it also speeds the composting process. It should be added that small-scale aerobic compostors are also available commercially.

Sheet composting is the method whereby organic materials are spread over the soil surface and then worked into the soil by spading or use of a rotary tiller. Sheet composting is often used in the fall when autumn leaves and garden residues are abundant. Green manuring is in essence a form of sheet composting. Regardless of the source of the organic contents, the addition of a nitrogen fertilizer will speed decomposition. When sheet composting is used it is not necessary to construct a compost pile. On the other hand, the area being composted must be kept out of production until the decay process is completed. Sheet composting differs only in degree from mulching. When organic residues are spread over the soil surface as a mulch they will gradually compost, much as occurs in nature. The **Ruth Stout method** of gardening employs just this kind of mulch. The subject of mulching is discussed in Chapter 8.

Fumigation

The soil contains a myriad of organisms, most of which are beneficial, or, at worst, harmless. In certain cases, however, the buildup of harmful organisms or other entities—including nematodes, insects, pathogenic fungi, weed seed—may be so great that soil fumigation is essential. **Soil fumigation** is essentially a process of soil sterilization, and it is important to recognize that fumigants may destroy beneficial, as well as harmful, organisms. Most of the soil fumigants convenient for home use are liquids or solids that, when applied to the soil, change to gases that diffuse for some distance from the point of application. The conversion to gaseous form greatly extends the zone of fumigant activity, and, fortunately, allows for the fumigant to escape from the soil over a two- to three-week period. The soil is then safe for normal use and soon becomes recolonized by beneficial organisms carried in by wind action or by untreated soil. The introduction of untreated soil can, of course, recontaminate a treated area. Use of clean garden tools and carefully selected planting stock, combined with care in cultural manipulations can prevent or greatly delay reinfestation.

All soil fumigants are somewhat toxic to humans and other animals and should be applied with caution. Generally, however, if fumigants are used outdoors in a slight breeze, and as directed by the manufacturer, they are quite safe. Finally, it should be added that cultural practices such as crop rotation, working the soil at the proper time to destroy germinating weed seed, and timely planting may reduce or eliminate the need for fumigation. Soil fumigation is not a substitute for skillful cultural practices; instead, it is an additional tool that may be needed to produce abundant, high-quality vegetables.

GARDEN LAYOUT—RAISED BEDS, FLAT BEDS, OR NO BEDS

A majority of gardeners lay their gardens out in rows, much as is done in large-scale farming operations. While this approach works well, the trend to more intensive gardening has brought about a shift toward layouts consisting of beds and pathways. These beds may be at ground level or they may be raised so that they appear as flattened mounds when seen in profile (Fig. 7-12). Beds not exceeding four feet in width can be worked conveniently from the pathways to either side. Generally, the area within beds can be cropped more intensively than is possible using the traditional row-type layout.

The raised bed, in particular, has been the subject of much interest in recent years. Actually, raised beds have been used for centuries. They are still common in parts of China, where this form of garden layout is said to have originated. Proponents of the raised-bed approach claim the following advantages for their system: (1) raised beds warm up faster in the spring and can be cropped earlier; (2) raised beds have improved drainage; (3) raised beds are permanent and easier to work; (4) it is never necessary to walk in the planting area, so the soil does not get packed down; (5) watering and fertilizing can be confined to bed areas and are therefore more economical; (6) working with plants in a raised bed is easier, because less stooping is required; (7) less

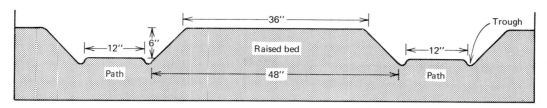

FIGURE 7-12 Profile of a raised bed. (From Ray Wolf, *Be a Gardener—Not a Small Scale Farmer, Organic Gardening and Farming*, Emmaus, Pa. 18049, October 1976, reprinted by permission of Rodale Press, Inc. Copyright 1976.)

weeding is required because the beds are cropped intensively; (8) the use of beds facilitates record keeping; and (9) raised beds display order and symmetry and are pleasing to the eye. Many of the advantages stated above apply equally to flat beds, of course. Interestingly, raised beds are now being used in some large-scale operations. In parts of Florida, for example, the commercial production of vegetables is made possible by forming raised beds on shallow soils overlying caliche.

The use of beds, especially of the raised kind, also presents some disadvantages. In general, they are more labor intensive, they restrict or inhibit the use of power equipment, and they take specific soil areas—the pathways—out of production permanently. The handsome raised bed garden shown in Fig. 7-13 has 43 percent of its total area devoted to permanent pathways. Some gardeners form raised beds with vertical sides retained by railroad ties, boards, or rocks. In general this practice is to be discouraged, as the border areas make ideal habitats for garden pests such as slugs.

Each kind of garden layout presents advantages and disadvantages. The kind to be chosen by individual gardeners depends almost entirely on personal preference; rarely is a specific layout mandatory. In this regard it should be added that, contrary to popular belief, the numerous planting systems discussed in a later section lend themselves, with few exceptions, to any kind of garden layout.

PLANTING THE GARDEN: WHEN TO PLANT

The future success of your garden depends in no small part on when and how your vegetables are planted, or transplanted. The "when" of planting depends on the climate of your area, the hardiness of the vegetables you wish to plant, and other factors, such as the characteristics of your garden soil. The "how" of planting depends on whether you are using seeds or transplants. Transplants can either be grown at home or purchased at garden stores. In this section we explore the "when" and "how" of planting vegetables, with the goal of enhancing our knowledge of these crucial areas of garden activity.

The Frost-Free Period

A major determinant in growing vegetables is the climate of your locality, and in particular, the length of the frost-free period. This period can be established in a general way by use of Figs. 6-12 and 6-13, which give the average dates of the last 32°F temperature in spring and the first 32°F temperature in fall. To estimate the frost-free period for your area find your location on the maps and then, the solid line that comes nearest to it. Two numbers are given for each line, the first representing the month, the second the day. In central Kansas, for example, the following figures are found:

FIGURE 7-13 A handsome garden formed of 10 raised beds, each measuring 4 by 25 feet. The long pathways are nearly 18 inches wide. The 4-foot-wide central path extending across the garden is mainly aesthetic. (From Ray Wolf, *Be a Gardener—Not a Small Scale Farmer, Organic Gardening and Farming,* Emmaus, Pa. 18049, October 1976, reprinted by permission of Rodale Press, Inc. Copyright 1976.)

Last 32°F temperature in spring = 4/20

First 32°F temperature in fall = 10/20

The interval between April 20 and October 20—183 days—is the frost-free period. The length of the frost-free period for certain localities in each province and state is also given in Appendix D. Most of our garden activities, including planting, are carried out within or just preceding this period. As stated in Chapter 6, the frost-free interval—which generally represents the growing season—must be long enough for any vegetable in question to complete its growth successfully.

The frost dates given in Figs. 6-12 and 6-13 also are used to determine planting dates for both spring and late summer plantings of vegetables. To determine spring planting dates do the following:

- Find the date of the last 32°F temperature for your locality using Fig. 6-12. Record this date.
- Then, using Table 2-2, you will find listed in columns 1 and 2 the cool-season vegetables, some of which can be planted four to six weeks before the frost-free date, others two to four weeks. You will also find, in columns 3 and 4, the warm-season crops that can be planted on the frost-free date, and those that should be planted at least one week later.

Certain precautions are in order, however. First, the frost-free date is computed from long-term averages and can fluctuate in any given year. Planting on the frost-free date gives a 50-50 chance of escaping future spring freezes, whereas planting earlier increases the chances and planting later reduces them (see Table 6-3). The actual frost-free date for a given locality may be two to three weeks later than the average date of the last 32°F temperature. Second, a given plant, say the garden pea, can be planted over only a limited span of time. This is because most cold-tolerant crops like peas actually do much better in cooler weather than hot weather. Thus, they must be planted late enough to escape severe cold, but early enough to avoid higher summer temperatures. Many vegetables that do poorly when planted in late spring, because of hot summer weather, can be planted in late summer to capitalize on the cooler weather of fall. Table 7-5 summarizes the earliest planting, or transplanting, dates for many vegetables, and also gives the *range of dates* for successful spring planting. The best time for planting is on or near the earlier date for both cool- and warm-season crops. Can you explain why this is so?

Determining Dates for Fall Planting

The period for fall plantings in your locality can be ascertained much as for spring planting. The fall gardening season represents roughly the interval between July 25 and the first 32°F temperature of fall (Fig. 6-13). Compare that interval to the number of days from planting to harvest for the vegetables listed in Table 7-2. If the interval is long enough, your fall garden should be successful. Keep in mind that some crops can tolerate light freezes, others hard freezes, but that for all vegetables, growth slows as the season progresses.

HOW TO PLANT

In times past the vast majority of North American gardens were planted as single-row plantings (with some hill planting), much as is done in large-scale agricultural operations. In re-

cent years, however, the press of population and shortage of gardening space has brought challenges to this time-honored system. Indeed, a major strategy of modern-day gardening is aimed at maximizing the yield per square foot of garden area by the use of innovative planting and cropping techniques. Before we examine the actual planting procedures used for seeds and transplants it is essential that we explore briefly some of these space-stretching techniques.

Planting and Cropping Patterns

The major planting and cropping patterns of use to home gardeners are the following:

Single-row Planting

The method where you make an indentation—a furrow or trench—in the soil and plant in a single line (Fig. 7-14a). This is a method used widely in large-scale agriculture operations and in gardening. Single-row planting lends itself well to the use of power equipment, but is less space intensive than many other methods.

Double-row Planting

A modification of single-row planting, where two rows of vegetables are planted in close proximity and separated from other two-row units by wider spacings (Fig. 7-14b). Double-row planting is used commercially in growing onion, root crops, and pineapple, and enjoys limited use by gardeners. Like single-row planting, this system lends itself to the use of power equipment, while at the same time allowing more intensive cropping. The system is often used commercially in conjunction with furrow irrigation.

Wide-row Planting

A relatively new, innovative planting approach where seed are broadcast in rows 10

TABLE 7-5 Earliest Dates, and Range of Dates, for Safe Spring Planting of Vegetables in the Open

Crop	Planting dates for localities in which average date of last freeze is						
	Jan. 30	Feb. 8	Feb. 18	Feb. 28	Mar. 10	Mar. 20	Mar. 30
Asparagus[a]					Jan. 1–Mar. 1	Feb. 1–Mar. 10	Feb. 15–Mar. 20
Beans, lima	Feb. 1–Apr. 15	Feb. 10–May 1	Mar. 1–May 1	Mar. 15–June 1	Mar. 20–June 1	Apr. 1–June 15	Apr. 15–June 20
Beans, snap	Feb. 1–Apr. 1	Feb. 1–May 1	Mar. 1–May 1	Mar. 10–May 15	Mar. 15–May 15	Mar. 15–May 25	Apr. 1–June 1
Beet	Jan. 1–Mar. 15	Jan. 10–Mar. 15	Jan. 20–Apr. 1	Feb. 1–Apr. 15	Feb. 15–June 1	Feb. 15–May 15	Mar. 1–June 1
Broccoli, sprouting[a]	Jan. 1–30	Jan. 1–30	Jan. 15–Feb. 15	Feb. 1–Mar. 1	Feb. 15–Mar. 15	Feb. 15–Mar. 15	Mar. 1–20
Brussels sprouts[a]	Jan. 1–30	Jan. 1–30	Jan. 15–Feb. 15	Feb. 1–Mar. 1	Feb. 15–Mar. 15	Feb. 15–Mar. 15	Mar. 1–20
Cabbage[a]	Jan. 1–15	Jan. 1–Feb. 10	Jan. 1–Feb. 25	Jan. 15–Feb. 25	Jan. 25–Mar. 1	Feb. 1–Mar. 1	Feb. 15–Mar. 10
Cabbage, Chinese	(b)	(b)	(b)	(b)	(b)	(b)	(b)
Carrot	Jan. 1–Mar. 1	Jan. 1–Mar. 1	Jan. 15–Mar. 1	Feb. 1–Mar. 1	Feb. 10–Mar. 15	Feb. 15–Mar. 20	Mar. 1–Apr. 10
Cauliflower[a]	Jan. 1–Feb. 1	Jan. 1–Feb. 1	Jan. 10–Feb. 10	Jan. 20–Feb. 20	Feb. 1–Mar. 1	Feb. 10–Mar. 10	Feb. 20–Mar. 20
Celery and celeriac	Jan. 1–Feb. 1	Jan. 10–Feb. 10	Jan. 20–Feb. 20	Feb. 1–Mar. 1	Feb. 20–Mar. 20	Mar. 1–Apr. 1	Mar. 15–Apr. 15
Chard	Jan. 1–Apr. 1	Jan. 10–Apr. 1	Jan. 20–Apr. 15	Feb. 1–May 1	Feb. 15–May 15	Feb. 20–May 15	Mar. 1–May 25
Chervil and chives	Jan. 1–Feb. 1	Jan. 1–Feb. 1	Jan. 1–Feb. 1	Jan. 15–Feb. 15	Feb. 1–Mar. 1	Feb. 10–Mar. 10	Feb. 15–Mar. 15
Chicory, witloof					June 1–July 1	June 1–July 1	June 1–July 1
Collards[a]	Jan. 1–Feb. 15	Jan. 1–Feb. 15	Jan. 1–Mar. 15	Jan. 15–Mar. 15	Feb. 1–Apr. 1	Feb. 15–May 1	Mar. 1–June 1
Cornsalad	Jan. 1–Feb. 15	Jan. 1–Feb. 15	Jan. 1–Mar. 15	Jan. 1–Mar. 1	Jan. 1–Mar. 1	Jan. 1–Mar. 15	Jan. 15–Mar. 15
Corn, sweet	Feb. 1–Mar. 15	Feb. 10–Apr. 1	Feb. 20–Apr. 15	Mar. 1–Apr. 15	Mar. 10–Apr. 15	Mar. 15–May 1	Mar. 25–May 15
Cress, upland	Jan. 1–Feb. 1	Jan. 1–Feb. 15	Jan. 15–Feb. 15	Feb. 1–Mar. 1	Feb. 10–Mar. 15	Feb. 20–Mar. 15	Mar. 1–Apr. 1
Cucumber	Feb. 15–Apr. 15	Feb. 15–Apr. 1	Feb. 15–Apr. 15	Mar. 10–Apr. 15	Mar. 15–Apr. 15	Apr. 1–May 1	Apr. 10–May 15
Eggplant[a]	Feb. 1–Mar. 15	Feb. 10–Mar. 15	Feb. 20–Apr. 15	Mar. 10–Apr. 15	Mar. 15–Apr. 15	Apr. 1–May 1	Apr. 15–May 15
Endive	Jan. 1–Mar. 1	Jan. 1–Mar. 1	Jan. 15–Mar. 1	Feb. 1–Mar. 1	Feb. 15–Mar. 15	Mar. 1–Apr. 1	Mar. 10–Apr. 10
Fennel, Florence	Jan. 1–Mar. 1	Jan. 1–Mar. 1	Jan. 15–Mar. 1	Feb. 1–Mar. 1	Feb. 15–Mar. 15	Mar. 1–Apr. 1	Mar. 10–Apr. 10
Garlic	(b)	(b)	(b)	(b)	(b)	Feb. 1–Mar. 1	Feb. 10–Mar. 10
Horseradish[a]	(b)	(b)	(b)	(b)	(b)	(b)	Mar. 1–Apr. 1

192

Kale	Jan. 1–Feb. 1	Jan. 10–Feb. 1	Jan. 20–Feb. 10	Feb. 1–20	Feb. 10–Mar. 1	Feb. 20–Mar. 10
	Mar. 1–20					
Kohlrabi	Jan. 1–Feb. 1	Jan. 10–Feb. 1	Jan. 20–Feb. 10	Feb. 1–20	Feb. 10–Mar. 1	Feb. 20–Mar. 10
	Mar. 1–Apr. 1					
Leek	Jan. 1–Feb. 1	Jan. 1–Feb. 1	Jan. 1–Feb. 15	Jan. 15–Feb. 15	Jan. 25–Mar. 1	Feb. 1–Mar. 1
	Feb. 15–Mar. 15					
Lettuce, head[a]	Jan. 1–Feb. 1	Jan. 1–Feb. 1	Jan. 1–Feb. 15	Jan. 15–Feb. 15	Feb. 1–20	Feb. 15 Mar. 10
	Mar. 1–20					
Lettuce, leaf	Jan. 1–Feb. 1	Jan. 1–Feb. 1	Jan. 1–Mar. 15	Jan. 1–Apr. 1	Jan. 15–Apr. 1	Feb. 1–Apr. 1
	Feb. 15–Apr. 15					
Muskmelon	Feb. 15–Mar. 15	Feb. 15–Apr. 1	Feb. 15–Apr. 15	Mar. 15–Apr. 15	Mar. 15–Apr. 15	Apr. 1–May 1
	Apr. 10–May 15					
Mustard	Jan. 1–Mar. 1	Jan. 1–Mar. 1	Feb. 15–Apr. 15	Feb. 10–Mar. 15	Feb. 10–Mar. 15	Feb. 20–Apr. 1
	Mar. 1–Apr. 15					
Okra	Feb. 15–Apr. 1	Feb. 15–Apr. 15	Mar. 1–June 1	Mar. 10–June 1	Mar. 20–June 1	Apr. 1–June 15
	Apr. 10–June 15					
Onion[a]	Jan. 1–15	Jan. 1–15	Jan. 1–15	Jan. 1–15	Jan. 15–Feb. 15	Jan. 15–Feb. 15
	Feb. 15–Mar. 15					
Onion, seed	Jan. 1–15	Jan. 1–15	Jan. 1–15	Jan. 1–Feb. 15	Feb. 1–Mar. 1	Feb. 10–Mar. 10
	Feb. 20–Mar. 15					
Onion, sets	Jan. 1–15	Jan. 1–15	Jan. 1–15	Jan. 1–Mar. 1	Jan. 15–Mar. 10	Feb. 1–Mar. 20
	Feb. 15–Mar. 20					
Parsley	Jan. 1–30	Jan. 1–30	Jan. 1–30	Jan. 15–Mar. 1	Jan. 15–Mar. 1	Feb. 15–Mar. 15
	Mar. 1–Apr. 1					
Parsnip			Jan. 1–Feb. 1	Jan. 15–Feb. 15	Jan. 15–Mar. 1	Feb. 1–Mar. 1
	Mar. 1–Apr. 1					
Peas, garden	Jan. 1–Feb. 15	Jan. 1–Feb. 15	Jan. 1–Mar. 1	Jan. 15–Mar. 1	Jan. 15–Mar. 1	Feb. 1–Mar. 10
	Feb. 10–Mar. 20					
Peas, black-eye	Feb. 15–May 1	Feb. 15–May 15	Mar. 1–June 15	Mar. 10–June 20	Mar. 10–June 20	Apr. 1–July 1
	Apr. 15–July 1					
Pepper[a]	Feb. 1–Apr. 1	Feb. 1–Apr. 1	Mar. 1–May 1	Mar. 15–May 1	Mar. 15–May 1	Apr. 1–June 1
	Apr. 15–June 1					
Potato	Jan. 1–Feb. 15	Jan. 1–Feb. 15	Jan. 15–Mar. 1	Jan. 15–Mar. 1	Jan. 15–Mar. 1	Feb. 1–Mar. 1
	Feb. 20–Mar. 20					
Radish	Jan. 1–Apr. 1	Jan. 1–Apr. 1	Jan. 1–Apr. 1	Jan. 1–Apr. 15	Jan. 1–Apr. 15	Jan. 20–May 1
	Feb. 15–May 1					
Rhubarb[a]							
Salsify	Jan. 1–Feb. 1	Jan. 1–Feb. 1	Jan. 15–Feb. 20	Jan. 15–Feb. 15	Jan. 15–Mar. 1	Jan. 15–Mar. 1
	Feb. 1–Mar. 1					
Shallot	Jan. 1–Feb. 1	Jan. 1–Feb. 1	Jan. 1–Feb. 20	Jan. 15–Feb. 20	Jan. 15–Mar. 1	Jan. 15–Mar. 10
	Mar. 1–15					
Sorrel	Jan. 1–Mar. 1	Jan. 1–Mar. 1	Jan. 15–Mar. 1	Jan. 15–Mar. 1	Jan. 15–Mar. 1	Feb. 1–Mar. 10
	Feb. 15–Mar. 15					
Soybean	Mar. 1–June 30	Mar. 1–June 30	Mar. 10–June 30	Mar. 20–June 30	Mar. 20–June 30	Apr. 10–June 30
	Apr. 20–June 30					
Spinach	Jan. 1–Feb. 15	Jan. 1–Feb. 15	Jan. 1–Mar. 1	Jan. 1–Mar. 1	Jan. 1–Mar. 1	Jan. 15–Mar. 15
	Feb. 1–Mar. 20					
Spinach, New Zealand	Feb. 1–Apr. 15	Feb. 1–Apr. 15	Mar. 1–Apr. 15	Mar. 15–May 15	Mar. 15–May 15	Mar. 20–May 15
	Apr. 10–June 1					
Squash, summer	Feb. 15–Apr. 15	Feb. 15–Apr. 15	Mar. 1–Apr. 15	Mar. 15–May 15	Mar. 15–May 15	Mar. 20–May 15
	Apr. 10–June 1					
Sweet potato	Feb. 15–May 15	Mar. 1–May 15	Mar. 20–June 1	Mar. 20–June 1	Mar. 20–June 1	Apr. 1–June 1
	Apr. 20–June 1					
Tomato	Feb. 1–Apr. 1	Feb. 20–Apr. 10	Mar. 1–Apr. 20	Mar. 20–May 10	Mar. 20–May 10	Apr. 10–June 1
	Apr. 20–June 1					
Turnip	Jan. 1–Mar. 1	Jan. 1–Mar. 1	Jan. 10–Mar. 1	Jan. 20–Mar. 1	Jan. 20–Mar. 1	Feb. 10–Mar. 10
	Feb. 20–Mar. 20					
Watermelon	Feb. 15–Mar. 15	Feb. 15–Apr. 1	Feb. 15–Apr. 15	Mar. 15–Apr. 15	Mar. 15–Apr. 15	Apr. 1–May 1
	Apr. 10–May 15					

TABLE 7-5 (continued)

Crop	Planting dates for localities in which average date of last freeze is						
	Apr. 10	Apr. 20	Apr. 30	May 10	May 20	May 30	June 10
Asparagus[a]	Mar. 10–Apr. 10	Mar. 15–Apr. 15	Mar. 20–Apr. 15	Mar. 10–Apr. 30	Apr. 20–May 15	May 1–June 1	May 15–June 1
Beans, lima	Apr. 1–June 30	May 1–June 20	May 15–June 15	May 25–June 15			
Beans, snap	Apr. 10–June 30	Apr. 25–June 30	May 10–June 30	May 10–June 30	May 15–June 30	May 25–June 15	
Beet	Mar. 10–June 1	Mar. 20–June 1	Apr. 1–June 15	Apr. 15–June 15	Apr. 25–June 15	May 1–June 15	May 15–June 15
Broccoli, sprouting[a]	Mar. 15–Apr. 15	Mar. 25–Apr. 20	Apr. 1–May 1	Apr. 15–June 1	May 1–June 15	May 10–June 10	May 20–June 10
Brussels sprouts[a]	Mar. 15–Apr. 15	Mar. 25–Apr. 20	Apr. 1–May 1	Apr. 15–June 1	May 1–June 15	May 10–June 10	May 20–June 10
Cabbage[a]	Mar. 1–Apr. 1	Mar. 15–Apr. 1	Mar. 15–Apr. 10	Apr. 1–May 15	May 1–June 15	May 10–June 15	May 20–June 1
Cabbage, Chinese	(b)	(b)	(b)	Apr. 1–May 15	May 1–June 15	May 10–June 15	May 20–June 1
Carrot	Mar. 10–Apr. 20	Apr. 1–May 15	Apr. 10–June 1	Apr. 20–June 15	May 1–June 1	May 10–June 1	May 20–June 1
Cauliflower[a]	Mar. 1–20	Mar. 15–Apr. 20	Apr. 10–May 10	Apr. 15–May 15	May 10–June 15	May 20–June 1	June 1–15
Celery and celeriac	Apr. 1–20	Apr. 10–May 1	Apr. 15–May 1	Apr. 20–June 15	May 10–June 15	May 20–June 1	June 1–15
Chard	Mar. 15–June 15	Apr. 1–June 15	Apr. 15–June 15	Apr. 20–June 15	Apr. 15–May 15	May 1–June 1	May 15–June 1
Chervil and chives	Mar. 1–Apr. 1	Mar. 10–Apr. 10	Mar. 20–Apr. 20	Apr. 1–May 1	Apr. 15–May 15	May 1–June 1	May 15–June 1
Chicory, witloof	June 10–July 1	June 15–July 1	June 15–July 1	June 1–20	June 1–15	June 1–15	June 1–15
Collards[a]	Mar. 1–June 1	Mar. 10–June 1	Apr. 1–June 1	Apr. 15–June 1	May 1–June 1	May 10–June 1	May 20–June 1
Cornsalad	Feb. 1–Apr. 1	Feb. 15–Apr. 15	Mar. 1–Apr. 1	Apr. 1–June 1	Apr. 15–June 1	May 1–June 15	May 15–June 15
Corn, sweet	Apr. 10–June 1	Apr. 25–June 15	May 10–June 15	May 10–June 1	May 15–June 1	May 20–June 1	
Cress, upland	Mar. 10–Apr. 15	Mar. 20–May 1	Apr. 10–May 10	Apr. 20–May 20	May 1–June 1	May 15–June 1	May 15–June 15
Cucumber	Apr. 20–June 1	May 1–June 15	May 15–June 15	May 20–June 15	June 1–15		
Eggplant[a]	May 1–June 1	May 10–June 1	May 15–June 10	May 20–June 15	June 1–15		
Endive	Mar. 15–Apr. 15	Mar. 25–Apr. 15	Apr. 1–May 1	Apr. 15–May 15	May 1–30	May 1–30	May 15–June 1
Fennel, Florence	Mar. 15–Apr. 15	Mar. 25–Apr. 15	Apr. 1–May 1	Apr. 15–May 15	May 1–30	May 1–30	May 15–June 1
Garlic	Feb. 20–Mar. 20	Mar. 10–Apr. 1	Mar. 15–Apr. 15	Apr. 1–May 1	Apr. 15–May 15	May 1–30	May 15–June 1
Horseradish[a]	Mar. 10–Apr. 10	Mar. 20–Apr. 20	Apr. 1–30	Apr. 15–May 15	Apr. 20–May 20	May 1–30	May 15–June 1
Kale	Mar. 10–Apr. 1	Mar. 20–Apr. 10	Apr. 1–20	Apr. 10–May 1	Apr. 20–May 10	May 1–30	May 15–June 1
Kohlrabi	Mar. 10–Apr. 10	Mar. 20–May 1	Apr. 1–May 10	Apr. 10–May 15	Apr. 20–May 20	May 1–30	May 15–June 1
Leek	Mar. 1–Apr. 1	Mar. 15–Apr. 15	Apr. 1–May 1	Apr. 15–May 15	May 1–May 20	May 1–15	May 1–15

Lettuce, head[a]	Mar. 10–Apr. 1	Mar. 20–Apr. 15	Apr. 1–May 1	Apr. 15–May 15	May 1–June 30	May 10–June 30	May 20–June 30
Lettuce, leaf	Mar. 15–May 15	Mar. 20–May 15	Apr. 1–June 1	Apr. 15–June 15	Apr. 15–June 15	May 10–June 30	May 20–June 30
Muskmelon	Apr. 20–June 1	May 1–June 15	May 15–June 15	June 1–June 15			
Mustard	Mar. 10–Apr. 20	Mar. 20–May 1	Apr. 1–May 10	Apr. 15–June 1	May 1–June 30	May 10–June 30	May 20–June 30
Okra	Apr. 20–June 15	May 1–June 1	May 10–June 1	May 20–June 10	May 20–June 30	June 1–20	
Onion[a]	Mar. 1–Apr. 1	Mar. 15–Apr. 10	Apr. 1–May 1	Apr. 10–May 1	Apr. 20–May 1	May 1–30	May 10–June 10
Onion, seed	Mar. 1–Apr. 1	Mar. 15–Apr. 1	Mar. 15–Apr. 15	Apr. 1–May 1	Apr. 20–May 15	May 1–30	May 10–June 10
Onion, sets	Mar. 1–Apr. 1	Mar. 10–Apr. 1	Mar. 10–Apr. 10	Apr. 10–May 1	Apr. 20–May 15	May 1–30	May 10–June 10
Parsley	Mar. 10–Apr. 10	Mar. 20–Apr. 20	Apr. 1–May 1	Apr. 15–May 15	May 1–June 1	May 10–June 1	May 20–June 10
Parsnip	Mar. 10–Apr. 10	Mar. 20–Apr. 20	Apr. 1–May 1	Apr. 15–May 15	May 1–June 1	May 10–June 1	May 20–June 10
Peas, garden	Feb. 20–Mar. 20	Mar. 10–Apr. 10	Mar. 20–May 1	Apr. 1–May 15	Apr. 15–June 1	May 1–June 15	May 10–June 15
Peas, black-eye	May 1–July 1	May 10–June 15	May 15–June 1				
Pepper[a]	May 1–June 1	May 10–June 1	May 15–June 10	May 20–June 10	May 25–June 15	June 1–15	
Potato	Mar. 10–Apr. 1	Mar. 15–Apr. 10	Mar. 20–May 10	Apr. 1–June 1	Apr. 15–June 15	May 1–June 15	May 15–June 1
Radish	Mar. 1–May 1	Mar. 10–May 10	Mar. 20–May 10	Apr. 1–June 1	Apr. 15–June 1	May 1–June 15	May 15–June 1
Rhubarb[a]	Mar. 1–Apr. 1	Mar. 10–Apr. 10	Mar. 20–Apr. 15	Apr. 1–May 1	Apr. 15–May 10	May 1–June 1	May 15–June 1
Rutabaga			May 1–June 1	May 1–20	May 1–20	May 10–20	May 20–June 1
Salsify	Mar. 10–Apr. 15	Mar. 20–May 1	Apr. 1–May 15	Apr. 15–June 1	May 1–June 1	May 10–June 1	May 20–June 1
Shallot	Mar. 1–Apr. 1	Mar. 15–May 1	Apr. 1–May 1	Apr. 1–May 1	Apr. 20–May 10	May 1–June 1	May 20–June 1
Sorrel	Mar. 1–Apr. 15	Mar. 15–May 1	Apr. 1–May 15	Apr. 15–June 1	May 1–June 1	May 1–June 1	May 10–June 1
Soybean	May 1–June 30	May 10–June 20	May 15–June 15	May 25–June 10	May 25–June 10	May 20–June 10	May 20–June 10
Spinach	Feb. 15–Apr. 1	Mar. 1–Apr. 15	Mar. 20–Apr. 20	Apr. 1–June 15	Apr. 10–June 15	Apr. 20–June 15	May 1–June 15
Spinach, New Zealand	Apr. 20–June 1	May 1–June 15	May 1–June 15	May 10–June 15	May 20–June 15	June 1–15	
Squash, summer	Apr. 20–June 1	May 1–30	May 1–30	May 10–June 10	May 20–June 15	June 1–20	June 10–20
Sweet potato	May 1–June 1	May 20–June 10	May 20–June 10		May 25–June 15		
Tomato	Apr. 20–June 1	May 5–June 10	May 10–June 15	May 15–June 10	May 25–June 15	June 5–20	June 15–30
Turnip	Mar. 1–Apr. 1	Mar. 10–Apr. 1	Mar. 20–May 1	Apr. 1–June 1	Apr. 15–June 1	May 1–June 15	May 15–June 15
Watermelon	Apr. 20–June 1	May 1–June 15	May 15–June 15	June 15–July 1	June 15–July 1		

Courtesy of U.S. Department of Agriculture.

[a]Plants.

[b]Generally fall planted.

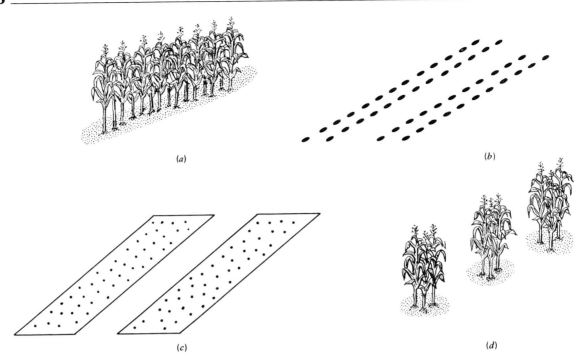

FIGURE 7-14 Some planting patterns: single row (*a*); double row (*b*); wide row (*c*); and "hills" (*d*).

or more inches wide, separated by pathways (Fig. 7-14*c*). To use this method prepare a smooth seedbed and mark off rows (a steel garden rake works well for marking because it is about 10 inches wide). Then, broadcast seed in the row areas and cover. When seedlings are one-half to one inch tall, thin by dragging an iron rake slowly across the rows. This system is useful for gardeners wishing to crop intensively, as several times more produce can be grown than in single rows using the same amount of space. The system has other advantages too. The closely growing plants in wide rows shade the ground creating a shade mulch that restricts the growth of weeds and conserves moisture. Crops that are grown successfully using wide rows include beans, beets, carrots, chard, Chinese cabbage, collards, dill, lettuce, onion (both from seed

and from sets), peas, rutabaga, spinach, and turnip.

Hill Planting

A method of planting where plants are grown in clusters, known as "hills" (Fig. 7-14d). Space-consuming crops such as pole beans, corn, cucumbers, melons, and squash are often planted this way allowing their roots to radiate out from a central point. If plants are grown from seed, sow 7 to 10 seeds in a 12- to 18-inch circle, and thin to leave three or four of the best seedlings.

Mound Planting

A method of planting similar to hill planting, but where plants are grown in mounds. This method works well for vegetables such as the cucurbits that require large quantities of water but may be affected adversely if water covers their crowns.

Staggered Planting

A method of planting where plants are staggered in consecutive rows to create an equal spacing in all directions. Use of the staggered arrangement allows more plants to be grown in a given area without decreasing the space between plants (Fig. 7-15). By staggered planting, and by reducing spacing between plants to a minimum, an enormous number of plants can be grown in an area no larger than 100 square feet (Table 7-6). This planting method offers great potential for those wishing to garden intensively.

Interplanting or Intercropping

A cropping method where two crops are grown in a given area at the same time, either by alternating plants within rows or growing them in separate rows (Fig. 7-16). The principle behind interplanting is to combine a small, quick-growing vegetable with a larger, slow-growing crop. In essence, the quick-growing vegetable will mature and be harvested before the slow-growing crop needs the space, thus allowing you to raise two crops in the same area without crowding either. Intercropping is not exactly a new idea. The American Indians grew mixed plantings of corn and beans long before the Pilgrims stepped ashore at Plymouth. Indeed, in early gardening, intercropping of a less precise nature was a common and widespread practice. Intercropping is, of course, the rule in nature, and allowed early hunting-gathering people to take advantage of a rich and varied diet.

Intercropping can be regarded as a form of companion planting. In **companion plantings** plants that complement each other are grown together. For example, one crop may help repel insects harmful to another crop. In successful intercropping the plants that are grown together also complement each other, in growth habit and time to maturity, and this allows for effective space utilization. Many plant combinations work well together, whereas others do not, often for reasons that are poorly understood. The following are some combinations that have proven successful:

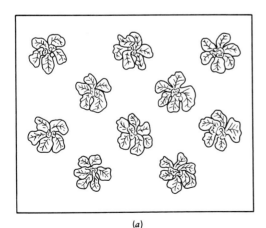 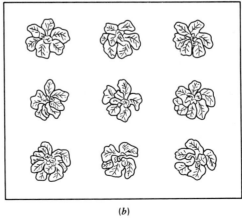

(a)　　　　　　　　　　　(b)

FIGURE 7-15 Through staggered planting (a) more plants can be grown in a given area than when planted in even rows (b). Significantly, plant density is increased without decreasing the spacing between plants.

TABLE 7-6 Intensive Planting Guide

Plant spacing (in.)	Planting centers (per 100 sq ft)	Plant spacing requirements
1	14,400	Radish
2	3,600	Carrot
3	1,600	Beet, onion, peas (bush)
4	900	Beans (bush), mustard greens, spinach, turnips
6	400	Peas (pole), rutabaga
8	225	Beans (pole), lettuce (leaf)
10	144	Lettuce (head)
12	100	Broccoli, cabbage, corn, peppers

From Ray Wolf, "Garden the French-intensive Way," Organic Gardening and Farming, *Emmaus, Pa. 18049, February 1976. Reprinted by permission of Rodale Press Inc. Copyright 1976.*

FIGURE 7-16 Intercropping. Here bronze and green leaf lettuce are interplanted between onions grown for bulbs. (Courtesy of National Garden Bureau, Sycamore, Illinois.)

Pea and carrot
Cabbage and lettuce
Carrot and radish
Onion and radish
Parsnip and radish
Cabbage and radish
Cabbage and onion sets

Corn and bean
Corn and spinach
Corn and lettuce
Tomato and spinach
Tomato and radish
Tomato and lettuce
Chard and pepper

A number of combinations not listed also work well. Simply remember that your goal is to combine a small, quick-growing crop with a slower-growing crop that requires wider spacing. Through experimentation you can determine those combinations best suited to your situation.

Succession Planting

A widely used cropping system where one crop is followed by another in a given space. Succession planting or cropping also refers to situations where a particular crop is planted at time intervals to provide a continuous supply of a vegetable throughout the season (Fig. 7-17). In succession cropping gardeners are taking advantage of the fact that vegetables take different lengths of time to mature, and also that they grow best at particular periods

FIGURE 7-17 Succession planting. Beans planted at 10-day intervals will provide a continuous supply throughout the growing season.

of the growing season. Succession cropping is often referred to as **double cropping** or **triple cropping,** indicating, and correctly so, that by this method it is possible to harvest two, or even three, crops a year in a space. When we make fall plantings of the same cool-season vegetables planted in early spring we are taking advantage of one aspect of succession cropping. Let us look at others.

Through succession cropping we can more effectively use our garden space, as is illustrated in Table 7-7. For example, one crop of carrot follows another in the same space, and the same is true for cabbage. Note that the second plantings of these cool weather crops were not made until the hot days of summer were past. You may question at this point if it was wise to follow carrot with carrot, or cabbage with cabbage: Should not rotation have been practiced? No, not necessarily! Rotation is a critical feature to employ from year to year, but is generally not required within a given season. Note also that two plantings of carrot, chard, zucchini, green beans, and peas were made, and three plantings of corn were made, with the later plantings being only a month or less after the first. In each case this procedure spread the harvest over a longer period of time, making for more effective crop utilization. Similar results could have been achieved using cultivars of a crop having different times to maturity.

The succession cross-reference chart illustrated is for southwestern Illinois, but gardeners everywhere who use this cropping pattern should construct a similar chart. Once established, the chart will be a valuable source of information. In principle, any "in" crop can follow any "out" crop if the times do not overlap. Thus, the sequence carrot-Swiss chard-cabbage would be possible, but the sequence cabbage-okra-zucchini would not work. It is important to remember, however, that with intense cropping fertility needs are greater. Before planting a second crop in an area add ¾ to 1½ pounds of 5-10-10 fertilizer, or its organic equivalent, per 100 square feet, and work into the soil.

Vertical Planting

Plantings, such as vegetables that climb or can be trellised and vegetables planted in vertical containers, constitute vertical plantings (Fig. 7-18). This planting method is discussed in detail in Chapter 14.

In summary, we have discussed several planting systems. A gardener may use a number of these in a gardening program to maximize yields. Through intensive efforts large quantities of high-quality vegetables can be grown in an area of limited size. Not all gardeners, however, are concerned with obtaining maximum yields from their garden. Indeed, few of us ever obtain maximum yields; but for all of us gardening should mean both food and fun.

Sowing Seed in the Garden

Seed Sources

The seed is a crucial stage in the life cycle of vegetable crops; without seed, most species

TABLE 7-7 Succession Cross-reference Chart for Southwest Illinois

	March	April	May	June	July	Aug.	Sept.
Beets	IN ▬▬ OUT						IN ▪
Carrots	IN ▬▬ OUT	IN ▬▬ OUT			IN ▬▬▬▬		
Swiss chard		IN ▬▬ OUT	IN ▬▬▬ OUT				
Lettuce	IN ▬▬▬ OUT						
Peppers			IN ▬▬▬▬▬▬▬▬▬▬▬				
Tomatoes			IN ▬▬▬▬▬▬▬▬▬▬▬				
Pumpkins				IN ▬▬▬▬▬▬▬▬▬			
Cabbage	IN ▬▬ OUT IN ▬▬▬▬▬▬▬▬▬						
Okra			IN ▬▬▬ OUT				
Zucchini			IN ▬▬▬ OUT	IN ▬▬▬ OUT			
Corn		IN ▬▬▬▬▬ OUT	IN ▬▬▬▬ OUT	IN ▬▬▬▬▬▬ OUT			
Broccoli	IN ▬▬▬▬ OUT						
Radishes	IN ▪ OUT	IN ▪ OUT					
Green beans			IN ▬▬▬▬ OUT	IN ▬▬▬▬▬ OUT			
Peas	IN ▬▬▬ OUT	IN ▬▬▬▬ OUT					

From Ellen E. Jantzen, "A Successful Planting Chart," Organic Gardening and Farming, *Emmaus, Pa., 18049, February 1977. Reprinted by permission of Rodale Press Inc., Copyright 1977.*

would not survive. Thus, a reliable supply of clean, viable, disease-free seed is of great importance. Gardeners now have available seed from several sources, including that from their own gardens, and seed offered commercially, either on racks at retail outlets, or through seed catalogs (Appendix E). In many cases it does not pay to save seed from the home garden, except in instances where a particular cultivar is not available from commercial sources. Even then, the gardener must be knowledgeable about the crops from which he or she plans to save seed. For example, the gardener needs to know such things as the breeding characteristics of the crop in question, proper time for harvesting seed, how to harvest and clean seed, and how to best store harvested seed.

FIGURE 7-18 Vertical plantings make effective use of limited space. (Courtesy of the National Garden Bureau, Sycamore, Illinois.)

Sowing Seed

Sowing seed is not difficult, but adherence to the following steps will assure greater success.

- Start with a well-prepared seedbed and high-quality seed. The only tools needed are those to mark off straight rows (two stakes and a roll of twine usually) and an implement to make furrows for planting, or **drilling,** seed.
- Mark off straight rows at recommended distances (Table 10-3), and make furrows for the seed. Shallow furrows can be made using a hoe handle or stick (Fig. 7-19a), and

deeper furrows using the edge of the hoe blade (Fig. 7-19b). Some very small seed (one-sixteenth of an inch or smaller) can be successfully sown right on the seedbed surface and covered with a paper-thin mulch layer.

- Sow seed in the row using the recommended spacing and depth (Table 10-3). As a general rule, seed should be buried to a depth three to four times their average diameter: some variability is permissible with larger seed, such as that of bean, corn, and pea. Small seed, if sown too deeply, may exhaust their food reserves before reaching the soil surface. In sandy soils or in dry weather seed can be sown a little deeper than usual.
- Cover seed to the level of the surrounding soil. Be sure to firm—not pack—the soil over the seed by gently tamping the soil with your hands, or with the back of an upright hoe or rake. Then, the garden should be watered, unless the soil was very moist to begin with. Excessive watering, or rainfall, can cause your soil to crust, thereby retarding the entrance of air into the soil and preventing seedling emergence. To prevent this, watch your garden closely until the seedlings have emerged, and break up any crust that may form. Covering the soil with a thin layer of mulch after planting is a practice that greatly reduces, or prevents, soil crusting.
- Finally, it is a good idea to attach empty seed packets to the stakes at their respective row ends. If you record planting dates on these packets you will have an additional record not only of what you have planted, but also when you planted. These packets are then available for handy reference right in the garden when you have questions on the time till seedling emergence, how to thin seedlings, or other cultural manipulations.

Seed Packets Versus Seed Tapes

One dilemma that confronts gardeners at planting time is the use of seed in packets

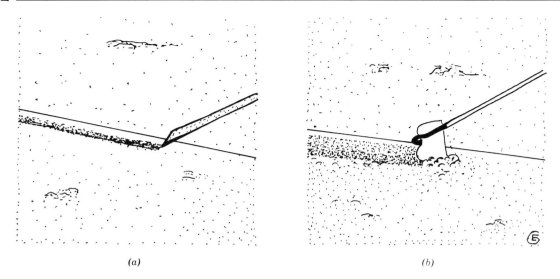

FIGURE 7-19 Making furrows. Shallow furrows can be made using a stick or hoe handle (*a*), and deeper furrows using the hoe edge (*b*). (Courtesy of Esther McEvoy.)

versus seed on tapes. Seed tapes are available for a number of small-seeded vegetables that have very high seed germinability. While the tapes are a somewhat more expensive source of seed, their use assures uniform plant spacing and virtually eliminates the need for thinning. If you sow seed by hand, instead of using tapes, remember that there is a tendency to sow too thick, especially when the seed is small. This wastes seed, causes competition among seedlings, and creates major thinning chores. More desirable seed spacing can often be achieved by mixing small seed with soil and then spreading the mixture uniformly down the row. Also, pay attention to stated germination rates. If the rate is high, as is the case with radish, then seed can be placed at desired plant spacing, thus eliminating the need for thinning.

Planting by the Moon

Another area in which gardeners frequently have questions pertains to planting in relation to phases of the moon. For centuries the belief has persisted that yields will be better when crops are planted during certain "fruitful" days, or that yields will be depressed if the crops are planted during certain "barren" days. Nearly all of us can recall a grandparent stating opinions such as do not plant with the moon in the fourth quarter, or plant approximately two days prior to the full moon. The authors do not presume to understand all the influences of our moon, or planets, on plant behavior. Perhaps there are subtle influences that scientific investigations have not yet clarified. It has been our experience, however, that other variables—soil quality, abundance of nutrients, availability of water, and general garden care—have an enormous influence on garden yields. If cosmic forces can affect crop yields, this influence is vastly overshadowed by these other well-understood aspects of gardening.

Storing Garden Seed

Many times gardeners wonder if they should save garden seed, and if so how it can best be stored. Seed vary in their longevity naturally. Longevity is also influenced by storage conditions, especially temperature and seed moisture content. Reduced temperature invariably lengthens the storage life of seed. The situation with respect to moisture is less clear. Some short-lived seed lose viability if their moisture content becomes too low, but most medium and long-lived seed must be dry for long storage life. The profound influence of temperature and moisture on seed longevity are illustrated by the following statements:

- For each 10° drop in storage temperature the life of the seed is approximately doubled.
- For each 1 percent decrease in seed moisture the life of the seed is approximately doubled.

Experimentally, onion seed, a normally short-lived seed, loses its viability in one week when stored at 90°F with a 14 percent moisture content. If the seed is dried to 6 percent moisture, however, it retains high viability even after 20 years.

Storage Times

The approximate storage life for vegetable seed in **open storage** (storage without moisture or temperature control) in the eastern United States is as follows:

One year—sweet corn, onion, parsley, parsnip.

Two years—beet, pepper.

Three years—asparagus, bean, celery, carrot, lettuce, pea, spinach, tomato.

Four years—cabbage, cauliflower, eggplant, okra, pumpkin, radish, squash.

Five years—cucumber, endive, muskmelon, watermelon.

Storage Methods

To encourage more efficient storage of seed the National Garden Bureau asked Dr. James Harrington, a seed storage specialist at University of California, Davis, to devise an inexpensive method for storing leftover garden seed. The method he proposed involves putting seed in glass jars with a "desiccant" (Fig. 7-20), and storing in a refrigerator. The procedure is carried out in the following manner:

1. Unfold and lay out a stack of facial tissues.

2. Place two heaping tablespoons powdered milk on one corner. The milk must be from a freshly opened pouch or box to guarantee dryness.

3. Fold and roll the facial tissue to make a small pouch. Secure with tape or a rubber

FIGURE 7-20 Storing seed in the refrigerator in tightly closed glass jars containing a desiccant can extend their storage life significantly. (Courtesy of National Garden Bureau, Sycamore, Illinois.)

band. The tissue will prevent the milk from sifting out and will prevent seed packets from touching the moist desiccant.

4. Place the pouch in a widemouth jar and immediately drop in packets of leftover seeds.

5. Seal the jar tightly using a rubber ring to exclude moist air.

6. Store the jar in the refrigerator, not in the freezer.

7. Use seed as soon as possible. Discard and replace the desiccant once or twice yearly. Dried milk is "hygroscopic" and will quickly soak up moisture from the air when you open the bottle. Therefore, remove seed packets quickly, and recap the jar without delay.

The lower moisture content provided by the desiccant and the lower temperature of the refrigerator lengthens significantly the storage life of vegetable seed. Most garden seeds can also be stored in the home freezer. With proper storage we can slow the decline in vigor, thus extending—often for several years—the storage life of garden seed.

Using Transplants in the Garden

In our discussion of using transplants in home gardening we will explore the following topics: (1) why we use transplants, (2) obtaining transplants—the alternatives, and (3) moving transplants to the garden.

Why Use Transplants?

Since virtually all vegetables can be started from seed why should we bother with the more cumbersome approach of using transplants? There are several reasons why we may choose to use transplants. In more northern sections of the country it is difficult, or impossible, to grow many vegetables directly from seed because of the short growing season. Through the utilization of transplants it is possible to initiate plant growth indoors (or in a region of milder climate) several weeks before the crops could be started outdoors. Tomato seed normally cannot be sown outdoors until after the frost-free date. By sowing seed indoors several weeks earlier, large, ready-to-grow seedlings can be transplanted out at the same time seed would be sown, thus allowing them adequate time to maturity. In essence, through the use of transplants the gardener can add several weeks to the gardening season.

In some regions, slower-growing, cool-season crops benefit from the additional cool period growing time provided by use of transplants. And, by using transplants, a crop can be harvested earlier. This means not only is there a longer harvesting season but early harvest also frees a space for planting to another crop. Finally, for certain crops, such as celery, it is difficult to establish good stands from seed sown directly in the garden, whereas use of transplants assure a uniform, quality crop. Virtually all vegetables could be handled as transplants. In fact, some gardeners do plant flats of crops, such as corn and bean, to use in filling empty spaces in rows that may occur in outdoor plantings. In general, however, transplants should be used only in those situations where the advantages of use override the additional expenditure of time, effort, and money they entail.

Transplants are most commonly used for the solanaceous crops (eggplant, pepper, tomato), the cucurbits, or vine, crops (cucumber, melons, squash), members of the cabbage family (broccoli, cabbage, cauliflower, and so on), celery, and lettuce. A number of other vegetables are grown from transplants on a less frequent basis (Table 10-3). Growing vegetables from transplants is not a single process, but involves a sequence of processes, or steps. The gardener can employ any or all of these steps, ranging from starting seed indoors for future transplants to buying transplants at the garden store or supermarket. Let us look at the alternatives.

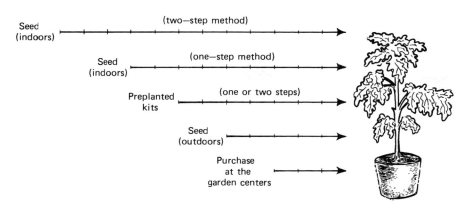

FIGURE 7-21 Some alternatives in starting vegetables from transplants. The length of arrows reflects the amount of time and effort required for each method.

Obtaining Transplants—The Alternatives

Transplants can be obtained in several ways, as is illustrated in Fig. 7-21. Each offers advantages and disadvantages of which the gardener needs to be aware.

Starting transplants indoors from seed. Starting transplants indoors from seed is the most time consuming and laborious method for getting transplants. On the other hand, this method offers the gardener a greater choice of cultivars than is usually available at garden centers, and a large number of transplants can be grown inexpensively. Gardeners starting seed indoors can use any one of several procedures, the most common of which are the one step and the two step methods (Fig. 7-22). Several variations of the two methods are possible. For example, seed can be sown in a tray or flat, as shown in Fig. 7-22, and then thinned to a spacing of three to four inches apart, rather than being moved to larger containers.

Gardeners frequently wonder how best to determine the time at which seed should be sown indoors to yield proper size seedlings at the correct time. With frost-sensitive plants such as tomato, timing is determined by counting backward from the frost-free date, as follows:

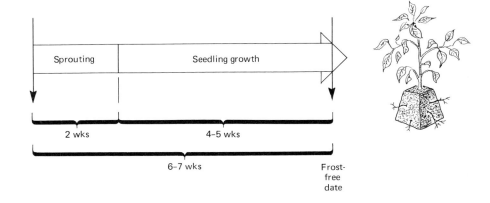

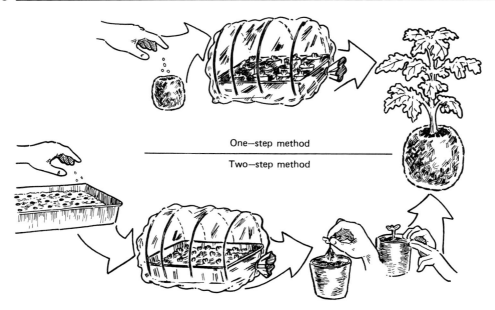

One—step method

Two—step method

FIGURE 7-22 Two common methods for starting plants indoors from seed. In the one-step method seed are sown directly into containers, cultured, and transplanted directly to the garden. In the two-step method seed are sown in flats, then moved to individual containers for further growth before transplanting to the garden. (Courtesy of Guerry Dean.)

By taking into consideration the time required for sprouting and subsequent seedling growth the planting time—six to seven weeks before the frost-free date—can be ascertained. Hardy seedlings, such as those of members of the cabbage group and lettuce, can withstand light frost and thus can be started and transplanted outdoors earlier. As a guide to beginning gardeners we present the approximate number of weeks needed for starting the most commonly used transplants indoors, as follows:

Cabbage family	four to six weeks
Cucurbits	two to four weeks
Lettuce (head)	four to five weeks
Onion	two to four weeks
Tomato family	six to seven weeks

It is important that gardeners use proper timing in starting plants indoors, because plants started too early give stunted plants, or plants that become too large and leggy. In either case future yields are adversely affected.

A multitude of growing media and containers are available for starting plants indoors. Vermiculite, or a mixture of vermiculite and peat moss, is probably the most widely used seed-starting medium. Since the nutrient value of these media is low, young seedlings must either be fed nutrients regularly, or transplanted to another medium of higher fertility. The synthetic soil mixes are also widely used, particularly as the potting mix in the two step method (Fig. 7-22). All of the above media are sterile when purchased, a matter of considerable importance in preventing damping-off. **Damping-off** is a seedling disease caused primarily by the fungus, *Phythium*. This, and other soil-borne fungi attack the succulent plant stem at the soil line. Stems shrivel at the point of infection and the seedlings fall over and quickly die. Once seedling stems are somewhat woody, damping-off is not a problem.

A wide range of containers is also available for starting seed indoors (Fig. 7-23). The Jiffy-7 pellet serves as an example, and illustrates strikingly the innovation occurring in container manufacturing (Fig. 7-24). Jiffy-7 pellets come in compressed form, and can be purchased in boxes containing 12, 24, 50, or more pellets. To prepare the unmoist pellet for planting, simply add water. Within minutes pellets expand to about seven times their

FIGURE 7-24 Jiffy-7 pellets in compressed, unmoistened form; after soaking in water; and with lettuce seedling ready to plant in the garden. (Courtesy of U.S.D.A.)

original size and emerge as net-covered pots of sterile, fertilized potting soil. Seed can be sown directly into these containers and seedlings grown until they are ready to plant in the garden, at which time you transplant the whole unit. Through use of containers of this sort vigorous, healthy seedlings can be grown with a minimum of effort. Many other kinds of containers, including peat pots, can normally be planted, pot and all, thus eliminating much of the shock associated with transplanting. The subject of containers and planting mixes is discussed in Chapter 14.

When starting seed indoors several environmental parameters must be considered. First, temperatures must be favorable for germination and seedling growth. Normal house temperatures of 70 to 72°F during the day, and 8 to 10 degrees lower at night, are adequate for these developmental processes. Second, once seedlings have emerged they need abundant light. This can be provided, at least in part, by placing the seedlings near a window with an unobstructed southern exposure or in an enclosed sun porch. Generally, however, supplemental light is required to produce stocky, vigorous seedlings. This can best be provided by growing seedlings under a bank of fluorescent lights left on 14 to 16 hours per

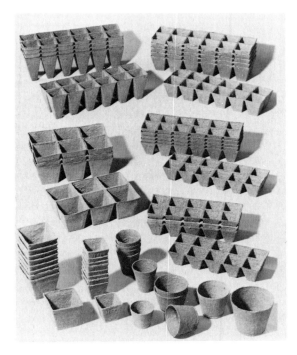

FIGURE 7-23 Some of the many containers available for starting plants indoors. (Courtesy of Jiffy Products, Chicago, Illinois.)

day. Third, the humidity around seedlings should be maintained at high levels. This can be accomplished by enclosing the seedlings in a plastic film. Finally, seedlings should be properly watered and fertilized. Details on the care of vegetables growing indoors are presented in Chapter 14.

Transplants from preplanted kits. Another relatively new and effective way to start vegetables indoors is through the use of preplanted kits offered for sale by several seedsmen. These kits are easy and convenient to use in that they come complete with individual growing compartments filled with a sterile growing medium, and seed of the cultivar in question. Seedlings grown in these kits may be thinned to one plant per growing compartment and grown to garden size. It is usually better, however, to give them more growing room by moving individual seedlings to larger containers after the first true leaf is formed (Fig. 7-25).

Starting "transplants" outdoors. Two additional alternatives in starting vegetables from transplants are suggested in Fig. 7-21 and need our attention. The first method—starting seed outdoors—is a valuable method that accomplishes the goals of transplanting, without actually using transplants. By this method seed is sown directly in place in the garden but protected by polyethylene seed row covers or homemade open-ended boxes. The boxes, for

FIGURE 7-25 Seedling at the proper size to move to a larger container. In moving, hold the seedling gently by one of its leaves, not by the stem. (Courtesy of U.S.D.A.)

example, can be placed over hill plantings, and glass plates placed over the boxes. After seedling emergence the glass plates can be partially, or completely, removed during the day, but must be returned to their original positions at night. To facilitate germination and early seedling growth a layer of composting materials or manure can be placed a few inches beneath the seed. The heat generated in its decomposition gives a hotbed effect, stimulating seedling growth. In areas of milder climates many gardeners report that success in starting plants by this method equals that achieved using tranplants.

Purchasing transplants. Purchasing transplants at the garden center or supermarket is no doubt the easiest and most widely used way of obtaining transplants. Purchased transplants are generally more expensive than those raised at home from seed, and varietal selection is more restricted. Nevertheless, buying commercially grown plants is the best route for many gardeners to take, especially if time and space considerations are important.

The selection of high-quality transplants depends on the gardener having basic knowledge of the features that constitute *quality* in transplants. Ideally, transplants should be stocky and of good color. Avoid spindly plants or those with off-color leaves. Select medium-sized plants, having from four to eight leaves. Avoid larger plants, or plants having flowers or partially developed fruits. Contrary to popular belief, larger transplants do not do better. Study after study has shown the superiority of smaller plants. In one Cornell University test, for example, the performance of small 37-day-old tomato transplants was compared to that of much larger 72-day-old plants. From plots of equal size the smaller plants, which had from five to seven leaves when transplanted, produced more than twice as much fruit as the larger transplants. Transplants should also have a vigorous, crisp appearance, and be free of disease and insects.

Examine each plant carefully for the presence of whitefly, mealybug, and aphids, for example, and reject infested specimens. The problems of transplant selection can be minimized by buying from trusted dealers early in the season while selections are good. And speaking of selection, remember to buy only named cultivars of transplants adapted to your area and growing conditions, just as you buy only named cultivars of garden seed.

Moving Transplants to the Garden

Moving plants from the indoor environment to the garden involves not one, but two distinct operations. First, sensitive seedlings must be exposed gradually to the rigors of the outdoor environment, a process known as **hardening off,** or simply as **hardening.** Second, the seedlings must be shifted, in as gentle a manner as is possible, from the culture medium of the container to the soil of the garden, a process termed **transplanting.** Since the future success of our seedlings depends in large part on the skill with which we carry out the two operations, they merit our close attention.

Hardening off. Young plants need to be conditioned for survival and growth in the garden. On plants grown indoors the waxy coating, the cuticle, that covers leaves and stems is too thin to protect seedlings against the drying effects of sun and wind. Furthermore, young seedlings are too tender and succulent to withstand the rigors of low temperatures, or the whipping action of wind. Exposing seedlings to environments intermediate between those of the house and the garden will firm or harden their tissues, and condition them for survival when transplanted outdoors.

The treatments used to adapt plants to outdoor conditions include (1) exposing them to temperatures too low for good growth, (2) allowing the soil of the planting medium to become somewhat dry, and (3) combining these two operations. A common practice employed by gardeners is to set their seedlings outdoors during the day and bring them back indoors before sundown. Outdoors, the seedlings are exposed to greater temperature variations, more direct sunlight, the drying action of wind, and greater moisture stress. This routine, if continued daily for 10 to 14 days will suitably harden seedlings for transplanting.

Another way in which seedlings can be hardened is through the intermediary of a coldframe. A **coldframe** is essentially an enclosed, unheated frame or box covered by glass or plastic (Fig. 7-26). Traditionally, the wooden frame covered by glass has been most popular, but in recent years frames covered by a layer, or layers, of plastic film have increased greatly in popularity. This popularity is largely justified, because the plastic covered frames are less expensive to construct, lightweight and portable, less easily broken or damaged, and commonly have translucent rather than solid sides.

The coldframe, which really is not that cold, has been aptly described as a plant's "halfway house." Placed outdoors in a sunny location, it is warmed during the day by sunlight. At night, as cooling occurs, the glass or plastic cover can be put back over the frame, slowing the loss of heat and preventing injury from frost. Care must be taken, particularly with the glass frames, to see that they are opened adequately each day to prevent excessive heat buildup. The plastic-covered frames normally allow for some ventilation, but they should be opened too. In colder regions some form of artificial heat is commonly needed in the coldframe. This heat can be provided through the installation of buried electric, lead, or plastic-coated heating cables, one or more 25-watt light bulbs (allow one bulb for each two square feet of space), or through the decomposition of organic materials. The use of decaying organic materials for the generation of heat is relatively new to most North Americans but has been practiced in northern China for cen-

FIGURE 7-26 The coldframe—a plant's halfway house. Coldframe can be constructed using wood and glass, or wood and plastic film as shown here. (Courtesy of Peter Chan.)

turies. One skilled gardener of our acquaintance simply digs a large hole beneath where he will later situate his coldframe and puts a bale of hay, or straw, into the hole (compost materials or manure work equally well). He then covers the material with three to four inches of soil and places the coldframe in position. Within a few days his coldframe has become a hotframe, or a hotbed.

The coldframe has other uses aside from that of hardening plants. A coldframe is useful for starting fall garden plants such as broccoli and parsnips and can even be used as a protected garden, supplying lettuce and other crops well past the frost-free season. In many areas root crops can be kept through the winter in a coldframe when covered with hay or straw. The coldframe can also serve as a miniature greenhouse; a place in which to start cuttings and other vegetative propagules.

Certain precautions related to hardening plants are in order. First, plants cannot be conditioned to withstand conditions for which they are not genetically equipped. Thus, while cool-season crops, such as cabbage, onion, and lettuce can be conditioned to withstand freezing temperatures, warm-season crops cannot. Their seedlings can be conditioned to have slightly more tolerance to cold, but not to withstand freezing temperatures. Second, plants can be overhardened. This results in delayed growth when the plants are set out, retards fruiting, and may reduce yields. Third, the goal in hardening off plants is to adjust them *gradually* to the outdoors growing environment. While this adjustment results in a temporary slowing of growth it sets the stage for successful adaptation in a more hostile environment. Fourth, any adequate system of hardening plants will prepare them for garden conditions within 10 to 14 days. The future success of these hardened plants depends in no small way, however, on the skill with which they are transplanted.

Transplanting. The following steps are provided as a guide to successful transplanting.

1. Start with a well-prepared seedbed and stocky, disease-free plants with good color and proper size.

2. Transplant preferably on a cloudy day or in the late afternoon or evening.

3. Thoroughly water plants an hour or two before transplanting. Plants in flats should be blocked out in such a manner as to get as much soil with each plant as possible.

4. Carefully mark and dig holes where transplants are to go. Dig holes deep enough to *completely bury* degradable pots, or the root ball. Containers made of peat that have been allowed to become dry and hard should be removed before transplanting, because root growth through them will be retarded.

5. Working quickly, carefully remove a plant from its container (where applicable) and place into hole. (Do not pull plants from pots by their stems; instead, tip pot and allow plant to slide out, tapping the pot if necessary.) Most transplants should be set just deep enough to cover their root ball or container, but tomatoes can be set almost to the level of the first leaf, because roots will form along the buried stem.

6. When a plant is in place in the hole cover the roots with soil and firm the soil around the root ball and stem.

7. Water in to settle the soil and remove air pockets that cause roots to dry out and die.

8. Fertilization is important. Either add solid fertilizer—organic or chemical—at step 4, or as a starter solution at step 8. When using solid fertilizers be sure to separate the fertilizer from the root ball by an inch or more of soil. Starter fertilizers are available commercially. They have complete solubility and a high phosphorus content, 10-50-10 for example, giving transplants a fast start.

9. Protect new transplants from excess heat, wind, or cold for a few days. This can be done using bushel baskets, inverted paper

bags, cedar shakes, old plastic jugs, or any of the commercial protectors available (Fig. 7-27). These protectors should allow for ventilation and should be removed as soon as possible.

In transplanting, it is important to remember that the roots are the most vulnerable part of the plant. Regardless of the care taken in moving plants many fragile roots and root hairs will be destroyed, and, as a consequence, the ability of seedlings to gather water and minerals will be diminished. To counteract this diminished ability gardeners often remove a portion of each leaf, say the outer half, to reduce the requirements for water and minerals imposed on the root system. This allows plants to adjust more gradually to growth in the garden soil.

Once the planting of seed and transplants is finished, the new garden is on its way. Much work remains to be done, however, before crops are harvested. Thinning, weeding, cultivating, watering—these are just a few of the many tasks that lie ahead. Care of the growing garden is discussed in Chapter 8.

Questions for Discussion

1. Define the following terms:
   - (a) Hardy vegetable
   - (b) Indore method
   - (c) Dolomite
   - (d) Companion planting
   - (e) Intercropping
   - (f) Succession planting

2. Say that you have found an area of approximately 200 square feet in which to plant your first garden. What factors would you take into consideration in deciding what vegetables to grow?

3. In a small garden, such as that of Question 2, what planting and cropping patterns could you employ to maximize yields?

4. Why do the authors emphasize record keeping as an important part of gardening?

5. What factors should be taken into consideration in evaluating and choosing vegetable cultivars?

6. Discuss the reproductive characteristics of common vegetable cultivars employing the terms open-pollinated, pureline, and hybrid in your discussion.

7. Where does the beginning gardener obtain information on the best vegetable cultivars for his or her area and purposes?

8. Describe to a classmate or friend the best way to till soils. Be sure to take textural characteristics of the soil into consideration.

9. Outline a fertilization program for your garden including timing of application, types of fertilizers to use, and rates of application.

10. Comment on the statement "All garden soils should be limed at least once every five years."

11. Why do gardeners go to the trouble of making compost?

12. Draw to scale a plan showing how you would lay out raised beds in a garden measuring 12 × 15 feet.

13. Determine to earliest dates, and range of dates, for the safe planting of beet, sweet corn, pea, and tomato in the open in your area.

14. What do you think of the practice of planting "by the moon"? Explain your answer.

15. Describe techniques for storing garden seed for one or more years.

16. How would you determine when to start transplants of cabbage, cucumber, and tomato indoors from seed?

FIGURE 7-27 Plastic containers can be used to protect young transplants from harsh weather.

Selected References

Baggett, J. 1976. Waking up the vegetable garden. *Horticulture,* 54:36–37.

Baggett, J. 1978. Saving your own vegetable seed: A pollination primer. *Horticulture,* 56:18–25.

Chan, P. and S. Gill. 1977. *Better Vegetable Gardens the Chinese Way.* Graphic Arts Center Publishing Co., Portland, Ore.

Faust, J. 1975. *The New York Times Book of Vegetable Gardening.* Quadrangle, New York.

Harrington, J. 1976. How to store your seeds alive until you want to use them. *Vegetable Growers News,* 31(4):3, P.O. Box 2160, Norfolk, Va.

Hawthorn, L. and H. Pollard. 1954. *Vegetable and Flower Seed Production.* Blakiston, New York.

Hayes, J. (ed.). 1977. *Gardening for Food and Fun.* USDA Yearbook of Agriculture.

Heiser, C. 1969. *Nightshades. The Paradoxical Plants.* Freeman, San Francisco.

Johnston, Jr. R. 1976. *Growing Garden Seeds.* Johnny's Selected Seeds, Albion, Maine.

Lunt, H. (ed.). 1956. *Handbook on Soils.* Brooklyn Botanic Garden, Brooklyn, N.Y.

Rothenberger, R. 1977. Keep your garden producing in fall. *Flower and Garden,* 21:54–65.

Vandemark, J., M. Shurtleff, and R. Randall. 1973. *Illinois Vegetable Gardening Guide Circular 882.* Urbana, Ill.

Ware, G. and J. McCollum. 1975. *Producing Vegetable Crops.* The Interstate Printers and Publishers, Danville, Ill.

Webster, R. 1975. Growing vegetables in the home garden. *USDA Home and Garden Bull. No. 202.*

LEARNING OBJECTIVES

By the time you have finished this chapter you should be able to:

1 Reveal the importance of proper thinning, and describe thinning methods.

2 Give two or three reasons why cultivation can be beneficial.

3 Define "dust mulch" and explain its potential value to the gardener.

4 List a number of benefits derived from the use of mulches.

5 Choose and use mulching materials to enhance vegetable production.

6 State potential problems related to the use of mulches.

7 Explain why water has been termed the "hazardous necessity" as relates to its use in the garden.

8 Describe the main elements of a drip irrigation system for the home garden.

9 Explain why the level of soil fertility is closely tied to the efficiency of water use by plants.

10 Describe a method whereby the percentage of available moisture in garden soils can be determined.

11 Given soil type and watering depth, calculate the approximate amount of water (in inches and gallons) required to bring the soil from the permanent wilting percentage to field capacity.

12 Discuss the use of "graywater" in the home garden.

13 Explain the principle involved in the use of protective structures to prevent frost damage.

8

CARE OF THE GROWING GARDEN

"Our England is a garden, and such gardens are not made
By singing—'O how beautiful!' and sitting in the shade."

Rudyard Kipling

In Chapter 7 we discussed planning and preparing the vegetable garden, and establishing healthy, vigorous seedlings. Employing the concept outlined in Chapter 6, that our garden is in essence a solar energy retrieval system, we can regard our young seedlings as a retrieval system in its infancy. Naturally, we want each of our retrieval units—the individual plants—to develop and operate at, or near, peak efficiency; but many factors can hinder these processes. The cultural manipulations discussed in this chapter will make our gardens more effective in capturing and storing solar energy.

Care of the garden requires effort, but this effort can be a source of great enjoyment. Garden work is best done during short, frequent visits. Such a schedule keeps gardening fun by reducing the tedium that may accompany repetitious tasks. Also, problems that could become serious if discovered too late

can be quickly corrected when addressed early in their development. Digging by animals, insect problems, and soil incrustation are examples of problems that can be corrected easily if the garden is inspected often.

On an early visit to the growing garden, the gardener will witness a miracle of rebirth—the emergence of seedlings, in rows, with their small cotyledons already opened and absorbing solar energy (Fig. 8-1). This tells the gardener that the first act of plant husbandry will soon be needed, that is, thinning.

THINNING

Why Thin?

Thinning consists of removing small or young plants from a row to provide the remaining plants with more space to grow and develop.

FIGURE 8-1 Spinach seedlings that need to be thinned. (Courtesy of National Garden Bureau, Sycamore, Illinois.)

Many gardeners, and especially the inexperienced, sow seed too thickly and are reluctant to thin. However, surplus plants must be removed before the competition for water, light, nutrients, and space becomes detrimental to growth. Thinning is especially important in root crops where failure to thin results in plants with large tops and poorly developed roots. The production of high yields of quality produce begins with judicious thinning.

How to Thin

How is thinning best accomplished? To begin with, thinning is not a one-time event, but a gradual process that should be completed over a period of weeks. For example, if seedlings of a vegetable such as spinach emerge at a spacing of an inch or less, but the desired spacing is four inches, they should first be thinned to a spacing of about two inches (Fig. 8-2). The first thinning should remove the weakest plants and should be done at the second or third true leaf stage, or by the time the plants are two inches tall. This allows for the removal of plants before their roots become too intertwined. Generally the plants spaced two inches apart should be thinned to the desired four-inch spacing when they have grown to the point where their leaves begin to overlap. The later thinnings of spinach, beet, lettuce, carrot, and certain other crops make outstanding contributions to salads and other dishes. Plants from the first thinning can also be used, as transplants to new rows or areas of row having poor stands. In short, gradual thinning assures that stands of desired density will be obtained and at the same time optimizes use of thinned plants, either as transplants or as nutritious early spring greens.

Thinning can be done by hand or by pulling a steel rake across the rows (Fig. 8-3). Thinning is best done when the soil is moist, and in the evening or on a cloudy day when transpirational demands on the root system are

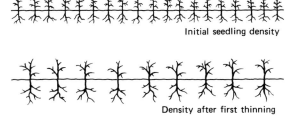

Initial seedling density

Density after first thinning

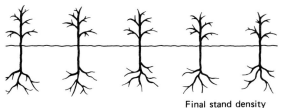

Final stand density

FIGURE 8-2 Crowded seedlings should be thinned gradually to the desired spacing.

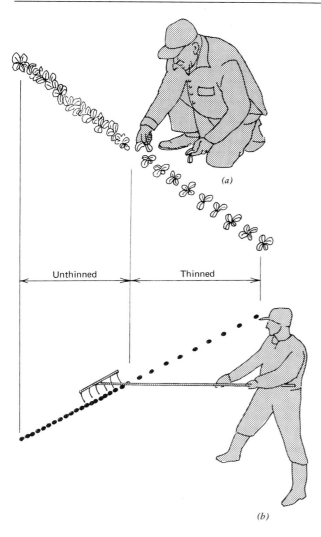

FIGURE 8-3 Thinning can be done by hand (*a*), or using a garden rake (*b*).

Unthinned   Thinned

(*a*)

(*b*)

reduced. The plants that remain after thinning should be gently pressed into the soil to assure that their roots have good soil contact. Watering can also be used to settle the soil and help remove air pockets which are detrimental to roots.

How to Reduce Thinning Chores

The thinning process can be reduced or facilitated in a number of ways. The use of seed tapes, where possible, virtually eliminates the need for thinning. Thinning chores can also be reduced by careful spacing of seed at planting time, a skill that comes with experience. Perhaps the biggest thinning problems are associated with vegetables like carrot whose seeds germinate slowly. By the time seedlings emerge they may be in a veritable forest of weedy seedlings and it may even be difficult to identify rows, much less carrot seedlings. Here thinning can be facilitated by planting seed that germinate rapidly, such as those of radish or lettuce, along with the carrot. The presence of these "marker plants" at intervals helps to highlight the rows and makes the carrot seedlings easier to find and thin.

CULTIVATING

Purposes of Cultivation

In Chapter 7 we defined tillage as including all of the operations involved in working the soil. Cultivation in the growing garden is one aspect of tillage. Cultivation is done for a number of reasons, the most important of which are (1) weed control, (2) improvement of soil aeration, and (3) enhancement of moisture retention.

Cultivating to Control Weeds

Beyond doubt, the primary value of cultivating in the home garden is weed control. Weeds compete with growing vegetables for water, minerals, and sunlight. Furthermore, weeds often harbor disease organisms, insects, and other pests. To illustrate, curly top in tomato and sugar beet is a serious virus disease carried by leafhoppers from border weeds to these crops. (Weeds also harbor insects and nematodes that can reinfest garden crops each year.) The control of weeds thus conserves moisture and nutrients for our vegetables and can reduce disease and pest problems. All of these measures have important effects upon yield. In one Cornell University study scientists compared yields of vegetables in noncultivated and cultivated plots. In all cases, the plots in which weeds were allowed to grow had strikingly lower yields (Table 8-1).

Gardeners often use cultivation rather than chemicals, as is frequently done in commercial vegetable production, to control weeds. Most of the chemicals used to control weeds can be used only for specific crops. Because of the diversity of crops grown within a relatively small space the gardener generally cannot effectively use these chemical agents, knows as **herbicides,** but rather must rely on cultivation to control weeds.

When and how should cultivation be done to control weeds? Generally it is best to cultivate just as weed seedlings emerge from the soil, because at this point the fragile seedlings quickly wilt and die, particularly in hot weather. When cultivating near vegetable seedlings take care not to cover them with soil because this can prove fatal. After each rain or irrigation new weed seedlings will emerge and these too should be quickly dealt with. Care must be taken not to cultivate heavy soils when they are wet, however, as the exposed soil chunks will later become hard clods which will not

TABLE 8-1 Effect of Cultivation on Yield of Selected Vegetables

Vegetable	Not cultivated	Average yield of edible portion of crop (pounds/plot)	
		Cultivated	
		Half of season	All season
Beet	45.6	239.7	240.3
Cabbage	129.1	234.6	233.6
Carrot	27.9	506.4	505.3
Onion	3.6	69.6	67.7
Potato	52.7	150.4	148.3
Tomato	23.3	166.6	164.0

Courtesy of Cornell University Agricultural Experiment Station.

crumble when broken up. Larger weeds are best removed when the soil is somewhat moist, as they tend to come out easier and cause less disruption to adjacent crop plants. Many larger weeds easily root again under cool, moist conditions, however. To prevent this, shake excess soil from their roots and remove the plants from the garden area. Remember, weeds, like other organic materials, are likely candidates for the compost pile.

Shallow cultivation is preferable to deep cultivation. Just scrape the surface of the soil with rototiller, wheel hoe, or hand hoe. A regular sharp, flat-bladed hoe works well for cultivating between rows, but a triangular-pointed hoe may be better for the closely spaced vegetables within rows.

As vegetables get larger the gardener should hoe less of the area between the rows to avoid injury to growing plants. Additionally, once crop plants are established the competitive threat from weeds is lessened considerably. In fact, weeds that begin growth late in the season have little or no effect on crop yields (Table 8-1), but their control may still be important for general garden sanitation.

Some weeds are difficult to control through cultivation. This group includes the herbaceous perennial weeds, such as Canada thistle, *Cirsium arvense* (Fig. 8-4), and field bindweed,

Convolvulus arvensis, that spread by means of rhizomes and roots. Simple cultivation has little effect on these aggressive competitors, because they soon send up new shoots from their underground structures. A mistake many gardeners make when dealing with these noxious weeds is to cut off new shoots as soon as they emerge from the soil. The new shoots continue to draw food from the root system for at least two weeks after emergence, after which time the new leaves are big enough to send food back into the roots. Thus, subsequent cultivations should be at approximately 14 to 18 days after new shoots emerge. In this way the maximum food reserves are drained from the storage organs and over a period of time the plants die by starvation. In most cases, however, herbicides give faster and more effective control.

Cultivating to Improve Soil Aeration and Moisture Retention

Cultivation also improves soil aeration and moisture retention by breaking up surface crusts and physically stirring to expose more soil volume to air. Cultivation can result in moisture conservation in a number of ways. First, moisture is conserved through the removal of weeds. In addition, cultivation in-

FIGURE 8-4 Canada thistle, showing the extensive, spreading root system from which new shoots can arise. (Reprinted with permission of Macmillan Publishing Co., Inc. from John H. Martin and Warren H. Leonard, *Principles of Field Crop Production.* Copyright © 1967, Macmillan Publishing Co., Inc.)

creses the permeability of soil to the entrance of water and thereby reduces or prevents surface runoff. Also, cultivating creates a loose layer of surface soil, known as a **dust mulch,** that slows evaporation of water from the soil surface by blocking its upward capillary movement.

The goals of cultivating to improve soil aeration and enhance moisture retention may be opposed to each other. For example, cultivation to introduce air can expose moist soil surfaces to the drying action of this air. Cultivating soon after a rain, and particularly a light rain, is likely to cause moisture loss because of the greater evaporative surface exposed to the air. Cultivation that disrupts a surface mulch will also accelerate moisture loss. In summary, cultivation should be done to remove weeds, to break up surface crusts, and perhaps to create a dust mulch. When none of these conditions are met, cultivation is generally not necessary and is often detrimental.

We have seen that the primary objectives of cultivation are the control of weeds, the conservation of soil moisture, and the improvement of soil aeration. While these objectives can generally be met through cultivation, they can also be met largely through another cultural practice—that of mulching.

FIGURE 8-5 Mulches create an insulating layer at the soil-air interface. Here an organic mulch is used to enhance plant growth and soil quality. (Courtesy of the Brooklyn Botanic Garden, Brooklyn, New York.)

MULCHING

What Are Mulches?

A **mulch** is created whenever the soil surface around crops is artificially modified. Typical mulching materials include straw, sawdust, paper, and plastic. Mulches form an insulating layer, or buffer zone, *between* the soil surface and the aerial environment (Fig. 8-5). Solar radiation no longer impinges directly on the soil, but rather on the mulch layer. Similarly, precipitation does not fall directly to the soil, but on the mulch layer too. The mulch can have profound effects on both the soil and the aerial environment, effects that can be manipulated to the advantage of the gardener.

Benefits of Mulching

Among the many benefits derived from the use of mulches are the following:

- Control weed growth.
- Conserve soil moisture.
- Regulate, or modify, soil temperature.
- Prevent soil crusting.
- Maintain soil structure and fertility.
- Encourage root growth in the topsoil.
- Reduce, or prevent, soil erosion.
- Eliminate root injury caused by cultivation.

- Reduce certain plant diseases.
- Give cleaner produce.
- Modify the aerial environment around plants.
- Hasten maturity.
- Improve yields.
- Allow for easier harvesting.
- Provide aesthetic appeal.

Obviously, no single mulch material will provide all of the above benefits in every situation. Instead, the precise nature of the effects depends on the kind of mulch used and the manner in which it is applied. Let us examine some types of mulches.

Choosing Mulching Materials

There are two basic types of mulching materials: (1) naturally occurring materials, having little or no human modification, and (2) synthetic materials. Examples of naturally occurring materials include:

Buckwheat hulls	Leaves
Compost	Manure
Crop residues	Peat moss
Gravel, stones	Sawdust
Ground corncobs	Straw
Hay	Wood or bark chips

Most of the naturally occurring materials are organic in nature and as such gradually break down. In practice these materials are often turned into the soil at the end of the gardening season. Examples of synthetic materials used as mulches include:

Aluminum foil	Polyethylene
Brown paper	(black plastic)
Newspaper	(clear plastic)

The paper products, brown paper and newspaper, are biodegradable, whereas the aluminum foil and polyethylene films are not and should normally be removed at the end of the gardening season.

Gardeners differ greatly in opinion when it comes to choosing between natural and synthetic mulching materials. In truth, each type of mulch material presents certain advantages and disadvantages. The choice should depend on the purpose of the mulch (i.e., how the gardener wishes to modify the environment), availability, cost, and preference. Many innovative gardeners find that a combination of mulches best fit their needs. Examples will illustrate these points.

Mulch materials form an interface between the atmosphere and the soil and as such receive incoming solar radiation. How this radiation is "handled"—reflected or absorbed—is of great consequence. As discussed in Chapter 6, reflected radiation is, for all practical purposes, the same as radiation never received. This is not so of absorbed radiation, which usually produces a rise in temperature through its conversion to heat energy. The nature of the mulch determines the extent of reflection and absorption. An aluminum foil (Fig. 8-6) or aluminum coated plastic mulch will have a high degree of reflectivity, and a low degree of absorptivity, relative to bare soil. Consequently the soil beneath this mulch will not warm up as much as if the mulch was absent. The reverse is true of a black plastic mulch. The temperature of this film soars on a warm day because much of the incoming radiation is absorbed. Some of the heat energy produced is reradiated back into the air above the plastic, and some is transferred to the soil below. As a consequence, soil temperatures increase up to 12°F, or more, over those of adjacent bare soils. The increase would be even greater were it not for the air pockets present between the plastic film and the soil. These pockets slow the transfer of heat because air is a relatively poor conductor of heat.

A combination of mulches could be chosen by a gardener who first wanted to warm the soil up earlier than normal, but who wanted later to protect the soil from excessively high summer temperatures. Black plastic is cred-

FIGURE 8-6 An aluminum foil mulch used with cucurbits. (Courtesy of National Garden Bureau, Sycamore, Illinois.)

ited for giving earlier maturity and increased yields of certain crops in cool areas. In summer, however, high soil temperatures may reduce nutrient availability and retard root growth. Covering the plastic with an organic mulch, such as sawdust, would lower soil temperatures. Painting the plastic a light color, to increase its reflectivity, is also a practice employed in areas where high summer temperatures present problems.

In addition to the temperature modifications caused by mulches, the gardener needs to consider the permeability of each mulch to water, to carbon dioxide and oxygen, and to light itself (Table 8-2). Knowledgeable selection of mulches will allow the gardener to dramatically modify the garden environment. However, the gardener must also know which vegetables will be benefited—or hindered—by such changes, and how to use the mulch in question to most effectively bring about these changes.

Mulches and Soil Temperatures

Beyond question, one of the major attributes of mulches is their regulation of soil temperature. Warm-season crops, and in particular the vine crops, show a marked increase in vigor, earlier yields, and greater total yields when mulched with black plastic. In a Virginia study comparing the yields of two muskmelon and two watermelon cultivars, plants mulched with black plastic grew much more vigorously and began spreading earlier than those on bare ground. Both early yields and total yields were greater on plastic than on bare ground (Fig. 8-7). Also, melons direct seeded through black plastic yielded 2.3 times as much as transplants on bare soil. While yields were greater, fruit quality was not changed, and the size, appearance, and flavor of the melons grown under the two systems were similar.

Mulching to Control Weeds

Mulches also control weeds and eliminate the need for cultivation by smothering weeds. Black plastic and aluminum foil are better for controlling weeds than are the organic mulch materials. Clear plastic is generally not recommended for use by home gardeners because light penetrates the film and encourages the growth of weeds beneath. However, if temperatures under the clear plastic consistently reach 90°F or more the growth of weeds is inhibited. To control weeds with organic mulches existing weeds need to be removed before the mulch is applied. The mulch layer should be deep enough so that weed seedlings cannot grow through it on their stored food reserves. Mulches of hay, leaves, and straw are frequently applied to a depth of three or four inches, whereas a layer of sawdust, bark chips or wood chips of about two inches is usually adequate. Perennial weeds are little

TABLE 8-2 Some Important Characteristics of Common Mulching Materials

Mulch	Soil temperature (°F)	Permeability to			Other comments
		Water	CO_2, O_2	Light	
Aluminum-coated plastic and foil	To −10	—	—	—	Surface high in reflectivity which keeps soil temperatures low. Research shows that the mulch also repels aphids. Expensive.
Black plastic	To +12	—	+	—	Early warming of the soil improves yields of many crops. Eliminates weed growth and protects fruit of vine crops from rots. A favorite.
Brown paper	To −8	—	+	—	Has a high degree of reflectivity, resulting in lower soil temperatures. Commonly has a thin plastic coating. Biodegradable. Stops weeds.
Clear plastic	To +20	—	+	+	Rays from sun penetrate and warm the soil. Benefits similar to black plastic, but may stimulate weed growth beneath plastic.
Organic mulches	To −10	+	+	—	Reduce soil temperatures and save water. Stops most annual weeds if applied thickly. Adds nutrients to soil as mulch decomposes. May cause temporary nitrogen shortages.

affected by organic mulches, but their control through removal is usually made somewhat easier.

Mulching to Conserve Moisture

Mulches conserve moisture in a number of ways. One obvious way, of course, is through their prevention of weed growth. Mulches also reduce evaporation by (1) lowering soil temperatures, (2) forming a layer impervious to water vapor, and (3) blocking off the drying effects of direct sun and wind. In addition, mulches tend to decrease the surface runoff

of water and increase the absorptive capacity of the surface soil. By slowing evaporation they also help to maintain a more even supply of moisture in the upper layers of soil.

Mulching and Soil Fertility

Another aspect of mulches that merits special attention is their role in maintaining soil structure and fertility. When organic mulches are used, a moist environment develops at the soil-mulch interface that encourages the growth of soil organisms. These organisms gradually decompose the organic materials providing a

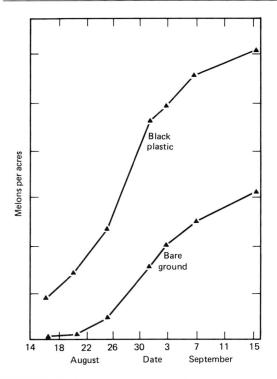

FIGURE 8-7 Relative yields of melons when grown under black plastic and bare ground in southwest Virginia. (Data from R. D. Morse, 1977. Virginia Polytechnic Institute and State University, Blacksburg, Va.)

isting weeds have been removed, the soil has warmed, and vegetable plants are large enough not to be smothered by the mulch. Some gardeners initially apply organic mulches between the rows and wait until the plants are several inches tall to cover the area within rows. Plastic mulches can be applied when the seedbed is prepared. Seed can be drilled or transplants set out directly through the plastic. When clear plastic is used seed can be sown even before the plastic is laid down, because the gardener can see the seedlings as they emerge from the soil and cut openings to allow for their continued growth. Plastic is best applied on a calm, windless day. Under ideal conditions a gardener can lay down plastic alone (Fig. 8-8), but generally things go better when one has the help of another person. While polyethylene is pervious to carbon dioxide and oxygen, it is impervious to water (Table 8-2). Soaker hoses or drip lines can be

slow but continuous release of plant nutrients and humus into the soil. The process of decomposition can be accelerated through the addition of a nitrogen fertilizer with the mulch. Eventually, as the mulch is turned into the soil, soil structure and fertility are even further enhanced, although temporary nitrogen shortages are likely to occur during the decomposition process.

How to Apply Mulches

The application of mulches depends on the kind used. Organic mulches are generally applied when seedlings are well established, usually after the first cultivation. At this time ex-

FIGURE 8-8 Black plastic mulch laid down before transplants were set out.

placed under the plastic, or plants can be watered (1) through the planting holes, (2) through special openings or slits cut in the plastic, or (3) along the edges of the plastic (if contiguous sheets do not overlap).

Problems Created by Mulching

The use of mulches presents some problems. Organic mulches can act as sources and havens for pests—weeds, disease-producing microorganisms, and rodents. Generally, these problems can be reduced by stirring the mulch occasionally. Also, to aid in disease control, the refuse of the crop being mulched should not be used as a mulch itself. Some mulch materials, such as hay and straw, present a fire hazard and should not be used close to buildings. Fresh leaves and lawn clippings, may form an impenetrable blanket that can smother plants. Nearly all organic mulches have a high carbon to nitrogen ratio. When turned into the soil mulches can bring about a temporary shortage of available nitrogen, because the organisms that decompose the organic matter compete with vegetables for the nitrogen in the soil. To prevent this shortage extra nitrogen should be added with the mulch or when the mulch is turned under. The symptoms of nitrogen deficiency are light green or yellowish foliage. Finally, frost damage can be greater to mulched crops than unmulched ones. Organic mulches, for example, reduce soil temperatures. On cool nights mulched surfaces will radiate less heat than adjacent bare ground. This lack of back radiation causes the temperature around mulched plants to be a few degrees cooler than around unmulched ones, and this difference can result in frost injury.

The Use of Winter Mulches

To this point our discussion of mulching has been confined to mulches applied during the period of active plant growth, that is, **summer mulches.** In addition to these, **winter mulches** are sometimes applied in late fall or winter to provide cold-weather protection to perennial crops. In contrast to the numerous benefits offered by summer mulches, mulching in winter is done for one principal reason—to provide protection against extremes in temperature. Winter mulches (1) form an insulating layer that reduces the loss of heat from the soil surface, (2) stabilize soil temperatures, thus preventing alternate freezing and thawing that disrupts and injures plant roots through soil heaving, and (3) slow the absorption of heat in the early spring through insulation and shading of the soil, thus preventing the early growth and flowering of certain plants during unseasonably warm spells, before the frost period has past. Winter mulches are usually applied after a light freeze or two so as not to delay dormancy. With strawberry, for example, a straw mulch two to three inches thick is commonly applied after the plants are dormant in the fall, but before heavy freezes have occurred. In spring the mulch material is moved to the middle of the rows.

Mulches—A Summary

Through the use of mulches the gardener is able to significantly modify the environment in which his or her plants are growing. For these modifications to be beneficial the gardener must clearly understand how a particular mulch modifies the environment, which crops will benefit from this modification, and how to correctly apply the mulch. Both naturally occurring and manufactured mulches are available, and both work well in specific instances. In recent years the use of black plastic mulches has increased markedly. More recently aluminum foil has been used as an effective mulch and, among other things, has been found to "repel" aphids by disrupting the aphid's plant-sensing mechanism (Table 8-2). As a result aphids fail to detect the plants so do not land. This principle has been tested in the Imperial Valley of California and was found to significantly increase yields of summer squash by reducing the spread of the watermelon mosaic virus carried by certain aphids.

WATERING

Water is essential for plant growth. It is the medium in which nutrients are transported throughout the plant. Water is the major constituent of the vacuolar sap that occupies the largest portion of mature, living plant cells. The average vegetable is from 75 to more than 90 percent water on a fresh weight basis.

The Importance of Proper Watering

If our vegetables do not receive an adequate supply of water they soon come under water stress. Some of the initial effects of water stress are a reduction in cell elongation, partial or complete closure of stomates, and lowering of photosynthetic rates. These effects cannot be seen, but they invariably precede wilting, the

major manifestation of severe water stress in plants (Fig. 8-9). Plants can survive wilting, and during hot periods may even wilt when water supplies in the soil are adequate. If wilting occurs in the hot part of the day, however, and is progressively more pronounced daily, it is a clear sign of water shortage. In wilted plants the stomates are closed; thus the exchange of gases necessary for photosynthesis cannot occur. Since respiration continues, however, the plant starts using its food reserves and begins to decrease in dry weight. In tomato the stressed leaves may draw water from developing fruits, contributing to the pathological condition known as **blossom end rot.** Prolonged wilting leads to plant death.

Too much water is not good either. You will recall from Chapter 4 that the soil consists about equally of solids and pore space (Fig. 4-9), and that the pore space is occupied by air and water in varying proportions. Thus, when the proportion of water in the soil is high the amount of air is low. Plant roots require air for growth. By overwatering, particularly on heavy soils, the air supply

FIGURE 8-9 Severely wilted leaves of squash plants in late afternoon of a hot day (*a*), and the same plants the following morning (*b*). The wilting was temporary because no water was added to the soil. (Used by permission of Holt, Rinehart and Winston from Carl L. Wilson, Walter E. Loomis, and Taylor A. Steeves, *Botany.* Copyright © 1971 Holt, Rinehart and Winston.)

is reduced and root growth slows or stops altogether. The longer air is lacking, the greater the damage. Damaged roots are susceptible to decay organisms and eventually, if adequate air is not made available, the plant will die.

Excesses of water have other less dramatic but nevertheless serious effects. When plants with well-developed root systems are growing on warm, moist soil under conditions that favor low transpiration rates they can absorb excessive water resulting in the occurrence of growth cracks. **Growth cracks,** the actual splitting of plant organs, are common in carrot and sweet potato roots, cabbage heads, and tomato fruits (Fig. 8-10).

While abundant quantities of water are required for good vegetable growth and yields, too little or too much can cause serious problems. One gardening professional described watering as the most misunderstood aspect of gardening. We will discuss three aspects of watering: (1) how to water, (2) when to water, and (3) how much water to apply.

How to Water

Several methods of watering are available to the gardener, including the use of watering cans, hoses, sprinklers, furrows.

Watering Cans and Hoses

Watering cans and hoses prove useful, and are sufficient for the small garden. They are ideal for watering vegetables in hanging baskets and other containers, as well as specific locations within the garden (Fig. 8-11). When purchasing a hose it is best to buy one of high quality reinforced with nylon. Better hoses will last for many years, do not kink or crack easily, and can be coiled even in cold weather. Also, if you have much watering to do a hose of larger inside diameter, $\frac{1}{2}$ or $\frac{5}{8}$ inch, is more efficient. A hose of $\frac{5}{8}$ inch diameter will deliver twice as much water per unit of time as one $\frac{1}{2}$ inch in diameter, and the $\frac{1}{2}$ inch about twice that of one $\frac{3}{8}$ inch in diameter. In general, when watering with a hose the stream of water should be directed around the base of the plant, not on the foliage. If the stream of water is strong enough to expose sensitive roots the force of the water can be reduced with shower head devices available in any garden or hardware store.

One problem gardeners face when watering with a hose is that of dragging the hose in and around, and sometimes over, vegetables. To eliminate this problem put stakes at strategic locations within the garden, and over each stake place a short piece of plastic pipe (Fig. 8-12). In this way the pipe turns on the stake allowing the hose to be pulled along easily, and also preventing it from causing damage to plants.

Soaker hoses are a specialized type of hose available to gardeners. These are perforated lengths of plastic or canvas tubing that release a slow, gentle flow of water into the soil around

FIGURE 8-10
Cracked fruits of tomato resulting from excess uptake of water by the plant. (Courtesy of U.S.D.A.)

FIGURE 8-11 Water delivered by hose to a submerged, perforated coffee can is an excellent way to water cucurbits. (Courtesy of the National Garden Bureau, Sycamore, Illinois.)

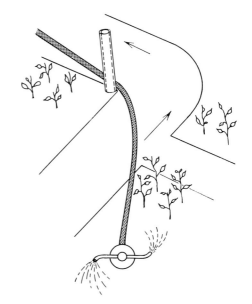

FIGURE 8-12 A short stake covered with a piece of plastic pipe allows the gardener to move the hose easily around the garden without causing injury to plants.

plants. The principal advantages of this watering method are that water is conserved because the flow can be directed into the ground near plants with little loss through runoff or evaporation. Also, regardless of which way the soaker is turned, its gentle streams of water cause little or no compaction of soil or splashing of muddied water onto plants.

Sprinkler Watering

Sprinklers provide another popular way for watering the garden. Many types are available giving a wide range of watering patterns and area coverages. In addition to providing a convenient means of watering, sprinklers have the advantage of lowering the temperatures around plants on hot midsummer days. On the negative side, evaporative losses through sprinkler watering tend to be high, particu-

larly on hot, windy days and if droplet size is small. Also, many sprinklers deposit less water as one moves outward from the source (Fig. 8-13). Because of this successive sprinkler settings should overlap. The newer oscillating sprinklers largely eliminate this problem and additionally can be easily adjusted to cover either square or rectangular areas. Other complaints registered against sprinklers are that they wet foliage, increasing the incidence of plant diseases and causing "burning" of leaves. While it is true that water on foliage may increase plant diseases the problem can be largely overcome by watering early in the day so that any excess moisture evaporates before evening. There is no truth whatsoever to the claim that water will cause foliage to burn, or that cold water hitting the leaves of plants will shock them and retard their growth. Finally, while sprinkler watering is well suited to cool-season crops grown primarily for their vegetative parts it may not be quite as good

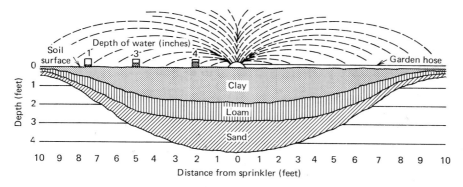

FIGURE 8-13 The pattern of water distribution with a common garden sprinkler. Note the shortage of water at the outer edges (perimeter) of the spray and the differences in the penetration of equal amounts of water in soils of different types. (Courtesy of University of California Cooperative Extension Service.)

for warm-season crops grown for their fruits. This is mainly because watering at the time of flowering can disrupt pollination and subsequent fruit development. Warm-season crops may also be more susceptible to foliage diseases such as mildew. Despite these objections, sprinklers remain the most popular way to water the garden.

Furrow or Surface Watering

Another way in which the garden can be watered is through furrows run between the rows (Fig. 8-14). Furrow (surface) watering is best adapted to gently sloping soils of medium to low water-intake rates where the surface soil is deep and uniform and the subsoil does not impede drainage. Soils having a steep slope or irregular contour, very sandy soils, and shallow soils are not suited to this method. Many professionals feel that in those situations where furrow watering can be used it is the best way to apply water to the garden. Generally evaporative losses are less than with sprinkler watering, and in addition the foliage is not wetted. On the negative side, time is required to make and maintain furrows, and there is a tendency for the soil in and along the furrows to crust and bake.

Drip Watering

Finally, a relatively recent, innovative watering method—drip irrigation—merits our attention. This system provides a constant, low volume release of water in exact location near plant roots. The conventional **drip**, or trickle, **system** consists of a main line, header, laterals, and emitters (Fig. 8-15). The **main,** commonly a garden hose, is connected at one end to the water supply and at the other end to the header. Since a supply of clean water at low pressure is required a strainer to remove foreign matter and a pressure-reducing valve are installed along the main line. The contaminant-free water flows under low pressure into the **header** and from there into the **laterals.** At any point along the laterals the gardener can install emitters. These **emitters** have minute orifices that allow water to drip slowly onto the soil surface near plants at rates ranging from one-half to four gallons per hour. Once in the soil the water moves by capillarity into the surrounding area. The system may be installed either above or below ground.

In using a drip system the gardener attempts to replace on an almost daily basis the water used by the plant. With a constantly renewed supply of water it is not necessary

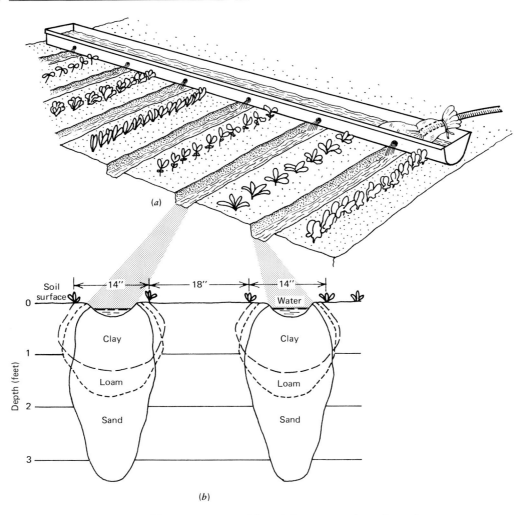

FIGURE 8-14 A system of furrow irrigation (*a*) and the penetration of equal amounts of water in furrows consisting of three soil types (*b*). (Modified from University of California Cooperative Extension Service.)

for the entire root system of a plant to be wetted. In general, researchers feel that if at least one-half of the root system of a plant is maintained at a favorable moisture level that water will not be a limiting factor in plant growth.

The advantages of the drip system of watering are many and include the following: (1) water is placed accurately in the root zone where it is needed; (2) it keeps leaves and fruit dry; (3) water is supplied at a slow rate so there is little waste; (4) only a small area around each plant is wetted allowing work in the garden even while watering is in progress; (5) much less water is used than with conventional watering methods; (6) weed growth is greatly reduced; (7) low quality water, including water high in salts, can be used; (8) ac-

encountered are the plugging of the small outlets of emitters and the tendency of rodents to chew on system parts. Much work is being done on the manufacture of more effective systems, and it is safe to say that both of these problem areas will soon be reduced or eliminated. In the meantime the clogging problem can be corrected by maintaining a clean strainer, by removing the plug from the downstream end of the drip line (Fig. 8-15) and flushing at full pressure, and by cleaning and/or replacing individual emitters as they become clogged.

Another method is to use, as laterals, any of the several kinds of perforated or porous plastic tubing that exude water along their entire length, thus eliminating the localized emitters. As a general rule, the types of tubing that allow water to trickle or ooze from their entire length can be used effectively for all plant spacings up to three feet. At spacings of greater than three feet the use of properly spaced individual emitters is regarded as a more effective method of watering.

The advantages of a properly installed and maintained drip system far outweigh the disadvantages. Continued advances in the area of drip technology, coupled with increased demands made on water worldwide, make it quite clear that drip irrigation will increase in popularity in the future, for the commercial grower and home gardener alike.

When to Water

When is the best time of day to water: morning, midday, or evening? Authorities differ in their answer to this question, but many feel that watering in the early morning hours is best, especially if water is applied by overhead sprinkling. Watering early in the day lets wetted foliage be dried by the hot sun of midday, and the chances of plants developing certain foliage diseases are reduced. Also, by watering in the early morning, or evening, the loss of water through evaporation is decreased. If

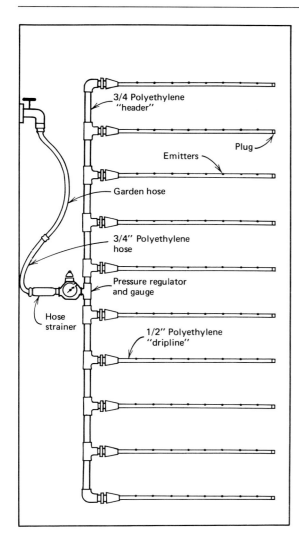

FIGURE 8-15 Basic elements of a drip irrigation system for the home garden.

cessories can be added that allow fertilizers to be put directly into the system; and (9) once the system is laid out little maintenance is required.

But there are disadvantages too. A drip system for the home garden costs more than other systems, but the extra cost may be offset by savings in the amount of water used and by improved yields. Other problems sometimes

other methods of watering are used the presumed advantages of early morning watering are reduced or eliminated entirely. To illustrate, watering using watering cans or hoses can conveniently be done in the evening after dinner, furrow watering can continue all afternoon or evening, and a common practice in drip watering is to let the system operate the entire night two, three, or even four nights a week.

In short, the time of day at which water is applied is of minimal importance from the standpoint of good garden practice. For most gardeners other factors are the major determinant to time of watering. For example, family schedules, water usage requirements or restrictions of the local water district, and convenience all influence the watering schedule. There is a matter of critical importance, however: how much water to apply. Too little water can lead to retarded growth, reduced yields, or even death. Similarly, too much water has adverse effects on plant growth. Determining how much water to apply and guidelines to tell when water is needed are the subjects of the next section.

How Much Water to Apply

Determining how much water to apply is difficult because amounts are affected by several variables including (1) weather conditions, (2) plant characteristics, and (3) soil properties.

Weather Conditions

In an earlier chapter we learned that all precipitation is not of equal value to plants; to be most useful it must be received during the growing season. Due to a favorable distribution of precipitation throughout much of the Midwest, for example, a gardener may only water the garden once or twice during the entire gardening season. In the Northwest, on the other hand, gardeners must water every week, despite an annual precipitation about twice that of the Midwest.

Not all growing season precipitation represents moisture available to plants. Some moisture is lost through evaporation and some through runoff, but the amounts lost vary from region to region. In general, moisture is most effective for plant growth in regions that are cool and where the humidity is high (Fig. 14-15). In much of the western third of the continent high summer temperatures and low humidity causes the evaporation of large amounts of water from soil and plant surfaces, resulting in a lower proportion of moisture available to plants. Within a given area, showers that deposit less than one-quarter inch of moisture on hot summer days are usually ineffective in promoting plant growth because of high evaporative losses. Similarly, a large proportion of the water from a heavy rain may be lost by runoff. In other words, the effectiveness of precipitation depends on the amount and the intensity, in relation to temperature, relative humidity, and the absorbing capacity of the soil. Only that water reaching the root zone can be used by plants.

As a rule of thumb, vegetable plants require the equivalent of at least one inch of rainfall per week throughout the growing season. In spring and early summer plants can draw on the water stored in the soil from precipitation of the previous fall and winter, but as the growing season progresses these reserves are depleted and additional water becomes essential. Moisture not supplied by rain must be provided by the gardener through a regular program of watering.

Plant Characteristics

Plant characteristics profoundly influence how much water to apply and the frequency of its application. To illustrate, the depth of rooting of vegetable crops in well-drained soils varied considerably (Table 2-4). In general, the more completely plant roots colonize the soil the more water they obtain, and the less frequently they need to be watered. This relationship is less precise than expected, however, because even deep-rooted crops obtain the majority of their water in the upper foot or so of soil (Fig. 8-16). In the home garden plants having widely differing root systems are grown in close proximity to each other, and it is therefore difficult to accommodate the water needs of individual crops. As a result it is recommended that watering be tailored to meet the needs of shallow-rooted crops. If these needs are satisfied, then we can be sure that those of more deeply rooted crops are met also.

Vegetable crops also vary in their need for water. A crop grown over a comparatively short period of time or in a cool period generally requires less total water than a crop grown over a longer period or during hot weather when transpirational losses are high. A more important consideration than the total amount of water used is, from a water efficiency standpoint, the amount of water transpired per pound of dry matter produced. Corn, with a high total water requirement, was found to be one of the more efficient crops in producing dry matter (Fig. 8-17). While the values given are typical, the actual amounts would

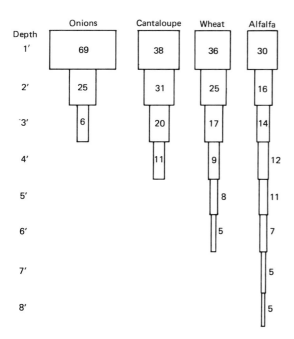

FIGURE 8-16 Percentage of water used from each foot of soil for several irrigated crops in Arizona. The shallow-rooted onions grew only in winter and used a total of 18 inches of water, whereas alfalfa grew the entire year and used a total of 74 inches of water. (Reprinted with permission of John Wiley & Sons from H. Foth and L. Turk, *Fundamentals of Soil Science.* Copyright © 1972 by John Wiley & Sons.)

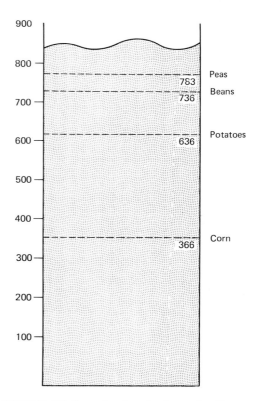

FIGURE 8-17 Pounds of water transpired by certain crops for each pound of aboveground dry matter produced at Arkon, Colorado. (Reprinted with permission of A. & L. Publications. Modified from Samuel R. Aldrich, Walter O. Scott, and Earl R. Leng, *Modern Corn Production.* Copyright © 1975 by A. & L. Publications.)

be higher in hotter, dryer regions, and lower in cooler, more humid regions. The values given also represent the *total* dry matter produced. Of special interest to the gardener is not the total amount of dry matter produced, but rather the proportion of edible dry matter. In endive, lettuce, and cabbage virtually all of the aboveground matter is edible, whereas in corn only a small percentage is consumed. Crops such as beet and turnip are efficient water users, regardless of their relative transpiration rates, because virtually the entire plant can be consumed.

Vegetables require an abundant supply of water at particular times. For fruit crops, such as corn, melons, snap beans, and tomato, the critical stage corresponds to the period of flowering and fruit development. In snap beans, for example, moisture stress during this period markedly depresses yields because of flower abscission and ovule abortion. In crops grown for their vegetative parts the critical period usually corresponds to the time of rapid development of the edible parts. Thus in broccoli, cabbage, cauliflower, and lettuce, moisture stress is most serious during head development, in onion during bulb enlargement, and in potato during tuber formation.

Soil Properties

Soils vary greatly in their water-holding ability; a property related directly to soil texture. The finer-textured soils, such as clays and clay loams, have a greater ability to store moisture, hold more available water, and also retain more water at the wilting coefficient than do coarse-textured soils (Fig. 4-20). To illustrate, for each foot of depth clay soils hold from 1¾ to 2½ inches of usable water, loams from 1 to 1¾ inches, and sandy soils from ¾ to 1 inch. Because of their greater ability to store water, the finer-textured soils can be watered less frequently. Thus, while a gardener with a sandy soil may need to water daily, a gardener having a clay soil may find weekly watering to be adequate.

In Chapter 5 we discussed the principle of limiting factors, particularly as it relates to the mineral elements required by plants. We saw that in general the level of crop production is restricted, or limited, by the most limiting of the essential growth factors. It is important to recognize that this principle applies to light, temperature, oxygen supply, and water as well as nutrients (Fig. 5-3). Thus, even when mineral nutrients are abundant growth can be limited by a shortage of water. Most gardeners are quite aware of this fact. What many gardeners do not realize, however, is that the reverse is also true: even when water is abundant growth will be limited by the shortage of a particular nutrient. The amount of water used by a high-yielding crop of beans, for example, is only slightly more than that used by a low-yielding crop. This is because water losses through transpiration (the major plant loss) and evaporation continue at about the same rates as when the plants are more productive.

Applying Water to the Garden

The soil water available for plant use is that held between field capacity and the wilting coefficient (Fig. 4-20). To assure optimum vegetable growth the gardener must apply water before the wilting coefficient is reached. The **irrigation point,** the moisture condition at which watering should start, varies somewhat depending on soil texture, type of crop and other factors, but generally it falls about midway between field capacity and the wilting coefficient. For effective watering the gardener needs to know how to estimate the percent of available water remaining in the soil.

Determining Soil Moisture Content

The percentage of available moisture in garden soils can be determined in several ways, including (1) using moisture meters, (2) keeping a weather balance record, (3) using the soil "balling" method, and (4) evaluating plant characteristics. Of these methods, the last two currently are the most useful to the home gardener. By use of the balling, or feel, method it is possible to determine the approximate amount of moisture in the root zone, as a percent of that present at field capacity (Table 8-3). Many soils also change color as the wilting coefficient approaches. Usually the soil is darker at the field capacity and becomes lighter as the moisture content decreases. To more fully appreciate the moisture characteristics of your soil try to correlate changes observed by the balling method with observed color changes.

Evaluation of plant characteristics is also a good index of the need for water. Watch for temporary wilting of shallow-rooted crops such as onion in the hot afternoons. Note also changes in the color of foliage and the amount or lack of young shoot growth. The older leaves at the base of cucumber, cantaloupe, and tomato plants, for example, may change slightly in color, becoming a darker green, often almost bluish or grayish, instead of the normal lighter green color as a result of retarded growth. Often small areas near the ends of rows, or in corners, do not get watered as thoroughly as the rest of the garden. Use these areas for comparison of growth rates in the rest of the garden.

Determining Water Application Rates

Once having determined the need to water, how much water should be applied to the soil? Enough should be added to wet the soil to a depth of *at least* two feet at each watering. Deep watering is essential, because most of the water in the surface four to six inches of soil is lost through evaporation. Additionally, vegetable plants as a group get the majority of their water from the upper two feet or so of soil and cannot achieve maximum growth unless this zone is kept moist (Fig. 8-18). But how do we tell when water has penetrated to

TABLE 8-3 Chart for Determining the Amount of Moisture in Medium-to-fine-textured Soils. With Sands and Sandy Loam Soils the Balls Are More Friable and Fragile Throughout the Whole Range.

Degree of moisture	Feel	Percent of field capacity
Dry	Powder dry	0
Low (critical)	Crumbly, will not form a ball	Less than 25
Fair (usual time to irrigate)	Forms a ball, but will crumble upon being tossed several times	25–50
Good	Forms a ball that will remain intact after being tossed five times, will stick slightly with pressure	50–75
Excellent	Forms a durable ball and is pliable; sticks readily; a sizable chunk will stick to the thumb after soil is squeezed firmly	75–100
Too wet	With firm pressure can squeeze some water from the ball	More than field capacity

From Strong, Sprinkler Irrigation Manual, *1956. Used with permission of Wright Rain, Ringwood, England.*

Here's Why Shallow Watering Is Bad

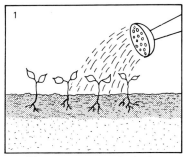

When you water to a depth of only an inch or two, the little rootlets stay up near the surface since they always seek water.

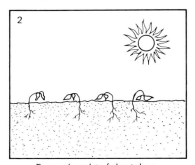

Even a drought of short duration can kill young plants by drying up the soil at the surface. This can happen very quickly.

FIGURE 8-18 The importance of deep watering. (Courtesy of Northrup King Co., Minneapolis, Minnesota.)

Here's Why Deep Watering Is Good

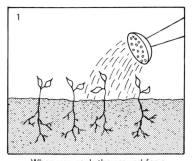

When you soak the ground for a considerable distance down, plants are able to sink deep, healthy roots.

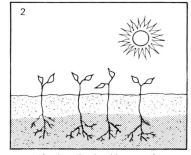

If a drought should occur, the plant's vigorous root system lets it draw on that deep "water bank" and keep on thriving.

a depth of two feet? The best way is to determine their water-holding capacity.

Some representative values for available water in soils are as follows:

Soil	Available Water (inches per foot)
Coarse sand	0.5
Very fine sand	1.2
Very fine sandy loam	1.9
Silt loam	2.1
Silty clay	2.6
Clay	2.8

Thus, a silt loam soil at the wilting coefficient would require 2.1 inches of water to bring the first foot of depth back to field capacity, or 4.2 total inches of water for two feet of depth; the minimum depth of water penetration desired. Of course, if only half of the available water had been removed from the soil, then 2.1 inches of water ($\frac{1}{2} \times 4.2$) would penetrate to the two-foot depth. One important point that should be made here is that, while plants can remove *any* proportion of the available water we allow them to remove, there is no way the gardener can return the soil to only one-half, three-fourths, or seven-eighths of field capacity. When water is added to the soil it moves downward (and to some extent laterally) through the force of gravity as a solid front. Thus, in our example of a silt loam soil at the wilting coefficient the addition of 2.1 inches of water would return approximately the first foot of soil to field capacity—it *would not* restore the first two feet half way to field capacity.

How do we determine when we have added the desired volume of water to our garden soil? As it turns out, one inch of water is equivalent to about 27,000 U.S. gallons per acre, or 0.62 gallons per square foot of garden space. There are simple ways to measure how much water you are giving your garden. One way is to place jars or cans under the sprinkler pattern, as illustrated in Fig. 8-13. Measure the depth of water in the containers at the end of the watering period, and take the average as the depth of water being applied. If the time of watering was noted, future waterings can be based on this time interval. Another way to compute water output is to measure the flow from your garden hose at a particular force. The output, in gallons, can be used to determine depth of water penetration in a given soil.

With the above information you should be able to answer questions such as the following:

Mr. Green has a 30 × 40 foot garden plot comprised of silty clay soil. How much water (in inches) will it take to bring the soil to field capacity to a depth of 30 inches if it is now at the wilting coefficient? How many total gallons of water will be required?

If you can answer this question you will be able to determine the water requirements for your own garden.

Water Conservation Ideas

Because of the increasing demands being made on water and water shortage problems sometimes encountered, water conservation has become an increasingly important topic. We have already mentioned a number of sound water conservation practices, including skillful cultivation and the control of weeds, mulching, and informed watering techniques. Sometimes, however, additional conservation ideas are required if one is to garden successfully. The experiences of a rural gardener of our acquaintance illustrate this point admirably. The drilled well on our friend's property produces, at best, only two to three gallons of water per minute, scarcely enough for household use, let alone for the garden. In spite of this limitation our friend has a large and beautiful garden each year. To accomplish this he has constructed a series of elevated, interconnected barrels into which water slowly trickles during the night when household water demands are low. As an added feature the barrels are painted a dark color allowing maximum absorption of solar energy and thereby more rapid warming of the cold well water. Outlets at the barrel bottoms allow the warmed water to be released into a drip system as needed. Using this system a luxuriant, high-yielding garden is grown each year, even though available water is in short supply.

Using "Graywater"

Throughout much of the West drought problems may become so severe that little or no fresh water is available for the garden. What do you do then? During such periods of crisis the dedicated gardener may have to use household wash water, also known as **"graywater,"** for his plants. While this water can be used for gardens, its use presents problems. The major problems relate to the fact that household water (1) may contain food residues and other organic matter that attract insects and rodents, (2) may contain bacteria and viruses that cause human illness, and (3) may contain chemicals that alter soil structure or are toxic to plants.

Bathwater and rinse water from dishes and laundry can be used for plants, but soapy wash water should not be used in the garden because of its probable chlorine and boron content. Chlorine can cause injury to plants, particularly if it touches foliage. The chlorine problem can be prevented by letting water stand for 24 hours before using on plants. Boron is present in many laundry products and, while essential for plant growth, could accumulate to toxic levels in the soil. This problem can also be prevented, by choosing laundry products low or lacking in boron. The phosphates present in detergents will not hurt plants and, as a matter of fact, act as plant fertilizer. If your home has a water softener your water will contain more sodium than normal, and over a period of many months the excess sodium could alter soil structure (see Chapter 4). This problem can be counteracted by mixing gypsum into the soil at a rate up to 25 pounds per 100 square feet. Remember, too, that the possible harmful effects of wash water can be reduced by dilution with water of higher quality. For health reasons, avoid letting household water touch edible plant parts during watering. Graywater should be used only during periods of severe water shortage.

FERTILIZING

The best time to fertilize is when preparing the garden soil for planting. In addition, it is often desirable to add more fertilizer after crops are growing. Most vegetables benefit from one or more applications of nitrogen during the growing season, particularly on more sandy soils and on soils where organic residues have been incorporated. A light green foliage color is characteristic of a nitrogen shortage, although a shortage of water can also cause poor leaf color. Other elements can be added too, and it is often convenient for the gardener to use the same complete fertilizer, or manure, that he or she used when preparing the garden soil.

Why Fertilize During the Growing Season?

By fertilizing during the growing season (1) plant growth rates can be maintained at high levels; (2) small quantities of fertilizer can be added (It is safer and more economical to apply fertilizers a small amount at a time.); and (3) fertilizer can be supplied (or withheld) selectively to crops to elicit specific growth responses. In this regard it should be remembered that, while nearly all vegetables benefit from the addition of extra nitrogen during the growing season, leafy vegetables such as cabbage, kale, lettuce, and spinach require more nitrogen than most other crops. The tuber and root crops, including beet, carrot, the potatoes, and turnip require a high percentage of potassium, and the fruit crops respond well to phosphorus additions. Remember also that too much of a fertilizer can reduce crop yields by producing excessive vegetative growth or through toxic reactions.

How to Apply Supplemental Fertilizers

The most common method of applying supplemental fertilizer is as bands along one or both sides of rows, a practice known as **sidedressing**. The method is similar to the banding procedure illustrated in Fig. 7-9 except that in sidedressing the bands are placed further away from the rows to avoid injury to the more extensive root systems of larger plants. Fertilizers can also be applied through irrigation systems. As mentioned earlier, accessories can be added to the home garden drip system that will allow the use of fertilizers such as ammonium nitrate, ammonium sulfate, calcium nitrate, and sodium nitrate without clogging or corrosion. In applying fertilizers through drip watering the manufacturer's directions should be followed closely. Fertilizers can also be applied in water solution to aerial plant parts, a practice known as

foliar fertilization. Commercial growers, however, have for years made foliar applications of fertilizers, particularly micronutrients, to orchard trees and vegetables with a high degree of success. Now, Dr. John Hanway, an Iowa State University scientist, has found that, by varying the kinds of fertilizer solutions used and the timing of their application, yields of soybean could be greatly increased, while causing only minimal leaf damage. Although much more research remains to be done in the area of foliar fertilization results such as these will no doubt find their way to the home garden before too long.

Timing and Rates of Supplemental Fertilizer Applications

The timing and rate of application of supplemental fertilizers will vary. A sidedress to leafy crops should occur after plants are well established and at least 4 to 6 inches high, to sweet corn when it is 12 to 15 inches high (earlier if deficiency symptoms are obvious), and other fruit crops such as cucumber, pepper, and tomato at about the time they begin to set fruit. Normally, a sidedress can be applied at about half the rate of preplanting fertilizer applications; that is, at 1 pound of 10-10-10 per 100 square feet, or 50 feet of row space when rows are 2 feet apart (Table 7-4). Nitrate of soda, ammonium nitrate, and urea are commonly used sidedress fertilizers. These high-analysis fertilizers (Table 5-8) usually provide nitrogen at a lower unit price but are more difficult to distribute evenly. As with preplanting fertilization, rates should be geared to nitrogen content. For those who prefer to use organic fertilizers for sidedressing, dried manure can be applied at a rate of 10 pounds of 1-1-1, or its equivalent, per 50 feet of row.

To sidedress a crop distribute the fertilizer in a band along one side of the row and about 4 to 6 inches from the plants, depending on size. Then hoe the fertilizer into the surface few inches of soil. If rainfall is not likely then water should be added to make the fertilizer more quickly available to plant roots. When adding a sidedress take care not to get chemical fertilizers on foliage as it may cause burning, particularly if the foliage is wet.

WEATHER-MODIFYING PRACTICES

Virtually all gardeners have, at one time or another, utilized some weather modifying, or weather circumventing, practice. These same kinds of practices must be employed on an almost daily basis by gardeners in more northern areas and at higher altitudes to assure successful vegetable production. In general, these practices are aimed at either (1) lengthening or circumventing short growing seasons, (2) moderating excessively hot or cold temperatures, or (3) providing protection from the wind. Using these techniques can mean the difference between success and failure in gardening.

The major ways the gardener can manipulate or circumvent the normal climatic limitations include:

- Choice of garden location. A wise choice of garden site is a major way of circumventing frost. Unfortunately, gardeners commonly have little flexibility in site selection.
- Informed cultivar selection. Use of rapid-maturing and hardy cultivars are among the best ways to overcome the limitations imposed by short growing seasons.
- Use of raised beds. Raised beds warm up earlier, can be planted earlier, and allow earlier harvests. In effect, the gardener can extend the growing season by employing raised beds.
- Use of transplants. Their use allows the gardener to lengthen the growing season by starting plants indoors, moving them to a coldframe, and then to the garden.

- Use of mulches. Mulching represents perhaps the most powerful tool available to gardeners for modifying the environment around plants. Through wise use of mulches the gardener can extend the growing season, and provide protection against extremes of hot and cold.
- By making fall plantings of the same cool season crops as were planted in spring. This practice provides the middle latitude gardener with two cool seasons per growing season. Through the use of late summer plantings the gardener can add months to his growing season. Some cool-season crops are grown more successfully in the late summer.

Protecting Plants Against Frost Injury

One frost protection practice is the application of water to foliage. Frost on foliage results from the freezing of dew. If the warm sun strikes leaves with frost still on them it can cause irreversible destruction of plant cells. To prevent this form of damage gardeners can spray foliage with water in the early morning. The applied water elevates the temperature of plants towards its own temperature, and thereby prevents the damage that can be caused by the sun's rays striking the plant. In addition, gardeners can keep their sprinkler systems going much of the night as a means of frost protection. The water applied throughout the night not only elevates plant temperatures, but also releases energy—in the form of heat—if it is converted from the liquid to the solid state. Both of these actions are effective in providing protection against cold injury.

On small plants the formation of frost can be prevented by the use of structures such as hot caps, cloches, and other protective coverings (Fig. 8-19). With all of these structures the principle of protection is the same: they either allow—or should be removed to allow—the penetration of short rays from the sun during the day to warm the soil. At night the protective covering traps the long rays leaving the soil, much like a layer of clouds or the eves of a house, thus creating a warmed zone around the plant. **Hot caps** are a widely used, commercially produced, protective covering for plants. These transparent structures, which come in a range of sizes, allow for the penetration of short rays from the sun while restricting the back radiation of long rays. Around the base of the cap a rim or lip is present for use in anchoring the cap to the soil. Small holes in the cap allow for needed ventilation, but caution must be exercised. On hot days excessive heat buildup may occur, requiring removal of the hot cap for a time. In any event, as plants enlarge and begin to touch the top of the cap it can be torn away, leaving only the sides for protection. When all danger of frost is past the entire remaining disc can be removed. While tents, cloches, A-frames, paper bags, and the like may not be quite as convenient to use as hot caps they work well.

The maintenance of abundant soil moisture is another means of cold protection. Water in the soil pore space enhances the movement of heat upward from lower layers by conduction. Protective structures such as hot caps also serve to keep moisture loss through evaporation to a minimum.

Enhancing Solar Radiation

Gardeners are sometimes faced with the problem of not enough solar radiation. One way of dealing with sites that receive only limited sun exposure is to plant crops such as the leafy vegetables that grow well in partial shade. If the entire garden is in limited sun, reflective panels of aluminum foil or metalized plastic will focus more solar energy to crops that need long daily exposure to full sun (Fig. 8-20).

(a)

(b)

(c)

(d)

(e)

(f)

FIGURE 8-19 A variety of simple structures that gardeners can use to protect plants from extremes of temperature and wind. (Courtesy of Guerry Dean.)

The sun's energy can also be captured and stored by use of water-filled inner tubes thereby promoting the growth of sun-loving crops like tomato (Fig. 8-20).

Sun and Wind Protection

Too much sun and wind present problems, especially for tender transplants. For these, A-frames, boards, shingles, or shakes can be conveniently placed to provide needed shade and also protection against wind (Fig. 8-19 *a, c*). For larger vegetables and young trees, shade cloth on a wooden frame, or a lath structure, can be used to effectively screen the foliage of plants from sun and wind. A properly located hedge or fence is, of course, the best form of permanent protection against wind.

FIGURE 8-20 Reflective panels and water-filled inner tubes can be used to channel solar energy to vegetables. (Courtesy of Guerry Dean.)

PLANT-SUPPORT TECHNIQUES

It may be necessary to provide support to plants that make use of vertical space. Such measures may also modify the environment around plants. For example, gardeners employ espaliering of fruit trees to conserve space and create interesting art forms (Fig. 14-11). But these are only two of the reasons why training trees to grow on a trellis against a wall is of value. A third, and very important reason, is that the reflected light and added warmth next to the wall enhance vegetative growth and fruit development. Similarly, cucurbits trained on a trellis or wall not only effectively utilize vertical space, create privacy, and afford beauty, they also thrive on the added warmth gained from their supported location (Fig. 14-9). This subject is also discussed in Chapter 10, which deals with the culture of individual vegetables, and in Chapter 14.

PESTS AND DISEASES

Because of the volume of literature on pests and diseases and the importance of plant protection in overall gardening success we have devoted Chapter 9 to the subject of protecting garden crops. We present the underlying principles of plant protection and follow this with discussions of plant diseases, insects and mites, weeds, and other pests. Additionally, we cite the most serious insect and disease problems of the individual crops as a part of discussion of those crops in Chapters 10 through 13.

Questions for Discussion

1. Define the following terms:
   - (a) Mulch
   - (b) Blossom end rot
   - (c) Emitters
   - (d) Irrigation point
   - (e) Graywater
   - (f) Hot cap

2. What are the main reasons for cultivating garden soils?

3. Prepare a set of guidelines on effective cultivation techniques for use by beginning gardeners.

4. Name four materials used for mulching and tell the advantages and disadvantages of each.

5. List several ways in which mulches enhance production of vegetables.

6. The authors indicate that mulches can be used either to lengthen or to shorten the growing season. How is this possible?

7. Describe at least three different methods of watering and indicate where each method would work most effectively.

8. Draw a plan for the installation of a drip system in your own garden, or in the garden shown in Fig. 7-2.

9. What options are available to the gardener to help in determining the percent of available water remaining in the soil?

10. In a soil initially at field capacity, growing plants are able to remove, for example, 50 percent of the available water. If the same soil is at the wilting coefficient it is not possible for the gardener to apply just enough water to bring it to the same 50 percent level. Explain.

11. How much water (in inches) would it take to bring a clayey soil now at the wilting coefficient back to field capacity to a depth of 30 inches? A silty loam soil?

12. Kelly Jones has a 15 × 20 foot garden plot comprised of a silt loam soil. How much water (in inches) will it take to bring the soil to field capacity to a depth of 3 feet if it is now at the wilting coefficient? How many total gallons of water will be needed?

13. You are preparing to start a garden on a plot that has not been gardened for years according to neighbors. Outline a long-range program of fertilization that should ensure a successful gardening effort.

14. The growing season in Timmons, Ontario, averages 93 days. What measures would you take to overcome the limitations imposed by the short growing season in this area?

Selected References

Aldrich, S., W. Scott, and E. Leng. 1975. *Modern Corn Production.* A & L Publishing, Champaign, Ill.

Chan, P. and S. Gill. 1977. *Better Vegetable Gardens the Chinese Way.* Graphic Arts Center Publishing Co., Portland, Ore.

Faust, J. 1975. *The New York Times Book of Vegetable Gardening.* Quadrangle/The New York Times Book Co., New York.

Frese, P. (ed.). 1957. *Handbook on Mulches.* Brooklyn Botanic Garden, Brooklyn, New York.

Gagnon, M. 1977. Mulches "foil" aphids. *Vegetable Briefs for California Farm Advisors,* Issue No. 192.

Janick, J. 1972. *Horticultural Science.* Freeman, San Francisco.

Janssen, K. 1977. Foliar fertilization, an old practice with possible new potential. *Vegetable Growers News,* 32(1):1, P.O. Box 2160, Norfolk, Va.

Johnstone, D. and E. Brindle. 1976. *Vegetable Gardening Basics.* Burgess Publishing Co., Minneapolis, Minn.

McGilvray, S. 1977. How to water your garden less and enjoy it more. *Horticulture,* 55:24–26.

Muzik, T. 1970. *Weed Biology and Control.* McGraw-Hill, New York.

O'Dell, C. and R. Morse. 1976. Improvement of muskmelon production in western Virginia. *Vegetable Growers News,* 31(3):2–3, P.O. Box 2160, Norfolk, Va.

Ray, R. (ed.). 1974. *Weather-Wise Gardening,* Ortho Books, San Francisco.

Spencer, E. 1957. *All About Weeds.* Dover Publications, New York.

Steffek, E. 1978. Do yourself and your garden a favor—Use mulches. *Flower and Garden,* 22:12–16.

U.S. Dept. of Agriculture. 1977. *Gardening for Food and Fun.* Government Printing Office, Washington, D.C.

University of Calif. 1977. Using household waste water on plants. Leaflet 2968.

Vandemark, J., M. Shurtleff, and R. Randell. 1973. *Illinois Vegetable Gardening Guide.* Circular 882, Urbana, Ill.

Ware, G. and J. McCollum. 1975. *Producing Vegetable Crops.* The Interstate Printers and Publishers Inc., Danville, Ill.

LEARNING OBJECTIVES

By the time you have finished this chapter you should be able to:

1 Name and describe several categories of living organisms that may cause damage to garden crops.

2 Define the term "pest" as used in the text.

3 Describe symptoms of plant damage caused by disease organisms and insect pests, emphasizing distinguishing characteristics of each.

4 Identify the pest control options available to the gardener, and comment on the advantages and disadvantages of each approach.

5 Discuss the computation and usefulness of LD_{50} values.

6 List three useful insects and state their value to the gardener.

7 Comment on measures you would implement to reduce the incidence of virus disease in the garden.

8 Discuss the status of fungi as plant pathogens using appropriate examples.

9 Describe the two distinct types of life cycles occurring among insects and tell how a knowledge of these may aid in pest control.

10 Insects vary in their feeding behavior. Explain how these variations might influence the type of insecticide used for control.

11 Explain what "caterpillars" are, and cite specific examples of plant damage caused by these pests.

12 List the four categories of weeds most commonly dealt with by gardeners, and indicate which categories, if any, present more difficult control problems.

13 Tell why herbicides are used rarely in home gardens.

14 Name three pest animals and cite measures effective in their control.

9

PROTECTING GARDEN CROPS

"An ounce of prevention is worth a pound of cure."

Wisdom of Many Ages

The harvest of vegetables from the garden can be reduced by a multitude of yield-reducing agents. Some of these agents are environmental in nature, as when yields are reduced due to nutrient deficiencies or mineral toxicities, lack or excess of soil moisture, diminished supply of oxygen, unfavorable soil pH, too low or too high temperatures, shortage or excess of light, and air pollutants. All of these environmentally caused plant disorders have been discussed in preceding chapters.

In addition to environmentally induced plant disorders, there are a large number of plant maladies caused by living organisms. In the process of growing on or burrowing into vegetables pests cause roots to grow abnormally, stems to rot, leaves to curl, and fruits to be misshapen and fall prematurely. In so doing they damage or destroy plants and reduce yields (Table 9-1). Because gardens are a collection of a few types of plants, often related, pests may multiply faster than in natural plant communities. Gardeners must, therefore, act quickly when garden pests build to damaging levels.

The organisms responsible for plant disturbances show great diversity. They may be of a microbial, plant, or animal nature, and they come in sizes ranging from invisible to the unaided eye to large, easily seen organisms such as weeds and rabbits. Some of the organisms are general in the plants they attack while others are quite specific. Moles, slugs, aphids,

TABLE 9-1 Estimated Average Annual Quantity Losses (in Thousands of Tons) Caused to Selected Vegetables by Insects, Diseases, and Weeds in the United States, 1951–1960

Crop	Actual production	Potential production	Losses due to Insects	Losses due to Diseases	Losses due to Weeds	Totals
Snap beans	665	944	79	164	46	289
Dry beans	1024	1536	205	174	133	512
Fresh market cucumbers	228	310	27	41	14	82
Sweet corn	2431	3189	462	53	243	758
Lettuce	1686	2157	127	217	127	471
Onion	1365	1952	246	273	68	487
Peas	587	821	23	135	76	234
Sweet potato	900	1083	49	110	24	183
Fresh market tomatoes	1095	1567	77	318	77	472

Courtesy of Farbenfabriken Bayer Ag. Leverkusen, Germany.

morning glory, and the damping-off fungi are examples of nonspecific organisms. The Colorado potato beetle and the apple scab and corn smut fungi, on the other hand, have a very restricted host range. Finally, organisms can be injurious at specific stages of plant growth. Thus, damping-off is essentially a disease of young seedlings, whereas adult corn rootworms feed on corn silks at a later stage of plant development.

For purposes of convenience garden pests may be divided into four groups, as follows:

- Viruses, bacteria, mycoplasma, fungi, nematodes.
- Insects and closely related organisms.
- Weeds.
- Larger animals, such as slugs, snails, rabbits, and deer.

These four groups are the subject of this chapter.

The organisms listed in group 1 are customarily regarded as disease agents, and the plant disturbances they cause as **diseases.** Plant pathologists also consider plant damage by nonliving agents as diseases. Boron deficiency, sunscald, and leaf burn due to excessive fertilizer, for example, are regarded as diseases.

Usage of the term "pest" presents some confusion. In scientific circles the insects and closely related organisms such as mites, ticks, and spiders are termed pests, thus differentiating them from diseases. The term is used in a somewhat broader way too, in that the word pesticide refers not only to insecticides, but to materials such as fungicides, herbicides, and rodenticides. Furthermore, larger animals are frequently referred to as pest animals.

What then are pests? We regard **pests** as living organisms that disturb our garden crops in any way that interferes with normal structure and function. Commonly these disturbances are manifest as reduction in yields, but they may also result in plant death. In essence, garden pests are unwelcome members of the garden community. If our gardening efforts are to be successful we must have a thorough understanding of each of the four groups of pests identified above. Our comprehension of these harmful organisms will be enhanced, however, if we first are aware of some basic principles of pest management.

PRINCIPLES OF PEST MANAGEMENT

Sound pest management requires that the gardener be able to determine the cause of plant problems. Once the causal agent is known it is then possible to begin a control program. Several methods of control are available to gardeners and understanding each one is prerequisite to a sound pest management program. Even when the causal organisms have been clearly identified and the control options are known, however, an important question remains: What level of control do we wish to attain? Pest control usually *does not* mean entirely eliminating the pest, although if the pest is a gopher, for example, it can be eradicated. More often, however, pest control involves reducing the number of pests to a nondamaging level.

Determining the Cause of Plant Problems

Identifying plant problems is the most fundamental part of pest management. Control measures are meaningless if they are not aimed at the causal organism.

Diagnosing pest problems is not an easy task. In general, the damage caused by disease organisms and insects is most difficult to diagnose correctly, but even the damage caused by larger organisms, such as slugs or bluejays, may be misinterpreted.

The disease organisms are invariably small and not easily seen (Fig. 9-6). Some are inhabitants of the soil, while others are carried from plant to plant by insects, wind, or rain. Generally, the presence of these organisms is not recognized until symptoms of disease are evident. Verticillium wilt disease of tomato, for example, is caused by soil-borne fungi. These organisms enter the plant roots and, by disturbing the vascular system, cause plants to wilt and die quickly. For diseases such as these, identification is somewhat academic; they must be controlled through prevention. Other diseases, such as the downy mildew and late blight fungal diseases begin as small areas on leaves of infected plants and spread. If identified early these damaging pests can be controlled.

Diagnosing insect problems is also challenging. These pests may be small, well camouflaged, feed only underground or at night, or be well concealed by foliage or soil. The cabbage looper, for example, resembles strikingly the foliage of cabbage in color. Furthermore, it feeds mainly on the underside of leaves or burrows into heads, and for these reasons is not easily found. Cutworms, which attack a variety of garden crops, commonly feed on foliage at night, burrowing into the soil by day. To find these pests it is usually necessary to carefully examine the soil around crops.

To complicate matters even further, both disease and insect pests may induce symptoms of a similar nature. Distortion in the shape of plant parts, loss or discoloration of plant parts, and plant death are common denominators of both types of pest damage. Proper identification comes from knowing the subtle differences in these general symptoms caused by each type of organism.

Diseases caused by bacteria and fungi usually produce sunken, discolored spots on foliage and fruits. These spots represent the initial points of infection, and on leaf surfaces turn gray, brown, or black in color. The infected areas are also characterized by one or more zones of different color surrounding the central darkened spot. A typical infection, for example, may have a darkened center of dead tissue, bordered by a circle of yellow or tan dying tissue, and that surrounded by a region of pale green tissue representing the area into which the disease is spreading. As an infection spreads the individual infections may merge, forming large, irregularly shaped areas of dead and dying tissue. Viral infections cause numerous symptoms, including (1) distortion or dwarfing of plant parts, (2) yellow streaking, blotching, or mottling of leaves, flowers, and fruits, (3) a reddish discoloration of green foliage, and (4) the formation of abnormally small, nonfruit-producing flowers. Some of the more common symptoms of bacterial, fungal, and viral infections are illustrated in Fig. 9-1. Additional aspects of disease identification are discussed later in the chapter.

Insects cause damage to plants by sucking, chewing, boring, and inducing the formation of galls. Foliage discoloration caused by sucking insects appears as yellowish or bronze-colored areas without the pattern of zonation referred to above. The density of the discolored areas depends on the severity of infestation.

Chewing insects leave holes that may be small or large in diameter, go partially or entirely through leaves, and appear at leaf margins or on the leaf surface. Unlike disease organisms, chewing insects commonly leave excreta in the vicinity of feeding areas. Boring insects leave entrance holes on foliage parts, and in leaves their pathways through internal tissues may be clearly visible. Gall-forming insects induce the formation of characteristic growths on leaves. Manifestations of these types of insect damage are shown in Fig. 9-2. Insect damage symptoms are also discussed later in the chapter.

Armed with the kinds of information outlined above it is often possible to determine the organism causing plant problems. Even when the causal agent cannot be determined, however, all is not lost. Your local agricultural agent or farm advisor is equipped to diagnose plant problems, either directly or by forwarding specimens on to the trained specialists at the provincial or state agricultural college. In either case, most pest problems can be quickly diagnosed and control recommendations provided.

Methods of Pest Control

For any pest to thrive certain conditions are necessary, as follows: (1) the pest must be present; (2) conditions must be favorable for it to multiply and attack plants; and (3) the plants must be susceptible to attack by the pest. In essence, these three conditions represent links in the chain of plant pest damage (Fig. 9-3). As such, if any link is weakened—altered to a pest's disadvantage—a measure of pest control is realized.

The pest control measures available to gardeners each work at disrupting one or more of the links in the chain of pest damage to plants. For simplicity, these control measures can be divided into five categories, as follows:

- Pesticides.
- Resistant crop cultivars.
- Cultural practices.
- Biological controls.
- Mechanical methods.

As we discuss the values and limitations of each of these control categories see if you can determine which link, or links, in the chain of pest damage are being altered.

Pesticides

Pesticides are defined broadly as materials that kill or control pests. While numerous kinds of pesticides are recognized, those most commonly used by gardeners include:

- Bactericides—control bacteria.
- Fungicides—control fungi.
- Herbicides—control weeds.
- Insecticides—control insects and related pests.
- Nematicides—control nematodes.
- Predacides—control pest animals.
- Rodenticides—control rodents.

Pesticides carry out their actions in several ways and can be grouped according to their modes of action. **Contact pesticides** kill pests simply by contacting them. **Stomach poisons**

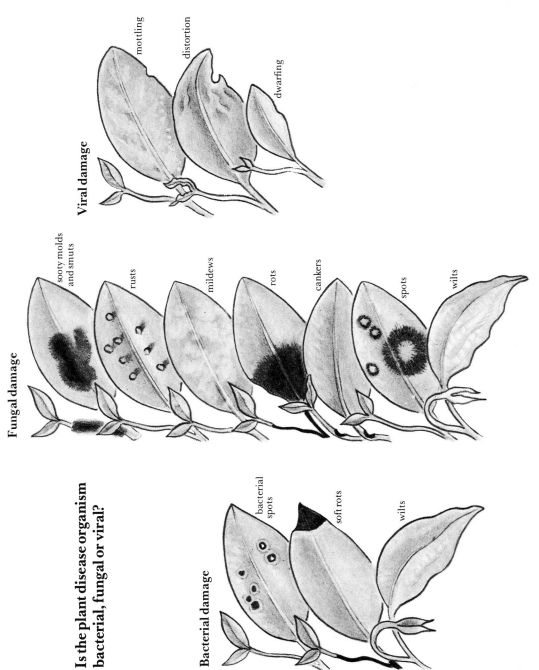

Is the plant disease organism bacterial, fungal or viral?

Viral damage

mottling
distortion
dwarfing

Fungal damage

sooty molds and smuts
rusts
mildews
rots
cankers
spots
wilts

Bacterial damage

bacterial spots
soft rots
wilts

FIGURE 9-1 Characteristic symptoms of bacterial, fungal, and viral infections of foliage. (Courtesy of James F. Gauss.)

Is the plant pest organism a sucking, chewing, boring or gall-making insect?

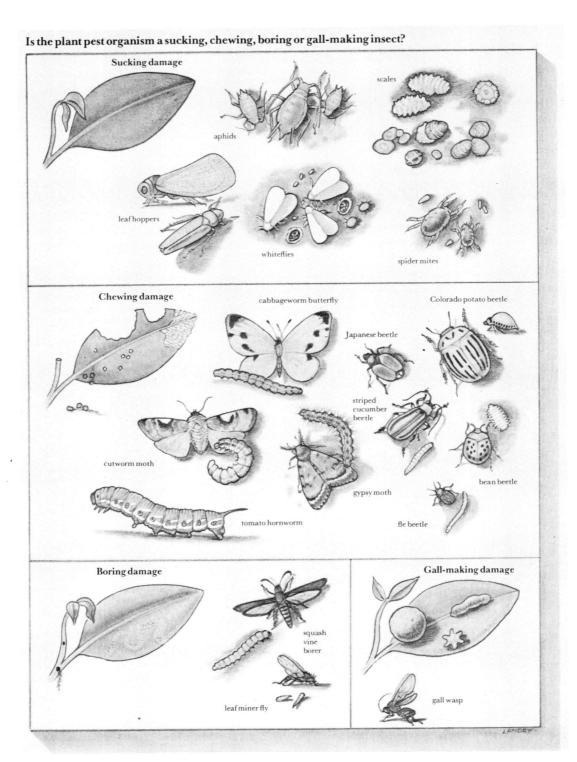

FIGURE 9-2 Examples of plant damage caused by insects. (Courtesy of James F. Gauss.)

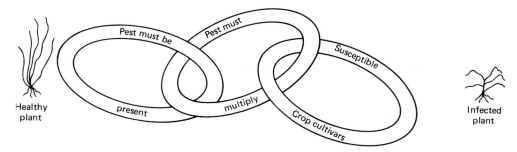

FIGURE 9-3 Three links in the chain of plant pest damage. Measures that weaken any link give a degree of pest control.

are pesticides that kill when swallowed. **Systemic pesticides** kill pests by being taken into the sap of a plant on which the pest is feeding. **Translocated herbicides** kill plants by being absorbed by leaves, stems, or roots and moved (translocated) throughout the plant. **Fumigants,** a category of pesticides discussed in Chapter 7, are gases that kill when they are inhaled or otherwise absorbed by a pest.

A given pesticide may act in more than one way. When a systemic insecticide is applied to the soil around vegetables, for example, the granules may kill pests with which they come in contact, or ingested granules may kill the pest that devoured them. Once the granules are dissolved in the soil solution, however, and the toxic ingredient taken up by plants the pesticide acts in a systemic manner, either killing or debilitating pests that feed on plant parts containing the toxic ingredient.

Another way in which pesticides may be viewed is on the basis of their selectivity. A **selective pesticide** carries out its actions against certain kinds of plants or animals without causing serious injury to others. The herbicide 2,4-D, for example, causes severe injury to many broadleaved species while leaving monocotyledons such as grasses more or less untouched. Thus, this herbicide may be put to use in removing field bindweed or Canada thistle from corn. **Nonselective pesticides,** on the other hand, injure or kill most of the plants or animals with which they come into contact.

Said in simple terms, they can't tell the "good guys" from the "bad guys." The soil fumigants display this type of action. The nonselectivity of many pesticides is a major drawback to their use.

Pesticides vary greatly in their toxicity. Toxicity is most often expressed by means of LD_{50} values. Such a value is a statistical estimate of the oral dosage necessary to kill 50 percent of a large population of test animals, usually white laboratory rats, under standarized conditions. The lethal dosage is stated in terms of milligrams of pesticide per kilogram of body weight of the animal. The higher the LD_{50} value of a pesticide the more it takes to be lethal. Said another way, the higher the LD_{50} value the lower the toxicity. Some representative LD_{50} values are given in Table 9-2. To give perspective to the meaning of LD_{50} values we have included in our list two compounds found in every home, aspirin and table salt. While LD_{50} values are useful for comparative purposes, it should be kept in mind that humans may vary in susceptibility to these chemicals as compared to the rat.

All pesticides should be used with caution. This warning applies not only to the synthetically manufactured materials, but also to those derived from naturally occurring organic sources. For example, a number of insecticides are derived from plants or plant parts. These insecticides, known as **botanical insecticides,** or simply as **botanicals,** include ro-

TABLE 9-2 The LD$_{50}$ Values for Some Commonly Used Substances

Substance	Oral LD$_{50}$
Aldrin	39
Aspirin	750
Chlordane	335
2,4-D	375–666
DDT	217
Diainzon	250
Dibrom	250
Malathion	1375
Nicotine	83
Pyrethrum	200
Rotenone	132
Sevin (carbaryl)	850
Table salt	3332
Toxaphene	90
Vydate (oxamyl)	5.4

Courtesy of Oregon State University Cooperative Extension Service.

tenone, pyrethrum, nicotine sulfate, saba-dilla, and ryania. While certain of these botanicals are less toxic than the modern synthetic pesticides this is not always the case. The botanical insecticides pyrethrum and rotenone are far more toxic than "Sevin," a manufactured inorganic pesticide. In short, all pesticides, regardless of their origin, should be handled carefully, following exactly the directions given by the manufacturer on the label of the pesticide to be applied.

Pesticides also vary greatly in their longevity in the environment. Some, such as DDT, persist and accumulate in the environment and thereby harm or contaminate plants and animals—including humans—that were not intended as target organisms. The general public was first alerted to this problem in 1962 by Rachel Carson in her then startling book *Silent Spring.* A major problem pointed out by Miss Carson was that these chemicals may persist in the environment and build up to higher levels in each stage of the **food chain** (series of living organisms connected to each other as sources of food or nutrition for each successive member). This phenomenon, known

as **biological magnification,** is illustrated in Fig. 9-4. Fortunately, the feeding patterns of humans are such that we are normally protected from accumulating lethal doses of pesticides (or heavy metals). Nevertheless, as concerned gardeners, we should be fully aware of the potential long range effects that may result from the indiscriminate use of pesticides. This does not mean that pesticides should be excluded from our pest control arsenal; only that they should be used judiciously.

Resistant Crop Cultivars

One of the most reliable and economical means of pest control in the home garden is through the use of disease and insect resistant cultivars. Indeed, host resistance is the primary means of control for many important vegetable problems including those caused by viruses, vascular pathogens (such as the wilt-causing fungi), root rots, and the rust diseases of cereals. Cabbage yellows, onion smudge, common mosaic of bean, black wart of potato, and fusarium wilt of tomato are among the many pests controlled by taking advantage of host resistance. The use of resistant cultivars, where available, is always recommended but is especially important when a gardener has had vegetable losses due to a particular pathogen.

Vegetables vary in their resistance to diseases and insects, just as they differ in their response to environmental conditions. A particular crop cultivar may be **immune** to a pest; that is, it is not attacked by the pathogen even under the most favorable conditions. Although this is a rather uncommon situation it does occur with diseases like cucumber scab where a plant is either totally immune or totally susceptible. More often, however, crops show various degrees of resistance ranging from almost immunity to complete susceptibility. Crop cultivars are said to be **resistant** when they are capable, to some degree, of enduring an attack by a particular pathogen.

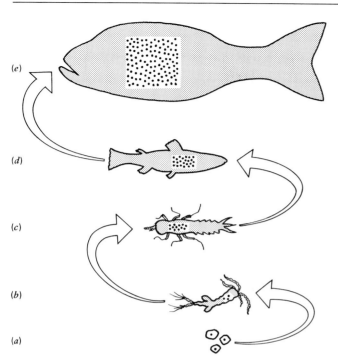

(e)

(d)

(c)

(b)

(a)

FIGURE 9-4 Biological magnification: (a) Some individuals of a single-celled plant species at the bottom of a food chain have picked up a small amount of a stable non-excretable chemical: (b) *Cyclops*, a small crustacean, incorporates into its own tissues the chemical from all the infected plants it eats; (c) a dragonfly nymph stores all the chemical acquired from the numerous *Cyclops* it eats; (d) further magnification occurs when a minnow eats many of the infected dragonfly nymphs; (e) when a bass, the top predator in this food chain, eats many infected minnows, the result is a high concentration of a chemical that was much less concentrated in the organisms lower in the chain. (Reproduced from William T. Keeton, *Biological Science,* Third Edition. Illustrated by Paula DiSanto Bensadoun, with the permission of W. W. Norton & Company Inc. Copyright © 1980 by W. W. Norton & Company, Inc.)

They are **tolerant** if they are moderately susceptible but capable of surviving and producing acceptable yields when attacked by a pathogen. **Susceptible** cultivars are easily subject to injury or damage from attack by a particular pathogen and as a result give poor yields or none at all. Normally a resistant or tolerant crop cultivar is easier to protect from damage by a particular pathogen than is a susceptible cultivar, when accepted control measures or cultural practices are used.

The nature of resistance in crop plants is, in many cases, poorly understood. It is known, however, that resistance is conditioned by a number of internal and external factors which operate to reduce the chance and degree of infection. It is also known that the resistance of a crop cultivar to a pathogen may be "lost" or "broken down." This happens when a change (genetic mutation) in a pathogen results in a new variant, known as a **biotype, strain,** or **race,** which can successfully attack

a heretofore resistant cultivar. Thus, the loss of resistance in the plant cultivar is generally due to a change in the pathogen, *not* the host plant.

The loss of resistance of crop plants to particular pathogens has been accentuated in recent years through the use of pesticides. When DDT was first brought to the market, for example, it was hailed as the miracle insecticide; a means of stamping out once and for all insect-borne diseases (such as malaria) and winning the farmers' war against crop destroyers. Soon, however, many insects developed strains that were resistant to DDT. Thus, not only may pesticides linger in the environment and cause injury to organisms not intended as targets, they may also, when used repeatedly over a period of time, select for the development of more virulent forms of a pathogen. Indeed, even weed species develop biotypes resistant to particular herbicides. The development of pest forms resistant to particular pesticides is yet another reason for their judicious use.

Cultural Practices

Many cultural practices serve as effective means of pest control. Scientists in the United States Department of Agriculture and various state universities suggest the following measures to help reduce losses caused by plant pests.

- Use fertile, well-drained soil and high-quality fertilizer.
- Purchase high-quality, disease-free seed or plants.
- Treat seed with chemicals to protect against decay and damping-off.
- Plant crops that are suited to the soil and climate.
- Follow recommended planting dates.
- Purchase disease-free plants; make sure they are the proper size to give good growth (see Chapter 7) and are free of pest problems.
- Keep down weeds.
- Water plants early in the day so the moisture on the plants will dry off before dew appears.
- Destroy plants of each annual crop as soon as harvest is completed.
- Avoid the unnecessary use of insecticides that may kill beneficial insects.
- Practice crop rotation.

To this list could be added the point made in an earlier chapter pertaining to record keeping. Carefully kept records of crops grown, cultivars used, planting dates, fertilizer additions, pest problems and control measures, yields, and rotation schedules provide an informed base for current and future pest control measures.

Biological Controls

A number of biological methods are employed to control harmful insects and related organisms. In general, these methods fall into either of two categories—the use of natural predators or companion planting.

Natural Predators. A broad range of natural predators can aid the gardener in the fight against crop pests. Perhaps the first organisms that come to mind are larger animals such as birds and bats. Birds such as the towhee, wren, swift, thrush, and purple martin feed almost exclusively on insects. Efforts to increase the number of birds in your locality by building bird houses, baths, and feeders are strongly recommended. Many bats are insect eaters, and are capable of capturing enormous numbers of flying insects. Frogs and toads also have a place in the garden. The major portion of a toad's diet, for example, consists of aphids, cutworms, spiders, ants, caterpillars, squash bugs, slugs, and the like.

Perhaps more important as predators are the several species of insects which prey on insects destructive to vegetable crops. A few of these predaceous insects are illustrated in Fig. 9-5, and an expanded list is presented in Table 9-3. Certain of these insects, such as the

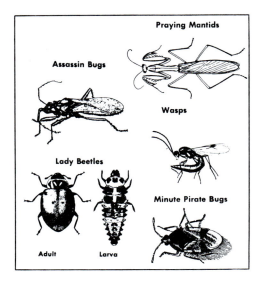

FIGURE 9-5 Some important beneficial insects. (Courtesy of U.S.D.A.)

TABLE 9-3 Some Beneficial Insects Including Distribution and Common Prey

Organism	Distribution	Prey species
Ant lion (doodlebug)	Many parts of United States; most abundant in South	Ants and other insects
Aphid lion (lacewing larvae)	Throughout the United States	Larvae feed on aphids, scales, thrips, and mites
Damsel bug	Throughout the United States	Aphids, fleahoppers, and small larvae of insects
Assassin bug	Throughout the United States	Immature forms of insects
Ground beetles	Throughout the United States	Adults and larvae feed on caterpillars and other insects
Lady beetles	Throughout the United States	Aphids, spider mites, scales, and mealybugs
Minute pirate bugs	Throughout the United States	Adults and larvae feed on small insects such as mites; they also feed on eggs and larvae
Praying mantids	Throughout the United States; most numerous in the northeast	Young mantids feed on aphids and other small insects; older mantids devour many kinds of larger insects
Spiders and mites	Many species occur throughout the United States	Aphids, beetles, caterpillars, cyclamen mites, flies, leafhoppers, spider mites, thrips, larvae, and eggs of many insects
Syrphid flies	Throughout the United States	Larvae eat insects
Wasps	Throughout the United States	Numerous mature insects and young caterpillars

Courtesy of U.S. Department of Agriculture.

praying mantids and lady beetles, may be purchased in bulk for release in the garden. The effectiveness of this method of control is reduced somewhat by the migratory habits of these insects, and by the fact that they may themselves serve as food for other organisms of prey.

The use of bacterial cultures is effective in controlling certain caterpillars. Interestingly, the bacteria involved, *Bacillus thuringiensis* and *B. popilliae*, reproduce only in the digestive tract of caterpillars and are completely harmless to other organisms. The bacteria are mass produced commercially and sold as a wettable powder—under the trade names of Bio-trol, Dipel, and Thuricide—which can be sprayed on plants. Application of this micro-bial agent has given moderate success in con-trol of cabbage loopers, the imported cabbage worm, and other larval pests.

Companion Planting. Companion planting is a form of interplanting in which plants are grown in specific combinations to help con-trol—by repelling or attracting—insects and other garden pests. The use of companion plantings to control pests is somewhat contro-versial because, while many gardeners have reported outstanding success using the method, the scientific community has been slow to doc-ument these successes or to determine the chemical agents responsible for reputed con-trol actions. Regardless of these controversial aspects, it is quite clear that many, if not all, plants emit powerful chemical substances into the air around them and from their roots into the soil solution. Some of these plant extracts have been shown to contain antibacterial or antifungal agents, and some have the ability to retard or inhibit the germination of seed of other plant species. An extract from cau-liflower seed, for example, has been found to inactivate the bacteria causing black rot.

Some of the plants that have been used in companion planting with vegetables for pest control purposes are listed in Table 9-4. The list is by no means complete, and through your own gardening efforts you may determine other effective combinations.

Mechanical Methods

While it would be impossible to list all of the mechanical controls employed by gardeners to control pests, some of the most popular and/or effective are as follows:

- Hand picking of insects where their size is large (as with the tomato hornworm) and their numbers small.
- Placing collars of cardboard or metal around the base of young plants to protect against cutworms.
- The use of mulches (Chapter 8) to control weeds and the pests they harbor.
- Employing aluminum foil and certain other mulches to repel aphids, Mexican bean bee-tles, and other insects.
- Using shallow pans or specially designed containers of stale beer to capture slugs and snails. These harmful night feeders are at-tracted by odors to the beer, fall in, and drown. Whether they fall in accidentally or

TABLE 9-4 Plants Used for Companion Planting with Vegetables and Reported Pest Control Actions

Plant	Reported actions
Marigold	Repels nematodes, bean beetles, and possibly Japanese beetles
Nasturtium	Deters aphids, squash bugs, and striped pumpkin beetles
Rosemary	Deters cabbage moth, bean beetles, and carrot fly
Garlic	Repels aphids and Japanese beetles; discourages mice
Soybean	Shields corn from chinch bug
Mint	Deters white cabbage moth and ants
Chive	Repels Japanese beetles
Onion	Repels many pests
Sage	Deters cabbage moth, carrot fly
Thyme	Repels cabbage worm

due to intoxication has not been determined with certainty!

- The use of fencing, netting, traps, noise-makers, and the like to control larger pest animals.

In summary, we have seen that several options are available to the gardener for use in combating many plant pests, including the use of (1) pesticides, (2) resistant cultivars, (3) good cultural practices, (4) biological controls, and (5) mechanical methods. No single approach will effectively solve all pest problems. The organisms that cause bacterial wilt of corn and bacterial blight of bean, for example, cannot be effectively controlled by crop rotation because these pathogens are carried on seed or by insects that spread them in the field. In short, the control methods used must be tailored to meet specific pest problems. Factors that influence which approach, or approaches, to pest control will be used include the nature and abundance of the pest, whether we are seeking to prevent pest problems or correct them, and the level of pest control desired.

A Philosophy of Pest Management

What constitutes a sound philosophy of pest management? An emerging philosophy among modern gardeners is that of preventing pest problems where possible, and reducing losses to acceptable levels in those situations where treatment is required. It is generally agreed that in most cases the total eradication of a plant pest is neither practical nor necessary. It is important only that we reduce pest numbers to a level where satisfactory yields will be obtained. After all, a few chewed or discolored leaves, some damaged fruits, and scattered weeds are of little consequence in an otherwise healthy garden.

Many of the preventative aspects of pest management are simply a matter of good sense. The cultural practices outlined above should

always be employed, and resistant cultivars should be used whenever available. In other cases more drastic preventative measures may be required. For example, many of the fungal diseases that affect leaves and fruits can only be prevented by applying the proper fungicide at the right time. Similarly, insect infestations, like those of radish by root maggots, can be prevented by properly timed applications of insecticide.

While prevention is unquestionably the cornerstone of pest management, there are always a number of pest problems that arise during the gardening season and require prompt treatment. Here too the gardener is presented with a number of options. A first line of defense should be the implementation of mild treatment procedures such as hand picking of insects, biological controls, or the use of live traps to capture pest animals. The use of pesticides should be employed only where other methods may not work or have not worked.

MANAGING VEGETABLE DISEASES

The Disease Agents

Plant diseases are caused by viruses, bacteria, mycoplasma, fungi, and nematodes. These organisms are small and generally not visible to the naked eye (Fig. 9-6). In spite of their small size, these pathogens have the potential to seriously disrupt plant growth and development.

Viruses

The viruses are submicroscopic, crystalline particles comprised of a protein sheath surrounding a central core of nucleic acids. Viruses possess many of the attributes of living organisms, including the ability to reproduce and to cause diseases. Of interest, the reproduction of viruses can take place only *within*

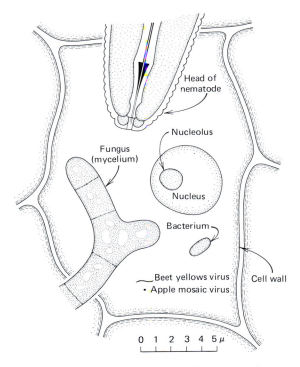

FIGURE 9-6 Sizes and shapes of common disease organisms in relation to a plant cell. (Reprinted with permission of Academic Press and George N. Agrios from *Plant Pathology* by George N. Agrios. Copyright © 1969 by Academic Press.)

living cells. Unlike cellular organisms, the viruses do not themselves divide; instead, they induce the host cells to form more viruses. If a given virus cannot multiply within host tissue infection will not occur.

Virus enter plant cells through openings caused by mechanical injury, through introduction by an insect vector, or through introduction into ovules by infected pollen grains. In susceptible plants the virus may multiply rapidly, and infections may be spread to other plants by aphids, mealybugs, mites, or other pests. Other common vectors are contaminated garden tools and human hands. Tobacco mosaic virus, for example, may be spread to susceptible plants by the hands of a cigarette smoker. During the period when annual vegetables are absent from the garden, viruses reside in perennial plants, soil, and in seed. These materials thus serve as reservoirs of virus, ready to infect other plants when introduced by a suitable vector.

The most common symptom of viral infection and sometimes the only symptom, is a reduced growth rate of the plant, resulting in varying degrees of dwarfing or stunting. Frequently, however, more obvious symptoms are visible on aerial parts. These symptoms include the formation of **mosaics** (light green, yellow, or white areas intermingled with the normal green of foliage) and **ringspots** (chlorotic or necrotic rings or patterns on leaves, stems, or fruits). Other symptoms include the clearing of leaf veins, leaf curling, and the formation of leaves that are more leathery than normal (Fig. 9-7). Infected flowers may appear streaked or fail to open properly, and fruits may be small or poorly shaped. Regardless of symptoms, almost all viral diseases cause some reduction in yields, and the life of infected plants is usually shortened.

Vegetable diseases caused by virus include common bean mosaic, cucumber mosaic, curly top, potato mosaic, and tomato bushy stunt. Most viruses have a wide host range.

FIGURE 9-7 Vein clearing of leaflets and stipules of a garden pea plant caused by bean mosaic virus. Note color of uninfected leaflets at lower right. (Courtesy of Dr. Frank P. McWhorter, retired. Oregon State University.)

Bacteria

Bacteria are small, usually single-celled microorganisms visible under an ordinary microscope. While the vast majority of bacteria are highly beneficial to humans, a relatively small number have been found to cause diseases in plants. These pathogenic bacteria penetrate plants through wounds or natural openings. Once in plants they live and reproduce either in the intercellular spaces or within cells. Bacteria have the ability to multiply within the plants at an astonishingly rapid rate, a feature that is correlated directly with their severity as pathogens. While bacterial infections can occur at almost any time, they are favored by warm, moist conditions. When unfavorable conditions prevail most bacteria have the ability to form thick-walled spores that can survive over a long period of time.

As with viral infections, the symptoms of bacterial infections are numerous. A common symptom with certain diseases is the appearance of darkened, water-soaked spots or streaks on leaves and stems. These spots may turn gray, brown, or black, and eventually fall out leaving small shot-holes in the leaf. Other bacteria cause soft rots of fruit, root, and storage organs, while still others invade the plant's vascular system, usually starting at the roots, causing wilts (Fig. 9-8). Symptoms of wilt disease include drooping or yellow foliage, usually starting at the top of the plant. Eventually leaves shrivel and dry, and they may drop off. Severely infected plants usually die, but less susceptible plants may show only slowed growth, sometimes accompanied by excessive blossoming and branching. Certain bacteria cause swelling or outgrowths of organs resulting in galls and tumors. Sometimes bacterial infections are difficult to distinguish from those caused by fungi. Rapid spread through-

FIGURE 9-8 Bacterial wilt of cucumber caused by *Erwinia tracheiphila*. (Courtesy of U.S.D.A.)

out the plant and the absence of fungal growth over infected areas help distinguish bacterial from fungal diseases.

Numerous vegetable diseases are caused by bacteria. Examples include fire blight of apple and pear, bacterial blight of bean, bacterial wilt of tomato, and soft rot of potato and other vegetables. In all, about 200 species of bacteria have been found to cause diseases in plants.

Mycoplasma

Mycoplasma are essentially bacterial cells without a normal, rigid cell wall. Aster yellows is a very serious disease caused by a mycoplasma and known to infect plants from more than 40 families. Among the many garden plants affected by aster yellows are aster, carrot, celery, lettuce, onion, potato, spinach, and tomato (Fig. 9-9).

Fungi

Fungi are small, usually microscopic plants lacking chlorophyll. Their bodies are comprised of numerous threads or filaments, known as **hyphae.** The hyphae form compact masses called **mycelia** (Fig. 9-10). Because fungi lack chlorophyll they must depend entirely on other organisms for their food supply. Many fungi are **saprophytic;** that is, they utilize dead organic material as food. Other fungi have a **parasitic** mode of existence, living on or in another living organism, the **host,** from which they obtain their food. Some fungi that cause plant diseases live exclusively as parasites, whereas others have both parasitic and saprophytic phases in their life cycles. Regardless of these nutritional differences, fungi enter plant tissues—their source of food—through wounds, through natural openings such as stomates, or directly through the epidermis.

In their evolution as plant pathogens the fungi have developed complex life cycles. The apple scab fungus, for example, overwinters on dead leaves on the ground (Fig. 9-11). Dur-

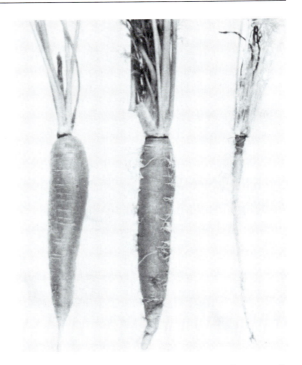

FIGURE 9-9 Characteristic symptoms of aster yellows infection of carrot. Note the bushy tops of the severely infected (right) and moderately infected (middle) specimens as compared to a healthy plant (left). (Courtesy of the late S. S. Ivanoff, Mississippi Agricultural Experiment Station.)

ing winter and early spring spores mature in these leaves, and are released into the air. They germinate on young apple leaves, penetrate the cuticle, and develop within the tissues of the host plant. Soon scab lesions appear on infected leaves and from these lesions enormous numbers of spores are produced, which infect still more plants. As fall approaches the infected leaves again fall to the ground where they overwinter. Thus, we see that the apple scab fungus has both a parasitic and saprophytic phase in its life cycle. More importantly, we see that a measure of control can be effected by a simple sanitary measure—that of destroying leaves in the fall.

Symptoms of fungal infections include the

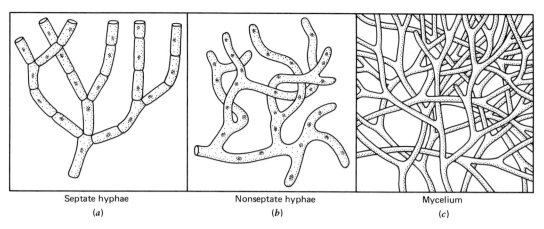

Septate hyphae
(a)

Nonseptate hyphae
(b)

Mycelium
(c)

FIGURE 9-10 The fungal body is composed of septate (*a*) or nonseptate (*b*) hyphae that, in mass, form the vegetative body, the mycelium (*c*). (Reprinted with permission of John Wiley & Sons from T. Elliot Weier, C. Ralph Stocking, Michael G. Barbour, and Thomas L. Rost, *Botany: An Introduction to Plant Biology*. Copyright © 1982 by John Wiley & Sons.)

formation of scabs and lesions as occur in apple scab. Other symptoms include the damping-off of young seedlings, root rot, basal stem rot, and a rapid browning of foliage. Some fungal infections result in the swelling and distortion of plant parts as exemplified by curling of leaves, profuse branching, gall formation, and enlarged roots appearing as clubs, a condition known as clubroot (Fig. 9-12). Other fungi invade the vascular systems of plants causing wilting, while still others form visible mycelia over plant surfaces, a condition known as mildew.

In all, more than 8000 species of fungi can cause diseases in plants. Some such as the late blight of potato, Dutch elm disease, and the smut and rust diseases of cereals have drastically changed the course of human history. Common vegetable examples include the fusarium and verticillium wilts of tomato, blackleg of brassicas, and anthracnose of cucumbers and melons. Peach leaf curl, corn smut, early blight of potato and tomato, downy mildew of grapes and onions are also common garden problems. Most fungal diseases are controlled by use of resistant cultivars or fungicides.

Nematodes

Nematodes, sometimes called eelworms, are small, wormlike animals that are common inhabitants of all soils. In size they range from 0.3 to 4.0 mm in length, but due to their small diameter are invisible to the naked eye. They are easily observed under the microscope, however. Nematodes are generally most abundant in the uppermost six inches of soil and in or around the roots of susceptible plants. Of the several thousand species of nematodes occurring in nature only a few hundred cause damage to plants.

Nematodes cause only minor injury to plants by the direct mechanical injury associated with their feeding. Instead, a major portion of damage is caused by the injection of saliva into host plants. Powerful enzymes in the saliva aid in the penetration of plant cell walls, in the digestion of cellular contents, and disruption

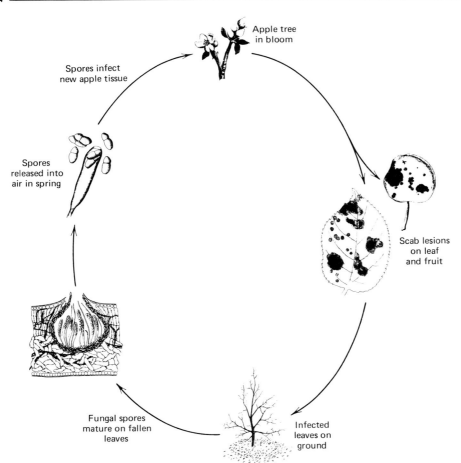

Spores infect
new apple tissue

Apple tree
in bloom

Spores
released into
air in spring

Scab lesions
on leaf
and fruit

Fungal spores
mature on fallen
leaves

Infected
leaves on
ground

FIGURE 9-11
Disease cycle of *Venturia inaequalis*, the fungus that causes apple scab. Destroying leaves in the fall is an excellent method of disease control. (Reprinted with permission of Academic Press and George N. Agrios from George N. Agrios, *Plant Pathology*. Copyright © 1969 by Academic Press.)

of normal cell behavior. Nematodes can also spread bacterial, fungal, and viral diseases.

Symptoms of nematode infections appear on both the roots and aboveground portions of plants. The most common root injury symptoms are root knots or root galls, lesions, excessive branching, injured root tips, and root rots. Swellings, caused by root-knot nematodes, for example, greatly distort roots (Fig. 9-13). Needless to say, plants displaying such symptoms show reduced growth. They may also display symptoms of nutrient deficiencies such as yellowing of foliage, show excessive wilting on hot days, and, more often than

not, give reduced yields of produce of lower quality.

Important diseases are caused by nematodes. The root-knot nematodes attack nearly all crops. Stem and bulb nematodes attack onion, potato, strawberry, and many of the bulb-forming flowers (Fig. 9-14). The cyst-forming nematodes, which are confined mainly to southeastern United States, attack many kinds of legumes and also some nonleguminous plants. The ubiquitous lesion nematodes attack vegetables such as tomato, potato, and carrot, fruit trees such as apple, peach, and cherry, and numerous ornamentals.

FIGURE 9-12 Clubroot, a serious fungal disease of the cabbage family. (Courtesy of U.S.D.A. and Pennsylvania State University.)

Disease Control Measures

Disease control measures begin before symptoms are observed. We have already alluded to many of these preventative measures. The use of resistant cultivars will prevent many diseases caused by virus, fungi, and nematodes. Purchase of disease-free seed and plants is also beneficial. Additionally, seed can be treated with chemicals or by other methods to protect against decay and damping-off. In certain cases soil sterilization may be necessary to destroy bacterial spores, fungi, and nematodes, as well as weed seed.

Keeping the garden and border areas clean is also an important preventative measure.

Weeds harbor many viruses as well as the insects that transmit them. Both perennial weeds and insects serve as hosts for carrying disease-causing organisms over winter. Crop residues should be disposed of promptly, as they too serve as havens for disease organisms. Diligent thinning, fertilization, mulching, and watering all contribute to a disease-free garden. The implementation of the abovementioned preventative measures as they apply to specific vegetables are outlined in Table 9-5.

Many fungal diseases can be prevented or controlled by fungicides (Table 9-5). Fungicides are not new. Sulfur was used to control mildew in the days of Homer, 1000 years before the Christian era. However, widespread

FIGURE 9-13 Nematode galls on roots of cucumber. (Courtesy of Harold Jensen, Oregon State University.)

FIGURE 9-14 Strawberries planted where infected red clover was grown are often heavily damaged by bulb and stem nematodes. Distorted leaves and fruiting stalks are characteristics most striking early in the season. (Courtesy of Harold Jensen, Oregon State University.)

use of fungicides began with the development of the Bordeaux mixture (copper sulfate and lime) in 1885 to control downy mildew of grapes in France. This was followed by an era of inorganic compounds of sulfur, copper, and mercury. Organic compounds were developed in the 1930s and have since dominated the fungicide field. One group of organic fungicides, the carbamates, are frequently used by gardeners. The fungicides maneb, zineb, and ziram referred to in Table 9-5 belong to this group. Note the many diseases controlled by each of these fungicides. Maneb, for example, will control most leaf spot diseases.

In general, best results are obtained if the fungicide is applied before the disease appears, but in some cases application can be delayed until disease symptoms are evident. Sprays usually give better control than do dust preparations. In either event, when applying a fungicide *all* aboveground plant surfaces must be thoroughly covered, because only covered portions are protected from the disease. Applications must be repeated whenever the fungicide coating wears off, commonly at 7- to 10-day intervals. In all cases the timing of fungicide applications should coincide with the stage when plants are susceptible to a particular pathogen. Fungal infections are usually most severe during humid or wet weather and fungicides may need to be applied more frequently when these conditions prevail. When applying any fungicide always follow closely the instructions given by the manufacturer on the label.

There are relatively few home remedies or biological controls for plant diseases. Some gardeners report limited success using spray preparations of camomile, chive, garlic, horseradish, hyssop, or onion to combat bacterial and fungal diseases. The use of **antibiotics** (chemical compounds made by an organism to protect it against disease-producing organisms) is an area that appears to offer hope for the future, however. To illustrate, the fungus *Penicillium notatum* makes and secretes the antibiotic penicillin for protection against its bacterial enemies. Plant scientists hope that just as this powerful antibiotic kills bacterial invaders in nature, other antibiotics can be applied to crop plants to kill bacterial and/or fungal pathogens. Experiments to date have shown the merits of this concept. Streptomycin, for example, has been used to control bacterial diseases of pepper and tomato. Some antibiotics are systemic in action; that is, when absorbed they move throughout the plant giving protection to all tissues of shoot and root. Furthermore, once within the plant they cannot be washed off by irrigation water or rainfall.

MANAGING VEGETABLE INSECTS

No aspect of gardening is of greater interest than insect control. With dramatic swiftness these creatures can invade a garden, quickly multiply, and in a short time turn to shambles the work of a season. Many insects are easy to see, but most are difficult to manage. With so many different insects attacking garden

TABLE 9-5 Common Vegetable Diseases and Their Control

Crop	Disease	Control measure
Asparagus	Rust	Apply fungicide containing zineb[a] after harvest. Make 5 applications at 10-day intervals.
Beans	Mosaic	Plant resistant cultivars.
	Leaf and pod diseases	No fungicide recommended. Do not cultivate, weed, or harvest beans when plants are wet. Plant certified, western-grown seed.
Beets, Swiss chard, Spinach	Leaf diseases	Apply fungicide containing maneb[b] or zineb[a] at 10-day intervals. Start when plants are 6 to 8 inches high.
Cabbage, Broccoli, Brussels sprouts, Cauliflower, Chinese cabbage, Kale, collards, Kohlrabi, Mustards, Rutabaga, Radish, turnip	Yellows	Plant resistant cabbage cultivars.
	Blackleg	Buy only hot-water-treated seed.
	Black rot	Buy only hot-water-treated seed.
	Clubroot	Apply one cup of transplanting solution containing pentachloronitrobenzene (Terraclor, PCNB) around the roots of each plant. The solution is made by mixing 3 level tablespoonfuls of 50-percent wettable Terraclor in 1 gallon of water.
Carrots, Parsnips	Leaf diseases	Apply fungicide containing maneb[b] or zineb[a] when spots first appear.
	Yellows	Control leafhoppers, which transmit the mycoplasm. Destroy infected plants.
Cucumbers, Pumpkins, Squash, Gourds	Bacterial wilt	Control cucumber beetles, which spread the bacteria from plant to plant. Remove infected plants.
	Scab	Plant resistant cucumber cultivars. Buy hot-water-treated seed.
	Mosaic	Plant resistant cultivars.
	Leaf and fruit diseases	Apply fungicide containing zineb,[a] or maneb[b] at 7- to 10-day intervals. Begin after vines start to spread. If control is needed before vines start to spread, use ziram[c] or captan.[d]
Eggplants	Fruit rot	Apply fungicide containing maneb,[b] zineb,[a] or ziram[c] at 7- to 10-day intervals. Begin when the first fruits are 2 inches in diameter.
	Verticillium	Plant resistant cultivars.

TABLE 9-5 (continued)

Crop	Disease	Control measure
Muskmelons (Cantaloupes) Honeydew melons, and Watermelons	Fusarium wilt Bacterial wilt Leaf and fruit diseases	Plant resistant cultivars. See Cucumbers. See Cucumbers.
Onions, Garlic, and Chives	Leaf diseases	Apply fungicide containing maneb[b] or zineb[a] at weekly intervals. Begin when leaf spots are first noticed. Add 1 tablespoonful of powdered household detergent or 1 teaspoonful of liquid detergent to each gallon of spray solution.
	Smut	Plant disease-free onion sets. Smut only attacks onions grown from seed. Treat seed with thiram before planting.
Peas	Fusarium wilt Root rots	Plant resistant cultivars. Plant early and use a seed treatment.
Potatoes	Tuber diseases	Buy certified seed potatoes. Plant uncut tubers. Grow cultivars resistant to scab and late blight.
	Leaf diseases	Apply fungicide containing maneb[b] or zineb[a] at 5- to 10-day intervals. Start when plants are 10 inches high.
Sweet potatoes	Black rot, scurf, foot rot	Buy certified plants. Use 3- or 4-year rotation.
	Wilt, root-knot, soil rot	Plant resistant cultivars.
Tomatoes, Peppers	Fusarium wilt, verticillium wilt	Plant immune or resistant cultivars.
	Leaf and fruit diseases	Apply fungicide containing maneb[b] or zineb[a] at 5- to 10-day intervals. Begin when the first fruits are 1 inch in diameter.

Courtesy of University of Illinois Cooperative Extension Service.
[a]Zineb fungicides such as Dithane Z-78, Stauffer Zineb, Ortho Zineb 75 wettable, Chipman Zineb, Niagara Zineb, Penwalt Zineb, Black Leaf Zineb, etc., contain zinc ethylenebis (dithiocarbamate).
[b]Maneb fungicides such as Manzate, Manzate D, Dithane M22, Dithane M-22 Special, Kilgore's Maneb, Black Leaf Maneb, and Penwalt Maneb contain manganese ethylenebis (dithiocarbamate).
[c]Ziram fungicides such as Zerlate Ziram Fungicide, Karbam White, Corozate, Orchard Brand Ziram, Ortho Ziram, Stauffer Ziram, etc., contain zinc dimethyldithiocarbamate.
[d]Captan fungicides such as Orthocide 50W, Captan Garden Spray, Captan 80 Spray-Dip, Orthocide Garden Fungicide, and Captan 50W contain N-(tricholoromethylthio)-4-cyclohexene-1, 2-dicarboximide.

crops it is easy to appreciate why insect management in the garden is a major concern of gardeners everywhere.

Knowing Vegetable Insects

Garden Entomology Basics

The small, mobile creatures that gardeners refer to in general terms as "insects" belong to two closely related but distinct groups—insects and arachnids. Members of the two groups differ from each other in several respects. The **insects** have three distinct body regions—head, thorax, and abdomen—and three sets of legs. **Arachnids,** on the other hand, have only two body regions but four pairs of legs. In contrast to insects, arachnids have no wings. The beetles (including weevils), moths, butterflies, grasshoppers, earwigs, and bees are examples of insects. Spiders, ticks, scorpions, mites and daddy longlegs, although sometimes regarded as insects, all represent specific kinds of arachnids. In our discussion we will often use the term "insect" in a broad sense to refer to both insects and arachnids.

Two distinct types of life cycles occur among insects and related pests. One type, known as **gradual metamorphosis,** is characterized by the fact that the physical changes leading to adulthood are spread more or less evenly throughout the entire life. In gradual metamorphosis the young, known as **nymphs,** are almost identical in appearance to adults, except for their smaller size. In **complete metamorphosis** the offspring go through four distinct stages of development, during which their appearance is altered drastically (Fig. 9-15). The process begins when an adult female lays **eggs,** which subsequently hatch into wormlike **larvae.** After a period of activity the larvae change into quiescent **pupae** (which may or may not be inside a cocoon). From the pupae stage emerges the adult.

An understanding of life cycle variations has important practical applications as it relates to pest control measures. Insects that undergo gradual metamorphosis, for example, can usually be controlled by the same methods throughout their life cycle. Generally whatever eradicates adults will also kill nymphs. In contrast, pests that undergo complete metamorphosis may not be vulnerable to the same control measures during all stages of their life cycle. Young insects of both types are usually softbodied creatures and as such are more vulnerable than adults to adverse weather, predation by birds and by other insects, diseases, and control measures implemented by the gardener.

Insects cause reductions in the yields of garden crops through their numerous activities, which include inducing the formation of galls, boring, chewing, and sucking on crop plants. Not only do they pirate plant juices and eat plant parts, but often their feeding or egg-laying activity causes plant parts to become curled, discolored, or abnormal in other ways. Also, insects can spread plant diseases during feeding. Leafhoppers and aphids are especially important as agents for transferring viruses from weeds to garden crops.

Many insects feed on a large number of different plants. Examples of such general feeders include aphids, cutworms, earwigs, flea beetles, spider mites, and grasshoppers. These pests may be found on nearly every kind of plant in the garden. In contrast, many other insects are quite specific in their feeding habits, including the asparagus beetle, bean weevils, and the cabbage worms.

Insects can locate host plants by specific compounds in the plant. The cabbage butterfly, for example, is attracted to cabbage and closely related crops by the pungent odor of mustard oils given off by these plants. Eggs laid by this butterfly develop into damaging cabbageworm larvae. Other insects use a combination of senses (i.e., smell, touch, color, taste) to locate their preferred vegetables.

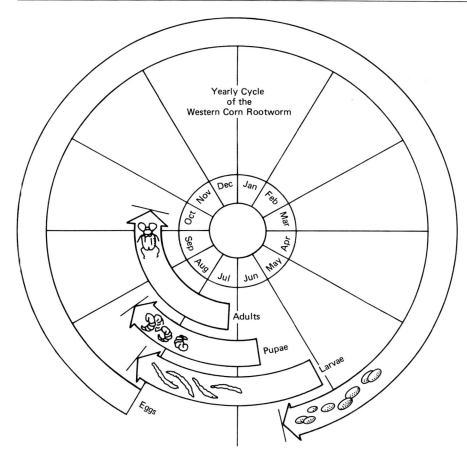

FIGURE 9-15 Life cycle of the western corn rootworm, an insect that displays complete metamorphosis. (Reprinted with permission of A. & L. Publications. Modified from Samuel R. Aldrich, Walter O. Scott, and Earl R. Leng, *Modern Corn Production*, 1975.)

Classifying Insects

Insects can be classified into two groups based on their feeding behavior. One group includes those insects with biting mouth parts, examples of which are illustrated on Fig. 9-16. Insects with biting mouth parts are further subdivided on the basis of the plant parts on which they feed. Accordingly, the following four groups are recognized: (1) stem and leaf feeders, (2) root feeders, (3) stem borers, and (4) fruit, seed, and fleshy organ feeders.

Can you think of insect examples in each of these categories? Perhaps you can assign the pests shown in Fig. 9-16 to their appropriate feeding category.

The second broad group of insects based on manner of feeding include those insects with piercing and sucking mouth parts. Examples of this grouping are illustrated in Fig. 9-17. As you can see, some of the more common plant pests—both indoors and out—belong in this group.

Pest control measures, especially insecticide

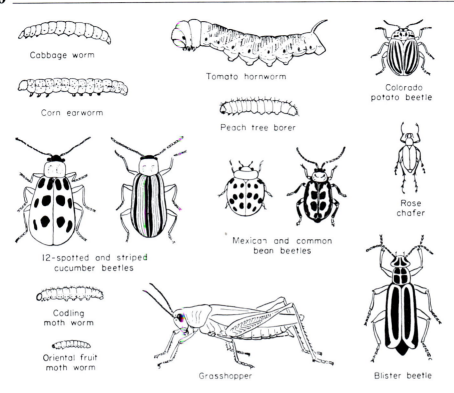

Cabbage worm

Corn earworm

Tomato hornworm

Peach tree borer

Colorado potato beetle

Rose chafer

12-spotted and striped cucumber beetles

Mexican and common bean beetles

Codling moth worm

Oriental fruit moth worm

Grasshopper

Blister beetle

FIGURE 9-16 Examples of insects with biting mouth parts. (Reprinted with permission of McGraw-Hill Book Company from J. B. Edmonds, T. L. Senn, F. S. Andrews, and R. G. Halfacre, *Fundamentals of Horticulture*, 4th ed. Copyright © 1975 by McGraw-Hill, Inc.)

application, are closely tied to the feeding behavior of insects. To illustrate, piercing and sucking insects such as aphids feed by removing plant sap through a flexible, needle-like beak inserted into leaves or stems. Pests of this type are quite susceptible to contact or systemic insecticides, but are left virtually untouched by stomach poisons. Many biting insects, on the other hand, are quite vulnerable to stomach poisons because they feed on plant parts and thereby ingest pesticides. Biting insects are also susceptible to insecticides that kill on contact. Note that informed pest management requires several pieces of information. We must know the identity of the insect and have a clear knowledge of its life cycle characteristics and feeding behavior. Only then

can we consistently achieve success in our pest management efforts. Space limitations prevent giving detailed information on each insect pest. Instead, we discuss briefly a few of the more important groups of garden insects.

Aphids. Aphids, also known as plant lice, are small insects that display gradual metamorphosis (Fig. 9-17). While generally greenish or whitish in color, they may also be gray, brown, black, or red. Both winged and wingless forms exist, depending on species and stages of development. Aphids injure plants primarily by sucking sap out of them. In the process of feeding, however, aphids secrete a substance known as **honeydew.** The sweet, sticky honeydew may attract ants and also

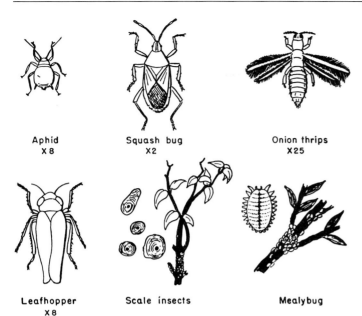

Aphid
X8

Squash bug
X2

Onion thrips
X25

Leafhopper
X8

Scale insects

Mealybug

FIGURE 9-17 Examples of insects with piercing and sucking mouth parts. (Reprinted with permission of McGraw-Hill Book Company from J. B. Edmonds, T. L. Senn, F. S. Andrews, and R. G. Halfacre, *Fundamentals of Horticulture*, 4th ed. Copyright © 1975 by McGraw-Hill, Inc.)

provide a medium for the growth of undesirable fungi.

Aphids are probably the most numerous of garden pests; a feature related to their reproductive behavior. During the gardening season *unfertilized* females produce generation after generation of young. Usually about 10 young are born at a time, and a generation may be born every two weeks or oftener. These young females can also multiply without fertilization. The whole process allows aphids numbers to increase with alarming rapidity. Only in the fall are aphid eggs fertilized. These eggs carry over the winter and provide a starting generation for next year's crop of aphids.

Beetles. Beetles, a category that includes weevils, come in literally thousands of forms that collectively comprise the largest of all insect groups. Adult beetles vary in size from less than 1 millimeter long to more than 100 millimeters. Shape and coloring also vary greatly; the eastern species of the cucumber beetle, for example, is striped, the western species spotted (Fig. 9-16). Beetles undergo complete metamorphosis, with the great majority producing only a single generation per year. Plant damage is caused by both larvae and adults, with both usually having similar feeding habits. The adult body is normally hard and compact with chewing-type mouthparts.

Adult beetles normally lay their eggs in the spring or early summer. Within about one to two weeks these hatch into larvae, known as worms or grubs. The larvae are voracious feeders on roots and shoots and attain full size during the summer. They pupate in the soil and within a few weeks adults emerge. The adults mature throughout late summer and autumn and hibernate during the winter. In spring these adults emerge, lay eggs to start the cycle again, and then die soon after egg laying is completed.

Borers and caterpillars. These two kinds of pests are discussed together because the species of each kind that damage temperate region vegetable crops belong to the same insect order, the Lepidoptera. Members of this order—the moths and butterflies—are perhaps more familiar to gardeners than any other insects. The Lepidoptera have complete metamorphosis with the often beautifully colored adults having generally two pairs of wings and greatly reduced mouthparts. Most larvae of the Lepidoptera are aggressive plant feeders, either feeding externally on foliage, mining the inside of leaves, or boring inside stems, trunks, or roots. Larvae of a great many species attack cultivated crops causing serious losses; thus the order is one of great economic importance.

While the distinction between moths and butterflies is not always sharp, moths are in general nocturnal whereas butterflies are diurnal in habit. A more important distinction is in the nature of their antennae. In moths the antennae are usually threadlike in appearance and not thickened or swollen at the tips. In butterflies, in contrast, the antennae are generally thickened or knoblike at their ends, as exemplified by the familiar swallowtailed butterflies.

Borers. The term "borer" is used in a general way to describe larvae that burrow into plant tissues. Larvae displaying boring characteristics are found in several kinds of insects including the beetles, moths, and less commonly the butterflies. Those that cause damage to vegetable crops, however, belong almost exclusively to the moth group. Familiar examples of borers are the southwestern corn borer, squash vine borer, peach tree borer, and the raspberry crown borer (Fig. 9-18). Larvae of some species attack stored grain and vegetables.

Borers begin their life as eggs deposited on foliage of weeds or crop plants. The eggs hatch into tiny larvae that feed first on foliage and later burrow into plant organs, often in preparation for hibernation. These pests may overwinter as eggs (common stalk borer) or more commonly as larvae or pupae (south-

FIGURE 9-18 Raspberry crown borer. Stems infected by the crown borer (*a*) as evidenced by the presence of frass. Longitudinal section through the stem (*b*) exposes a borer larva. An insecticide crown drench applied early in the season provides effective control, and at the same time avoids killing the valuable egg parasite of the borer that is present in the fall. (Courtesy of Joseph Capizzi, Oregon State University.)

western corn borer). Regardless of timing, adults eventually appear and egg laying begins anew.

Caterpillars. These damaging plant pests represent the larval stages of moths or butterflies. The list of harmful species is almost endless and includes (1) cabbage caterpillars that attack many members of the Brassicaceae, (2) the tent caterpillars of fruit and other trees, (3) corn earworms that attack cotton, corn, tomato, and other crops, and (4) numerous armyworms, cutworms, cornworms, and leaf rollers (Fig. 9-16).

Leafhoppers. The leafhoppers are closely related to the aphids, and like the latter have piercing-sucking mouthparts, gradual metamorphosis, and both winged and wingless forms. In general they are similar in size to aphids, and most are slender and nearly parallel sided (Fig. 9-17). There are literally thousands of species of leafhoppers and also an extremely large number of individuals. Frequently entomologists find that leafhoppers are gathered in greater numbers than any other insect when making random sweeps in the field. Leafhoppers attack and injure a number of garden crops, not only by direct damage from feeding but also through the transmission of plant diseases. The beet leafhopper, for example, transmits the virus that causes curly top of sugar beet, and the plum leafhopper transmits the destructive virus that causes peach yellows.

Maggots. The term "maggot" is used to designate the larva of true flies. Most are moisture loving, with food habits much different than those of adults. Many of these legless creatures establish their residences inside living plant tissues. Maggots of the cherry fruit fly, for example, tunnel into cherry fruits and apple maggots burrow into apples. The majority of maggots attacking garden crops are subterranean, however, developing from eggs laid in the garden soil by flies. Most notable are the cabbage, onion, and seed-corn maggots that attack sprouting seed, stems and roots, and storage organs of several crops including bean, broccoli, Brussels sprouts, cabbage, cauliflower, kale, onion, pea, radish, rutabaga, and turnip. Severe infections may result in the failure of seedlings to emerge, the wilting, yel-

lowing, and eventual death of plants, and serious damage to root crops.

Insect Control Measures

While a majority of insect control measures are implemented after a pest has been observed in the garden, some are of a preventative nature. Prevention, as with diseases, should start with the purchase of pest free seed and plants. Plant scientists have recently determined that plants have the capacity to kill or inhibit pathogens by secreting chemicals, known as **phytoalexins,** at the point of attack or invasion by a pathogen. Modern corn hybrids, for example, have considerable tolerance to the European corn borer, because the first injury by a borer causes the plant to secrete a chemical that is distasteful to them. Crop cultivars of the future will, no doubt, show greater insect tolerance, now that plant breeders are more fully aware of these remarkable defense mechanisms in plants.

Many additional cultural practices can help prevent damage as illustrated by the following examples. By chemically treating seed or transplants injury by maggots can be avoided. Timely planting can also reduce or eliminate maggot damage. Planting lettuce in sheltered areas near buildings or hedges can avoid damage from the six-spotted leafhopper because they prefer open areas. Removing and destroying crops heavily infected by aphids can prevent or slow their spread to other plants. Cleaning up the garden area in the fall will remove the food material the sweet potato weevil needs in order to survive over the winter. Keeping the garden area free of weeds destroys insect breeding places and removes overwintering habitats. Last, but not least, the rotation of crops prevents the buildup of soil organisms harmful to a particular crop or crops.

Mechanical controls are also effective preventative measures. Mulches, a topic discussed in Chapter 8, can be employed to reduce or eliminate weed growth, form a barrier between the aerial portion of the plant and the soil, and in some cases repel insects directly. Paper or metal collars can be used to protect young seedlings against cutworms. Borers, such as the squash vine borer, can be punctured with a needle or knife. Larger insect specimens, such as carrot caterpillars, corn earworms, Colorado potato beetles, harlequin bugs, and especially tomato hornworms can be removed from plants manually. In other cases, egg masses can be crushed or removed from foliage, or a stream of water can be used to wash insects from plants

Some preventative measures involve the use of insecticides. Diazinon and other chemicals can be applied to infested soils prior to planting to protect bean, beet, carrot, celery, lettuce, onion, potato, sweet potato, turnip, and other crops from damage by wireworms, the destructive larvae of the click beetle. Similarly, chemicals can be used on soils prior to planting to avoid damage by mole crickets, a pest problem in the Southeast, and to protect a large number of vegetables from attack by the widely distributed root maggots.

The implementation of certain of the above-mentioned preventative measures as they apply to specific crops are presented in Table 9-6.

In addition to the preventative measures outlined above, a large number of insect control measures can be put into effect once a pest is found in the garden. Many of these measures involve the use of standard insecticides or biological controls, as detailed for specific crops in Table 9-6. Numerous biological controls as they relate to specific pests are given in Table 9-3.

Other corrective controls used by gardeners are less well established. The so-called **kitchen insecticides** (repellent sprays prepared from garlic, onion, and other plants) have found wide acceptance among gardeners, but little substantiation in the scientific community. In fact, in some cases these sprays have appeared to act as seasoning agents; making plants even more palatable to pests! The use of sprays prepared from liquified pests have proven to be more effective. Although this area is only beginning to be explored, initial results show that when insect pests or slugs are liquified with water in a blender and the mixture sprayed on plants certain pests are repelled. A modification of this approach calls for the use of diseased pest specimens in the mixture. In either event, it is not known with certainty why such mixtures aid in insect control. One explanation is that **pheromones,** a group of naturally produced insect scents that influence sexual and other behavior, are released into the mixture and act as insect repellents.

The use of pheromones offers exciting possibilities for effective insect control. Already chemists have learned to synthesize sex pheromones that will attract male insects to traps, thereby preventing them from mating with receptive females. Other scientists have synthesized a chemical almost identical to one produced by aphids in trouble to warn other aphids of their plights. When a mixture containing this chemical is sprayed on garden crops aphids flee. These two examples serve only to illustrate the possible application of pheromones in insect control. Beyond question, this important group of chemicals will find widespread garden use in the future.

MANAGING WEEDS IN THE GARDEN

Weeds are plants that compete with garden crops for moisture, nutrients, and growing space. Weeds also harbor numerous garden pests. In short, weeds are unwanted plants that the gardener must deal with in a decisive manner. In the words of Dr. Tom Muzik, a

TABLE 9-6 Common Vegetable Insects and Their Control[a]

Insect	Crop	Dust Formula	Spray Formula	Remarks
Aphid	Cabbage Cucumbers Melons Peas Potatoes Tomatoes	5-percent malathion	2 Tsp. 50-57-percent emulsifiable malathion *or* 1 Tsp. 64-percent emulsifiable dibrom in 1 gal. water	Apply on foliage when aphids appear. Repeat weekly as needed.
Blister beetle	Potatoes Corn Tomatoes Beans	5-percent sevin	2 Tb. wettable sevin in 1 gal. water	
Cabbage worms	Broccoli Cabbage Cauliflower Greens	4-percent dibrom	1 Tsp. 64-percent emulsifiable dibrom in 1 gal. water *or* *Bacillus thuringiensis*, 1½ Tsp. 3.2-percent wettable powder	Thorough treatment is necessary. Repeat weekly as needed. Begin treatment when worms are small.
Corn earworm (⅔ nat. size)	Sweet corn Tomatoes	5-percent sevin *Bacillus thuringiensis* (Thuricide, Dipel, Biotrol) on tomatoes	Inject ½ medicine dropperful of mineral oil into silk channel as silks start to dry *or* 2 Tb. wettable sevin in 1 gal. water	Dust or spray silks with sevin every other day for 10 days. Dust or spray tomatoes with sevin 3 to 4 times at 10-day intervals; begin when first fruits are small.
European corn borer	Sweet corn	5-percent sevin *or* 5-percent sevin granules	2 Tb. wettable sevin in 1 gal. water *or* 2 Tb. 25-percent diazinon in 1 gal. water	Apply insecticide four times at 5-day intervals beginning with egg hatching near mid-June. Avoid early spring plantings. On late corn dust as for corn earworm.
Striped cucumber beetle	Cucumbers Melons Squash	5-percent sevin	2 Tb. wettable sevin in 1 gal. water	Treat as soon as beetles appear. Repeat when necessary.
Cutworm	Most garden crops		2 Tb. 25-percent diazinon in 1 gal. water	At transplanting, wrap stems of seedling cabbage, pepper, and tomato plants with newspaper or foil to prevent damage by cutworms.

[a]Insects are about natural size except where otherwise indicated. Where two drawings are shown, the smaller one is natural size. One pound of dust or 3 gallons of spray should be sufficient to treat 350 feet of row. Tb. = tablespoon. Tsp. = teaspoon.

TABLE 9-6 (continued)

Insect	Crop	Dust Formula	Spray Formula	Remarks
Flea beetle	Most garden crops	5-percent sevin	2 Tb. wettable sevin in 1 gal. water	Apply as soon as injury is first noticed. Thorough application is necessary.
Grasshopper	Most garden crops	5-percent sevin	2 Tb. wettable sevin in 1 gal. water	Treat infested areas while grasshoppers are still small.
Hornworm (½ nat. size)	Tomatoes	5-percent sevin *or* *Bacillus thuringiensis* (Thuricide, Dipel, Biotrol)	2 Tb. wettable sevin in 1 gal. water	Ordinarily hand-picking is more practical in the home garden.
Leafhopper	Beans Carrots Potatoes Cucumbers Muskmelons	Use sevin dust or 5-percent methoxy-chlor dust	2 Tb. wettable sevin in 1 gal. water	Spray or dust once a week for 3 to 4 weeks, beginning when plants are small. Apply to underside of foliage.
Mexican bean beetle	Beans	5-percent sevin	2 Tb. wettable sevin in 1 gal. water	Apply insecticide to underside of foliage. Also effective against leafhoppers on beans.
Potato beetle	Potatoes Eggplant Tomatoes	5-percent sevin	2 Tb. wettable sevin in 1 gal. water	Apply when beetles or grubs first appear and repeat as necessary.
Squash bug	Squash	5-percent sevin	2 Tb. wettable sevin in 1 gal. water	Adults and brown egg masses can be hand-picked. Trap adults under shingles beneath plants. Kill young bugs soon after they hatch.
Squash vine borer	Squash	5-percent sevin	2 Tb. wettable sevin in 1 gal. water	Dust or spray once a week for 3 to 4 weeks beginning in late June when first eggs hatch. Treat crowns of plants and runners thoroughly.

Courtesy of University of Illinois Cooperative Extension Service.

weed specialist from Washington State University,

> Weeds are as important in home gardens as they are in any other agricultural operation. They reduce yields, affect crop quality, deplete fertility, serve as hosts for insects and diseases, and reduce the attractiveness of the surroundings.

To effectively deal with weed problems we need to understand weed biology and weed control.

Weed Biology

Nearly all of our weeds are herbaceous rather than woody plants. Many were introduced into America from Europe or Asia, and a majority of the species fall within only a limited number of plant families. Of the approximately 700 weed species imported into eastern North America from Europe, for example, more than half are found in just five plant families (Table 9-7). Many of our weeds were imported as ornamentals, forage and grazing plants, medicinal plants, or even as vegetables or herbs. Others were introduced inadvertently as impurities in seed, on implements, or by other means.

Life Cycles of Weeds

Weeds can be classified as annuals, biennials, or perennials. Annual weeds complete their life cycle from seed to seed in one year or less.

Two distinct kinds of annual weeds are recognized. Winter annuals germinate in the fall, live through winter, and produce seed in early summer. Summer annuals, on the other hand, germinate in spring and produce seed in summer or autumn of the same year. Summer annuals, such as the well-known pigweed and lambsquarters, are more serious in home gardens, because cultivation for the spring-planted garden destroys most fall germinating winter annual weeds.

Biennial weeds complete their life cycle over two years. The first year they produce a well-developed root system and a rosette of leaves. The cold winter period following the first year's growth is needed for flower initiation. In spring of the second year these plants flower, set seed, and die. Biennials characteristically occur in temperate areas where they receive the required cold stimulus. Only a few biennials behave as weeds, including burdock, bull thistle, wild carrot, and mullein. In general, biennial weeds are not serious garden pests because they are removed before they set seed, thus diminishing their reservoir of seed in the garden soil.

Perennial weeds live for more than two years and may live almost indefinitely. Some perennials, such as dandelion, dock, and plantain reproduce naturally only by seed, but many others spread aggressively by means of vegetable structures (Fig. 8-4). In contrast to biennials, perennial weeds of the garden are numerous and difficult to control. Young

TABLE 9-7 Examples of Plant Families Containing a Large Number of Weed Species

Family	Scientific name	No. of species
Sunflower	Asteraceae	112
Grass	Poaceae	65
Mustard	Brassicaceae	62
Mint	Lamiaceae	60
Pea	Fabaceae	54

plants can be controlled by cultivation, but once established they are difficult to eradicate.

Broadleaves and Grasses

Within the three life-cycle types outlined above two distinct taxonomic groups of plants—broadleaves and grasses—are recognized. Thus, weed specialists commonly classify important weeds as annual or perennial broadleaves, and annual or perennial grasses. Examples in each of these categories are as follows:

Annual Broadleaves	Perennial Broadleaves
Cocklebur	Horse nettle
Giant ragweed	Canada thistle
Annual morning glory	Common milkweed
Jimson weed	Field bindweed
Lambsquarters	
Pigweed	
Purselane	
Smartweed	

Annual Grasses	Perennial Grasses
Barnyardgrass	Johnson grass
Crabgrass	Nutsedge
Foxtail (giant, yellow, green)	Quackgrass
Wild cane	

In controlling weeds, particularly through the use of herbicides, it is helpful to know to which of these groups a particular weed species belongs.

Reproductive Behavior

Virtually all weeds reproduce by means of seed, but annuals and biennials reproduce naturally only by seed. Weeds, like other plants vary in the amount of seed they produce. Some weeds produce enormous amounts of seed (Table 9-8). There is no marked difference

TABLE 9-8 Number of Seeds Produced per Plant by Various Weeds

Common name	No. of seeds per plant
Barnyardgrass	7,160
Dock, curly	29,500
Lambsquarters	72,450
Mullein	223,200
Nutsedge, yellow	2,420
Oats, wild	250
Pigweed, redroot	117,400
Plantain, broadleaf	36,150
Purselane	52,300
Shepherdspurse	38,500
Spurge, leafy	140
Thistle, Canada	680

Reprinted by permission of John Wiley and Sons From Glenn C. Klingman, Weed Control: As a Science. *Copyright 1961 by John Wiley and Sons.*

in the amount of seed produced by annuals, biennials, or perennials. Nor is there any correlation between the amount of seed produced and the seriousness of a given weed pest.

Seed dormancy is uncommon in crop plants because plant breeders have selected crops for their lack of dormancy to assure uniform germination and growth. Weeds, on the other hand, have been unwittingly selected for those which germinate unevenly over a period of time. This has happened because a weed whose seed germinate uniformly would be easily eliminated by cultivation, whereas weeds whose seed germinate over a long period of time would survive and flourish. Due to seed dormancy the seed of weeds may remain buried in the soil for long periods of time, germinating only when conditions are favorable. This reservoir of seed in the soil, which may reach many hundreds of pounds per acre, complicates the task of weed control, because every time the soil is disturbed a new crop of seed is brought close enough to the surface to sprout and grow.

Vegetative or asexual propagation is important in the spread of many perennial weeds. Vegetative structures contributing to the spread and overwinter survival of weeds include stolons, rhizomes, bulbs, tubers, and roots. Control measures must take into account these vegetative structures, some of which may be some distance underground.

Weed Control

Weed control is an easy idea to grasp; simply remove unwanted plants and prevent others from growing in their place. In practice, however, garden weeds are difficult to control because they are so well suited to their life-style. Many produce vast numbers of seed that fall to the soil or are disseminated by water, wind, animals, or other agents. Others spread aggressively by means of vegetative structures as well as by seed. The methods of weed control available to gardeners fall into four categories, as follows: (1) mechanical, (2) cultural, (3) biological, and (4) chemical.

Mechanical Methods

These methods include plowing, rototilling, hoeing, and hand weeding. For most gardeners these procedures, especially hoeing and hand weeding, are the major ones used to control weeds. The main reason for the popularity of mechanical methods is their effectiveness; even the most persistent weed can be eradicated by repeated cultivation. They have the added advantages of being immediate, nonpolluting, and requiring no elaborate equipment or special applicators license. And your neighbors will not be worried about chemical drifting across the property line.

For weed control purposes it is best to plow or spade the garden in fall and rework again in spring to stimulate the germination of weed seed. Cultivation, as detailed in Chapter 8, should be started as soon as crops are discernible in rows, and continue at regular intervals as long as weeds are a problem. Many gardeners stop weeding as their crops mature. This practice generally has little adverse effect on total yields (Table 8-1), but it allows weeds to go to seed thereby adding to the reserves of weed seed in the soil. If weedy plants are not allowed to go to seed in the garden there will be a gradual reduction of weeds over a period of several years.

Cultural Methods

The major cultural technique employed to control weed growth is the use of mulches. Indeed, a major reason for the mulching is to control or inhibit the growth of weeds. However, if you use a mulch to control weed growth in your garden, be sure that the other effects derived from the mulch are commensurate with your gardening goals.

To illustrate this point, many kinds of mulches will inhibit weed growth. Of these, some will cause soil temperatures to be strikingly lower than normal, others much higher, and still others will have little effect on soil temperature. All other factors being equal, the main criterion in your choice of mulches should be the temperature influence on the crops in question. The reason for this is clear. You will achieve equally good weed control regardless of which mulch is chosen, but crop growth will vary greatly due to the temperature influence. Since the major goal of weed control measures is to enhance yields we will choose the mulch material whose temperature influence maximizes our returns.

Biological Methods

The growth of many weeds can be controlled through the use of biological agents, most notably insects. A 1974 Weed Science Society of America report lists a large number of weed species that can be controlled by specific pathogens. While it is impossible to present the entire list here, some examples are the following:

Weed	Biological Control Agent(s)
Curly dock	Rust (*Uromyces rumicis*)
Klamath weed	Leaf beetles (*Chrysolina* spp.)
Prickly pear	Moth (*Cactoblastis cactorum*)
Tansy ragwort	Seed fly (*Hylemya seneciella*); cinnabar moth (*Tyria jacobaeae*)
Canada thistle	Flea beetle (*Haltica cardworum*)

The saga of weed control through the use of insects and other pathogens includes some outstanding successes. Perhaps the most notable of these is the control of the Klamath weed (*Hypericum perforatum*) by goatweed beetles. This weed was at one time a serious pest throughout much of the far West. In an attempt to control this weed, goatweed beetles were imported from Australia to California and released into fields in 1945. Both the larvae and the adults of this beetle feed exclusively on Klamath weed, and as a result weed numbers were greatly reduced in just a few years. Studies show that the beetles aid in control not only by killing plants, but also by reducing their ability to compete and to survive adverse conditions, such as dry summers.

Although biological controls have been used to control weeds over broad areas of their range, this method is not generally applicable to an individual garden site. Nevertheless, the gardener is indirectly involved in and enjoys successes from the broadscale implementation of this method if his or her garden is in areas where biological controls are effected.

Chemical Controls

The use of chemicals to control weed growth is widely practiced in commercial agricultural operations, but seldom employed in home gardens. One reason for this is that in home gardens—in contrast to large-scale agriculture—a wide diversity of crops is grown in a small area. Under these confined conditions it is hard to find a herbicide that will selectively remove specific weeds while leaving crops untouched. Then, too, even if herbicides are available, their selective application requires attention to details that the gardener may be unwilling or unable to provide. Finally, as compared to the hoe and hand weeding, the use of herbicides is more costly, gives slower results, and carries the ever-present possibility of damaging the environment.

We do not wish to imply from the above that herbicides cannot be used in the garden. Indeed, some gardeners may find it challenging to safely and effectively control weeds through the use of herbicides. For these enthusiasts we call attention to the detailed table of herbicide usage for vegetables presented by Ware and McCollum. Local agricultural advisors and herbicide manufacturers can also provide information on the use of herbicides in home gardens.

MANAGING PEST ANIMALS

The pest animals comprise an unusual and unique group of garden pests. Some are far less widespread than the majority of garden pests discussed in previous sections. The armadillo, for example, which causes problems for gardeners over much of the south central United States, is a totally unfamiliar animal to most of us. If any generalization can be made about the distribution of pest animals as a group it is that they are more numerous in rural areas than suburban areas.

Damage caused by pest animals is also quite variable. Some, such as deer and raccoon, may invade the garden only once or twice during a season and then cause only minimal damage to crops. Others, such as gophers, moles, and rabbits may take up residence in or near the garden site and, if not controlled, raise havoc with many crops. Determining the source of animal damage may at times be difficult, because many of these raiders are nocturnal in habit and nearly all are elusive by nature.

Control measures for animal pests also vary greatly. For some there are no significant means of control, and for others control may not be desirable. To illustrate, ring-necked pheasants invade our garden each fall to feed on corn. We try to encourage these animals by planting more corn than we plan to harvest and leaving our corn stalks standing in the garden over winter. Even where control measures can be put into effect their implementation may be expensive and laborious. In most cases, some understanding of the behavioral characteristics of the pest animal is necessary for successful control.

Among the more serious and widespread of the pest animals are birds, deer, moles and gophers, rabbits and raccoons, and slugs and snails. Many others cause damage on a regional basis. Measures to aid in the control of these can be obtained from local sources of agricultural information.

Animal Damage Control

Birds

A variety of birds, including blackbirds, crows, sparrows, and starlings, can cause garden damage. Generally, bird damage coincides with two distinct stages of plant development: (1) seed planting and early stages of germination, and (2) fruiting. Between these times plants are relatively immune to damage by birds.

Certain control measures can reduce bird-related losses. Care should be taken when planting not to leave uncovered seed on the surface as these will attract birds. Planted seed of crops such as corn can be covered with mulch of hay or straw, the idea being that by the time the seedlings emerge from the mulch they are so large that birds will no longer be interested in them. Planting losses can also be reduced by treating seed with chemical repellents or by putting netting over newly planted rows. Preventing fruit losses is more difficult. Netting can, of course, be placed over bushes or small trees. In general, birds prefer tart, wild fruits and berries to our cultivated ones. Some gardeners believe that if ample supplies of wild berries are available, domesticated plants will suffer less damage. Scarecrows and a variety of noisemakers can be used to keep birds away but most are ineffective, expensive, or both.

An important aspect to keep in mind when considering bird-control measures is that most birds, including those that damage crops, consume large numbers of insects and other pests, to say nothing of how they enrich our lives. Before implementing control measures, the balance between positive and negative attributes of birds should be carefully weighed.

Deer

In earlier times deer numbers were kept in check by several natural predators, including wolves and cougars. Now, with these predators almost eliminated, deer numbers have soared. Improved browse and cover, wise herd management by game agencies, and the natural ability of deer to coexist with humans have also contributed to this increase. A large deer population and the fact that deer regard tender vegetables and fruits as delicacies results in occasional damage to garden crops.

Several control measures can reduce damage caused by deer. A fence at least four feet high around the garden may deter these graceful animals, although they are capable of leaping somewhat higher. Some gardeners have found that fencing, coarse rock, or, to a lesser extent, black plastic, spread flat on the ground at the perimeter of the garden is more effective than an upright fence. If deer enter your garden only at certain points—and they probably do—then these horizontal barriers will probably be effective if placed only at points of entrance. Other control measures involve the use of repellents. A variety of these are on the market for use in spraying shubbery.

bery. Dried blood, human hair, or perspired clothing hung in bags around the garden are also reported to act as repellents. Another reliable, mobile, and sometimes vigilant repellent, which it would be difficult to praise too highly, is, of course, the family dog.

Moles and Gophers

These two groups of mammals are widely distributed over temperate North America. Moles occur almost everywhere except in the Rocky Mountains and the arid Southwest, whereas gophers are found virtually everywhere except the northeastern part of the United States

and extreme western Canada. In many areas the ranges of these two pest animals overlap. Both moles and gophers lead largely subterranean lives and, consequently, are rarely seen. You can easily determine which of these two pests you have, however, by examination of the characteristic mounds formed by each animal. Mole mounds are more or less hemispherical with no indication of the centrally located entrance to the burrow. Gopher mounds, in contrast, are fan shaped with a clearly visible peripheral entrance to the burrow (Fig. 9-19).

The eating habits of moles and gophers differ greatly. Moles are basically carnivorous

MOLE

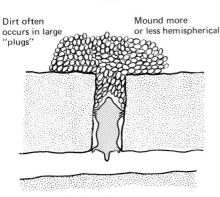

Dirt often occurs in large "plugs"

Mound more or less hemispherical

Dirt "cores" are shoved up the vertical shaft, and simply fall over as the mound is formed.

The top of the shaft is more or less at the center of the bottom of the mound, which is roughly circular when seen from the top:

Top view

POCKET GOPHER

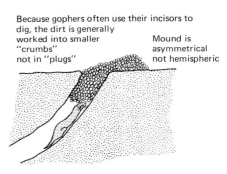

Because gophers often use their incisors to dig, the dirt is generally worked into smaller "crumbs" not in "plugs"

Mound is asymmetrical not hemispheric

The dirt is shoved up a slanting shaft, and the gopher shoves it away from the opening with vigorous pushes using the nose and face

Since dirt is shoved away from the opening in more or less the same direction as the tunnel, the opening is at the edge of the mound

Top view

FIGURE 9-19 Characteristics of mounds formed by moles and gophers. (Courtesy of Richard Forbes, Portland State University.)

creatures, feeding mainly on insects, earthworms, and the like, although some are reported to consume small amounts of vegetable matter. All moles are voracious eaters, some eating as much or more than their own weight daily. Gophers, on the other hand, are strict vegetarians, feeding on roots and tubers as well as some surface vegetation. Gophers have cheek pouches (pockets) on either side of their mouth into which they place food materials, hence the name pocket gophers. Both moles and gophers are active throughout the year.

How can moles and gophers be controlled? The general consensus is that moles are here to stay, and, although their numbers can be reduced in specific areas, extermination is impractical and probably not desirable. Within a given area moles can be successfully trapped if the intricacies of their runway systems are understood. Traps properly set along well-used main runs may yield literally dozens of moles. The use of chemical deterrents, such as lye, creosote, or tar, is probably a better control measure for home gardeners. To use this method effectively the gardener needs to determine the location of deeper runways entering the garden site. Deterrents placed in the runways at these points will keep moles from entering the garden area.

Gophers can also be successfully trapped. Traps may be set either in main runways or in the lateral runways that lead to the surface mounds, depending on trap design. Normally only one gopher occupies a single runway system. Gophers can also be controlled by placing toxic baits in the lateral runways, which are visited on an almost daily basis. Details on the control of these two animal pests are available through most agricultural agencies.

Rabbits and Raccoons

These animal pests, like those discussed above, are widely distributed over temperate North America. Except for crowded suburban areas, rabbits occur virtually everywhere from the Mexican border to the Arctic Circle. Raccoons, while widely distributed in the continental United States, have extended their range only short distances into Canada. Rabbits may be seen almost any time of day, but most do their feeding in the interval between early evening and late morning. Raccoons are chiefly nocturnal but may occasionally be seen during the day.

The eating habits of the two animals are of interest. Rabbits are strict vegetarians, feeding on green vegetation in summer and bark and twigs in winter. Due to their feeding habits and large numbers, rabbits can do considerable damage to gardens, including injury to shrubs and small trees. Raccoons are omnivores, eating fruits, nuts, grains, insects, frogs, bird eggs, and many other items. Raccoons may occasionally cause damage in the garden, particularly as related to their liking for ripe sweet corn.

It has been said that rabbits are better at multiplication than gardeners are at subtraction. Nevertheless, a number of measures can help to control these prolific garden pests. Beyond doubt, fencing is the first line of defense. A four-foot, poultry wire fence around the garden, or the trunks of trees, will keep rabbits out. Other measures include sprinkling dried blood meal around the garden, noisemakers, kitchen sprays containing garlic and/or fish emulsion, and last, but not least, the family dog.

Raccoons can be somewhat more difficult to keep from the garden area. Ordinary fences, such as those used to deter rabbits and other animals, prove noneffective because of the raccoon's ability to climb. A properly installed electric fence is effective but costly. Be sure to check local ordinances before employing this option. Other methods used by gardeners to reduce or eliminate losses include timely harvest, growing beans or cucumbers with corn, lighting and/or playing a radio in the garden at night, and dogs. Before employing any of these measures each gardener should determine if the damage caused outweighs the pleasure of having these animals in the vicinity.

Slugs and Snails

These widespread garden pests belong to the same phylum of animals as the familiar clam, oyster, squid, and octopus. The two can be distinguished in that snails form protective shells, whereas slugs are without visible shells. A major feature of both organisms is that they are easily dessicated in dry, hot weather by the rapid loss of water across their body surface. Consequently, slugs are restricted in distribution to areas of moderate to high rainfall. Snails can survive under somewhat harsher conditions because of their protective shells.

An understanding of the habits and habitats of these pests provides clues to their successful control. Both animals are nocturnal, hiding during the day in dark, damp places. Under conditions of high humidity and favorable temperatures, however, they may be active during the day. In colder regions slugs and snails hibernate during the winter. They also go into a resting state, known as **estivation,** during the summer in hot, dry climates. Slugs and snails feed on a variety of vegetation, including members of the cabbage family, carrot, lettuce, strawberry, tomato, and turnip. Reproductive characteristics of the two bisexual pests are similar. Both lay large clutches of eggs that within a period of a few weeks develop directly into young animals. Eggs laid in fall overwinter and hatch in spring.

Several effective control measures are available. First and foremost is garden cleanliness—remove anything that provides a hiding place for these pests. In addition, provide temporary hiding places such as shingles or boards between the rows or along the garden border. Slugs will collect under these and can be easily removed. Other methods that provide varying degrees of success include handpicking animals, putting out pans or saucers of beer into which the animals fall and drown, and the use of barriers of coarse sand, cinders, crushed oyster shells, or the like over which these pests have difficulty traveling. Finally, a number of toxic baits are available commercially, but care should be taken in using these because they may be harmful to pets and children.

Questions for Discussion

1. Define the following terms:
   (a) Pest
   (b) Herbicide
   (c) Systemic pesticide
   (d) Biological magnification
   (e) Larvae
   (f) Pheromones

2. What do the authors mean when they say that a plant is diseased?

3. Distinguish between pest control and pest eradication. Why is this distinction important?

4. Diagnosing the agent causing plant damage is often difficult. What guidelines would you offer to a group of 4-H students to help them to distinguish damage caused by disease organisms as opposed to that caused by insects?

5. Determine the LD_{50} values for two or three garden or household products that you have used in the past. Explain what these values mean in terms of toxicity.

6. Outline several aspects of resistance in vegetable cultivars.

7. Write a short essay on the use of biological controls in home gardens. Be sure to identify clearly the advantages and disadvantages of this method of pest control.

8. Identify at least three groups of organisms responsible for plant diseases and list several diseases caused by members of each group.

9. What methods would you employ to control plant diseases caused by fungi?

10. What is the practical importance of understanding insect life cycles as relates to pest control measures?

11. Comment on the importance of borers and

caterpillars as plant pests, giving specific examples and control measures.

12. Give three examples of mechanical methods that can be used to control specific insect pests in the garden.

13. Four methods of weed control are outlined in the text. Identify these methods, outline the merits of each, and reveal which methods you have used in your own garden.

14. How would you distinguish damage caused by moles from that caused by gophers? How would you control each of these pest animals?

Selected References

Agrios, G. 1969. *Plant Pathology.* Academic Press, New York.

Burt, W. and R. Grossenheider. 1964. *A Field Guide to the Mammals.* Houghton Mifflin, Boston.

Carson, R. 1962. *Silent Spring.* Houghton Mifflin, Boston.

Cox, J. 1976. An insect control method "too good to be true." *Organic Gardening and Farming,* 23:62–65.

Cox, J. 1977. The bug juice method: How safe? How effective? *Organic Gardening and Farming,* 24:164–172.

Edmond, J. et al. 1975. *Fundamentals of Horticulture.* McGraw-Hill, New York.

Gauss, J. 1979. Putting pests in their place. *Horticulture,* 57:60–67.

Goeden, R., L. Andres, T. Freeman, P. Harris, R. Pienkowski, and C. Walker. 1974. Present status of projects on the biological control of weeds with insects and plant pathogens in the United States. *Weed Science,* 22:490–495.

Holm, L. 1977. *The Worlds Worst Weeds: Distribution and Biology.* University of Hawaii, Honolulu.

Hyman, L. 1967. *The Invertebrates: Mullusca I.* McGraw-Hill, New York.

Muzik, T. 1970. *Weed Biology and Control.* McGraw-Hill, New York.

Rodale, J. (ed.). 1975. *The Encyclopedia of Organic Gardening.* Rodale Books, Inc., Emmaus, Pa.

Ross, H. 1956. *A Textbook of Entomology.* John Wiley, New York.

Southwick, C. 1967. *Ecology and the Quality of Our Environment.* Van Nostrand, New York.

U. S. Department of Agriculture. 1975. *Insects and Diseases of Vegetables in the Home Garden.* Agricultural Information Bulletin No. 380.

Ware, G. and J. McCollum. 1975. *Producing Vegetable Crops.* The Interstate Printers and Publishers, Danville, Illinois.

LEARNING OBJECTIVES

By the time you have completed this chapter you should be able to:

1 Identify the more popular garden vegetables and suggest reasons for their popularity.

2 List the types of beans and so-called beans available to gardeners and tell of their similarities and differences.

3 Describe the three types of harvest that may be secured from broccoli.

4 List two or three crops that can be left in the ground and harvested as needed during the winter.

5 Describe how to determine the proper timing of harvest for sweet corn, eggplant, pea, and watermelon.

6 Identify the major types of onions and tell which ones are best suited for winter storage.

7 Name the three types of peas that are available for planting in the home garden.

8 Tell why caution must be exercised in applying nitrogen fertilizer to vegetables such as sweet potato and tomato.

9 Explain what causes knobby tubers to develop in the white potato.

10 Describe the major feature used to distinguish squashes from pumpkins, and list three members of each group.

11 Name five vegetables that are grown for their greens, and tell how each performs during hot summer weather.

12 Discuss successful cultural techniques for tomato, including planting, staking, and pruning.

13 Identify two garden crops that the average gardener should probably not try to grow and tell why this is so.

14 Describe the cultural characteristics of two important perennial vegetables.

10

GROWING GARDEN VEGETABLES

by Wesley P. Judkins, Emeritus Professor of Horticulture
Virginia Polytechnic Institute and State University

*"No culture is so delightful to me as the culture of the earth,
and no culture comparable to that of the garden . . . though an
old man, I am but a young gardener."*

Thomas Jefferson

In the three preceding chapters we have discussed practices that relate to the production of vegetable crops in general. We will now consider the unique characteristics, cultural methods, and values, of the most important vegetables that may be grown in the home garden.

The first question to arise in our discussion concerns which vegetables should be considered. A logical approach is to concentrate on the vegetables that are grown most commonly by gardeners. Two lists that provide guidance in this matter are presented in Table 10-1. One list is based on sales from seed displays in stores, and the other records a Gallup survey of home garden vegetable favorites. There are several obvious and understandable differences between the two lists. Several popular vegetable crops, such as tomato, broccoli, cabbage, eggplant, and pepper, are commonly purchased as young transplants from

a garden center or greenhouse, rather than from a seed display. Onion sets, as well as white potatoes and sweet potatoes, are not displayed on seed racks in stores and thus were not included in the national ranking by seed companies.

The selection of vegetables to be planted by home gardeners seems to depend on the edible appeal of the harvested product, and the adaptability of the different vegetables for production under a wide range of climatic conditions. The susceptibility to damage by diseases or insects seems to be of less concern to gardeners than might be expected. This may be due in part of the availability of disease resistant cultivars, and the relative effectiveness of insect control measures that can be employed.

After carefully considering the many vegetables grown by home gardeners, and the material in Table 10-1, we have selected the

TABLE 10-1 The "Top 20" Vegetables as Seen from Two Points of View[a]

National ranking among seed companies of seeds sold through store displays[b]	Favorite vegetables and percentage of vegetable gardeners growing each[c]	
1. Lettuce	Tomato	92
2. Cucumber	String bean	63
3. Beans	Onion	59
4. Carrot	Cucumber	58
5. Squash	Pepper	57
6. Tomato	Radish	51
7. Beet	Lettuce	50
8. Corn, sweet	Carrot	47
9. Radish	Corn, sweet	45
10. Onion	Beet	39
11. Pea, green	Pea	38
12. Melons (except watermelon)	Cabbage	37
13. Turnip	Squash, summer	36
14. Watermelon	White potato	36
15. Pepper	Squash, winter	15
16. Okra	Pumpkin	14
17. Mustard greens	Eggplant	14
18. Cabbage	Spinach	13
19. Spinach	Broccoli	10
20. Chard, Swiss	Sweet potato	9

[a]Popular herbs such as parsley and dill, and perennial vegetables such as asparagus, have been deleted from these lists.
[b]Courtesy National Garden Bureau, Sycamore, Illinois.
[c]Courtesy Gardens for All, The National Association for Gardening, 1974.

"Top 20" that seem to be most important. A second group contains the "Next 10." A third group is composed of 12 miscellaneous crops that seem to be of relatively limited importance. A final group includes the most popular perennial vegetables. These vegetable groups are listed in Table 10-2.

A REVIEW OF PRODUCTION PRACTICES FOR HOME GARDENS

The basic production practices are similar for many of the vegetables that are grown in the home garden. These have been discussed in previous chapters. For convenience, however, we are reviewing a few important points that should be applied to the specific vegetables listed in Table 10-2.

Vegetable Planting Guide

Planting, germination, and time-to-maturity information for each of the more important vegetables is provided in Table 10-3. The planting guide contains information which is needed for planning and planting the home garden. This material has been assembled alphabetically by crops and listed in tabular form for easy reference. To avoid repetition, this information will usually not be repeated in the discussions of the individual vegetable crops.

TABLE 10-2 The Best Vegetables for Home Gardens

The top 20		The next 10	Miscellaneous
Bean	Onion	Brussels sprouts	Chinese cabbage
Beet	Pea	Cauliflower	Celeriac
Broccoli	Pepper	Swiss chard	Celery
Cabbage	Sweet potato	Collard and kale	Endive
Carrot	White potato	Kohlrabi	Garlic
Sweet corn	Pumpkin	Mustard	Leek
Cucumber	Radish	Okra	Peanut
Eggplant	Spinach	Parsnip	Popcorn
Lettuce	Squash	Rutabaga	Salsify
Melons	Tomato	Turnip	Shallot
			Soybean
			New Zealand spinach

Some perennial favorites

Artichoke (globe)
Artichoke (Jerusalem)
Asparagus
Horseradish
Rhubarb

Season of Planting

The planting date for vegetables depends on the hardiness of the particular crop. The average date of the last spring frost is the best guide to use in determining when to plant. Hardy vegetables may be planted from four to six weeks before the average last frost date in your area. Half-hardy crops may be planted from two to four weeks before the last frost date, tender crops at about the last frost date, and very tender crops about one week after the last frost date. The average date of the last frost in your locality may be secured from the Extension agent in your county or city, or may be determined from Fig. 6-12, or from the information provided in Appendix D.

Planting Patterns

Numerous planting and cropping patterns, including single row, double row, wide row, hill, succession, and intercropping are dis-cussed in Chapter 7. Certain of these planting methods can be used to (1) enhance yields, (2) provide a continuous supply of fresh vegetables over a longer period of time, and (3) fit the growing needs of particular crops. For example, onion, leek, and the smaller root crops can be planted in double rows or in wide rows to secure higher yields. Successive plantings at about three-week intervals can be made for vegetables that have a short harvest period such as snap beans, beets, cabbage, sweet corn, lettuce, and radishes. This will provide a supply over a much longer period than would be possible from only one planting.

Planting Distances

The rows in the garden should be spaced as close together as possible to secure high yields, and yet far enough apart to allow space for the plants to grow and room for cultural manipulations. A more complete utilization of soil moisture and nutrients will usually

TABLE 10-3 Vegetable Planting Guide

Vegetable	Plant hardiness	Seeds or plants/ 50 feet	Depth to plant[b] (inches)	Planting distance (inches)		Approx. days to germinate	Approx. no. days ready for use
				Between row	Between plant		
Bean, bush lima	Very tender	4 oz	1½	24	3	7	65–80
Bean, bush snap	Tender	4 oz	1½	24	3	6	45–60
Bean, pole snap	Tender	4 oz	1½	36	6	6	60–70
Beet	Half hardy	½ oz	½	18	3	8	50–60
Broccoli[a]	Hardy	⅛ oz	¼	30	18	10	60–80
Brussels sprouts[a]	Hardy	⅛ oz	¼	30	18	10	90–100
Cabbage[a]	Hardy	⅛ oz	¼	30	18	10	60–90
Carrot	Half hardy	¼ oz	¼	18	2	8	70–80
Cauliflower[a]	Half hardy	⅛ oz	¼	30	18	10	70–90
Celery[a]	Half hardy	⅛ oz	⅛	24	6	21	125
Chard, Swiss	Half hardy	½ oz	½	24	4	8	45–55
Collard	Hardy	⅛ oz	¼	24	12	10	50–80
Corn, sweet	Tender	2 oz	1½	30	8	7	70–90
Cucumber[a]	Very tender	¼ oz	1	48	12	7	50–70
Eggplant[a]	Very tender	⅛ oz	¼	30	12	10	80–90
Endive	Half hardy	¼ oz	¼	18	18	10	60–90
Kale	Hardy	⅛ oz	¼	24	8	10	50–70
Kohlrabi	Hardy	⅛ oz	¼	24	8	12	55–75
Leek	Hardy	¼ oz	¼	30	4	10	

Lettuce, leaf[a]	Hardy	1/8 oz	1/8	18	8	7	40–50
Lettuce, head[a]	Hardy	1/8 oz	1/8	18	12	7	7–75
Muskmelon[a]	Very tender	1/4 oz	1	48	12	7	85–100
Mustard	Hardy	1/8 oz	1/4	18	6	9	30–40
Okra[a]	Very tender	1 oz	1/2	30	12	10	55–65
Onion[a]	Hardy	1/4 oz	1/4	18	4	10	90–120
Parsnip	Half hardy	1/4 oz	1/4	24	6	18	120–170
Pea, English	Hardy	8 oz	1 1/2	24	1	8	55–90
Pepper[a]	Very tender	1/8 oz	1/4	24	18	10	60–90
Potato	Half hardy	3 lb	3	36	12		
Pumpkin[a]	Very tender	1/4 oz	1	72	24	7	75–100
Radish	Hardy	1/2 oz	1/4	18	1	6	25–40
Rutabaga	Half hardy	1/4 oz	1/4	24	6	9	80–120
Salsify	Half hardy	1/2 oz	1/2	1/2	24	8	150
Spinach	Hardy	1/2 oz	1/2	18	4	8	40–60
Squash, summer[a]	Very tender	1/4 oz	1	48	30	7	50–60
Squash, winter[a]	Very tender	1/4 oz	1	72	24	7	85–100
Sweet potato	Very tender	50 plants	3–4 in. of base	36	12		
Tomato[a]	Very tender	1/16 oz	1/4	48	18	8	70–90
Turnip	Hardy	1/4 oz	1/4	24	4	7	30–60
Watermelon	Very tender	1/4 oz	1	72	24	8	80–100

[a]Frequently grown from transplants.
[b]Plant shallower in heavy clay soils and deeper in sandy soils. Cover seeds deeper when planting the fall garden in midsummer.

be secured when plants are spaced evenly along the row rather than in hills or mounds. The normal minimum suggested spacings between rows and plants is presented in Table 10-3.

Depth of Planting

Many gardeners plant seeds too deep. They need only be sown deep enough to obtain moisture from the soil for germination. In good loamy soil, most vegetable seeds should be covered to a depth three to four times their average diameter. In sandy soils, plant seeds deeper and in clay soils, shallower. During the summer, when the soil is warm and dry, plant the seeds deeper. Approximate depths for spring planting in loam soil are given in Table 10-3.

Firm the soil along the row after planting to ensure adequate movement of moisture to the seed. The emergence of young seedlings from a heavy soil will be improved if the seeds are covered with sand or vermiculite to conserve moisture and reduce crusting.

Rate of Seeding and Thinning Seedlings

To ensure a full stand of plants and high yields sow two or three times more seed than needed to obtain the distance between plants suggested in Table 10-3. Large seeds, like those of bean and corn can be dropped about twice as close together as suggested for final plant spacings.

When seedlings are about two inches tall begin thinning. Thinned seedlings need not always be thrown away. Young seedlings may be transplanted, and larger seedlings of lettuce, Swiss chard, collard, kale, or beet may be used as salad or greens. Remember, crowded plants are more susceptible to diseases, and will not produce as large a crop of high-quality vegetables.

Using Transplants

The earliest crop of broccoli, cabbage, cucumber, eggplant, muskmelon, pepper, and tomato can be secured by setting young seedlings into the garden rather than planting seed. Transplants can be grown at home by any of the several methods discussed in Chapter 7, or purchased at a garden center or supermarket. In either case, transplants of desirable size, which are adequately hardened and properly set out, will give best results. As with garden seed, purchase only transplants of identified cultivars.

Late Summer and Fall Gardening

Planting a fall garden can lengthen greatly the period over which fresh vegetables are produced. The planting dates for your fall garden are easy to determine. Vegetables may be planted at any time which allows them to mature before they are killed by frost. To establish planting time, determine the number of days required for maturity as cited on seed packets, in seed catalogs, or in Table 7-2. Plant hardy crops not later than that number of days plus two weeks before the average date of the first fall frost in your locality. Plant half-hardy crops the number of days to maturity plus about three weeks before the first expected frost. Plant tender crops the number of days to maturity plus four weeks before the first expected frost. The average date of the first frost in your area can be secured from the Agricultural Extension agent in your county or city, or may be determined from Fig. 6-13, or from the information provided in Appendix D.

The best crops for the fall garden are those that reach maturity relatively fast and have a short harvest period, so that the entire yield can be secured before the plants are killed by cold. Hardy and half-hardy vegetables that continue to grow in cool weather are particularly desirable. Some of the best vegetables for fall planting are snap bean, beet, broccoli, Brussels sprouts, cabbage, Chinese cabbage, carrot, cauliflower, collard, kale, lettuce, radish, spinach, and turnip.

Furrows for planting seeds in the fall garden may be made without plowing or rototilling if the soil is mellow. Cover the seeds about twice as deep as you did in the spring.

Dry soil may be a problem during the midsummer planting period. To ensure germination of seeds, apply water along the row after the seeds are planted. Use a sprinkling can with small holes in the nozzle. Repeat daily as needed to maintain soil moisture until the young plants are well established and making rapid growth.

Selection of Cultivars

Specific vegetable cultivars for planting in the home garden are not included in this text. Such recommendations are omitted because (1) the list of cultivars is too long for practical inclusion, (2) new and improved cultivars are appearing at a rapid rate, and (3) cultivars must be selected by each gardener to satisfy his or her particular preferences and climatic conditions.

In brief, select the vegetables you like best with special attention to such features as length of time to maturity, disease-resistance, superior hybrids, and All-America Selections. To assist you, secure the latest list of recommended cultivars from the Extension agent in your county or city. Also, consult seed catalogs. These are reliable sources of information on gardening.

Harvest for Optimum Quality

Harvest your vegetables as soon as they reach edible size to secure maximum yields and high quality. One of the principal causes of poor quality vegetables is overmaturity. In particular, do not allow broccoli, cucumber, kohlrabi, summer squash, and sweet corn to become too large before they are picked.

GROWING THE TOP 20 VEGETABLES

Beans

Beans are among the oldest and most widely grown food crops. The common or garden bean (*Phaseolus vulgaris*) originated in Central America. Dry beans of this species were grown by the Indians for many centuries before the first settlers came to America from Europe.

The snap bean is the type of common bean that is most widely planted in home gardens (Fig. 10-1). Green and wax podded and bush and pole cultivars are available. To reduce the problem of disease control, select cultivars that are resistant to mosaic and mildew. Snap beans may be harvested when the pods reach full size and the beans are very small, or the picking may be delayed until beans reach mature size as long as the pods are still tender.

FIGURE 10-1 Snap bean, a popular garden crop. (Courtesy of Burpee Seed Co.)

The horticulture or pinto types of common beans have large seeds and are excellent as green shell beans, or as dry beans for winter use. Many bean cultivars are also available for harvest as dry beans to be used as baked beans or in chili. Except for the horticulture type, it is usually more practical for the home gardener to buy dry beans at the store rather than attempt to raise her or his own.

Lima beans (*Phaseolus limensis*) have large seeds and are popular for green shell or dry use. The Sieva bean (*Phaseolus lunatus*) is a small-seeded lima type that is preferred by some gardeners. Because of the large cotyledons, lima beans may have difficulty forcing their way through a heavy clay soil during the germination process. Therefore, they will grow best when planted in a mellow sandy loam soil. Fertilize lima beans moderately. Excess fertilizer will cause vigorous vine growth and reduce the set of beans.

Snap and lima bean cultivars are available in pole or climbing types that have long twining stems that grow to a height of six to eight feet or more. These are useful in a home garden where space is limited. They need to be supported by a pole, teepee, stout cord, trellis, or fence. Pole beans continue to produce over a relatively long period during the summer and early fall, and successive plantings are not needed as is suggested for bush-type snap beans.

The broad or Fava bean (*Vicia faba*) is not a true bean, but is related to vetch. It is the oldest known bean and was grown in Asia and Europe several thousand years B.C. The chief value of the broad bean is its tolerance of cold weather. It may be planted several weeks before the average last spring frost date. Broad beans do not thrive in hot weather. They are used like lima beans but are not as good quality.

Common and lima beans are tender to cold and should be planted not earlier than one week after the average date of the last spring frost. If planted early during cold, wet weather the seeds may rot in the ground and not germinate. Make small plantings of snap beans every three weeks to provide a continuous harvest during the summer.

The most important insects that attack beans are the Mexican bean beetle, leaf hoppers, flea beetles, and red spiders. To reduce damage by diseases, select cultivars that are resistant to mosaic and mildew.

Snap beans are a good source of vitamins A, B_1, and C. They contain about 190 kilocalories per pound. Lima beans and green shell beans are an excellent source of vitamins B_1 and B_2 and a good source of vitamins A and C. They have about 595 kilocalories per pound.

Beet

The beet (*Beta vulgaris*, var. *crassa*) originated in the region around the Mediterranean Sea. The earliest known forms were leafy plants similar to Swiss chard, with no enlarged root. The first report of large-rooted beets seems to be in German literature around 1550. Beets are a relatively easy crop to grow, and usually have no serious insect or disease problems.

The most popular cultivars of beets are red, and mature in 60 to 70 days after the seeds are planted. As a novelty, you may wish to try a row of the golden beets in your garden (Fig. 10-2). The root has an attractive yellow color that does not bleed out into the water during cooking as does the anthocyanin pigment of red beets. The flavor of yellow beets is similar to that of red cultivars, but some people object to the somewhat dull yellow color when the beets are cooked.

Although the swollen structure of a beet, and also of a radish and turnip, is commonly called a root, it is actually an enlarged hypocotyl. This is the part of a seedling plant between the root and the stem.

Beets are half-hardy vegetables that may be planted about two weeks before the average date of the last frost in your area. Make successive plantings at about three- or four-week intervals to ensure continuous harvest during

FIGURE 10-2 Beets are easy to grow and generally have no serious insect or disease problems. (Courtesy of Burpee Seed Co.)

the summer and fall. To increase production in the garden, sow seeds in a band three or four inches wide. Beets should be grown rapidly on a fertile soil with a plentiful supply of moisture to secure the best quality.

Since each beet "seed" is actually a dried fruit containing several seeds, the plants will come up in bunches. When the first true leaves appear at the top, thin the plants to about one inch apart. In another three weeks, when the plants are six to eight inches tall, thin again leaving the plants about three inches apart. The young, succulent plants may be used for greens or salad.

Young developing beet roots are delicious when they are about one inch in diameter. This is a desirable size for canning or pickling. The preferred size for mature roots is two to three inches in diameter.

Beet greens are a good source of vitamin C. Beet roots are a good source of vitamin B_1, and have about 205 kilocalories per pound.

Broccoli

Broccoli (*Brassica oleracea*, var. *italica*) has been grown in Europe for several thousand years, but has only become popular in the United States during the last century. The word broc-

coli is derived from the Latin *brachium*, and means a branch or arm. The edible part of the broccoli plant is a cluster of flower buds on a thick green stem (Fig. 10-3).

Broccoli plants are hardy and should receive the same cultural treatment as cabbage. Plant broccoli seeds or set young plants in the garden about four weeks before the average date of the last spring frost. Plant seeds for the fall crop in late June or early July about three months before the average date of the first fall frost.

Three types of crop or harvest may be secured from broccoli plants. Some of the young leaves may be cut off and used like collard greens when the top cluster of buds is two or three inches across. Cut the leaf at the base of the expanded leaf blade and leave the leaf stalk or petiole on the stem of the plant. Do not remove more than 25 percent of the leaves because most of the foliage is needed by the plant to nourish the developing bud clusters.

FIGURE 10-3 Broccoli, a hardy vegetable that matures for harvest in 60 to 90 days. (Courtesy of Burpee Seed Co.)

The top bud cluster of broccoli is the part of the plant which constitutes the main commercial crop. This cluster should be cut off with a four- to six-inch stem when it is four to six inches in diameter, and the flower buds are still tight and green. If the buds open and the yellow petals are exposed the cluster is inedible.

After the top bud cluster is harvested, small side bud clusters will develop. These may be harvested for at least four to six weeks after the main cluster is cut. They are excellent for home freezing because of their small size.

The most troublesome insects that attack broccoli are the green cabbage worm, cabbage loopers, and the cabbage aphid.

Broccoli bud clusters and leaves are an excellent source of vitamins A and C, and are a good source of vitamins B_1 and B_2. Broccoli has about 170 kilocalories per pound.

Cabbage

Cabbage (*Brassica oleracea*, var. *capitata*) is a hardy crop that grows best in cool weather. It seems to have originated in western Europe from nonheading forms such as collards. The modern, firm-headed types were first recorded about 1536, and were brought to the United States in 1641 (Fig. 10-4).

The earliest crop of cabbage may be secured by setting plants in the garden about four weeks before the average date of the last 32°F freeze. Seeds may be planted about two weeks earlier. For the fall crop of cabbage make a second planting in late June, or about 100 days before the average date of the first fall frost.

Green cabbage is the most common type grown in home gardens. Select cultivars that are resistant to yellows. This fungus disease lives over in the soil and is easy to control if resistant types are planted. Red cabbage is a variation of the green type that has anthocyanin pigment in the epidermal cells of the

FIGURE 10-4 Cabbage, a biennial vegetable that is a perennial favorite. (Courtesy of The National Garden Bureau, Sycamore, Illinois.)

leaf. It adds interesting color when used as salad or cole slaw.

Savoy cabbage has wrinkled leaves (Fig. 2-21). It has tender texture and a pleasant mild flavor when served either raw or cooked. It may be boiled without producing the pronounced odor that is common with green or red cabbage.

The cabbage is a biennial plant that typically forms a leafy head or enlarged terminal bud during the first growing season. It then produces flowers and seeds the second season, after a dormant period during the winter. However, if young cabbage plants are exposed to temperatures below 50°F for two or three weeks, the flowering or bolting stage will occur early in the first season and no usable head will form. This is usually not a serious problem if the stem of the young plant is as large as a lead pencil, or larger, when the low temperature occurs.

Heads of cabbage may split open if allowed to remain in the garden after they mature, or if a heavy rain occurs after a period of drought. The splitting may be reduced or prevented by deep cultivation close to the plant, or by bending the plant to one side to break off the roots.

Cutworms and root maggots may cause serious damage to cabbage soon after it is planted. Cabbage worms and aphids may be destructive during the season as the plants develop. Be sure to select yellows-resistant cultivars to reduce the losses from this disease.

Cabbage is a good source of vitamin B_1 and provides small amounts of vitamins A, B_2, and C. It has about 130 kilocalories per pound. Cabbage provides an alkaline reaction for the body, roughage to aid digestion, and in the raw form is a good cleanser for teeth and gums.

Carrot

The carrot (*Daucus carota*, var. *sativa*) originated in southern Asia several thousand years ago. The early types had stringy roots that were quite inedible. The present form, with sweet, fleshy roots, was developed in Europe about 600 years ago (Fig. 10-5).

Carrots are an excellent crop for the home garden if a mellow, loamy soil is available. The plants are relatively free of insect and disease problems.

Heavy compact soils are not suitable for carrots. The small, weak seedlings will have difficulty emerging through the crust which commonly forms on such soils. The roots may be forked and irregular when grown in dense or rocky soil. The use of fresh manure also may cause forked roots.

Several shapes and sizes of carrots are available (Fig. 10-6). These include the round and blunt, half-long cultivars, as well as those with long, tapering roots. The half-long type is best for most home gardens. In general, the longer carrots have smaller cores and higher quality,

FIGURE 10-5 Carrot, a crop which, in many areas, may be left in the ground in the fall and dug as needed for use during the winter. (Courtesy of Burpee Seed Co.)

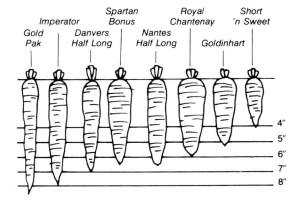

FIGURE 10-6 Carrots come in several shapes and sizes. The half-long type is best for most home gardens. (Courtesy of Burpee Seed Co.)

but are not recommended except for deep, mellow soils. Most cultivars require about 70 to 80 days to reach maturity after the seeds are planted.

Carrots are half-hardy and may be planted two to four weeks before the average date of the last spring frost. Plant in a band about three to four inches wide. The seed germinates quite slowly and it may be two weeks or more before the seedlings appear. A few radish seeds scattered along the row will help the carrots emerge, and provide an early crop before the carrots need the space. Thin the carrots to about two inches apart when the plants are two inches tall.

A second planting of carrots should be made in late June to produce a fall crop. During the dry summer weather, it will be necessary to sprinkle water along the row daily, or irrigate, to promote germination of the seeds.

In most areas carrots may be left in the ground in the fall and dug as needed for use during the winter. In the colder sections of the country, cover the row with leaves, old hay, or corn stalks to prevent the ground from freezing. The quality will be crisp and sweet when harvested during the cool fall or winter weather.

Carrots are delicious when eaten either raw or cooked. The roots are rich in carbohydrates and sugars, with about 205 kilocalories per pound. Carrots are a good source of vitamins A and B_1.

Sweet Corn

Sweet corn (*Zea mays*, var. *saccharata*) is one of the few vegetables that originated in the Western Hemisphere. Evidence indicates that this crop was grown in South America at least 2000 years B.C.

Corn requires a warm growing season. It should not be planted until about the average date of the last spring frost. Each of the desired hybrid cultivars of sweet corn have a harvest period of only 7 to 10 days duration.

Therefore, make small successive plantings every two or three weeks to provide continuous harvest during the summer. The last planting of midseason cultivars should be about 90 days before the average date of the first fall frost.

Plant corn with two or three rows together to ensure good pollination (Fig. 10-7). Sow seeds three or four per foot of row, and thin to a spacing of about 8 inches between plants. Select a fertile, well-drained soil and provide adequate moisture during the entire growing season. Sidedress liberally with a fertilizer containing nitrogen when the corn is about 12 inches tall. The growth of suckers around the base of corn stalks has no effect on the production of ears, and they need not be removed.

Harvest sweet corn when the kernels reach full size, and have developed their mature yellow or white color to the tip of the ear, but are still in the soft milk stage of growth. The ears are usually ready to harvest about three weeks after the silk first appears, when it becomes dry and dark colored.

After the corn has been harvested, cut the stalks into six- to eight-inch pieces with prun-

FIGURE 10-7 Corn, a popular warm-season crop. Plant several rows of corn close together to ensure good pollination. (Courtesy of The National Garden Bureau, Sycamore, Illinois.)

ing shears or a sickle. Leave the pieces on the ground during the winter and plow or rototill into the soil in the spring to improve the organic matter content. As soon as your cornstalks have been cut, set cabbage, broccoli, or cauliflower plants for the fall harvest to secure double cropping of your garden plot.

Sweet corn has the best edible quality if it is cooked and eaten immediately after it is harvested, when the sugar content is highest. If corn is held for several hours or longer under warm conditions, the sugars turn to starch, and the fine flavor is lost. Therefore, corn should be cooked at once, or cooled down rapidly after harvest, if it is to retain its good quality.

The European corn borer and corn ear worm may cause serious damage to sweet corn. Wilt or Stewart's disease is sometimes a problem but can be controlled by selecting resistant cultivars.

Sweet corn is a good source of vitamins A, B_1, and B_2, and has about 490 kilocalories per pound.

Cucumber

The cucumber (*Cucumis sativus*) has a long and interesting history. It is believed to be a native of southern Asia, probably in northern India in the region of the Himalaya mountains. Ancient literature reports its culture in western Asia more than 3000 years ago.

The cucumber has been grown in many countries, and is one of the few vegetables mentioned in the Bible. It was important in Egypt and was introduced into China from Persia about 100 B.C.

Columbus brought the cucumber to Haiti in the New World in 1494 and its culture spread rapidly in eastern North America. Indians were growing the crop in Canada as early as 1535. DeSoto found cucumbers in Florida in 1539, and Captain John Smith reported them in Virginia in 1584.

Cucumber plants are tender to cold. The seed should be planted about a week after the average date of the last spring frost. Select a well-drained, fertile soil high in organic matter. Best results are usually secured by planting seeds about 4 inches apart along the row and thinning the plants to about 12 inches. This will result in a better distribution of roots in the soil than when the crop is planted in hills or mounds.

The first flowers to appear near the base of a cucumber vine are male or staminate flowers, which will not develop into fruits. In about a week the female or pistillate flowers appear with the small cucumber at the base. These are pollinated by the male flowers and produce the edible crop. The new **gynoecious** hybrids produce only female flowers, and thus bear fruits earlier and closer to the base of the vine. They usually have about 10 or 12 percent of normal plants mixed in to provide pollination.

The two most common types of cucumbers are the medium to long slicing cultivars, and the short, somewhat chunky types, which are used for pickles. Some gardeners prefer the long burpless cucumbers that are suggested for people who develop stomach gas from eating regular cucumbers. Regardless of the type you plant, select disease-resistant hybrid cultivars to ensure productivity. Cultivars are now available that are resistant or tolerant of such diseases as anthracnose, leaf spot, mildew, mosaic, and scab.

Cucumbers may be trained on a trellis or fence to save space in a small garden (Fig. 10-8). In fact, this is an excellent way to raise fine, straight cucumbers even if space is not a limiting factor. Under favorable conditions the plants will reach the top of an eight-foot support by midsummer.

Space the plants about 12 inches apart along a fence or trellis made of woven wire cattle fencing, or stout cord such as binder twine, with openings about 6 inches square. As the cucumber stems elongate, the ends must be guided in and out of the openings in the woven

FIGURE 10-8 Cucumbers produce well when trained on a fence or trellis. (Courtesy of The National Garden Bureau, Sycamore, Illinois.)

support, or spiraled around individual cords. Tendrils will form along the stem and hold it firmly to the support. The vines will not climb the support by themselves as will pole beans and tall peas.

Harvest cucumbers when the fruits attain the desired size, and before they start to swell and lose their dark green color. If fruit is left on the vine to mature it will stop the setting of new fruit.

The striped cucumber beetle and aphids may attack cucumbers. Disease problems may be reduced by selecting resistant cultivars and controlling insects that spread diseases.

Cucumbers have a low kilocalories content of about 65 per pound, and are useful as a food for weight watchers. Cucumbers have negligible quantities of vitamins.

Eggplant

The eggplant (*Solanum melongena*) may have originated in southern Asia in the region of Burma or India, but its early history is somewhat confused. It has also been reported to be native to South America, as are three other vegetables of the Solanaceae family: potato, tomato, and pepper. The name eggplant relates to the fact that varieties that were popular many years ago produced small, egg-shaped fruit.

The eggplant is strictly a warm-season vegetable that should not be planted until at least a week after the average date of the last spring frost. To ensure an early harvest, start plants indoors about six to seven weeks before they are to be set in the garden.

If eggplants are grown on a fertile soil, and receive an abundance of moisture, they will continue to produce fruit until the plants are killed by frost (Fig. 10-9). Plants with a heavy crop of fruit may need to be staked.

Flea beetles, and the Colorado potato beetle may cause serious damage to eggplants, un-

FIGURE 10-9 Eggplant, a strictly warm-season vegetable commonly started indoors. (Courtesy of Burpee Seed Co.)

less they are controlled. Eggplant wilt may also become a problem if the plants are grown in the same location in the garden year after year.

Harvest eggplant fruits at a young stage of development when they are from one-third to two-thirds full mature size. Eggplant fruits should have a glossy color. When the color becomes dull they are past the stage of best edible quality. Harvesting is easier with pruning shears because the stem is quite woody.

Eggplants do not provide significant quantities of vitamins to the diet. They contain about 130 kilocalories per pound.

Lettuce

Lettuce (*Lactuca sativus*), like many other vegetables, is native to southern Asia and the region around the Mediterranean Sea. It was grown in the gardens of Persian kings at least 500 years B.C. Lettuce was brought to America by Columbus, and is now one of the most important crops in home gardens.

The best lettuce for most home gardeners is the leaf type, because it is easier to grow and has high quality. Leaf lettuce has an open, nonheading growth habit. The mature leaves may be frilled, crumpled, or deeply lobed. Select a heat-resistant cultivar that is slow in going to seed.

The butterhead or Bibb-type lettuce forms a loose head of thick, succulent, high-quality leaves. The outer leaves are green or brownish in color, whereas the inner leaves become cream colored.

The crisphead or iceberg lettuce is the typical head lettuce sold in grocery stores. It does not grow well in hot or rainy weather, and is therefore, more difficult to produce in the home garden. It may break down or become bitter if it matures during hot midsummer weather conditions.

Cos or romaine lettuce forms an erect head about eight or nine inches tall (Fig. 10-10). It is increasing in favor for home gardens because of its distinctive flavor.

FIGURE 10-10 The romaine or cos types of lettuce (background) is increasing in favor in home gardens. (Courtesy of The National Garden Bureau, Sycamore, Illinois.)

Lettuce is a hardy crop that grows best in cool weather. Leaf lettuce may produce satisfactory yields in midsummer if the heat tolerant, slow-bolting cultivars are selected. The plants will tolerate partial shade during hot weather if adequate soil moisture is available.

Start planting lettuce four to six weeks before the average date of the last spring frost. Make successive small plantings every four weeks to provide a continuous harvest during the summer and fall. Start thinning the plants when they are not over 2 inches tall, leaving them about 3 inches apart. As soon as the plants reach a young edible size, continue removing the larger ones for use as salad, until a final spacing of about 8 inches is secured.

Head lettuce plants should be thinned to a 12-inch spacing when the plants are not over 2 inches tall. Butterhead and cos lettuce should be thinned to a spacing of about 8 or 9 inches. Harvest head lettuce as soon as a suitable size is attained.

Lettuce usually has no serious disease problems. Aphids and cutworms may cause some damage in early spring.

Leaf lettuce is an excellent source of vitamin A and a good source of vitamins B_1 and C. Head lettuce is a good source of vitamin B_1. Lettuce has about 85 kilocalories per pound.

Melons

Three species of melons should be considered for planting in the home garden: muskmelon (*Cucumis melo*, var. *reticulatis*); honeydew, casaba, or winter melon (*Cucumis melo*, var. *inodorus*); and watermelon (*Citrullus lanatus*). All melons are members of the gourd family, which also includes cucumbers, gourds, pumpkins, and squash.

Melons are tender plants that require a warm growing season, and should not be planted until at least one week after the average date of the last spring frost. They require a fertile, sandy loam soil, with adequate moisture during the growing season.

The best flavor will develop when melons ripen during warm, dry weather. A bland, unsweet flavor may be the result of ripening under cold rainy conditions.

Muskmelons

Muskmelons originated in southern Asia, and have been important as a food crop for at least 5000 years. Sketches of muskmelons have been found on the walls of Egyptian tombs dating back to about 2400 B C. The name musk comes from the Persian language and means a kind of perfume.

A common mistake is to call a muskmelon a cantaloupe. The cantaloupe is actually a warty fruit of inferior quality, which is found in northern Africa and is not grown in the United States.

The striped cucumber beetle may cause damage to muskmelons, and also spread bacterial wilt. The most serious diseases are fusarium wilt, leaf blight, leaf spot, and mildew. Select resistant cultivars to reduce the problem.

Muskmelons should be harvested at the full-slip stage, when the stem separates easily from the fruit. At this time the basic green color will become somewhat yellow.

Some home gardeners believe that the poor flavor of their muskmelons is caused by the cucumbers that are growing in the same garden. This is not true because muskmelons cannot cross pollinate with cucumbers, pumpkins, squash, or watermelons. As stated previously, muskmelons must ripen under warm, dry weather conditions to secure sweet, aromatic flavor.

As a food, muskmelons are a good source of vitamin C and have 125 kilocalories per pound.

Watermelons

Watermelons seem to have originated in tropical Africa. Dr. David Livingstone, the English explorer, discovered large quantities growing wild in that area.

Watermelon vines grow quite vigorously and may require more space than is available in the home garden. The new icebox-size cultivars have smaller vines, and smaller fruit, and are a more desirable selection (Fig. 10-11). Avoid using excess nitrogen fertilizer that may result in a vigorous vine, with luxuriant foliage and few fruit.

The striped cucumber beetle is the most troublesome insect on watermelons. The most serious diseases are anthracnose and fusar-

FIGURE 10-11 Watermelon. The cultivar shown is one of the new icebox-size cultivars that produces smaller vines and fruits. (Courtesy of Burpee Seed Co.)

ium wilt. The selection of resistant cultivars will help to keep them under control.

It is somewhat difficult to determine the stage of maturity at which watermelons should be harvested. The best indicator is probably the change in color of the bottom of the melon where it rests on the ground. When the watermelon reaches edible maturity, this spot usually changes from white to a yellow or yellow-brown color. The tendril near the stem of the melon turns brown and becomes dry and shriveled about 7 to 10 days before the full ripe stage is reached. Also, ripe watermelons have a duller sound when thumped than do immature melons.

Watermelons may be stored for about a month after harvest, in a cool dry place. The fruit is a good source of vitamin C, and has about 140 kilocalories per pound.

Honeydew, Casaba, or Winter Melons

The honeydew melon requires a very long growing season of 120 to 150 days, and needs high temperature and low humidity. Mildew and wilt may be very serious under high humidity conditions.

Honeydew melons have a delicious, sweet, refreshing flavor. They have a hard rind, and will keep for two or three months in a cool, dry storage.

Onions

The onion (*Allium cepa*) is one of the most prized vegetables of the gourmet cook (Fig. 2-1). It originated in Asia where it has been grown since prehistoric times. It was brought to America by Spanish explorers and soon became an important crop for both the Indians and the colonists.

Onions develop their leafy top during cool weather and form bulbs in warm weather. The bulbing process of American types such as Ebenezer and Globe occurs under long photoperiod or daylength conditions of 14 hours or longer. The bulbing process of the Bermuda type of onion occurs under short light periods of 10 or 11 hours.

The most important onions for home gardeners are the Globe or Ebenezer cultivars, and the Bermuda or Spanish types. As a general rule, the Globe and Ebenezer types form bulbs under long daylength conditions and are more adapted to the central and northern states. The Bermuda or Spanish types form bulbs when days are shorter and grow best in the southern states. The brown Globe or Ebenezer type onions will usually keep better in winter storage.

Onions are hardy plants that may be planted four to six weeks before the average date of the last spring frost. Onions may be grown from seeds, small plants, or sets. The use of sets is most satisfactory for home gardeners. The brown Globe type onions may be grown from seeds if planted about a month before the last frost date and may keep somewhat better in storage than onions grown from sets. Any of the common cultivars of onions may be used in the green bunching stage by harvesting when the bulbs first start to develop.

Onions grown from seed should be planted in a band about three to four inches wide and thinned to a spacing of about four inches. Sets should be planted in a double row with the rows four inches apart, and sets about four inches apart in each row. Select sets that are about one-half to five-eighths inch in diameter. Larger sets are more likely to produce a seed stalk that results in a thick stem through the center of the bulb and poorer keeping quality.

Onions from the garden may be harvested at any time from the green bunching to the full bulb stage of development. For storage during the winter, pull the onions when three-fourths of the tops have fallen over. Pull in the morning of a sunny day and allow the bulbs to lie in the full sun for the remainder of the day . Remove the roots from the base and cut the top off one inch above the top of the bulb (Fig. 10-12). Place the onions in a

FIGURE 10-12 Harvesting onions. (Courtesy of The National Garden Bureau, Sycamore, Illinois.)

well-ventilated garage or shed for several days or weeks to continue the drying process. Some authorities suggest the onions may be left on the ground for several days to dry. Under such conditions they become moist with dew each morning, and drying in a garage or shed is preferred.

Store onions for the winter in a cool, dry, dark basement room. Store in shallow trays not more than two or three onions deep. Inspect and sort your onions every few weeks to remove those that become soft or start to rot. If this procedure is followed, the brown winter onions will keep in excellent condition until spring.

There are usually no serious disease problems with onions in the home garden. The onion maggot, and onion thrips may cause damage under some conditions.

Onions are a good source of vitamins B_2 and C and contain about 220 kilocalories per pound.

Peas

Three types of peas are available for planting the home garden. The green or English pea (*Pisum sativum*) is the most common. The edible-podded, sugar, or snow pea (*Pisum sativum*, var. *macrocarpon*) is somewhat of a novelty in which the entire young pod is eaten like snap beans. The southern, black-eye, or Crowder pea (*Vigna sinensis*) is an edible form of cowpea that is adapted to southern climatic conditions. All peas are legumes and support nitrogen-fixing bacteria in root nodules.

Green or English Peas

The common pea seems to have originated in Asia and countries around the Mediterranean. Seed of primitive types have been found in the lake mud beneath the homes of prehistoric Swiss lake dwellers, dating back about 5000 years to the Bronze Age. Peas were grown in Greece and Rome before the Christian Era. The English designation has been applied to peas because many fine cultivars were developed in England during the last century.

English peas are hardy and grow best in cool weather. They should be planted four to six weeks before the average date of the last frost. They will not grow well under hot, midsummer conditions. Peas should be planted in a fertile, well-drained soil, in a band or double row about 4 inches wide. Thin the plants to leave about 12 or 15 per foot of row. The dwarf cultivars, which are 24 inches tall or less, may be grown without support. Taller types should be grown on a chicken wire fence or other support to improve the yield and make harvesting easier.

The pea aphid is the most important insect pest of peas. Root rot is the most common disease. The damage from rot may be prevented or reduced by planting in well-drained soil, and promoting vigorous growth with liberal fertilization.

Peas should be harvested when the seeds are tender and sweet, just before they reach full size. At this time, the pods will be plump, with a fresh, velvety, bright green color. Pods become dull green or yellow-green when they are past the ideal stage for best quality. To retain the highest quality, peas should be shelled as soon as possible after harvest, and then cooked immediately or placed in a refrigerator to reduce the conversion of sugars to starch.

English peas are an excellent source of vitamin B_1 and a good source of vitamins A, B_2, and C. They supply about 460 kilocalories per pound.

Edible-podded, Sugar, or Snow Peas

Edible-podded peas are an interesting crop for the home gardener (Fig. 10-13). The plants are hardy, and may be grown in the same way as English peas.

Edible-podded peas are harvested when the seed first become evident in the pod, before they begin to enlarge. The entire pod is then cooked and eaten like snap beans.

Southern, Black-eye, or Crowder Peas

The southern pea is an edible type of cowpea. The plants are very tender to cold, and seed should not be planted until about a week after the average last frost date. Southern peas are more disease resistant than beans in the warm climatic conditions of the South.

Southern peas may be harvested in the green-shell stage when the seed are fully developed but not hard, or they may be allowed

FIGURE 10-13 Edible-podded peas are hardy and may be grown in the same way as English peas. (Courtesy of Peter Chan.)

to ripen and then stored and used as dried peas.

Pepper

The pepper (*Capsicum annuum*) belongs to the Solanaceae or nightshade family along with potato, tomato, and eggplant. The pepper is native to tropical America and may have originated in Brazil. It apparently was never found growing wild in Asia which is the source of many cultivated crops. There is no reference to peppers in ancient Roman, Greek, Hebrew, or Chinese literature.

The sweet bell pepper (*Capsicum annuum,* var. *grossum*) is the type most commonly grown in home gardens (Fig. 10-14). The long, hot peppers (*Capsicum annuum,* var. *longum*) are suggested for those who use them in seasoning chili or other hot dishes. These peppers are not related to *Piper nigrum,* from which we obtain the dry, ground black and white pepper, which is used as seasoning.

The pepper is a very tender plant that grows best in a somewhat limited range of warm weather conditions. Blossoms will drop when the temperature falls much below 60°F, or goes much above 80°F. Fruit setting will resume when the temperature returns to the more desirable range. The small-fruited cultivars are more tolerant of high temperature conditions than are the larger types.

Pepper seed may be planted directly in the garden but most home gardeners will find it more satisfactory to secure young seedling plants from a garden center or greenhouse. Fertilize sparingly until the plants become well established and start to set fruit. The liberal use of nitrogen will cause excess vegetative growth and dropping of flower buds. As soon as fruits start setting, a light side dressing may be applied.

Practically all sweet peppers are dark green when immature and turn bright red at maturity. They are generally used in a green condition. They retain their fine flavor when fully ripe and red, and at that stage are very attractive in a tossed salad.

Peppers are an excellent source of vitamins A and C, and a good source of vitamin B_2. They contain 155 kilocalories per pound.

Sweet Potato

The sweet potato (*Ipomoea batatas*) is a member of the morning glory family that originated in the tropics, probably in Mexico or South America. Columbus mentioned this plant in the records of his fourth voyage and took plants back to Spain. It is reported that sweet potatoes were grown in Virginia in 1648.

The sweet potato is a swollen storage root (Fig. 3-6). It is not a tuber such as the white potato, which is a swollen, underground stem.

Sweet potatoes will thrive only in a warm climate, with an average temperature above 70°F during July and August. The growing season must have at least five frost-free months. Select a sandy loam soil because the roots will be long and stringy in heavy soils.

Sweet potatoes are started from slips or sprouts that arise from roots. These roots are

FIGURE 10-14 The sweet bell pepper is the type most commonly grown in home gardens. (Courtesy U.S.D.A.)

buried under two or three inches of sand in a hotbed or coldframe (Fig. 7-26). Set the slips or sprouts in the garden at least one week after the average last frost date.

Adequate fertilizer should be applied but not excessive nitrogen, which will stimulate vine growth and poor root development. Too much water will cause roots to be elongated and not the desired blocky shape. Excess moisture prior to harvest may cause the roots to crack.

In preparation for harvest, the tops of the sweet potato vines should be cut off. The roots need to be dug very carefully to avoid cutting or bruising, which will allow rot to develop. The roots should be dried in the sun, placed carefully in crates or baskets, and stored in a warm, well-ventilated basement where the temperature will not go below 50°F.

The sweet potato is one of the most complete, nutritious foods available. It was the principal item in the diet of the early settlers in the southern states. Some South Sea Islanders live almost exclusively on sweet potatoes. The roots are an excellent source of vitamin A and a good source of vitamins B_1 and C. They provide about 565 kilocalories per pound.

White or Irish Potato

The white potato (*Solanum tuberosum*) seems to have originated in Central or South America, probably in the Andes mountains of Peru. Potatoes were carried back to Spain by explorers in 1550, and from there through countries of Europe, during the next several decades. It arrived in Ireland in 1590 and rapidly became a major food source of a hungry people. As a result, it became known as the Irish potato. The potato was brought to the United States from Ireland in 1718 by Presbyterian immigrants.

The potato now exceeds all other vegetables of the world in terms of tons produced and value of the crop. The value of potatoes in the United States alone is over one billion dollars annually. The largest producers are the Soviet Union and Poland, which together produce about one-half of the world crop.

The potato is an enlarged underground stem called a tuber (Fig. 3-4). The improved cultivars of potatoes do not develop true to name from seeds. Therefore, they are propagated asexually by planting "seed pieces" about 1½ inches square, which are secured by cutting large tubers into sections (Fig. 10-15). Each seed piece must contain one or two eyes or bud clusters, from which the stems of the new plant develop. Seed pieces should be cut about one week before they are planted to allow time for the cut surfaces to dry and heal over. The seed pieces are planted about three inches deep with the cut side down and the eye or bud on top. Seed pieces should be cut only from certified disease-free potatoes.

White potatoes grow best on a well-drained, well-aerated loam soil. They will not make satisfactory growth and good yields on heavy clay soils. A constant supply of moisture is needed during the growing season to promote continuous growth and avoid the development of knobby tubers. Cool nights are needed to promote good tuber formation. Under hot weather conditions, the plants make a large vegetative top and form very few potatoes.

As the potato plants grow during the early part of the season, soil should be mounded up along each side of the row. This will make digging easier when the potatoes are harvested. It will also keep the tubers covered with soil and protected from the sun. If tubers are exposed to the sun they turn green, have a bitter flavor, and are mildly poisonous. If such potatoes are to be eaten, the green portion should be peeled off and discarded.

The most serious insect attacking the potato is the Colorado potato beetle. Damage may also be caused by flea beetles and leaf hoppers. Late blight and mosaic are the most serious diseases. The use of resistant cultivars will reduce losses from these diseases.

Harvest potatoes whenever the tubers attain a satisfactory size. Maximum size will be

FIGURE 10-15 The white potato is propagated asexually from "seed pieces." Each seed piece should have at least one eye and should weigh about $1\frac{1}{2}$ ounces. (Courtesy of the University of Illinois Cooperative Extension Service.)

reached when the tops start to die. One plant will yield from six to eight pounds of potatoes. The home gardener should dig only the quantity needed for immediate use. Potatoes will keep better in the soil in the garden unless a cool basement is available. The best storage temperature for potatoes during the winter is 36 to 40°F. If potatoes are held at lower temperatures nearer the freezing point, the starch may change to sugar and reduce the edible quality.

Potatoes have a pleasing, bland flavor that allows them to be used frequently in the menu without objectionable monotony. They are a good source of vitamins B_1 and C and contain about 386 kilocalories per pound.

Pumpkin and Squash

Pumpkins and squash originated in tropical America, probably in Mexico or Central America. Seed and fragments of stems have been found in the ruins of ancient cliff dwellings in the southwestern United States.

Pumpkins and squash are being discussed in the same section of this chapter because there is so much confusion in the use of the botanical and common names of these crops. Both belong to the *Cucurbita* genus. The difficulty arises in the use of the species names.

Cucurbita pepo is the correct botanical name for the most common pumpkin cultivars. The most conspicuous characteristic of this group, which can be observed easily by home gardeners, is the five-ridged stem that connects the pumpkin fruit to the vine from which it is growing. This is very evident in the sugar or pie pumpkins, as well as the 'Connecticut Field' and other cultivars which are used for jack o'lanterns.

The problem arises from the fact that many types that are commonly called squash are really pumpkins. These include the acorn, scallop, zucchini, and the yellow straight and crookneck varieties. These are really pumpkins because they belong to the *Cucurbita pepo* species, and have the five-ridged stem on the fruit. The acorn is a winter type that can be stored for several months. The others are summer types that are harvested and cooked when immature.

Cucurbita moschata has a hard, ridged fruit stem and should also be called a pumpkin. However, the most popular cultivar of this type, the 'Butternut', is commonly called a squash (Fig. 10-16). It is a winter type that may be stored for several months.

Cucurbita maxima is the true squash (Fig. 10-16). The stem of the fruit is round in cross section, and usually somewhat spongy when mature. Examples of true squashes include 'Hubbard', 'Buttercup', and 'Boston Marrow'. These are winter-type squashes that should be harvested when fully mature and will keep for several months in a cool basement.

FIGURE 10-16 *Cucurbita maxima is* The true squash. The cultivar shown, 'Boston Marrow', is one of the oldest squash cultivars in use. (Courtesy of The National Garden Bureau, Sycamore, Illinois.)

All species of pumpkins and squash grow best in a fertile, well-drained loam soil, in a sunny location. The long vine types may extend for 10 or 12 feet or more and are not well adapted to small gardens. The bush-type summer squash such as zucchini can be grown in a relatively small amount of space. Bush-type acorn squashes and the new bush-type pumpkins also have small space requirements. Use the spacings suggested in Table 10-3 for the summer- and winter-type squashes.

The most troublesome insects of pumpkin and squash are vine borer, squash bug, and striped cucumber beetle. The plants usually have no serious disease problems.

Summer squashes should be picked when they are young and tender. The rind should be soft and the seeds underdeveloped. Winter squash and pumpkin varieties should be fully mature before harvest to develop the best flavor and good storage quality. The winter types should be cut from the vine with pruning shears

leaving one inch or more of stem attached to each fruit.

Pumpkin and squash are excellent sources of vitamin A, and have 160 to 200 kilocalories per pound.

Radish

The radish (*Raphanus sativus*) probably originated in Asia or China, but the exact location is unknown. It is an ancient vegetable that has been cultivated since prehistoric times.

Radishes are usually the first vegetable that can be secured from the home garden. A delicious, succulent crop will be ready for harvest in three to four weeks after the seed are planted (Fig. 10-17). Although commonly called a root, the radish, like the beet and turnip, is actually a swollen hypocotyl, which is the part of a seedling plant located between the root and the stem.

Radishes are hardy and grow best during cool weather, in a fertile, moist soil. They develop a hot, pungent flavor and a pithy texture when grown in hot weather. Under long day-length conditions of 15 hours or more and warm temperatures, the plants mature rapidly and develop flowers and seedstalks. Radishes are a good source of vitamin C and have only 100 kilocalories per pound.

FIGURE 10-17 Radishes will be ready for harvest in three to four weeks after the seed are planted. (Courtesy of U.S.D.A.)

Plant radishes in narrow rows, or bands about three or four inches wide, starting four to six weeks before the average date of the last spring frost. Plant in short rows only a few feet long, and make successive plantings every 8 to 10 days during the cool early spring season. Start plantings for the autumn crop about six weeks before the average date of the first fall frost.

Thin radish plants to at least one inch apart as soon as possible after the seedlings emerge. This is essential to reduce competition between plants and to obtain plump, succulent radishes. The plants will develop too many leaves and stunted roots if they remain too thick in the row or are planted in the shade.

The most common insect problem with radishes is the cabbage root maggot.

Spinach

Spinach (*Spinacia oleracea*) is a popular crop in the home garden for use as greens. It is an easy and quick vegetable to grow during cool spring or fall weather (Fig. 10-18).

Spinach is native to southwest Asia and probably originated in Iran and adjacent areas. The earliest evidence of its use as a food crop seems to be in Chinese records, which report that it was introduced from Nepal in 647 A.D. It reached Spain from North Africa in 1100, and was cited in 1390 in an English cookbook. The early colonists brought spinach to America, but commercial production did not start until about 200 years later in 1806.

Spinach is a hardy vegetable that grows best under cool weather conditions. It will withstand cold temperatures as low as 10 to 12°F below freezing. It may be planted in the spring a month or more before the average date of the last frost.

Spinach plants demonstrate photoperiodic response by initiating flowers under long-day conditions of 14 hours or more. This development is increased by low temperature when the plants are young, and by high temperature as they mature. Long-standing or slow-bolting strains are available that should be planted instead of the older cultivars. The long-standing cultivars go to seed more slowly than other types, and have a longer harvest period. This information relative to the photoperiodic response of the different cultivars is included in the description that is found in seed catalogs.

Plant spinach in a well-drained, fertile loam soil. The entire plant may be harvested, or individual leaves may be picked, when the desired size is reached about 40 to 50 days after planting. Use the spacings for planting and thinning, which are suggested in Table 10-3. Spinach does not grow satisfactorily during continually wet weather. If your soil is somewhat heavy, plant on low ridges to provide extra drainage.

Spinach goes to seed rapidly during the hot weather of early summer. This reduces the yield and results in a poorer quality product. Swiss chard or New Zealand spinach may be preferred in the home garden because they will continue to grow during warm weather, and may be harvested several times.

The fall crop of spinach should be planted about 40 days before the average date of the first autumn frost. The seed do not germinate

FIGURE 10-18 Spinach, a popular crop in the home garden for use as greens. (Courtesy of U.S.D.A.)

well in warm weather. Sprinkling water along the row or using sprinkler irrigation to provide moisture and cool the soil will be beneficial.

There are two common types of spinach. The savoyed or crumpled leaf type is used mostly for home gardens and fresh market. The flat or smooth leaf sorts are more commonly used for commercial canning. Insects are easier to control on the smooth leaf cultivars.

The aphid is the most serious insect pest of spinach. Losses from mosaic or blight may be reduced by selecting resistant cultivars and controlling aphids that spread virus diseases. Downy mildew may be a troublesome leaf disease in wet seasons.

Spinach contains several essential nutrients that are beneficial in the diet. A half cup of cooked spinach provides two to four times the daily required amount of vitamin A for adults. Spinach is also an excellent source of vitamin C, and a good source of vitamins B_1 and B_2. It contains about 110 kilocalories per pound.

Tomato

The tomato (*Lycopersicon esculentum*) is by far the most popular crop in the home garden. It originated in South America, probably in the Peru-Ecuador-Bolivia area. Tomatoes were used as food by the Indians for many centuries before Europeans came to America.

Tomatoes are a good source of vitamins A, B_1, and C. They contain about 105 kilocalories per pound.

Types of Tomatoes

Many types and cultivars of tomatoes are available for the home garden. Consult your favorite seed catalog for pictures and descriptive information. The Extension agent in your county or city can suggest the best cultivars for your area. Regardless of the cultivars you plant, try to select those with the VFN des-

ignation after the name. This means the cultivar is resistant to verticillium wilt, fusarium wilt, and nematodes.

The large red tomatoes are the most common type grown by home gardeners (Fig. 10-19). They are excellent for use fresh, cooked, or for juice. Most cultivars are somewhat acid. If you prefer a subacid tomato try the yellow or orange varieties.

Paste tomatoes are mild in flavor and are produced on vigorous, productive plants. The fruit has an oval shape, is about three inches long, and has a firm, meaty texture.

There are also many cultivars of cherry, pear, or plum tomatoes that bear small red or yellow fruit on dwarf plants. These dwarf types are especially adapted for growing in containers.

FIGURE 10-19 The large, red tomatoes are the most popular in home gardens. (Courtesy of The National Garden Bureau, Sycamore, Illinois.)

Culture of Tomatoes

Tomato plants are very tender and should be planted in the garden not earlier than one week after the average date of the last spring frost. They need a well-drained soil but will grow better on heavy soils than most other vegetable crops.

The earliest harvest will be obtained by growing or buying transplants that are set in the garden after the danger of frost is past. Such plants should be about six or seven weeks old. Tomato seed may be planted directly in the garden for the midseason and late crop. When the young transplants are used, they should be set quite deep in the ground, with the lowest leaf just above the surface of the soil. If the plants are tall and leggy, they may be laid down with soil covering the lower part of the stem, and only the top six inches remaining above ground. These deep-planted tomatoes will form roots along the stem and make vigorous growth.

Tomatoes, like many other crops in the home garden, will benefit if a mulch of sawdust, leaves, or other organic material is used around the plants. The mulch will help keep the soil cool, conserve moisture, and reduce or eliminate weed growth. The mulch will add organic matter and improve the soil when it is plowed or rototilled into the ground when the garden is prepared for planting in the spring.

Tomato plants set fruit best when the night temperature is between 60 and 70°F and the day temperature is not above 85°F. If the temperature is 55°F or below at night, or 95°F or higher during the day, the flowers will usually drop and not set fruit. Excess nitrogen fertilizer early in the growing season, which causes vigorous vegetative growth, will also prevent fruit set. Therefore, wait until the lower flower clusters have set fruit before side dressing tomato plants.

Blossom end rot is a common problem with tomatoes. It is a brown spot of dead cells at the end of the tomato fruit, which may later enlarge into a rotted area. It is caused by a combination of water stress within the plant, and a lack of calcium. Blossom end rot may be reduced by maintaining a constant moisture supply in the soil by irrigating or mulching, and applying ground limestone to the soil around the plants.

Sunscald and cracking of tomato fruit may be increased by sparse foliage as a result of poor growth on an infertile soil or heavy pruning. Some cultivars are more resistant to cracking than others and should be selected for planting.

Staking and Training Tomatoes

The compact, low-growing, determinate type tomatoes, and the dwarf cultivars, may be grown by allowing them to spread over the ground. If this method is followed it is very important to use an organic or plastic mulch to keep the fruit off the ground and reduce the problem of rot.

The tall, indeterminate types of tomatoes will give best results if they are trained up off the ground. One of the best methods is to train the plants to one or two stems, tied to a single stake. As the stem or stems elongate, the suckers or little shoots that develop in the axils of the leaves are removed when they have developed two leaves (Fig. 10-20). The stem is tied loosely to the stake with coarse string or binder twine, at about eight-inch intervals (Fig. 10-20).

Tomato stakes should be about seven or eight feet tall to allow about eight inches to be driven into the ground and still have six feet or more exposed. The lower portion of the stake that extends into the soil should be treated with a wood preservative such as copper naphthenate each spring. Wooden stakes treated in this way will last for many years if stored in a garage or shed during the winter. Do not use creosote or pentachlorophenol as preservatives because they may seriously damage your tomato plants.

Tomatoes trained to a stake usually will have ripe fruit ready to harvest earlier than plants

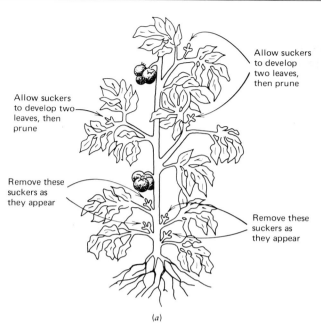

Allow suckers to develop two leaves, then prune

Allow suckers to develop two leaves, then prune

Remove these suckers as they appear

Remove these suckers as they appear

(a)

FIGURE 10-20 For best yields of quality fruit, tomatoes must be properly pruned (a) and staked (b). (Courtesy of The University of Illinois Cooperative Extension Service.)

To stake a tomato plant, tie the string tightly around the stake and loosely around the plant. Tie the knot just below a leaf so that the plant cannot slide down.

(b)

allowed to spread on the ground. Each plant will produce a smaller crop of fruit than larger plants on the ground or in cages. However, the plants can be set 15 inches apart in rows 30 inches apart and produce as large a yield for the same space in the garden as the spreading plants that must be set farther apart.

Another popular way of training tomatoes is to grow them in wooden or wire cages (Fig. 10-21). The supports are made in the form of a cylinder about 15 inches in diameter and five feet tall, using heavy 6 × 6 inch wire mesh such as is used to reinforce concrete. The cages are placed over the tomatoes after they are transplanted to the garden. As the plants grow, the side shoots are tucked back inside the cage, and not allowed to project through the wire. The tomatoes are harvested easily by reaching through the openings in the wire fencing.

FIGURE 10-21 Tomatoes can also be trained in wooden or wire cages.

A cagelike support for tomatoes may be constructed by driving five- or six-foot stakes on four corners around each plant to form a square about 14 inches on each side. Binder twine, or other strong cord may be wrapped around the stakes at about 5-inch intervals. The tomato plants are trained up inside this framework as in the wire cage. The stakes that are used to form this structure take much less storage space during the winter than the bulky wire cages.

Tomatoes may be harvested whenever they attain a desired size, and develop their typical red or orange color. Fruits may be harvested anytime after they have started to turn red and will continue the ripening process off the plant. Green tomatoes may be used to make pickles or relish.

Insects and Diseases of Tomatoes

Insects may be a serious problem on tomatoes. Cutworms may destroy plants soon after they are transplanted into the garden. Flea beetles and the Colorado potato beetle may devitalize and damage tomato plants. Other pests that may be troublesome include aphids, tomato hornworm, leaf hoppers, and tomato fruit-worm.

Fusarium and verticillium wilt are serious diseases of the tomato. Nematodes, which are very thin eelworms in the soil, may also be a problem. These may be controlled by selecting the resistant VFN cultivars.

Early and late blight may severely damage or kill tomato plants. As yet, no satisfactory cultivars are available that are resistant to these diseases.

GROWING THE NEXT 10 VEGETABLES

The garden will become more interesting, and your menu more varied, if you try some vegetables other than those listed as the most popular 20. Each of the 10 to be discussed here has certain cultural or culinary qualities that you may wish to consider.

Brussels Sprouts

Brussels sprouts (*Brassica oleracea*, var. *gemmifera*) was developed about 400 years ago from a Mediterranean strain of cabbage, by gardeners of Brussels, Belgium. The plants produce lateral buds like small cabbages, on the side of a thick stem 18 to 24 inches tall (Fig. 10-22).

Brussels sprouts develop the best flavor as a fall crop when the buds mature under cool conditions. Plant seed in late June about 100 or more days before the average date of the first autumn frost. Sprinkle water along the row to promote germination, if the soil is dry.

Pinch off the tip of the stem when the sprouts commence to form in the axils of the leaves near the base of the stem. This will promote a more rapid and uniform growth of the buds. Harvest by snapping the sprouts sideways, starting at the base when the buds are about one inch in diameter.

Although not recommended except for the cool northern states, Brussels sprouts may be grown as a late spring or early summer crop. The young transplants should be set in the garden at least four weeks before the average date of the last spring frost. A good crop may be harvested if the weather is cool in early summer. If the weather is hot when the sprouts are picked they will lose their delicate flavor.

Brussels sprouts are attacked by the same insects and diseases which are a problem when growing cabbage.

Brussels sprouts are a good source of vitamins B_1, B_2, and C, and contain about 260 calories per pound.

FIGURE 10-22 Brussels sprouts, a hardy vegetable that produces enlarged lateral buds. (Courtesy of Burpee Seed Co.)

Cauliflower

The cauliflower (*Brassica oleracea*, var. *botrytis*) is native to Europe and western Asia where it seems to have been cultivated before the Christian Era.

Cauliflower is a hardy plant that is grown best as a fall crop like Brussels sprouts. It may be grown as a spring crop in northern states that have cool weather in early summer when the crop matures. The heads will be loose and of poor quality if they mature during hot weather.

Cauliflower should be planted on well-drained, fertile soil. An adequate water supply must be available to promote vigorous, rapid growth and high quality. When the white head or curd starts to form, the leaves should be tied together, or one or two leaves should be broken over to protect the head from the sun. Harvest cauliflower when the heads reach full size, but before the clusters separate or turn brown.

Cauliflower is an excellent source of vitamin C, and a good source of vitamins B_1 and B_2. It contains about 140 kilocalories per pound.

Swiss Chard

Swiss chard (*Beta vulgaris*, var. *cicla*) is a botanical variety of beet that is an excellent crop for use as greens. It does not produce the swollen, edible storage organ which is found in beets. Swiss chard is native to southwest Asia, and the region around the Mediterranean, where it has been cultivated for several thousand years.

Plant Swiss chard as you would beets. Thin the plants to about two inches apart as soon as the first true leaves appear. Thin again to about four inches apart when the plants are about six inches tall. Use the plants you pull out for greens or salad.

Only one planting of Swiss chard is necessary to supply greens for the entire season. Harvest by cutting the leaves off about 2 inches above the soil, when not over 12 inches tall. New leaves will develop rapidly and repeated harvests may be obtained throughout the summer. The plants grow vigorously in hot weather and do not bolt and go to seed as is the case with spinach and leaf lettuce.

Swiss chard has tender texture and mild flavor like spinach when used as cooked greens. The young leaves may be used in a tossed green salad, or in a sandwich, as a substitute for lettuce. The leaf stalks or petioles may be cooked and used like asparagus.

Swiss chard is an excellent source of vitamin A, and is a good source of vitamins B_2 and C. It contains about 150 kilocalories per pound.

Collard and Kale

Collard and kale are different forms of the same genetic plant, *Brassica oleracea*, var. *a-cephala*. The botanical name means "cabbage without a head." In fact, collards are probably a primitive form of cabbage. Vegetable historians report it has been used as food for over 4000 years, and has been grown in its present form for about 2000 years.

Both collard and kale are hardy and can be planted four to six weeks before the average date of the last spring frost. They grow best in cool spring or fall weather. They will survive as a winter vegetable in southern states where the ground does not freeze. Collard will tolerate more summer heat than kale but may develop an unpleasant flavor if harvested during warm weather.

The leaves of collard are relatively flat, like young leaves of cabbage or broccoli. The leaves of kale are more crinkly and puckered. Both crops should be harvested when the leaves are only 9 or 10 inches long. Old leaves become tough and stringy.

Collard and kale are excellent sources of vitamin A, and good sources of vitamins B_1, B_2, and C. The leaves have about 225 kilocalories per pound.

Kohlrabi

Kohlrabi (*Brassica oleracea*, var. *caulorapa*) is a relatively modern vegetable that has been known for only about 500 years. It developed from wild cabbage in northern Europe.

Kohlrabi develops an enlarged stem just above the ground, which looks like a small turnip with leaves coming out from the sides (Fig. 10-23). This stem has a crisp, white flesh, which tastes like a very mild turnip. It may be eaten fresh or cooked.

Kohlrabi is a hardy plant that grows best in cool weather as a spring or fall crop in the garden. Sow seed in single or band rows, and thin plants to a distance of four inches apart. If they are left crowded in the row they will not develop the swollen, edible stem. The plants must receive ample fertilizer and water to promote rapid growth. They will tolerate

FIGURE 10-23 Kohlrabi develops an enlarged stem just above the ground. (Courtesy of U.S.D.A.)

partial shade in the garden better than most vegetable crops.

Harvest kohlrabi when the stems are about two inches in diameter. The stems become less succulent when larger and are of poor quality. Kohlrabi is an excellent source of vitamin C and has about 165 kilocalories per pound.

Mustard

Mustard (*Brassica juncea*) is a half-hardy, leafy vegetable cultivated in China for 4000 years. It should be grown as a spring or fall crop because it develops flowers and seeds quickly in hot weather.

Mustard is ready to harvest in 40 to 50 days after planting and is an excellent source of vitamins B_2 and C. It has a pronounced and distinctive flavor when used as greens, or in a tossed salad. The plants must be grown on fertile soil with an abundance of moisture to produce tender, succulent leaves. Mustard greens should be picked when the leaves are only three to four inches long to secure the best quality.

Okra

Okra (*Abelmoschus esculentus*), also known as gumbo, is a native of north Africa. It is a very tender plant which is adapted to the southern states as a summer or fall crop (Fig. 10-24). Okra is relatively free of insect and disease problems.

To promote rapid germination, soak okra seed in water for 24 hours before planting. Sow the seed in a sandy loam soil, and thin the young plants to 12 to 15 inches apart.

FIGURE 10-24 Okra, a southern favorite. (Courtesy of U.S.D.A.)

Harvest okra every two or three days when the young pods are tender and not over 3 inches long. If the stem becomes hard and resists cutting with a knife, the pods are past their peak of flavor and quality. The plants will stop bearing if pods are allowed to ripen on the stems. Okra is a good source of vitamins B_1 and C and contains about 155 kilocalories per pound.

Parsnip

The parsnip (*Pastinacea sativa*) originated in the Mediterranean or European region, possibly in Germany. The plants are quite free of insect and disease troubles.

Parsnips are half-hardy and may be planted about four weeks before the average date of the last spring frost. The seed germinate very slowly and two or three weeks will elapse before the seedlings emerge. Select a deep, well-drained soil because the roots grow 12 to 15 inches long. They will have irregular shape and be difficult to dig in heavy or rocky soil.

Parsnips must be exposed to temperature that is near freezing or below, to cause the starch in the root to change to sugar and produce the desired sweet, nutlike flavor. The roots are hardy and may be left in the ground and dug as needed during the winter. In northern states the row may be covered with a deep mulch to reduce freezing. If digging in midwinter is difficult, the roots may be left in the ground and dug in the spring.

Parsnips are a good source of vitamins B_1 and C, and contain about 380 kilocalories per pound.

Rutabaga

Rutabaga (*Brassica napobrassica*) is native to northern Europe, and is considered to be a natural cross between a cabbage and a turnip. It is cultivated widely as a food crop in countries around the world which have a cool climate. The edible part, which is commonly called a root, is actually a swollen hypocotyl as described for the radish.

The rutabaga has smooth, waxy leaves. The interior color or the root of most cultivars is yellow, although white-fleshed cultivars are available. The plants require 90 to 100 days to reach edible maturity. The best quality is developed if the crop matures in the fall under cool temperature conditions. Harvest rutabagas when they are four inches or more in diameter. They may be left in the ground and dug as needed during the winter, or stored in a cool, humid cellar.

Root maggots may cause serious damage to rutabagas. Aphids may also be a problem. Rutabagas are a good source of vitamins B_1 and C and contain about 185 kilocalories per pound.

Turnip

The turnip (*Brassica rapa*) is an ancient crop that appears to have originated in western Asia and the Mediterranean region. The edible part of the plant is commonly called a root but is a swollen hypocotyl, as we have reported for beets, radishes, and rutabaga. In this text we have also used popular terminology, and called these plant structures roots, even though this is botanically incorrect. Most turnip roots have white flesh, but some yellow cultivars are available.

The turnip is a fast-maturing crop that is ready to harvest in 40 to 60 days, and grows best under cool climatic conditions. Plant in very early spring to secure an early summer harvest. The best quality will be attained by a fall crop that matures during the cool autumn weather.

Sow the seed in single rows, or in a band about four inches wide. Thin the plants to a four-inch spacing, and use the thinnings as greens, as is done with beets. Harvest the roots when they are two to three inches in diameter. Larger roots become pithy, fibrous, and bitter.

Root maggots and aphids may cause damage to turnips.

Turnip greens are an excellent source of vitamins A and B$_2$, and a good source of vitamins B$_1$ and C. The roots contain about 155 kilocalories per pound.

GROWING SOME
MISCELLANEOUS VEGETABLES

In addition to the more popular vegetables that have been discussed previously in this chapter, there are about a dozen more that should be considered. Some of these crops have a unique quality or culinary use that may add a new or special interest to the home garden. In some cases we have suggested reasons why they should *not* be planted.

Chinese Cabbage

Chinese cabbage (*Brassica pekinensis*) is a half-hardy, erect-growing, loose-headed cabbage. It should be grown only as a fall or early winter crop because it will go to seed and not make firm heads when it matures in the summer.

Sow seed directly in the garden about 100 days before the average date of the first fall frost. Thin plants to about 12 inches apart and cultivate as you would cabbage. Harvest when the heads are fully mature. Remove the outer coarse leaves and use the tender center portion as a fresh salad, or as a cooked vegetable.

Cabbage worms, aphids, and cutworms may attack Chinese cabbage.

Celeriac

Celeriac (*Apium graveolens*, var. *rapaccum*) is also known as knob celery, or turnip-rooted celery. It apparently originated from the same wild plant species as celery but developed a swollen root rather than the enlarged succulent leaf petioles as in celery.

Celeriac has the same exacting climatic and cultural requirements as celery, and is, therefore, not an easy crop to raise in the home garden. The roots are harvested when about four inches in diameter. They may be used raw or cooked in salads, soups, or stews. Celeriac, like celery, is not a particularly good source of vitamins, but it does provide needed roughage and contains about 250 kilocalories per pound.

Celery

Celery (*Apium graveolens*, var. *dulce*) probably originated near the Mediterranean, in wet marshy areas. The plants are quite exacting as to climatic and cultural conditions, and are not recommended for the average home garden. Celery is adapted to the northern states for the summer and fall crops, and to the southern states for winter and spring crops.

Celery requires a long, cool growing season, a very fertile loam or muck soil, and an abundant supply of moisture. If the temperature remains above the desired 60 to 70°F range for even relatively short periods during the summer, the stalks may become tough and stringy, and have a strong flavor.

The quality of the older cultivars of celery may be improved by blanching. This is not practiced with the green types that are now popular. Celery is not a good source of vitamins. It does, however, provide desirable roughage in the diet, and is of interest to weight watchers as it contains only about 100 kilocalories per pound.

Endive

Endive (*Cichorium endivia*) was an important crop with the ancient Egyptians, Greeks, and Romans. It has been less popular in the United States because of its slightly bitter flavor.

The two most common types of endive are the broad leaf or escarole, and the curled or fringed type. Both are half-hardy and require about 90 to 100 days to reach edible size. They should be planted in a fertile loam soil. Adequate moisture is needed to promote rapid growth and produce a succulent, high-quality product.

Endive may be grown as an early summer crop if planted very early in the spring. However, the best quality and flavor will be secured by planting in late June for fall harvest. It may be grown as a winter crop in mild southern regions.

As the plants approach maturity, the outer leaves should be tied together over the center of the plant using string or an elastic band. This will cause a blanching of the center leaves that reduces the bitterness and causes them to become more tender. The plants should be dry when the leaves are tied, to reduce the possibility of rot developing in the enclosed center leaves. The leaves will be blanched and ready to eat in two to three weeks.

Endive is an excellent source of vitamin A, and a good source of vitamins B_1, B_2, and C. It contains about 110 kilocalories per pound.

Garlic

Garlic (*Allium sativum*) is used in such small quantities in cooking that it is usually purchased at the store rather than grown. For those who wish to produce it in their home garden, its culture is similar to that of onions.

Garlic grows as a cluster of cloves or bulblets, surrounded by a papery membrane or sac. Individual cloves are planted as you would onion sets, spaced about four inches apart in the row. Garlic is harvested, dried, and stored the same as onions.

Leek

The leek (*Allium porrum*) is a hardy plant of the onion group that requires 120 to 140 days to reach edible size after seed are planted. Leeks do not form an enlarged bulb like onions, but have an elongated base one to two inches thick and six to eight inches long, composed of tightly packed leaf bases.

Plant leek seeds four to six weeks before the average last frost date, and thin the plants to about four inches apart. In midsummer mound soil up on each side of the row to a height of six to eight inches, causing the lower edible portion to become white and tender.

Some growers sow leek seed rather thickly in a band type row three or four inches wide. When the plants are eight inches tall they are carefully dug and cut back to four inches. The pruned plants are then reset about four inches apart in a furrow five inches deep. As the plants grow, the soil along the sides of the furrow is gradually moved down to cover the basal part of the plant. Additional soil may be mounded up along the row to develop a longer white edible portion.

Leeks may be used when they reach edible size in the fall, or left in the ground and dug as needed during the winter. In central and northern states, the row should be covered with leaves, old hay, or other mulching material to reduce or prevent the freezing of the soil.

Leeks are delicious for use in soup, salad, or other culinary purposes where a mild onion flavor is desired. They are a good source of vitamin C and have about 205 kilocalories per pound.

New Zealand Spinach

New Zealand spinach (*Tetragonia tetragonioides*) may be used as a substitute for spinach in the home garden because the plants are tolerant of hot weather and will continue to produce edible leaves throughout the summer.

The plants of New Zealand spinach have a low, trailing growth habit, and form a spreading mat about three to four feet across (Fig.

10-25). The seed are slow to germinate and may be started indoors in peat pots, 6 to 8 weeks before the plants are to be set outdoors. Set plants in the garden about one week after the average date of the last frost. Set plants at least 15 inches apart, in rows three feet apart.

New Zealand spinach is harvested by cutting about three-inch stem tips with several leaves attached. New shoots will continue to develop, and repeated crops can be harvested throughout the summer and fall, until the plants are killed by frost. New Zealand spinach may be used raw or cooked like spinach.

Peanut

The peanut (*Arachis hypogaea*) is a tropical plant that originated in South America. It is adapted to southern states that have a warm growing season of 110 to 120 days. Peanuts do not grow well in northern and central states because the summer temperature is too low, even though the frost-free growing season may be long enough.

The Virginia-type peanut produces large seed, one or two per pod, with rich flavor, and requires a 120-day growing season to mature. The Spanish peanut is smaller with two or three seed per pod, with a sweet flavor, and requires about 110 days to mature.

Peanut plants are legumes and resemble yellow-flowered, low-growing sweet pea bushes. After the flowers wither, a stalk called a **peg** elongates and pushes the ovary or pistil of the flower into the soil to a depth of one or two inches. The pistil then develops into the pod containing the peanuts (Fig. 10-26). This explains why peanuts grow best on a light sandy loam soil.

The Virginia peanut grows quite vigorously and should be planted in rows about three feet apart, with plants 18 inches apart. Plant two seed in each location and thin to one plant per hill after the seed germinate.

Spanish peanuts are less vigorous and may be planted in rows two feet apart. Sow the

FIGURE 10-25 New Zealand spinach—not a close relative of spinach—can be used as a substitute for spinach. The plant will not tolerate low temperatures, but it will continue to produce during warm summer weather. (Courtesy of the National Garden Bureau, Sycamore, Illinois.)

FIGURE 10-26 Peanut, a nitrogen-fixing crop, grows best on a light, sandy loam soil. (Courtesy of U.S.D.A.)

seed about four inches apart, and thin to eight or nine inches apart when the plants are three inches tall. The seed will germinate in about seven days in warm weather, but may require two weeks under cool conditions.

The area around the young plants should be cultivated to a shallow depth until they are 12 inches tall. Then mound the soil around the plants as you would with potatoes. Mulching between the rows with straw or fresh sawdust will help conserve soil moisture and promote better growth.

Peanuts develop best in soils with a high level of calcium in the top three or four inches. To promote the maximum production of nuts, the plants and soil surface may be dusted with ground limestone during the blooming period.

The peanut crop should be harvested about 60 days after flowering when the plants turn yellow or are killed by frost. The inside of the shell should be brown, and the seedcoat will be light pink, and very thin. The seeds should have developed their maximum size. Overmature pods will break off the vine during the digging operation.

Lift the plants carefully with a garden fork, and shake off the loose soil. Dry the peanuts on the vine, or pulled off in trays, in a warm airy place. About four to eight weeks will be needed to cure the nuts before roasting, depending on the temperature and relative humidity of the air.

Popcorn

Popcorn (*Zea mays*, var. *praecox*) is an ancient and unique American crop that developed from primitive types of maize, probably in Mexico. It is a tender crop that grows best in the climatic conditions that are needed for sweet corn, and with the same cultural treatment.

Popcorn is not a good crop to grow in the home vegetable garden. If pollen from the tassel of popcorn, pollinates the silk of an ear of sweet corn, the sweet corn's edible quality, and especially the sweetness will be seriously reduced. Also, if popcorn is pollinated by sweet corn, its ability to pop will be reduced. Therefore, sweet corn and popcorn must be grown in separate locations, which usually is not possible in a home garden. Considering the rather small quantities of popcorn that are used by the average family, other vegetables in the garden will produce a greater return for the energy expended.

Salsify

The salsify or oyster plant (*Tragopogon porrifolius*) produces long slender roots with white flesh, which are about 1 inch in diameter at the top, and 8 to 10 inches long. It requires a deep, well-drained, sandy loam soil. Too many lateral roots may form in heavy soils, and forked or branched roots may develop if fresh manure is used.

As the name oyster plant suggests, the roots have a delicate flavor something like oysters. Salsify is half-hardy and requires a long growing season of 100 to 120 days. Seed may be planted two to four weeks before the average date of the last spring frost. Sow seed in a band about four inches wide, and thin to a spacing of two to three inches.

The roots may be harvested in the fall and stored in damp sand in a cool place. Leave one inch of top on each root to reduce shriveling. The roots may be left in the ground during the winter and dug as needed, as has been suggested for carrots and parsnips.

Salsify roots contain about 385 kilocalories per pound. They have a relatively low content of vitamins.

Shallot

The shallot (*Allium ascalonicum*) is a multiplier-type onion that grows in a cluster of bulblets like garlic, but they are not enclosed in a paper membrane. The plants are hardy and very resistant to diseases and insects. The bulbs are used in cooking or salads as you would mild-flavored onions.

Shallots are raised from sets or bulblets like onions or garlic. The hardy bulbs will live over the winter in the garden but best results will be secured by digging the clusters and re-planting individual bulblets each spring. The crop may be harvested and stored as suggested for onions.

Soybean

The soybean (*Glycine max*) is an ancient crop that is known to have been cultivated in China as early as 3000 B.C. The plants are tender and easy to grow, following the same cultural practices that are suggested for snap beans.

One important characteristic must be considered if maximum yields of soybeans are to be secured. The development of flowers on the plant is controlled by the length of the daylight and night portions of the 24-hour day. In general, short nights and long days delay flowering, whereas long nights and short days speed up flowering. Select cultivars that respond best to the night and daylength conditions in your locality. The Extension agent in your county or city should be able to recommend the best cultivars for your area.

Soybeans may be used in the green bean stage like lima beans or as dried beans or sprouted beans. As green beans, they are ready to harvest as soon as the pods are plump, and the beans are nearly full size but still green. As a rule, all of the beans on a plant ripen at about the same time. Harvesting may be easier if the entire plant is pulled up and the bean pods then are picked off. Blanching the pods in boiling water for five minutes will make shelling much easier.

PERENNIAL VEGETABLES FOR THE HOME GARDEN

Several perennial vegetables are useful additions to the home garden. They will persist and produce an edible harvest for several or many years after the planting is established. Locate perennial vegetables at the side of the garden area so they will not interfere in the spring with soil preparation for the annual vegetables.

Globe Artichoke

The globe artichoke is a type of thistle that is native to North Africa. The immature flower bud is the edible part (Fig. 10-27). The commercial crop is grown primarily in California, with small quantities in the Gulf states.

Globe artichoke plants are tender to cold and may be damaged when the temperature drops five degrees or more below freezing. Roots may survive 10 to 15 degrees below freezing if they are protected with a mulch 10 or 12 inches deep. This mulch should be removed about one month before the average last frost date in the spring.

A deep, well-drained, fertile loam soil is best for globe artichokes. Avoid soils with known nematode infestations. The plants may be started from seed but do not come true to name. The use of suckers is preferred if a vigorous, productive type is available.

The suckers that arise around the base of the plants are used to propagate the globe artichoke. When about 12 inches high, remove the sucker and a piece of the parent rootstock, with a spade or knife. A mature plant may produce 15 to 20 suckers. All but 5 or 6 suckers should be removed to secure the best production from the parent plant. Artichokes started from suckers should produce a few edible flower buds the first season.

Space globe artichokes about four feet apart each way in the garden. Fertilize heavily with about four pounds of 5-10-5 fertilizer per 100 square feet each spring. Irrigate as needed during dry weather. Plants usually will last about four years. A portion of the planting should be renewed each year to ensure continued production.

The edible flower buds of the globe artichoke are ready to harvest in July, when they

FIGURE 10-27 Globe artichoke. The immature flower buds should be harvested when they are two to four inches in diameter. (Courtesy of U.S.D.A.)

are two to four inches in diameter. The stem should be cut about one inch below the bud, before the bracts begin to separate. The base of each bract is thick and fleshy and is eaten along with the solid center of the bud.

Jerusalem Artichoke

The Jerusalem artichoke, girasole, or canadian potato, is a member of the sunflower family. This robust perennial grows to a height of 8 to 10 feet. The North American Indians have known and cultivated this crop for many centuries.

The edible part of the Jerusalem artichoke is a tuber, which is an enlarged underground stem like a potato (Fig. 3-4). This vegetable is used by diabetics since a considerable portion of its stored carbohydrates is in the form of inulin. The tubers may be cooked like white potatoes, or made into pickles or relish.

The Jerusalem artichoke is propagated by planting small whole tubers, or cutting large tubers into pieces about one and one-half inches square. The plants grow vigorously in fertile soils. They are very persistent and may become a troublesome weed.

The tubers of Jerusalem artichoke may be dug when frost kills the tops, or they may be allowed to remain in the ground and harvested as needed. The plant is hardy, even in northern states, and freezing of the tubers in the ground will not hurt them.

Asparagus

Asparagus is one of the best perennial crops for the home vegetable garden. When planted in a well-drained, fertile soil, it will produce a bountiful harvest of delicious spears for many years (Fig. 10-28).

Asparagus is a member of the lily family, which includes such plants as onions, chives, tulips, and hyacinths. Asparagus grows wild in many parts of the world but is thought to have originated in southern Asia. It has been cultivated in countries around the Mediterranean Sea for over 2000 years.

The Greeks apparently collected asparagus from the wild because their literature contains no cultural information. As early as 200 B.C. the Romans published detailed gardening instructions for this crop. The shoots were eaten not only during the harvest season but were dried for later use.

The home garden planting of asparagus may be started from seed or roots. The seed may be planted anytime from a month before to

FIGURE 10-28 Asparagus spears ready for harvest. (Courtesy of U.S.D.A.)

a month after the average last spring frost date in your area. The young plants, or one-year-old roots purchased from your seedsman or garden center, may be set in the permanent location in the fall or early spring.

The best planting procedure for asparagus is to dig a trench about a foot deep, wide enough so the roots can be spread out to their full length. Place about two inches of rich compost, well-rotted manure, or dried manure in the trench. Add four inches of soil, and four pounds of 10-10-10 fertilizer per 50 feet of row. Mix thoroughly and cover with two inches of topsoil.

Set the plants firmly in the prepared soil about 18 inches apart. Cover with rich loam soil to a depth of 4 inches over the crown. A 1-inch mulch of sawdust will conserve soil moisture and help suppress weed growth.

Do not cut asparagus the first season after setting the crowns. The plants need to become well established before harvesting starts. Spears may be cut for two or three weeks during the second season, and for six or eight weeks each

year thereafter. When the spears become spindly or crooked, stop cutting.

To secure a good crop of asparagus each year, the planting must receive special care to build up the reserve food supply in the roots. To do this you need a heavy growth of tops during the summer and fall.

As soon as harvesting has been completed, hoe the bed thoroughly to destroy all weeds. Apply two pounds of 10-10-10 per 50 feet of row and work into the top inch of soil. Renew the mulch by adding an inch of fresh sawdust, shredded leaves, or compost.

The best asparagus cultivar is 'Mary Washington'. It is rust resistant and produces high yields of thick, dark green spears that are suitable for use fresh, frozen, or canned.

Asparagus beetles may become a serious pest and if present should be controlled after the cutting season, and on new beds not yet old enough to harvest. Aphids may become a problem in midsummer. Rust and fusarium root rot are two diseases that may cause damage.

Cooked fresh asparagus is an excellent vegetable to serve with meat, fish, or poultry. It is also delicious with egg, cheese, or macaroni dishes. Or try tender, cooked asparagus spears covered with cheese sauce garnished with thin tomato slices, egg yolk, or strips of pimento.

A cup of cooked asparagus will provide two-thirds of the recommended allowance of vitamin C, one-third of the vitamin A, and about one-tenth of the daily iron supply needed for an adult. It is very low in sodium for those who need to restrict their intake of this element. For weight watchers, asparagus has a kilocalorie content of only 35 per cupful.

Horseradish

Horseradish is a hardy perennial that is native to eastern Europe from the Caspian sea north to Finland. It was commonly planted in Colonial gardens in America to be used as a condiment with meat and fish. It escaped from cultivation and flourished as a wild plant along streams and in other moist locations.

Horseradish plants produce a fleshy tap-root system with laterals arising from the top 8- to 12-inch portion. The main root commonly divides into several prominent fleshy roots. The lateral roots are used for propagation since the plant rarely produces seed.

A deep, fertile, moist, loam soil is needed to promote the development of thick, smooth, fleshy horseradish roots. Set the cuttings about 12 inches apart in rows three feet apart. Place the small end of the root cutting down, with the top or large end about two inches below the surface of the soil. Roots set in early spring will be large enough to harvest in the fall. The most rapid increase in root size occurs during late summer and early fall. Therefore, the best time to harvest horseradish is in October and November. Since the plants are hardy, unneeded roots may be left in the ground and harvested as needed during the winter.

The size and smoothness of the harvested roots may be improved by removing the laterals during the early summer. When the leaves of the plant are about one foot tall, remove the soil from around the top 8 to 10 inches of the tap root, and carefully cut off all lateral roots. Repeat this operation about one month later.

Rhubarb

Rhubarb is a rather unique crop. The edible part of this vegetable is a large succulent petiole, or stalk of a leaf. And yet, it is used more like a fruit in making pie and other types of desserts. In fact, it is commonly called the pie plant.

Rhubarb is a perennial plant that originated in the colder parts of Asia, probably Siberia. This crop tolerates severe cold in winter, and dry weather in summer. It is not adapted to the southern United States or to the warmer coastal areas.

The 'MacDonald' and 'Valentine' cultivars of rhubarb have desirable red-colored petioles. 'Victoria' is an old standard cultivar with green stalks shaded with red. All cultivars develop a more intense red color when grown under cool climatic conditions.

Like many other perennial crops, rhubarb does not come true to name from seed. The plant is propagated by dividing the crown or clump. The best divisions for planting are those that have a large piece of the fleshy storage root, and a prominent bud.

A well-drained fertile loam soil with a high content of organic matter and pH 6.5 is best for rhubarb. Plant divisions three feet apart, in rows four feet apart, and cover with three inches of soil. The planting may be done about a month before the average last frost date in the spring, or in the autumn a few weeks after the first frost has killed the foliage.

Your rhubarb crop needs liberal amounts of manure or chemical fertilizer to promote vigorous growth. Apply eight pounds of 5-10-5 per 100 feet of row before growth starts in the spring. Broadcast the fertilizer over the entire space between the rows and cultivate into the top two inches of soil.

The leaves of rhubarb should not be harvested during the first year after it is planted, and only a few should be pulled during the second summer. After the plants become well established, leaves may be removed for about two months each spring.

A healthy planting of rhubarb will continue to produce good crops for six or eight years. The petioles may become small and spindly when the plants are 10 or 12 years old. The crown should then be divided and replanted in a new location.

Rhubarb is easy to harvest by pulling the leaves to one side to detach them from the crown. Cut off the leaf blade leaving one-fourth inch to prevent splitting of the petiole. Only the petiole or stalk of the rhubarb leaf should be eaten. The blade has a high content of oxalic acid and may be quite poisonous.

Rhubarb wilts quite rapidly when held under room temperature conditions. It will retain its good, fresh, edible quality for three or four weeks under refrigeration at 34°F and high relative humidity.

The high acid content and unusual flavor influence the use of rhubarb in the diet. It is served primarily as desserts in the form of pie, tarts, sauce, or pudding. It may also be used for jam or jelly, and the cooked sweetened juice is refreshing as a chilled drink.

Rhubarb is quite low in calories, but desserts made from it may have high energy value because of the sugar used in their preparation. It supplies only moderate amounts of vitamins A and C.

Questions for Discussion

1. Define the following terms:
   (a) Fava bean
   (b) Cantaloupe
   (c) Slips
   (d) "Seed piece"
   (e) VFN-resistant cultivars
   (f) Gumbo

2. Tomato, bean, and onion are grown more often by gardeners than any other vegetables. What reasons can you advance to account for the popularity of this threesome?

3. Discuss late summer and fall gardening, including choice of vegetables, planting times, and cultural practices.

4. Identify some of the types of beans suitable for use in home gardens.

5. Broccoli, cabbage, and several other related vegetables belong to a group known as the cole crops. Identify other members of this illustrious group and list any cultural characteristics that you feel are applicable to the group as a whole.

6. You are invited to speak to a local garden club. The topic is cucumber. Make an outline for your talk and comment on its salient features.

7. List the several types of lettuce grown by gardeners, and indicate criteria the inexperienced gardener could employ to select the types most suitable for his or her first garden.

8. Discuss the culture of onions from planting through harvesting.

9. You live at approximately 47° N latitude and garden in an area measuring 12 × 16 feet. You are anxious to start your garden as near New Year's Day as possible. Identify three vegetables you would grow and approximate dates of planting.

10. If you could grow only three root crops in your garden what would be your choices? State the reasons for your choices.

11. You are on a sailing cruise in the South Pacific. Suddenly the captain announces that all must abandon ship and make for a nearby uninhabited island. He instructs each passenger to bring with them only their purse or billfold and one additional item. In your cabin you have both sweet and Irish potato that you were taking back home for planting. Which will you bring and why?

12. How can one distinguish clearly between pumpkins and squashes?

13. Several vegetables are biennials. List some of these and classify each as either cool-season or warm-season. Do you notice any unusual correlations? Explain.

14. Discuss methods for staking and training tomatoes.

15. Outline the cultural procedures for growing leeks.

16. Name three important perennial vegetables and indicate accepted methods of propagation for each.

Suggested References

Nissley, C. H. 1942. *The Pocket Book of Vegetable Gardening*. Pocket Books Inc., New York.

Ortho Books. Staff. 1973. *All About Vegetables*. Chevron Chemical Company, San Francisco.

Rodale, J. I. (Ed.). 1971. *The Encyclopedia of Organic Gardening*. Rodale Books Inc., Emmaus, Pa.

Ware, G. and J. McCollum. 1975. *Producing Vegetables Crops*. The Interstate Printers and Publishers, Inc., Danville, Ill.

LEARNING OBJECTIVES

By the time you have finished this chapter you should be able to:

1 Identify several of the characteristics used to distinguish between herbs and spices.

2 List the major herb-containing plant families.

3 Name five important perennial herbs.

4 Identify three herbs that can be grown in containers.

5 Discuss the ornamental values of herbs using appropriate examples.

6 Describe techniques to be used in harvesting and storing herbs.

7 Explain why herbs are sometimes described as "multipurpose plants."

8 Identify the most widely used herb in America and comment on its major use.

9 Describe the broad group of chemical substances that give the herbs their unique properties.

10 Describe some of the patterns that can be used in designing a herb garden.

11 List some of the major factors to be considered when planning and planting a herb garden.

11

GROWING CULINARY HERBS

"He causeth the grass to grow for the cattle, and herb for the service of man . . ."

Psalms 104:14

The herbs have been put to the service of humans since time immemorial, as attested to by the psalmist quoted above. Interest in herbs is also found in the ancient Greek and Sanskrit literature, and records of their many uses form a continuum between then and now. The search for herbs (and spices) has had a profound influence on civilization. The explorations of Vasco da Gama and Christopher Columbus for a new sea route to India were motivated by interest in these plants. Indeed, the Dutch developed a great empire through their control of the spice trade.

Different cultures have employed herbs in different ways; among the more important uses have been those related to medicine, flavoring, air fragrance and purification, cosmetics, and landscaping (Fig. 11-1). Each of these uses is based on one or more of the unique properties of herbs. Some authors have gone so far as to describe herbs as Nature's "miracle plants." But what are herbs? How are they grown? What are their uses to our modern technological age? Let us explore these questions and more.

HERBS AND THEIR CLASSIFICATION

WHAT ARE HERBS?

Before we can talk intelligently about herbs we need a working definition of what herbs are, and are not. To professional botanists **herbs** are plants whose stems and leaves die

FIGURE 11-1 Herb gardens are sometimes planted with intertwining borders called knot gardens. This decorative knot garden is at the Brooklyn Botanic Garden. (Courtesy of the Brooklyn Botanic Garden, Brooklyn, New York.)

back entirely or at least down to the ground each autumn; the plants may be annuals, biennials, or perennials. To the gardener or amateur horticulturist, however, the term "herb" has a more restricted usage, referring to the many plants whose leaves, stems, flowers, or other parts possess unique qualities that make them useful in medicine, cooking, perfumery, and so on. Most of these useful plants are herbaceous in habit, but a few are distinctly woody. **Potherbs** are flavoring herbs used for boiling in the pot as sources of natural flavoring.

It is not always easy to distinguish between herbs and other plants of similar character. For example, there is not an entirely clear distinction between herbs and spices. **Spices**, like herbs, are parts of aromatic plants used for their flavor-producing qualities. Unlike herbs, however, spices are usually of tropical

origin, are generally used in the dried state, and are likely to be more pungent. Spices are said to give their own distinctive taste to foods, whereas herbs add a delicate flavoring. Both can be regarded as **condiments**; that is, additives that give flavor to food. Examples of spices are cinnamon, vanilla, ginger, nutmeg, turmeric, cloves, and the capsicum peppers.

The distinction between herbs and vegetables is not entirely clear either. Chives and garlic are usually considered as herbs, but a related species, onion, is regarded as a vegetable. Parsley and dill are herbs, but another member of the same family, carrot, is a vegetable. In summary, the terms herb, spice, and vegetables are really culinary, not scientific terms. Their usage is tied to customs, and customs vary with different societies and with time.

Pronouncing "Herb"

Another problem related to herbs is simply how to pronounce the word "herb." Sophisticated Americans say "erb," and to say "herb" is regarded as a sign of one's educational neglect. On the other hand, in England, "herb" is regarded as correct and those who drop the "h" are considered provincial. Etymologically the Latin word *herba* came to England through the Old French word *herbe* or *erbe*. Until 1745 the word was "erb" both in spelling and pronunciation. At the start of the sixteenth century the "h" was rejoined but remained silent until about 1800. Since then, pronunciation of the "h" has become common in England, but most Americans still hold to the older pronunciation. We will leave it up to the reader to choose a preferred pronunciation.

Important Herb Families

A list of the favorite garden herbs, their family affiliation, and a number of pertinent characteristics is presented in Table 11-1. Note that the vast majority are found to be con-

TABLE 11-1 Some Common Culinary Herbs and Certain of Their Characteristics and Uses

Name	Family	Growth habit	Height (inches)	Propagation	Cultural directions	Uses in cooking
Angelica *Angelica archangelica*	Apiaceae	Biennial (perennial)	72–96	Seed (self-sows)	Moist, rich soil; partial shade; space 18 in. apart; plants mature in 70 days	Leaves— salads, with cooked rhubarb Stems—salads, preserves, cakes and puddings (candied form)
Anise *Pimpinella anisum*	Apiaceae	Annual	18–24	Seed (self-sows)	Moderately rich soil; full sun; space 6–8 in. apart; maturity about 75 days	Seeds— pastries, candy, cookies, beverages Leaves— salads, garnish
Balm *Melissa officinalis* (also known as lemon balm)	Lamiaceae	Perennial	24–36	Seed Runners Divisions	Thrives in moist, poor or better soils; full sun or partial shade; space 12–18 in. apart; matures in 60 days	Leaves—soups, meats, fish, salads, tea and summer beverages
Basil *Ocimum basilicum*	Lamiaceae	Annual	18–30	Seed	Prefers well-drained fertile soil; protected sunny site; space 10 in. apart; matures for use in 50–60 days	Leaves—soups, stews, omelets, salads, meats, sauces

TABLE 11-1 (*continued*)

Name	Family	Growth habit	Height (inches)	Propagation	Cultural directions	Uses in cooking
Borage *Borage officinalis*	Boraginaceae	Annual	24–36	Seed (self-sows)	Moist, average soil; full sun; space 15 in. apart; matures in 50 days	Leaves—salads, greens, garnish Flower sprays—summer drinks
Burnet *Sanguisorba minor* (also known as salad burnet)	Rosaceae	Perennial	12–18	Seed Divisions	Does well on dry, poor soil; full sun; space 12 in. apart; matures in 45 days	Leaves—salads, soups, cool drinks, tea
Caraway *Carum carvi*	Apiaceae	Biennial	24–36	Seed (self-sows)	Moderately rich soil; full sun or light shade; space 6 in. apart; start in fall to bear next year	Leaves—garnish Seeds—breads, cakes, soups, sauces, salads
Chervil *Anthriscus cerefolium*	Apiaceae	Annual	18–24	Seed (self-sows)	Moist, rich soil; partial shade, space 6 in. apart; matures in 75 days	Leaves—soups, salads, meat, poultry, eggs, garnish
Chives *Allium schoenoprasum*	Liliaceae	Perennial	8–18	Seed Divisions	Does well in rich soil; full sun; space 6–12 in. apart; mature in 80 days	Leaves—omelets, salads, soups, sauces, dips
Coriander *Coriandrum sativum*	Apiaceae	Annual	24–36	Seed	Does well in average soil; full sun; space 8–12 in. apart; matures in 90 days	Seeds—pastries, sauces, curries, salads, preserves, drinks

TABLE 11-1 (continued)

Name	Family	Growth habit	Height (inches)	Propagation	Cultural directions	Uses in cooking
Dill *Anethum graveolens*	Apiaceae	Annual	24–36	Seed (self-sows)	Does well in average or better soil; full sun; space 12 in. apart or sow in clumps; matures in 100 days	Seed heads—pickles, sauces, meats, salads, vinegar Leaves—soups, sauces, fish, lamb, cook with vegetables, salads
Fennel *Foeniculum officinale* (also known as sweet fennel)	Apiaceae	Perennial (treat as annual)	36–48	Seed	Moderately rich soil; full sun; sow in early spring, thin to 18 in. apart; don't sow near other umbels; matures in 45 days	Shoots—confections, vinegar Leaves—salads, soups, fish Seeds—salads, fish, bread, tea
Hyssop *Hyssopus officinalis*	Lamiaceae	Perennial	12–24	Seed (self-sows) Cuttings Divisions	Does well on poor or average soils; full sun; sow in early spring, thin to 10–12 in. apart; matures in 60 days	Leaves—salads, egg dishes, soups, poultry and fish, tea
Lovage *Levisticum officinale*	Apiaceae	Perennial	36–72	Seed Divisions Cuttings	Does best in a rich, fairly moist soil; full sun; space 15–18 in. apart; plants take about four years to reach full size, use after one year	Leaves and stems—celery flavor to soups and salads, omelets, meat and poultry

333

TABLE 11-1 (*continued*)

Name	Family	Growth habit	Height (inches)	Propagation	Cultural directions	Uses in cooking
Marjoram *Origanum majorana* (also known as sweet or knotted marjoram)	Lamiaceae	Perennial (treat as annual)	8–12	Seed Cuttings	Light, well-drained soil; full sun; space plants 6–12 in. apart	Leaves—salads, egg dishes, soups, dressings, meats, poultry, tomato dishes (milder than oregano)
Mint *Mentha* sp.	Lamiaceae	Perennial	12–60	Divisions Cuttings	Prefer moist, rich soils; partial shade; space 8–12 in. apart; est. plants mature in 30–45 days; prevent from going to seed by pinching	Leaves—cold drinks, tea, mint sauce, soups, meats, salads, desserts
Oregano *Origanum vulgare* (also known as wild marjoram)	Lamiaceae	Perennial	24–36	Seed Divisions Cuttings	Plant in average or better soil; full sun; space 12–18 in. apart; plants mature in 45 days; prevent seed formation	Leaves—Italian dishes, vegetable soup, Mexican casseroles, and in places where marjoram is used
Parsley *Petroselinum crispum*	Apiaceae	Biennial (treat as annual)	6–15	Seed	Prefers moist, moderately rich, medium-textured soil; partial shade; seed slow to germinate, space 6–8 in. apart; matures in 70–90 days	Leaves—garnish, salads, soups, sauces, omelets, vinaigrettes

TABLE 11-1 (*continued*)

Name	Family	Growth habit	Height (inches)	Propagation	Cultural directions	Uses in cooking
Rosemary *Rosmarinus officinalis*	Lamiaceae	Perennial	36–72	Seed Cuttings	Does well on light-textured, poor or average soil with ample lime; full sun; space 18–30 in. apart	Leaves— sauces, soups, stews, gravies, meat, poultry; add to water in cooking peas, potatoes, turnips
Sage *Salvia officinalis*	Lamiaceae	Perennial	18–24	Seed Cuttings	Does best in light-textured, well-drained soil; full sun; space 12–18 in. apart; plants mature in 75 days	Leaves—meats, fish, dressings, stuffing, egg and cheese dishes, gravies, tea
Savory *Satureja hortensis* (also known as summer savory)	Lamiaceae	Annual	18–24	Seed	Grows best in a moderately rich, well-drained soil; full sun; space 6–8 in. apart; plants mature in 60 days	Leaves— salads, soups, dressings, poultry and fish, rice, bean, pea, and cabbage dishes
Sorrel *Rumex scutatus* (also known as French sorrel)	Polygonaceae	Perennial	12–18	Divisions Seed	Grow best in light-textured, moderately rich soil; full sun or partial shade; space 12 in. apart; plants mature in 100 days; remove flower heads	Leaves— salads, omelets, souffles, soups, stews, ragout, and cooked vegetables

TABLE 11-1 (*continued*)

Name	Family	Growth habit	Height (inches)	Propagation	Cultural directions	Uses in cooking
Sweet cicely *Myrrhis odorata*	Apiaceae	Perennial	24–36	Seed (self-sows) Divisions	Grows best in moist, acid soils; partial shade; space plants 18 in. apart; plants mature in 60 days; sow seed late summer or fall	Seeds—mixed with other herb seeds, cabbage, certain liqueurs Leaves—salads, soups, stews Roots—boiled and served as a vegetable
Tarragon *Artemisia dracunculus*	Asteraceae	Perennial	18–24	Cuttings Divisions	Grows best in medium-textured, well-drained soil; full sun or light shade; space plants 10–12 in. apart; mature for use in 60 days	Leaves—salads, sauces, eggs, vegetables, vinegar
Thyme *Thymus vulgaris* (also known as garden thyme)	Lamiaceae	Perennial	6–12	Seed Cuttings Divisions Layering	Grows best in light-textured, well-drained soils; full sun; space plants 8–12 in. apart; est. plants mature in 50 days	Leaves—salads, soups, dressings, egg dishes, vegetables, meat, fish, gravies, cheese, bread

centrated in only a few of the numerous families of flowering plants (Table 11-2). It is beyond the scope of this text to give detailed information on the growth, maintenance, and uses of all of these herbs. Certain references cited at the end of the chapter present this information and are recommended for those who wish to pursue the subject in greater detail. We limit our treatment to the twelve most commonly grown herbs. These are suggested as a beginning point for less experienced gardeners. After the husbandry of these has been mastered we then encourage experimenting with additional herbs. We realize that the information presented on the "top 12" will overlap to some extent with the material presented in the summary table (Table 11-1), but we try to keep this duplication to a minimum.

THE TOP 12 HERBS

Basil
Ocimum basilicum
Family: Lamiaceae

Description

Basil is a handsome plant that reaches about two feet in height and has light green, rather broad, simple leaves (Fig. 11-2). Its flowers are small, white or purplish, and occur in spikes. Although basil is a perennial in tropical areas, we treat it as an annual. There are several kinds of basil in cultivation including:

O. basilicum   Sweet basil (herb and ornamental)

TABLE 11-2 Distinguishing Characteristics of Several Important Herb-containing Plant Families

Plant family	Distingishing characters
Borage family (Boraginaceae)	Plants with usually alternate, simple leaves; flowers bisexual, regular, corolla five-lobed, five stamens adnate to corolla; inflorescence a scorpioid raceme or cyme.
Composite family (Asteraceae)	Plants with alternate, opposite, or whorled leaves; flowers unisexual or bisexual; inflorescence a compact head (may look like a "flower").
Mint family (Lamiaceae)	Plants with square stems and opposite or whorled leaves; flowers bisexual; hypogynous, irregular, two-lipped sympetalous corolla; four stamens; inflorescence a spike or raceme.
Lily family (Liliaceae)	Plants with parallel veined leaves and a bulb or bulblike storage organ, or fleshy rootstocks; flower parts in threes or multiples thereof; perianth parts frequently colored and alike; inflorescence variable.
Parsley family (Apiaceae)	Plants with alternate, mostly compound leaves; flowers small, usually bisexual; inflorescence a simple or compound umbel.

FIGURE 11-2 Sweet basil. (Courtesy of U.S.D.A.)

O. minimum	Bush basil
O. sanctum	Holy basil
O. suave	Tree basil

Basil is sometimes called the "King of Herbs," and its Greek name *Basileus* means king. *O. sanctum* is the most sacred plant in the Hindu religion. The plant has a very long history. It apparently originated in India and spread from there into Europe where it has been cultivated for about 2000 years. Both Pliny and Dioscorides recorded basil in their first century writings.

Culture

Basil does best on a well-drained, fertile soil. It should be placed in a sunny position sheltered somewhat from the wind if possible. The plant grows easily from seed, but should be started indoors or planted outside after the danger of frost is past. Basil seeds germinate rapidly and plants mature for use in about 50 days. Pinch off selected shoot tips of growing plants to encourage branching and prevent flowering. As fall approaches plants can be cut back to one or two pairs of leaves, lifted from the soil, and brought indoors.

Harvesting and Uses

Green, healthy leaves can be picked and used fresh when plants are about six to seven weeks old. Harvested leaves can also be kept for a short time in polyethylene bags in the refrigerator. Leaves can be dried by hanging plant portions in a warm, dry place away from sunlight. (One excellent method is to put bunches of the herb in brown paper bags, tie the neck of the bag around the stems, and hang in an attic or other dry place.) When dry, the leaves should be crumbled, put into airtight jars, and labeled. Basil can also be preserved by covering freshly cut leaves with salt or olive oil, or by freezing them in water. The spicily scented leaves go well in soups, stews, sauces, salads, and with meats, fish, and shellfish. They are considered to go especially well with tomato dishes.

Chives
Allium schoenoprasum
Family: Liliaceae

Description

Chives are nothing more than small, dainty onions that grow in clumps that reach about a foot in height (Fig. 11-3). The plant is a hardy perennial and reproduces by bulbs as well as by seed. Leaves of the plant are cylindrical and hollow and come to a point at their distal end. The flowers have a delicate rose-purple color and occur in compact umbels.

Chives came to Europe from Asia. They are reported to have been used in China for at

FIGURE 11-3 Chives, a handy perennial (Courtesy of U.S.D.A.)

least 5000 years. Legend has it that Marco Polo spread the good news about chives throughout the Mediterranean region.

Culture

Chives do well in almost any soil, but a sunny location is recommended. Plants can be started from seeds or divisions. Seeds can be sown outdoors in the spring as soon as the soil can be worked. Established seedlings should be thinned to 6 to 12 inches apart. An easier way to establish plants is to use divisions. These can be planted outdoors any time after the last frost of the spring. Established plants require little care. Clumps must be dug up and divided after three or four years to prevent overcrowding. This can be done in spring or fall. The plants have a rather shallow root system and should be watered during dry spells. Regular weeding and fertilization are also important. Chives die back with the onset of fall frosts, but send up new growth in the spring.

Chives will flourish indoors in pots that can be placed on any sunny windowsill.

Harvesting and Uses

The management of chives depends on the intended use. If leaves are desired for cooking, plants should not be allowed to flower and, as a matter of fact, should be kept cut back to one or two inches above the ground. Fresh leaves can be harvested as needed and plants should be cut back regardless. Chives do not dry well, but leaves can be chopped, placed in a container, covered with water, and frozen. Leaves are commonly added to already cooked dishes and go well in omelets, salads, soups, sauces, spreads, dips, and cottage cheese. The attractiveness of chives also gives the plant ornamental value, especially for borders. When put to this use normal growth and flowering is desired.

Dill
Anethum graveolens
Family: Apiaceae

Description

Dill is an annual herb that grows to three feet in height and has bluish-green stems and finely divided, yellowish-green leaves (Fig. 11-4). Flowers are small, yellowish, and occur in umbels that may reach eight inches in diameter. Dill, like other members of the parsley family, produces a prodigious quantity of seed. By the time the seed are ripe the leaves of dill have taken on a reddish-purplish hue.

Dill is a native of the Old World and its historical record extends back to the Egyptian tombs. It was widely recommended by Roman and Greek physicians and its fame gradually spread throughout Europe. It is known to have been cultivated in England about 1570 and traveled with the Pilgrims to the New World.

FIGURE 11-4 Dill plants in bloom. (Courtesy of U.S.D.A.)

Culture

Dill does well in average or better soils with adequate drainage. Plants thrive in sunny spots that are somewhat protected from winds. Seed should be sown outdoors in place (seedlings are not easy to transplant) after the danger of spring frost is past and the soil is warm. For a continuous supply of foliage, plantings can be made at two- to three-week intervals. Young seedlings should be kept free of weeds to reduce competition and thinned to 12 inches apart. Plants mature for seed in about 100 days but the leaves can be harvested much earlier. If your plants shatter seed in the garden new seedlings will appear the following spring. If dill is planted too close to fennel or angelica cross-pollination may occur.

Harvesting and Uses

Leaves of dill can be harvested at any time in the later part of the growth cycle but are of best quality just before or as the flowers open. Leaves to be dried should be spread in a thin layer and put in a warm, dark place. When dry crumble the leaves, put in airtight containers, and label. To collect the seed, harvest plants as the flowerheads turn brown but before they are ripe enough to shatter their seed. Tie the plants in bunches and hang upside down in a warm or sunny place. When the plants are dry, shake out the seed and store in containers. Dill leaves go well with cucumbers, salads, green vegetables, soups, sauces, stews, and a variety of meat dishes. Seed are used in sauces, vinegar, salads, and meat dishes. Last, but certainly not least, flower heads are used in the time-honored way of preserving cucumbers—as "dill pickles" or dill pickled cucumbers.

Marjoram and Oregano
Origanum majorana and
Origanum vulgare
Family: Lamiaceae

Description

Marjoram, more accurately called sweet or knotted marjoram, is a tender perennial that reaches up to 15 inches in height and is characterized by small, ovoid, grayish-green leaves that are velvety to the touch (Fig. 11-5). Flowers form in clusters at regular intervals along the stem giving the plant a knotted appearance. The flowers are white or violet in color and display the familiar two-lipped corolla. Oregano, or wild marjoram, is the presumed parent stock from which sweet marjoram was developed. As expected, the plants have numerous similarities, but the wild form can be distinguished in that it reaches a height of 30 inches or more, has longer leaves of greener color, has pink to purple flowers that form in

FIGURE 11-5 Plant of sweet marjoram in flower. (Courtesy of U.S.D.A.)

loose clusters at the ends of longer stems, and is somewhat more hardy.

The marjorams are native to Europe and Asia. They have a long history as culinary and medicinal herbs, and were used extensively by the early Greeks and Romans. The Greeks gave them a name that literally translated means "joy of the mountains." In both cultures the plants enjoyed a high degree of symbolism, being a part of marriage and burial rites. These plants, like dill, made the long trip to the New World with the Pilgrims and have since become naturalized over large areas of North America.

Culture

The marjorams thrive in light, well-drained soils at sites where they receive full sun exposure and some protection from winds. They can be grown from seed, cuttings, or root divisions. Sweet marjoram, although technically a perennial, is best treated as an annual in northern areas. Seeds can be started indoors in March and the seedlings planted out after the danger of frost is past. A spacing of 8 to 12 inches is recommended. Oregano can also be started from seed planted indoors, or outdoors after the last frost of the spring. Once established, oregano is a vigorous grower and must be divided every couple of years to control spreading. Cuttings of the marjorams root most easily if taken from lush, early summer growth. Remember, leaves are the sought-after prize and preventing flowering will encourage growth of foliage.

Harvesting and Uses

Fresh leaves can be harvested at almost any time plant size permits. If leaves are to be dried the harvesting time is more critical and should correspond to the time of flowering. Tie harvested portions in small bundles and keep in a dark place with good air circulation. When dry, crumble the leaves and store them in properly labeled airtight jars.

The uses of the marjorams are legend. Sweet marjoram, which has a more delicate flavor than oregano, can be used with cooked vegetables, soups, salads, and egg and meat dishes. Oregano, of course, gives its distinctive flavor to pizza, spaghetti sauces, and other Italian, Spanish, and Mexican dishes. Oregano can, to some extent, subsitute for sweet marjoram but should be used more sparingly.

The Mints
Mentha sp.
Family: Lamiaceae
Description

There is an almost endless number of mints. Of necessity, we confine our discussion to the three that are said to be of most common usage:

M. rotundifolia	Apple mint or Bowles mint
M. piperita	Peppermint
M. spicata	Spearmint

All the mints display a number of features in common: They are hardy perennials; have white, pink, or purple flowers; spread—sometimes aggressively—by means of surface or underground runners; and have square stems with opposite, simple leaves. Apple mint grows to a height of 30 inches; has sessile, pubescent, oval leaves of up to 2 inches in length; and purple flowers in spikes up to 4 inches long. The variety *variegata* has variegated foliage. Peppermint has reddish stems that reach a height of 30 inches. Its dark green, petiolate leaves have toothed margins and may reach 3 inches in length. Flowers are usually purple and occur in spikes up to 3 inches long. Spearmint seldom reaches more than two feet in height; its leaves are sessile, coarsely toothed, pointed, and lighter green than those of peppermint (Fig. 11-6). The plant tends to have a rigid, crisp appearance.

The mints are among the oldest and most widely used of all the culinary herbs. Many believe them to have originated in Asia and traveled to Europe by way of North Africa.

FIGURE 11-6 Spearmint plant in flower. (Courtesy of U.S.D.A.)

Regardless of their origin it is clear that they have been used in Europe for more than two millennia. Spearmint was a valuable herb for the ancient Egyptians, Greeks, and Romans and is recorded in the writings of Dioscorides, Hippocrates, and Pliny as having numerous important uses. The herb is said to have been introduced into England by the Romans and has become a vital part of English culture. Many mints have been brought to North America over the years and some have become thoroughly naturalized.

Culture

The mints do best in a well-drained but moist soil in a semishady location. They can be started in early spring by means of divisions placed into a bed or spaced 8 to 12 inches apart in rows. The divisions should be covered with soil to a depth of about 2 inches and watered weekly. New plants will arise at nodes, and vegetative portions can be harvested from late spring until fall. New plants can also be grown from cuttings taken from established plants. The mints require watering during dry spells and should be controlled to promote vegetative growth and restrict flowering.

One problem with the mints is that they spread rapidly, crowding out less aggressive plants. Growing the plants in beds with well-defined boundaries both above and below ground will prevent excessive spreading. The different mints also freely cross-pollinate. For this reason plants produced from seed may not be uniform. Probably the major problem in mint growing, however, is the danger of a fungal disease, verticillium wilt, causing a stunting of plants and a concomitant bronze coloration of the foliage. This danger can be lessened by using resistant cultivars (such as Todd's Mitcham), by a regular rotation of beds, or by soil fumigation. If infection occurs, the mint should be harvested immediately and a new crop started in a fresh bed.

Harvesting and Uses

Fresh young leaves can be harvested any time plant size permits. These can be used directly or stored for a short time in a polyethylene bag in the refrigerator. Newly harvested leaves can also be frozen or dried. In freezing, wash the leaves, cut into smaller pieces, put into containers, cover with water, and freeze. The leaves can later be unfrozen and used as fresh. Mint leaves dry well. For drying, harvest plant portions as flowering begins and hang in small bunches or spread out on screens in a dry, dark place. When dry remove excess stem material and store the leaves and flowering shoots in properly labeled airtight containers.

There are many uses of mints. Foremost, perhaps, is their use to flavor cold summer drinks, in teas, and to make mint sauce. Mint also complements soups, salads, cooked potatoes, carrots, zucchini, and desserts. The commercial uses of mint are also worthy of mention. In 1975, 67,900 acres of peppermint were grown in the United States; 40,000 of this acreage in Oregon alone. The processed peppermint oil—historically more popular than spearmint—finds its way into our chewing gum, toothpaste, mouthwash, and confectioneries. The addition of mint essence to dentifrices is not exactly a recent innovation, having been practiced since about the sixth century. Menthol, as we know it, is a component of mint oil from *M. arvensis*. Major producers are Brazil and Japan. Corsican mint (*M. requieni*) is used horticulturally as a ground cover.

Parsley
Petroselinum crispum
Family: Apiaceae

Description

Parsley is a hardy biennial that reaches a foot or more in height and is grown for its finely divided and sometimes curly, dark green leaves (Fig. 11-7). Flowers, although usually not

FIGURE 11-7 Parsley plants. (Courtesy of U.S.D.A.)

present when treated as an annual, are small, greenish-yellow, and occur in compound umbels. The most commonly grown parsley is the curly leaved, dark green variety, but several are available, including the following:

'Extra Curled Dwarf'. 85 days. Compact plants with finely cut and curled, dark green leaves.

'Hamburg Rooted'. 90 days. Principally grown for the parsnip-like roots of white flesh.

Plain or Single (Italian). 72 days. Plain, dark green, deeply cut, flat leaves.

Parsley is thought to have originated in the Mediterranean region, and has been used in Europe for centuries. The ancient Greeks regarded it as a symbol of death and used it in funeral rites. The Romans, however, employed parsley in their diets as well as in other ways. Pliny was one of the first to extol the merits of parsley, supporting its use in fish ponds to improve the health and vitality of its inhabitants. The herb also gained the reputation of increasing the virility of males and the fertility of females. The curly leaved forms of parsley were popular in medieval England, and were introduced to America long ago.

Culture

Parsley does best in a moist, moderately rich, medium-textured soil. A somewhat shady location is preferable, but the plant also does well in full sun. Parsley is a cold-hardy plant that can be seeded in place in the garden several weeks before the frost-free date. The seed has a notoriously slow germination rate, sometimes taking up to eight weeks under natural conditions. Germination time can be shortened by soaking the seeds in water for a day or two before sowing, or by pouring boiling water over the seed before the rows are covered. Some gardeners prefer to start their plants indoors in flats and transplant after the soil has warmed. Regardless of technique, established seedlings should be spaced 6 to 8 inches apart in the row. Flowering should not be allowed. As fall approaches plants may be dug up with a good ball of soil, set in pots, and grown indoor in any sunny, cool spot. A winter supply of parsley is thus assured.

Harvesting and Uses

The attractive green leaves of parsley can be used fresh as soon as plant size permits. Always pick leaves from the outside of the plant, as these are the oldest. Newly harvested leaves can be stored in the refrigerator for several days. For longer storage, material can be frozen or dried. For freezing, gather bunches and put into plastic bags, remove as much air as possible, and put immediately into the freezer. When using frozen material it is easier to cut pieces to the desired size before they

are completely unthawed. Leaves are easily dried in a warm oven. When dry they can be crumbled and put into labeled containers for storage.

Parsley is beyond doubt the most widely used herb in America. Each year millions see it as a garnish with meals. Few eat it, however. This is unfortunate, since parsley is an excellent source of vitamins (especially A and C) and minerals (see Appendix F). Freshly picked and chopped parsley sprigs go well in salads and are a welcome addition to stews, soups, casseroles, and cooked vegetables. It is also used in preparing fish, poultry, and meat dishes. The roots of Hamburg parsley are used either fresh or dried for flavoring soups and stews. Horticulturally, parsley serves well as a border plant, thus taking on a dual role.

Rosemary
Rosmarinus officinalis
Family: Lamiaceae
Description

Rosemary is a half-hardy, long-lived (up to 30 years) perennial shrub that can grow to a height of six feet or more. The plant has narrow, needlelike leaves up to 1¼ inches long that are dark green and shining above, and light colored on the underside (Fig. 11-8). The leaves are rather leathery and have a spicy, resinous scent. The flowers of rosemary are a light blue in color, about ½ inch long, and occur on short axillary racemes. The plant flowers in early spring. The variety *prostratus*, as the name suggests, has a prostrate form of growth.

The Latin name *Rosmarinus* literally translated means "sea dew." This aptly describes the native habitat of rosemary for it is found abundantly all along the Mediterranean coastline. It is said that its fragrant odor carries far out to sea. Rosemary has been used for centuries in Europe and western Asia, and was grown on graves as a symbol of remembrance. It was introduced into Britain by the Romans

FIGURE 11-8 Rosemary, a handsome evergreen shrub. (Courtesy of U.S.D.A.)

Sunny locations are favored. The plant can be grown from seed but this is a slow process and several years may elapse before you have a productive bush. It is better to purchase small plants from a nursery or start your own from cuttings taken in the early fall. For cuttings, take a small side stem with a heel of old wood attached, root in a sand-peat mixture, and grow in a cool window location over winter. When plants are moved to the garden a spacing of about three feet is desirable. Established plants should be fertilized each spring and pruned back regularly to control their growth. Rosemary plants generally will not survive the winter in plant hardiness zones 1–6 (Fig. 6-11). The plant grows well in pots or tubs, however, and these can be moved indoors during the harsh winter period. Cuttings can also be taken as outlined above.

Harvesting and Uses

The fragrant leaves of rosemary can be harvested as needed when plant size permits. Since the plant is an evergreen and provides fresh leaves throughout the year there is little need to dry material for storage. Plant material dries well however, and can be dried and stored in the manner of other herbs.

Rosemary goes well in sauces, soups, stews, gravies, and with a variety of fish, poultry, and meat dishes. Some cooks add rosemary to water for cooking peas, potatoes, turnips, and other vegetables. In warmer climates rosemary is much used in landscaping. Its size makes it useful as a hedge plant. Rosemary oil is used as a fragrance in a number of products such as soap and candles.

Sage
Salvia officinalis
Family: Lamiaceae

Description

Sage is a hardy, perennial shrub that reaches about two feet high. Foliage of the plant, as

before the eleventh century and found wide use in its new home. The herb came to America with early settlers and was held in high regard as a seasoning agent, a reputation it still enjoys.

Culture

Rosemary does well on any well-drained soil of average fertility and high lime content.

well as aroma, is quite distinct. The leaves are one to two inches long, petiolate, grayish-green in color but darker above than below, and characterized by a rather unusual wrinkled texture (Fig. 11-9). The flowers, which appear in midsummer, are purple, blue, or white, and occur in racemes. *Salvia* is a vast genus with approximately 700 species recognized. *S. officinalis* is the most important from a culinary standpoint, and is sometimes called "true sage" or "garden sage." One variety has variegated foliage, another white flowers, and another ruby red flowers.

Sage is native to the Mediterranean region but is grown in most temperate climates of the world. The plant, like other herbs, has a long history. The ancient Greeks and Romans used extracts from it as a remedy for snake bites and as a general tonic. In the Middle Ages it was popular as an antidote for constipation, cholera, colds, and epilepsy. The name *Salvia* comes from the Latin *salvare* meaning "to save," and alludes to the presumed healing properties of the herb. In more recent times sage has gained in culinary popularity, but for centuries the Chinese have used it to make a tea.

Culture

Sage does best in a light, well-drained soil and in a sunny spot. The plant can be started easily from seeds, cuttings, or crown divisions. Seeds can be started indoors and the seedlings put outdoors as soon as the danger of frost is past. Because plants started from seed may be quite variable in type, propagation by cuttings is preferred. This method also gives faster results. Cuttings should be taken in late spring or early summer and planted directly into the desired site, or put into containers and transplanted later. After three or four years sage plants become woody and less productive. These plants can be divided, with younger crowns used to propagate new plants, and older wood discarded. Established sage plants need little care other than an annual spring application of a balanced fertilizer, weeding, pruning, and light watering during prolonged dry periods. Sage plants generally will not survive the winters in plant hardiness zones 1 to 5 (Fig. 6-11). Fortunately, the plant does well in containers and thus can be brought indoors in harsh climates.

Harvesting and Uses

Sage leaves can be used fresh or dried. New growth to be dried should be harvested before the plant flowers, tied in small bunches, and dried slowly in a warm, dry place away from sunlight. Once dry the leaves can be crumbled or stored whole in the usual properly labeled airtight container.

Perhaps the best known use of sage is in poultry stuffings. As one author puts it "sage and turkey stuffing are practically synonymous." The herb is also an important ingredient in sausage and other pork products, and in sauces, gravies, egg and cheese dishes, fish and meat dishes, and tea. The plant has some ornamental value and can be used in landscaping.

FIGURE 11-9 Plants of garden sage. (Courtesy of Brooklyn Botanic Garden, Brooklyn, New York.)

Savory
Satureja sp.
Family: Lamiaceae

Description

There are two kinds of savory that are grown as garden herbs. Summer savory, *S. hortensis*, is a tender annual that grows to a height of about 18 inches and has upright, pubescent stems and small, rather narrow, bronze-green leaves (Fig. 11-10). Flowers are small and white or lavender in color. Winter savory, *S. montana*, is a hardy perennial that grows to two feet in height and has shiny, dark green leaves that are shorter and stiffer in texture than those of summer savory. The flowers of winter savory are larger than those of the annual form but of about the same color range.

The savories, like so many other culinary herbs, are native to the Mediterranean region. The Romans used savory before the oriental spices were known and the poet Virgil (70–19 B.C.) advocated growing the plant near beehives. An early culinary use of savory was in combination with vinegar to make a sauce somewhat like present-day mint sauce. Later the Romans brought the savories to Britain where they were widely accepted. It is said that early British settlers brought the herb to America. In America the herb escaped from cultivation and has become naturalized over large areas of the eastern United States.

Culture

The savories do best in well-drained soils in full sun. Summer savory is started easily from seed. The seed can be sown indoors several weeks before the mean date of the last spring frost and seedlings transplanted to the garden after danger of frost is past. Or seed can be sown in place in the garden after the frost-free date. Established seedlings should be thinned to 6 inches apart in the row. Little care is necessary other than weeding, watering in dry weather, and perhaps fertilization. Plants mature in about 60 days.

Winter savory can be started from seed, cuttings, layering, and by division of old plants. The method used will depend on availability of material and preference. Propagation by seed can be handled as for summer savory. Cuttings are best taken in the late spring and can be planted in place in the garden. In layering, bend some of the stems to the side and cover partially with soil. When rooted, sever the stems from the parent plant and transplant them to a suitable site. Although a perennial, it is best to divide plants every three or four years. In the process the older central portion can be discarded and younger crowns used for propagation. Established plants should be fertilized each spring, kept free of weeds, and watered in dry weather. Winter savory can survive the winter in hardiness zone 4 (Fig. 6-11) and southward. The plant will be nearly evergreen, even in cold climates, if mulched well with straw in the fall.

Harvesting and Uses

The savory leaves have a strong, somewhat peppery taste. The summer form is generally

FIGURE 11-10 Summer savory. (Courtesy of Brooklyn Botanic Garden, Brooklyn, New York.)

considered a little milder than the winter form and is more widely used. The leaves can be used fresh or dried. For drying summer savory the leaves should be harvested just as the plant begins to flower. They can be dried and stored in the usual manner and will retain their flavor for a long time. The leaves of winter savory can also be dried but this is usually not necessary as fresh material is available nearly the year around.

Summer savory is known as the "bean herb," not because it resembles a bean plant, but because it is a delightful and essential addition to all bean dishes. The savories also go well in other vegetable dishes, pea and cabbage for example, and in salads, soups, and dressings. They also find use, often in combination with other herbs, in poultry, fish, and meat preparations. The perennial savory finds some use as a low hedge plant or accent plant.

Tarragon
Artemisia dracunculus
Family: Asteraceae

Description

Tarragon is an important member of the genus *Artemisia*, which, in addition, contains such notables as the western sagebrush, southernwood, wormwood, and dusty miller. In all, about 200 species are known, most of which occur in arid regions of the Northern Hemisphere. Nearly all are perennial, highly aromatic shrubs, and being composites they display the familiar head type of inflorescence. The true tarragon, also known as French tarragon, is a moderately hardy, herbaceous perennial that grows to a height of about two feet. The plant is quite different in appearance from other artemisias having slender, branching, upright stems with narrow, somewhat twisted green leaves that can be up to four inches long (Fig. 11-11). The lower leaves of the plants are divided at their tips. Heads (inflorescences) are whitish-green in color and

FIGURE 11-11 Tarragon at the flowering stage. Note the narrow, slightly twisted leaves. (Courtesy U.S.D.A.)

about one-eighth inch across. A related tarragon, *A. dracunculoides*, is a tougher, more robust plant reaching a height of six feet, and reported to have very little culinary value.

The tarragons are thought to be native to Siberia and southern Asia, and probably reached Europe during the Middle Ages. A thirteenth-century naturalist, Inbal Bayter, living in Spain, mentioned tarragon in his writings. French tarragon was widely grown in France and became an important ingredient in famous French sauces such as bearnaise and hollandaise. The plant was introduced into Britain about 1548 and subsequently found its way to the New World colonies.

Culture

Tarragon does best in a medium-textured, well-drained soil in either full sun or light shade. French tarragon does not set seed so it must be multiplied vegetatively. The tarragon seed occasionally offered by growers is Russian tarragon and is regarded by many, but not all, as not worth growing. Tarragon, like many other perennial herbs, is commonly propagated by divisions, but cuttings are also successful. Established plants should be divided in the early spring and the divisions spaced one to two feet apart in rows. Root cuttings can also be taken and planted in place at the same distance. Tarragon requires little care other than light fertilization, weeding, light watering in dry weather, and division every two or three years to control plant size and vigor. The plant is adapted in hardiness zones 4 to 10 but with a little care can survive in even harsher climates. One approach is to cut plants back to within a few inches of the ground before the first frost of the fall and mulch with straw or other material. Another approach is to grow plants in containers that can be moved indoors in inclement weather.

Harvesting and Uses

Leaves can be harvested from new shoots when the plants are about a foot high and progressively through the summer. If the leaves are to be dried they should be put in a warm, dry place away from strong light. When dry, crumble and store in the conventional manner. Fresh leaves can also be chopped into a container, covered with water, and frozen. When thawed this material can be used as fresh.

Tarragon leaves go well in salads, and with fresh tomatoes. Tarragon vinegar is made by immersing fresh young shoots for a time in wine vinegar. Many sauces and herb mixtures contain tarragon, and the herb is used with steak, poultry, and fish dishes. Asparagus and artichokes are delightful when served in combination with tarragon butter.

Thyme
Thymus vulgaris
Family: Lamiaceae

Description

Thyme is a low-growing, spreading, perennial herb that reaches to one foot in height. Its stiff, wiry stems bear small, oppositely arranged leaves that are oval to linear in outline and grayish-green in color (Fig. 11-12). Flowers, which appear in May and June, are small and lilac or purplish in color. There are about 50 species of thyme, which are regarded by most as having developed from *T. serpyllum*, the so-called "mother-of-thyme." *T. vulgaris*, known as "common thyme" or "garden thyme" is most important as a culinary herb, but "lemon thyme," *T. citriodorus*, is also popular and imparts a distinctly fruity flavor to foods and beverages.

The thymes are native to the Mediterranean region but are now widely distributed in temperate regions of the world. They, like so many other herbs, are steeped in symbolism. In ancient days thyme signified courage, eloquence, and energy. It is said that Roman soldiers bathed in water infused with thyme to give them vigor for pending battles. The herb also acquired culinary uses during the Middle Ages. One seventeenth-century recipe for soup had thyme and beer as ingredients. The soup, in addition to providing nutrition, was said to be a sure cure for shyness. No mention was made of the number of servings recommended! By 1721 this herb was found in gardens in North America.

FIGURE 11-12 Common thyme. (Courtesy of U.S.D.A.)

Culture

Thyme grows nicely in any light-textured, well-drained soil, preferably in full sun. The herb is propagated by seed cuttings, divisions, and layering. Seed can be started indoors several weeks before the last frost of the spring and moved outdoors after that date. The other methods of propagation can be carried out much as described for winter savory (see page 347). Several now familiar practices enhance the culture of thyme. First, the soil should be limed at the time initial plantings are made and a balanced fertilizer applied to established plants each spring thereafter. Second, new plants should be kept moist but established plants need watering only in dry periods. Third, plants should be clipped so as to promote bushiness and retard woodiness. Fourth, divide established plants every three or four years or they will become straggly and less aromatic. Finally, protect plants during the winter. The plant does well in hardiness zones 5 to 10 (Fig. 6-11) but should be mulched. The herb can be grown in portable containers, window boxes, and indoors.

Harvesting and Uses

In the first year of growth only one harvest should be made, but two or more harvests are possible in subsequent years. Thyme, like many other herbs, is best harvested just as the flower blossoms open. The leafy tops and flower clusters dry well and dried leaves are said to be much more aromatic than fresh ones. When dry, material should be crumbled and stored in properly labeled airtight containers. Thyme stores well, keeping most of its fine flavor for many years. Freshly harvested material can also be stored in polyethylene bags in the freezer for up to two months.

Thyme is used to flavor many dishes, especially salads, soups, stews, stuffings, and sausages, but also beef, fish, and game. It is used to flavor many kinds of vegetables, and lemon thyme adds a unique tang to fresh fruit dishes. Thyme's strong, sharp flavor easily overpowers those of more delicate herbs and thus needs to be used with caution. Common thyme finds use as an edging plant and in rock gardens. Mother-of-thyme enjoys a similar popularity.

THE HERB GARDEN AND MORE

In the preceding chapter we discussed growing garden vegetables. The vegetables were treated as a group because they share a number of features in common. Most are annuals or, if not, are treated as such. Most are grown together in a garden space and share in common cultural manipulations. All those vegetables are widely grown, frequently in a complementary manner. For example, two vegetables having different sizes and maturity dates can be grown in close proximity, one vegetable can follow another in a given space in a season, and so on. The herbs are also discussed as a group, and they too share features in common, albeit different ones. The annual herbs at least differ little in their cultural requirements from the vegetables, but in their uses the two are a world apart. The vegetables are mainstays of our diets, the herbs touches of flavor.

But what gives the herbs their unique properties as condiments? Much of the flavoring value and aroma of herbs is due to their content of **essential** or **volatile oils**. The word "essential" is not used in the sense of these oils being absolutely necessary to the plant, but rather as their having an essence; that is, being volatile and imparting a fragrance. Many essential oils belong to a class of chemical compounds known as **terpenes**, and most are **monoterpenes**. Terpenes are common natural products of plants. Other familiar terpenes, not all of which are essential oils, include turpentine (a mixture of terpenes), camphor, some alkaloids, carotenes, vitamin A, the resin acids, and natural rubber. The roles of these compounds in plants are problematic but probably include attracting and repelling predators and pollinators, inhibition of growth of competing plants, and antiviral activity. Some bitter substances in plants are known to have a terpenoid nature. Such include the bitter principles in cucurbits that are said to act as repellents for bees but attractants for cucumber beetles.

Because of their unique properties and concomitant uses the herbs have for centuries been grown in a separate garden area—the herb garden. Some of the earliest herb gardens were no doubt those of medieval monasteries. The real Golden Age of herb gardens—and herbs—however, began in Britain about the sixteenth century. Here formal gardens were shaped in designs resembling knots or figures (Fig. 11-1), the garden owner's coat of arms or initials. In more recent times circular gardens have become popular. This pattern developed from the practice of planting herbs between the spokes of old wagonwheels. The method was useful in that the rim and spokes of the wheel provided a continuous border for the garden as well as partitions between individual herbs (Fig. 11-13). The circular garden is still a favorite although most of us will have to use something other than a wagonwheel for our framework. Rectangular, ladder-shaped beds are also popular. In these the flanks and rungs serve as pathways (Fig. 11-13).

Several factors should be taken into consideration when planning a herb garden. Foremost among these are:

1. Location (site): soil properties, lighting, exposure, proximity to kitchen.

2. Plant characteristics; annuals, biennials, or perennials, height, aggressiveness in spreading, pollination, part(s) to be harvested, watering, fertilization, and other cultural requirements.

3. Size: generally just one or a few of a given herb is adequate for a family.

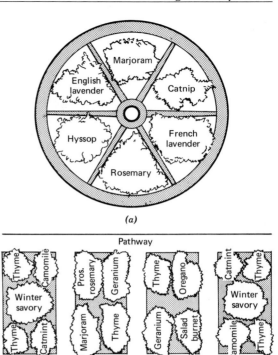

FIGURE 11-13 Examples of circular (a) and rectangular (b) herb gardens. (From Dorothy Hall, *The Book of Herbs*, reprinted with the permission of Angus & Robertson (United Kingdom) Ltd.)

4. Utilitarian features: flavoring properties, fragrance, value in landscaping, and as patio and indoor plants.

The reader can relate these many considerations to the respective herbs. Remember, design the herb harden that is best suited to your own particular herb needs. Such a garden will be a source of great pleasure and beauty.

Proper harvesting and preserving are critical aspects of herb gardening. In general, unless seed are the part to be harvested, the best time to harvest is at about the time the flower buds begin to open. At this time the foliage parts are said to contain the greatest quantity of essential oils and are of a general high quality. Herbs are a joy to use fresh, but for many, long-term storage is required. Drying is no doubt the most common method for preserving herbs, and the current popularity of food dryers will further enhance this technique. It is important to realize that the drying process removes water, thus concentrating the flavoring component of herbs. As a rule experienced cooks use two to three times as much fresh material as they would dried. They keep in mind that the key to cooking with herbs is moderation; that the goal is to enhance flavor, not to overwhelm it. Dried material should always be stored in properly labeled airtight containers.

Of necessity, we have limited our discussion of herbs mainly to their culinary aspects. Herbs have had in the past, and will continue to have, many other uses. Foremost among these are application as healing agents, and in cosmetics and perfumes (including potpourri). The references at the end of the chapter marked by asterisks expand on these important uses. Evidence suggests that in the future herbs will take on even more widespread roles in our lives. For example, a Rutgers University scientist has discovered and isolated a natural food preservative from rosemary and sage. Such preservatives could play an important role in reducing the amount of synthetic additives now put into our food, as well as open up new markets for specialty crops by farmers. Clearly, herbs have a promising future as a resource for improving the quality of our food and our lives.

Questions for Discussion

1. Define the following terms:
   (a) Spice   (d) Wild marjoram
   (b) Condiment   (e) Essential oil
   (c) Potherb   (f) Terpene
2. What does the term "herb" mean to the home gardener? To the professional botanist?
3. The important culinary herbs are grouped into just a few plant families. Can you offer reasons as to why this is so?
4. Prepare a list of general guidelines on the harvesting and storage of herbs to ensure peak quality.
5. You have only a small area for outdoor gardening and must limit the number of herbs you grow to three kinds. Which three would you choose? What factors did you consider in making your choices?
6. Identify four herbs that could be grown successfully indoors in containers.
7. List three important perennial herbs and tell how they are propagated.
8. What do you regard as the main attributes of herbs as relates to home gardening?
9. Identify three important herbs belonging to the plant family Lamiaceae, and indicate the culinary values of each.
10. What factors might you take into consideration when establishing a bed of mints?
11. You have friends coming for dinner and an evening of gaiety. Your guests have never used herbs to any extent in cooking and as a result you plan to surprise them with an herbal extravaganza. Draw up your menu for a five-course meal, and indicate the herbs that you use in each course.
12. As a library exercise, read about the genus *Artemisia*, paying particular attention to those that have culinary value. Present your findings in essay form. What do you conclude about the usefulness of the artemisias?

Selected References

Baily, L. H. 1949. *Manual of Cultivated Plants.* Macmillan, New York.

Clarkson, R. E. 1966. *Herbs, Their Culture and Uses.* Macmillan, New York.

Culpeper, N. 1959. *Culpeper's Complete Herbal.* Sterling, New York.

*De Baracli-Levy, J. 1971. *Nature's Children: A Guide to Organic and Herbal Remedies for Children.* Schocken, New York.

Fenaroli, G. 1971. *Handbook of Flavor Ingredients.* 2 Vols. CRC Press, Inc., Cleveland.

Fitzgibbon, T. 1976. *The Food of the Western World.* Quadrangle/The New York Times Book Co., New York.

Foster, G. B. 1973. *Herbs for Every Garden.* Dutton, New York.

Fox, H. M. 1972. *Gardening with Herbs for Flavor and Fragrance.* Macmillan, New York.

Grieve, M. 1971. *Culinary Herbs and Condiments.* Dover, New York.

*Hall, D. 1972. *The Book of Herbs.* Scribners, New York.

Howarth, S. 1976. *Herbs with Everything.* Holt, Rinehart and Winston, New York.

Lowenfeld, C. 1964. *Herb Gardening.* Faber and Faber, London.

*Lowenfeld, C. and P. Back. 1974. *The Complete Book of Herbs and Spices.* Little, Brown and Co., Boston.

Schuler, S. 1971. *Gardens Are for Eating.* Collier, New York.

Snyder, R. (Ed.). 1977. How to grow and appreciate herbs. *Flower and Garden*, 21:29–36.

Sunset Books (Eds.). 1972. *How to Grow Herbs.* Lane Books, Menlo Park, Calif.

LEARNING OBJECTIVES

By the time you have finished this chapter you should be able to:

1 Identify some of the merits of perennials in the home garden.

2 Name the broad categories of perennial food plants and list representatives of each group.

3 Discuss some of the unique problems associated with the growing of perennials.

4 Identify the major categories of cane fruits and discuss the place of each in the home garden.

5 Explain how to properly plant strawberries, including planting depth and bed layout.

6 Describe a commonly used technique for pruning and training grape.

7 Discuss the methods employed in the production of dwarf apple trees.

8 Compare fruit production in apple and cherry and explain how differences affect pruning procedures.

9 Describe the role of "heading back" and "thinning out" in controlling the growth of fruit trees.

10 Explain how you would train the growth of apple or peach, starting with the planting of a one-year whip and continuing on until the desired form is established.

11 Compare the nut trees to fruits such as apple and pear as relates to place in the home garden, maintenance, and utility.

12 Identify some common nut crops for use in the home garden.

13 Comment on the cultural requirements for three important nut crops grown in home gardens.

12

GROWING PERENNIAL FOOD PLANTS

by Mary Tingley Compton, Courtesy Professor of Horticulture
Oregon State University

*"Trees full of soft foliage; blossoms fresh with spring beauty;
and, finally, fruit—rich, bloom-dusted, melting, and luscious—
such are the treasures of the orchard and garden . . ."*

A. J. Downing (1845)

There is no one way to grow a plant. Gardeners have good results with many methods. Books have been written about each subject we are including in this chapter. Our purpose is to present, in limited space, information about most of our important temperate-zone perennial food plants and practices used to grow them successfully. Obviously much must be left unsaid.

Some of our choicest foods come from perennials: apples, strawberries, blueberries, brambles, cherries, peaches, and nectarines, pears, plums, and walnuts, to name a few. Most do not produce a crop the first season after planting, but require two to eight years to produce. These continue to bear several to many years. Thus, a perennial food garden is an investment for the future.

Perennial food plants may serve a dual purpose in the home landscape (Fig. 12-1). Utility with beauty is a good theme for any planting. Trees, shrubs, creepers, and even herbaceous perennials are available for landscape use (Table 14-1). Many are of exceptional beauty and interest. However, thought must be given not only to their contribution of beauty, but also to their maintenance. Can the necessary chores of pruning, grooming, fertilizing, watering, and harvesting be accomplished with ease and without injury to other plants or to buildings? What will be their impact on people? What about bees, pollen, scents, and sprays near living areas? Are sensitivities and property rights of family and neighbors respected? Are you prepared to give the year-round understanding and care that perennials require?

FIGURE 12-1 Perennials add beauty as well as utility. This apple tree graces an entrance and produces fruit too.

A plant that looks unloved does not add to a landscape picture. Some home gardeners successfully substitute food plants for more conventional landscape material, while others need separate food gardens screened from public view. Much depends on the industry and judgment of the gardener.

Growing perennials requires much the same knowledge as that required for growing the annual vegetables and herbs. Due to the unique nature of these plants, however, certain aspects of their growth and culture require special consideration. Before looking at our important temperate-region perennials we will examine the topics that are especially relevant to growing perennial food plants successfully.

FACTORS INFLUENCING PERENNIALS

Climate

Many factors affect success with perennial plants. Climate is tremendously important and largely determines the geographic range of a given cultivar. Aboveground parts must endure the year-round changes of weather and underground parts are subject to variations in temperature and water supply and sometimes to frost heaving. An area may be too cold, too warm, too wet, too dry, or too cloudy.

Degree and duration of low winter temperature often determine survival (Fig. 6-11). Almost any region may experience exceptional cold at times. In places ordinarily favorable for fruit and nut trees, a test winter may occur about once in 10 to 15 years. Some plants are killed outright, roots and all; others die to the ground, but produce new growth from the crown; still others show varying amounts of bark or wood killing.

Spring frosts may kill flowers or young fruits of early blooming plants such as apricots and peaches. Even apples and pears suffer crop loss at times. A slight slope toward the north can aid in avoiding frost injury by delaying growth and bloom and providing air drainage, while a location at the bottom of a hill may be a frost pocket. Mulches can also be employed to effectively delay the season (see Chapter 8).

Some areas are too warm for best growth. Why are apples not grown extensively in southern Florida or Hawaii? Those places lack low temperature. Many temperate-zone perennial plants have a rest period in winter. With the onset of cold weather, they become dormant, and their physiology also changes so that they will not grow even when all external conditions are favorable. Length of the rest period varies with species and cultivar. To break this rest and let the plant resume growth, a spell of cold weather is necessary.

Most apples and pears need six or more weeks of temperature a little above freezing. A few have little or no rest requirement, however, and will grow in parts of California and southern United States where most cultivars of these fruits do not thrive.

Resting plants are more resistant to winter injury than are growing ones. Most peaches have a rest period, but not so long as apples and pears. Peach trees that have finished their rest may start to grow during a warm spell in winter and be injured by a later drop in temperature.

Other weather conditions are also important. Cool, cloudy summers enhance the growth of pears, but inhibit the best development of peaches and some apple cultivars. Rootstocks successful in summer-dry areas of Washington may not thrive in the Willamette Valley or Oregon, or in New York, or Nova Scotia. As to microclimate, the author once knew of an apricot tree on the north side of a big New Hampshire barn. It bore annual crops until the barn burned down and it was exposed to full sun and night radiation. With earlier bloom and colder nights, it never bore fruit again.

Soils

Soil is not nearly as critical a factor as climate. Any good garden soil is suitable if it is well drained. A dependable, controllable water supply is essential. A soil that grows an assortment of vigorous weeds is likely to grow good garden plants. Ask the Agricultural Extension agent about the potential of your locality. Soil tests will show deficiencies of essential elements. Nitrogen is a yearly consideration whose need varies with the fertility of the soil and the time of year. Since perennial plants occupy their area several to many years, careful preparation before planting pays off. Adding organic matter, lime, phosphorus, and potassium fertilizers and any other nutrients lacking in the region or in the garden site is necessary for successful home gardening.

Fruitfulness

When we grow fruit plants for the crop, most important in production is **fruitfulness,** the ability to bear fruit. Fruitfulness is absolutely essential in all **fruit crops**—those perennials that by definition yield edible botanical fruits. Before fruit set can take place certain events must normally occur. Pollen from anthers must be transferred to the stigma(s) of a flower. Transfer of pollen, termed pollination, is, in our fruit plants, usually done by insects, chiefly honeybees. More than this, however, the transferred pollen must be of a compatible type, able to germinate, grow through the style, and produce viable male gametes that successfully fuse with egg cells. A cultivar that produces pollen able to fertilize a different cultivar is a pollinizer of it and is compatible with it. A bee pollinates; a compatible cultivar pollinizes. A cultivar able to set fruit without cross-pollinization is **self-fruitful.** Two cultivars able to set fruit with each other's pollen are **cross-fruitful;** each is a pollinizer of the other, and they are compatible with each other. Fruit plants may be wholly, partly, or not at all self-fruitful, depending on the genetic nature of the plant and sometimes on the environment.

Productivity is affected by a number of factors including previous treatment of the plant, culture, weather, open flowers, compatible pollen, and active bees. Fruitfulness, however, depends largely on the genetic makeup of the plant. Each cultivar has its own pattern of compatibility or fruitfulness. Most apples, for example, are self-unfruitful, but are cross-fruitful with most others in bloom at the same time. Triploid cultivars such as 'Baldwin' and 'Gravenstein' produce poor pollen, so they are not pollinizers; they may be more or less self-fruitful. 'Golden Delicious' and 'McIntosh' are compatible with most other cultivars and thus are good pollinizers (Table 3-3). Even in self-fruitful apple cultivars, however, best production may be achieved through use of a pollinizer.

Fruitfulness in certain pears varies in different growing areas. 'Bartlett' is self-fruitful in some of the southern sections of its range, but not dependably so elsewhere. A few other pears are somewhat self-fruitful, but most are more productive with a pollinizer. On the other hand, most peaches are self-fruitful, the best known exception being 'J. H. Hale', which produces no pollen (Table 3-3). While pollen sterility is a problem in certain plants, poor productivity is more often caused by unfruitfulness or incompatability.

If a plant is self-fruitful only one is needed for fruit set, whereas cross-fruitful plants require a compatible cultivar. Where space is limited a multiple-graft tree may be desirable. In theory, one tree grafted to several cultivars can provide early, midseason, and late cultivars of fruit without a pollinization problem. In practice this seldom happens with small trees. Vigorous grafts crowd and shade weaker ones or careless pruning removes one or more cultivars. In the author's garden, standard trees of apple, pear, plum, and cherry produce well from three or more grafts. Such trees make extra work, as each cultivar must be sprayed and harvested according to its own schedule of bloom and fruiting.

Flower (fruit) buds of apples, peaches, and other tree fruits form in the summer of the year before they appear as blooms. An important factor in flower bud formation is the amount of food available from photosynthesis. A tree producing a heavy crop of fruit uses much of its food for the growing fruits, but with a small crop, or none, food is available for other uses, such as the formation of flower buds and the development of resistance to cold. Thinning of fruit reduces demand by the crop for available photosynthate and generally enhances flower bud formation.

Certain apple cultivars characteristically have alternate or **biennial bearing,** producing many flowers and fruits one year and little or none the next. The same is true of some pecan cultivars. Loss of crops to frost, improper pruning, or other causes may push a tree into alternate bearing. Some normally alternate bearers can be kept producing more or less annually by removal of flowers, either by hand or by spraying, plus devoted thinning of fruit that sets in spite of treatment. Others are very hard to change. Most home gardeners should choose annual bearers rather than cope with the difficult ones.

Control of Insects and Diseases

Perennial food plants are susceptible to pests and diseases. Sometimes an all-purpose proprietary spray or dust gives adequate control. However, such preparations tend to be wasteful, as not all ingredients are needed. Always read the complete directions on the package carefully before proceeding. Never use more than directed. Sometimes less will do the job. Store sprays and dusts under lock and key separate from other material. Use extreme care with containers and utensils. Avoid inhaling the material. Protect skin from contact during application and wash thoroughly afterward. To protect children and pets, the ground under a treated tree may need to be hosed.

Under most conditions undisturbed ecology means wormy, blemished fruits. Controlling insects and disease isn't easy. Specific problems often need specific solutions. It is necessary to distinguish between an insecticide, which controls insects, and a fungicide, which controls diseases caused by certain nongreen plants. Examples of common, useful insecticides are diazinon, malathion, and nicotine sulfate. Captan, ferbam, liquid lime sulfur, wettable sulfur, and fixed copper are fungicides. A dormant season spray of liquid lime sulfur kills many insect eggs as well as many common fungi. Sevin is an effective insecticide that must be used with careful timing because it kills honeybees and useful predators.

Persons opposed to conventional methods of control have alternatives. Nicotine, pyrethrum, and rotenone preparations are useful against many insects. **Trap crops** (plants that are more attractive to pests than the garden plants) and hand picking are sometimes practical for insect pests. Soapy water is an old control for aphids. Detergent solutions discourage many pests. Hosing or sprinkling dislodges spider mites. Baited traps can reduce Japanese beetle and other populations. Predators such as ladybird beetles and praying mantids are available commercially and are sometimes released in gardens. Sanitation, the removal of all possible infectious and infested leaves, fruits, stems, weeds, and debris, is most healthful for all kinds of gardens and cannot be overemphasized. Resistant cultivars should be used whenever possible to control pests.

Selecting Cultivars

In this chapter we will normally avoid making cultivar recommendations, for the same reasons as outlined in Chapter 10 and elsewhere. Reliable information on those best suited to your area can be obtained from the sources mentioned below, or from specialized publications such as *Hortus Third*.

Where to Get More Information

For the amateur fruit gardener membership in the North American Fruit Explorers is a good investment. The organization is a source of seeds, scions, and information. Its publication *Pomona* is helpful to the grower of any kind of fruit.

Local home orchard societies are reviving interest in old cultivars and spreading information about new ones. These can be located through the Agricultural Extension Service

but are not part of its program. Such societies, museums, and individuals hold fall festivals, "apple squeezin's," fruit shows, and demonstrations and are encouraging the development of display orchards. Information, techniques, plant materials, and pleasant social occasions are available. Information about old cultivars and scions may be obtained from the Worcester County Horticultural Society, 30 Elm Street, Worcester, Massachusetts 01608.

The New York State Fruit Testing Association, Geneva, New York 14456, sells to its members apples on Malling and Malling-Merton rootstocks, scab-resistant apples, and virus-free cherries, strawberries, and raspberries, among other plant materials.

For dependable, up-to-date information consult the Agricultural Extension Service in your area (Appendix C). Acquire and read their bulletins and have your name placed on their mailing lists. Attend their demonstrations and also the clinics held by Master Gardeners, who are trained and sponsored by the Agricultural Extension Service.

Categories of Perennials

A few of our common perennial food plants are classified as vegetables, since the edible parts do not include the matured ovary of a flower. Included in this group are the artichokes, asparagus, horseradish, and rhubarb. These perennial vegetables are discussed in Chapter 10. The present chapter deals with those perennials that are grown for their fruits. Most, but not all, are long-lived, woody plants. Included in this grouping are the small fruits, tree fruits, and nut trees.

GROWING SMALL FRUITS

The **small fruits** grow on perennial plants of relatively small stature. Numerous crops are included in this group. We will discuss the more important of these, including blueberry, the brambles, strawberry, and grape.

Blueberries

Description

Blueberries were an important food for native Americans long before colonial times. Wild species varying in size from low to tall shrubs are found in much of North America and are still harvested in local areas. In eastern Maine and adjacent Canada, many acres of wild low-bush blueberries are fertilized, dusted, pruned by burning, harvested, and sold or processed, constituting an important industry. From zone 5 southward the highbush blueberry occurs and farthest south the rabbiteye is found (Fig. 6-11). In the Northwest pickers harvest pailfuls of "huckleberries" on the slopes of Mt. Adams in Washington and elsewhere. All these are species of *Vaccinium*, the blueberry, which in some localities is called "huckleberry." Also called huckleberries in botanical texts are the evergreen, or florists' huckleberry, *V. ovatum*, and the red huckleberry, *V. parvifolium*. However, the *Vacciniums* are not to be confused with the "true" huckleberries, species of the genus *Gaylussacia*, distinguished by berries with hard seeds and leaves with resinous dots. Both are members of the heath family, Ericaceae. Most blueberry cultivars have been bred from the highbush blueberry, *V. corymbosum*, and a few from *V. ashei*, the rabbiteye blueberry. Selections of wild plants of the latter are also grown. Rabbiteye cultivars withstand heat and lack of moisture better than others.

The blueberry is the only fruit domesticated entirely within the twentieth century. In 1908 Frederick V. Coville and Elizabeth White began a search for large-fruited, good flavored plants of *Vaccinium corymbosum*, which were used in a breeding program. A number of cultivars were produced, some of which are still home garden favorites. Later cultivars have been produced by Coville's successors and by others.

Most cultivars of *V. corymbosum* are medium to tall shrubs slightly spreading toward the top. The thin, simple, oval leaves are up to three inches long, reddish in spring, and turning bright yellow, orange, and red before leaf fall (Fig. 12-2). Clusters of small white or pinkish flowers with one-piece, urn-shaped corollas add interest in spring. The blueberry fruit, which develops from an inferior ovary, contains many soft seed. The tiny crownlike, five-pointed calyx remains attached to the top of the fruit and turns blue with it. Light blue berries with waxy bloom are usually preferred over darker, less coated ones, although these characters are not correlated with flavor.

Blueberries are usually propagated by hardwood or softwood cuttings set in a sand-peat medium. Bushes may also be divided and canes layered. Seed germinated after cold treatment do not come true, but may produce acceptable plants. Nursery-grown plants produce fruit after two to four or more years.

Culture

Highbush blueberries require an acid soil, well-aerated, well-drained, well-watered, and the more organic matter, the better. A pH of 4 to 5.2 is favorable. As blueberries are shallow rooted, soil less deep than for tree fruits is

FIGURE 12-2 Blueberries. A highbush blueberry plant showing leaves and fruits. (Courtesy of U.S.D.A.)

adequate. Preparation for planting should extend down about a foot, with the digging in of peat or other organic matter and potassium and phosphorus fertilizers as needed. Do not use lime, bone meal, ashes, eggshells, animal manures, or fertilizers containing nitrate nitrogen.

Where the soil is suitable, cultivars of *V. corymbosum* grow in zones 5 to 9 (Fig. 6-11). They need a growing season of about 160 days. Plants are about as hardy in wood as peaches and require some winter cold. In zones 3 to 4 on acid soil, the lowbush blueberry, *V. angustifolium,* may be grown as a ground cover, with little culture except the burning off of one-third of the area each spring before the ground thaws. Thus bushes are pruned and weeds controlled. The burned portion does not bear fruit the year of burning. Selections of cultivars of *V. ashei,* the rabbiteye blueberry, are grown from zone 7 southward. These are far less demanding about soil, water, and rest-breaking cold.

One- or two-year-old plants are set out five to six feet or more apart at the same depth as in the nursery. Growing at least three different cultivars is recommended, because blueberries are usually considered self-unfruitful and sometimes cross-unfruitful. Some highbush cultivars, however, are self-fruitful. Apparent unfruitfulness may be due to absence of bees.

Weeds are controlled and organic matter is constantly adding its residue to the soil, when a layer of sawdust, bark dust, or other mulch is applied each spring. Peat is excellent for mixing into the soil before planting but is not a suitable mulch, because it is hard to wet once it has dried out. Blueberries, being shallow rooted, are subject to injury from cultivation. Remove large weeds by hand and stir the surface mulch only enough to keep a crust from forming.

Usually blueberries need only a nitrogen fertilizer in spring: in many areas, ammonium sulfate; in milder areas, cottonseed meal. Before renewing mulch, addition of a complete rhododendron-type fertilizer to aid soil organisms may be useful.

Until bearing age is reached, remove only weak and interfering canes. Prune bearing bushes the same as many ornamental shrubs: remove older, unfruitful canes at ground level each spring (Fig. 12-3). Cut back remaining canes to induce branching. Long flower clusters characteristic of some cultivars, may be cut to one-half their length or be thinned out to improve fruit size and earliness.

Most home garden blueberries are not sprayed or dusted. Where the blueberry maggot occurs, the control program for the area must be followed. If canker and die-back occur on stems and branches, or if mites are present at harvest time, a dormant spray is beneficial. If birds are a problem, cover bushes with netting. A wild or cultivated cherry tree nearby will sometimes divert the birds.

Harvesting and Uses

Taste berries before harvesting the crop. A few kinds are sweet and good when they have turned blue, but most need a few more days to reach best flavor. Ripe fruits may be spicy, sweet, or tart. Market berries are often lacking in sweetness and flavor because they were picked too soon. They are not all they could be either as fresh fruit or cooked.

Homemade bread with milk and sun-ripened wild blueberries were a satisfying supper dish on a hot day in New England in the early part of the century. Often the fresh berries were available for any meal of the day throughout the berrying season, and were baked in pancakes, muffins, cake, pies, and puddings. Surplus berries were "stewed in their own juice"—nothing added, that is—in an open kettle on a wood-burning range, put into hot jars, and sealed at once. No further processing—a real saving of energy. These were used in winter, mostly in pies, with sugar, flour, and a very little nutmeg added. Berries in-

FIGURE 12-3 A four-year-old highbush blueberry plant before (a) and after (b) pruning. The plant was pruned heavily to produce an earlier crop of larger berries. (Courtesy of U.S.D.A.)

tended for sauce were sweetened while cooking. Today fresh-frozen blueberries are excellent in pies and other cooking, but as fresh fruit they have lost somewhat in texture and flavor. They are best when not quite thawed, and many people like them with cereal. Some find them more easily digestible than other fruits.

Where the soil is acid, blueberries are good dual-purpose shrubs. Besides the annual crop of blue fruit, they also contribute color of flowers, young leaves, autumn leaves, and winter stems besides a soft, medium texture of summer foliage that blends and contrasts well with other plants. They will grow in light shade, but produce less fruit. They are one of the best shrubs to attract birds. Berries usu-

ally hang on the bush until picked, so are not messy. In many places they need no spray or dust. Most cultivars are tall shrubs, but may be kept medium-sized by pruning. Where blueberries will grow, they are excellent shrubs for summer homes, since harvest time is July and August.

Bramble Fruits

Description

The **bramble fruits,** or simply **brambles,** include the raspberries, blackberries, and dewberries of the genus *Rubus* (Rosaceae). They have aggregate fruits made up of adhering drupelets, with or without a core of receptacular tissue, and perennial roots with biennial shoots termed **canes** (Fig. 12-4). Typically a cane grows from the root one season, produces flowers and fruits its second season, and then dies. Some cultivars, however, produce fruit in fall on canes of the current season. Canes may be stiff and freestanding as in the raspberries and northeastern blackberries, or procumbent or trailing as those of the eastern dewberry and the very vigorous western cultivars such as 'Logan' and 'Boysen.' The word dewberry has come to be synonymous with trailing blackberry, whereas they were formerly separated on the form of flower clusters and the presence of absence of tip-rooting.

Brambles are widespread throughout North America. Each growing section has its own native and garden varieties. Most cultivars are derivatives of wild North American species, native or naturalized, but the exact ancestry of each may not be traceable.

The important raspberries of North America belong in three groups: red raspberries, descendants of *R. idaeus* var. *strigosus,* distributed from Newfoundland southward into North Carolina, westward to the Rocky Mountains, and along the Pacific coast into Alaska; black raspberries, blackcaps, or thimbleberries, cultivars of native *R. occidentalis;* and purple raspberries or purple canes, hybrids between red and black forms. Cultivars of the European red raspberry, *R. idaeus* var. *vulgatum* were grown in North America until cultivars of the more productive native *R. idaeus* var. *strigosus* began to replace them in 1865. Present-day North American cultivars are all considered to be of *strigosus* origin.

Red raspberries are one of our hardiest fruits, their culture extending into zone 3 (Fig. 6-11). Most cannot tolerate heat and drought, so only selected cultivars are successful south of Maryland. Black raspberries, *R. occidentalis,* are often found in the same habitat as red ones but do not extend as far north, ranging from zone 4 south into Georgia and westward to the Rocky Mountains. Natural hybrids, *R. x neglectus,* are found where both species occur. Plants are vigorous and productive with variable fruit and plant characteristics. *Hortus Third* lists raspberry cultivars for various sections of North America.

Like the aggregate fruit of raspberries, the fruits of blackberries are made up of adhering drupelets; but unlike raspberries, the drupelets adhere to the receptacle, which breaks off at harvest, constituting the core of the aggregate, accessory fruit. A raspberry, having no core, is shaped like a thimble.

A European species with partly perennial stems, *Rubus laciniatus,* naturalized in the Pacific islands and in the American Northwest, became the Oregon evergreen blackberry. Today, Oregon's 'Thornless Evergreen' is the

FIGURE 12-4 A blackberry branch with a typical fruit cluster ready for picking. (Courtesy of U.S.D.A.)

most planted blackberry in Oregon and the most important commercial blackberry in North America. Productivity has not decreased in the thornless form. It is liked in home gardens for its ease of picking and generous yield.

The Himalaya berry also naturalized in the Northwest, is cultivated in northern California and has given genes to important cultivars. Other species genealogically important are *R. flagellaris,* and its cultivar 'Robibaccus,' and *R. ursinus.*

Scientific breeding programs at state universities have given us excellent blackberry cultivars. *Hortus Third* and others offer useful lists for the Northeast; for Maryland, Arkansas, Florida, and the Gulf states; for Texas and the Southwest; and for Washington and Oregon. Choose cultivars known to produce well in your area.

Culture

Brambles need full sun with protection against drying winds and extreme cold. Many lack resistance to heat and drought. Red raspberries, a home garden crop in zone 3, are the hardiest, except for the little northern bog plant, *R. chamaemorus,* known as cloudberry or baked-apple berry, adapted in zone 2 (Fig. 6-11). Most western cultivars, suited to zones 6 to 9 are usually not hardy east of the Rockies; fruit quality is poor and often does not set.

Any good garden soil is suitable, provided it is well drained, retentive of moisture, and has a continuous water supply. Water is especially important just before and during harvest. Cultivate soil before planting as for other perennial crops. Keep weed free.

Usually new shoots (canes) from the roots, referred to as suckers, are dug for replanting and each should be attached to a piece of the old root. Do not pull them out, if they are to be planted. Blackcaps, purple canes, and some trailing blackberries may also be tip layered. Tips of canes bend down and take root, or those inserted in a spade hole in August are ready for fall or spring planting (Fig. 3-8*b*). In cold areas spring is planting time.

Cut back starts, leaving about a foot of cane to be used as a handle. Set starts in a furrow or in spade-holes about as deep as they were growing, pressing the soil about the roots with a foot to remove air pockets. When planting is finished, cut off the canes at the ground. New canes will grow in spring and will bear the second year.

Thinning out excess canes and training pattern depend on the vigor of the cultivar and its place in the landscape picture. Often the easiest method is to grow plants in a row and tie them to one or two wires supported by strong posts. Plant red raspberries two to three feet apart in a row along wires about four feet aboveground (Fig. 12-5). At spring pruning save seven to eight strong canes from each

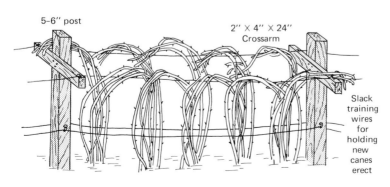

5-6″ post

2″ × 4″ × 24″ Crossarm

Slack training wires for holding new canes erect

FIGURE 12-5 A double row of heavy wire fastened to posts will give raspberry canes all the support they need. (Courtesy of Oregon State University Agricultural Extension Service.)

plant. Cut off tips of canes at five to six feet depending on vigor and tie loosely to the wire. Such a row is easy to prune and cultivate and fruit is easy to pick. Blackcaps, purple canes, and eastern blackberries may be trained the same way. The very vigorous west coast blackberries or dewberries need wires—one wire two feet and the other three and one-half feet or more above ground, and require additional space in the row, up to eight feet between plants (Fig. 12-6). Lay growing canes along the row to avoid trampling or other injury. It may be helpful to dab paint on the bases of bearing canes to make them easy to see after harvest when they are pruned out. In spring before growth begins, remove weak canes and divide the rest into two groups, one right and one left. Carry a group up, over and along the top wire, then down and back along the bottom wire. Tie loosely where support is needed. Some cultivars may need heading back of side branches to improve size of berries and to avoid tangles (Fig. 12-14).

After harvest cut out and remove all bearing canes at the ground except in very cold areas where they are needed to hold snow. These may be pruned out in spring when canes are thinned and trained.

Certain cultivars—called everbearing—produce fruit in fall on side branches from

tips of new canes. The next season they bear at the normal time (midsummer) on side branches from lower nodes. If only a fall crop is wanted, cut or mow off all canes after harvest.

Plants are shallow rooted, so cultivate only enough to keep the rows weed free, remove surplus suckers, which may be the worst weed present, and prevent compaction of the soil. An organic mulch is desirable. Each spring berries in rich soil, well prepared before planting, need a moderate application of complete fertilizer followed by a layer of mulch. Keep a bare area around bases of canes.

Brambles are subject to many diseases and insect pests. Plant only virus-free stock. If anthracnose cankers or orange rust appear, remove and destroy affected plants. Prompt pruning out and removal of canes after harvest helps in disease control. A dormant spray helps control fungous diseases and kills insect eggs. Where berry mite is a problem—it causes hard, red drupelets and sour, nonmaturing berries—a prebloom spray may also be needed. Blackcaps, being less resistant to viruses and other diseases, should not be planted near other cane berries, especially red raspberries. Do not set plants with lumpy growths on roots or canes. These are caused by crown gall bacteria and may spread in soil and destroy a planting. It may also attack fruit trees, roses, and other plants.

Harvesting and Uses

Pick fruits when fully ripe. Black or dark fruit types may be colored days before they reach best flavor and often are picked too soon. When ripe—in prime eating condition—berries show a slight softening or loss of lustre. Look, feel, and taste before harvest; and if fruit is acid or flat, wait. Most will improve. A quart of fruit from a yard of row is considered an average yield and often much more is produced. Harvest may last several weeks. The average life of a planting is five to eight years, but with

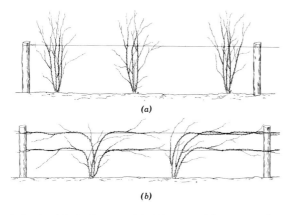

(a)

(b)

FIGURE 12-6 Training blackberries. (Courtesy of U.S.D.A.)

a good location and care, it may last many years.

All bramble fruits may be used fresh, or after freezing, offering a wide variety of flavor, color, and texture. They are also good cooked or canned. The red raspberry is one of our choicest fruits, fresh or frozen, and one of the most rewarding to grow at home. All bramble fruits make good jam, but none can surpass that of the blackcap. The latter is also one of the easiest fruits to dry. They are grown for jam and also for extraction of a purple dye used for stamping meat.

Brambles may be trained on various supporting structures that lend interest in the landscape or grown in large containers. The light glaucus stems of blackcaps often attract attention, especially in a winter landscape. The author's small untrained Himalaya thicket forms an impenetrable barrier six feet tall that supplies delicious fruit for harvest and for many birds, besides cover for upland quail and small animals and birds. *R. chamaemorus* is an excellent dual-purpose plant. Its upright stems arise from rhizomes and are less than a foot tall. These bear solitary, white, one-inch flowers followed by yellow to orange-red blackberrylike fruits that taste like baked apples. They are delicious fresh and they make beautiful amber jam. Plants are recommended for cool, moist rock or bog gardens.

Strawberries

Description

The strawberry may be the most widespread of North American fruits, since its culture extends throughout much of the United States and temperate Canada, into Alaska and parts of Mexico. *Fragaria,* suggesting fragrance, is a low, spreading genus in the rose family (Rosaceae), and is one of our most flavorful fruits (Fig. 12-7). Modern North American cultivars are designated as *Fragaria* x *Ananassa* in *Hor-*

FIGURE 12-7 Strawberries, one of our most flavorful fruits. (Courtesy of U.S.D.A.)

tus Third, indicating their hybrid nature. They are regarded as hybrids of *F. chiloensis* from Chile and *F. vesca* var. *virginiana* from eastern North America. All are of mixed ancestry. For the home gardener the word "strawberry" and the cultivar name identifies a cultivated variety, as the 'Marshall' strawberry or the everbearing strawberry 'Ozark Beauty.'

Fragaria vesca, long cultivated in Europe, was brought to North America by early settlers and cherished until replaced by the healthier and more productive North American species *F. vesca* var. *virginiana,* which was cultivated also in England as early as 1629. These species produced small, variable fruits. Large-fruited cultivars became available in the 1850s, after plants of *F. chiloensis,* native to the Pacific coast of both North and South America, had been taken from Chile to Europe and used in breeding. Since that time thousands of cultivars have been developed and new ones are released frequently. Old cultivars succumb to virus, red stele rot, or unfavorable conditions. New ones come disease free and often with characteristics for success in particular areas.

The strawberry stem is a short, woody structure with three-parted leaves arising from nodes at the top or crown. Each stem has several to many fibrous roots. Horizontal stems called runners grow from buds in the leaf axils, as do also compound cymelike inflorescences with flowers opening over several days (Fig. 3-1). The rosaceous flowers have five or more green sepals with attached bracts constituting the "hull," five white or pink petals, several to many stamens, and many simple pistils borne on a fleshy receptacle, which becomes the edible part of the aggregate, accessory fruit (Figs. 3-21, 3-23). Each pistil becomes a one-seeded fruit called an achene. Each achene contains a single seed. In a mature strawberry the achenes may protrude from the surface or may be depressed. Their position is useful in identification of cultivars.

Strawberry plants with normal stamens are self-fruitful. Cultivars without pollen have been grown in North America and may still be found in local areas, but none are important at present.

Cultivars come in two types: **Junebearing,** which produce fruit for a period in early summer; and **everbearing,** yielding berries in summer and also in fall. Some are genetically fixed in the everbearing habit, while others are influenced by climate, especially in California. In general, June bearers produce larger crops of better quality, but everbearers are valuable for prolonging the fresh fruit season, thus saving freezer space and electricity, and for producing a late crop if an early one has been lost to frost.

Choose cultivars developed for your area or known to succeed there. An old cultivar—'Marshall'—has long been considered the standard of quality, but is now hard to find. Accept only plants certified free of virus and red stele rot.

Culture

Strawberries need full sun, good air drainage, and dependable water supply. They produce on a wide range of soils (pH 5.5 to 6.5), but most cultivars do best on light loam, if water can be provided during the bearing season. Where only heavy or poorly drained soil is available, plant in raised beds or in containers (Figs. 7-12, 14-8). Saline soils are to be avoided. The ground should be prepared with phosphorus and potassium fertilizers and organic matter incorporated at least spade deep. Soil should be weed free and as friable as possible, so that new plants can take root.

Planting

Depth of planting is critical. All roots, but not the crown, must be covered, so plant with the crown at the soil line (Fig. 12-8). In friable soil two persons can plant strawberries rapidly. One thrusts a spade into the soil and moves it forward to make a hole large enough

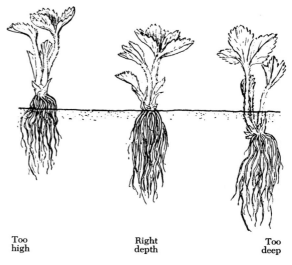

Too high   Right depth   Too deep

FIGURE 12-8 For good results, strawberries must be planted at the correct depth. (Courtesy of Oregon State University Agricultural Extension Service.)

to hold the roots. The other spreads the roots in the hole with the crown at the ground line, then with a foot presses the soil against the roots, being careful not to leave air pockets. Protect plants against drying out while handling. Use strong, young plants formed at the tips of runners, either purchased or home grown.

Allow everbearers to bear fruit their first fall, but not in June. From June bearers set out in spring remove flower stalks throughout the season. The next year expect a heavy crop and the third year a smaller crop. Start a new planting not later than the third spring and preferably, the second.

The two usual ways of growing strawberry plants are the **hill system** and the **matted row.** The latter is perhaps a little less demanding and produces a good crop of acceptable berries. Plants are spaced at one and one-half to two foot intervals in rows which are three to three and one-half feet apart. Runners are allowed to develop and set new plants making a row 18 inches or more wide. This method is often used with June bearers, which produce fruit only during a short period in early summer. The hill system is preferred for everbearers and may be used with June bearers. Using this system, plants are set 12 to 15 inches apart in rows that are two and one-half to three feet apart. All runners are cut off as they develop. Hill culture produces a heavier crop of larger, high-quality berries. A mulch of straw will conserve moisture, control weeds, and keep the berries clean. In cold areas a winter mulch may be needed to protect against freezing injury or sun damage.

About three weeks after setting out plants, stir into the soil one heaping teaspoonful of 5-10-5 fertilizer around each plant and water it. Repeat the application in early August and twice each year. Do not use sodium nitrate or any other sodium carrier; sodium tends to reduce yield in strawberries.

Strawberries have several important fungus and insect pests. Red stele rot is caused by *Phytophthora fragariae*, a soil-borne fungus. To avoid red stele rot, do not plant where strawberries have been grown recently. Also avoid places where potatoes, tomatoes, peppers, or eggplant have grown as verticillium root rot may be present. Several kinds of root weevils attack strawberries. The adult snout beetles eat at the edges of leaves at night, hiding in the soil or on other plants by day; the larvae live and feed on the roots. Control is usually spray or dusts when chewed leaf edges show that beetles are active. Different species of beetles are present at different times of year. If identification is necessary, stalk them at night with a flashlight. Your Extension agent will help you identify them (Appendix C). Follow control practice for your area. Prompt picking of ripe berries plus sanitation helps control gray mold, or botrytis blight, of the fruits, however, spraying or dusting may be necessary during rainy harvest periods. Plants may need protection against spring frosts at bloom-time and a cover of netting to keep out birds as berries ripen.

Harvesting and Uses

Pick ripe fruits every day—with hulls and stalks unless fruit is to be used at once. Fresh strawberries, sliced, sweetened, and served with cream or other topping, berries with hulls left on arranged around a little mound of powdered sugar, and shortcake of rich biscuit dough are favorite home and festival dishes. Put berries in or over homemade ice cream. Try a pie combining strawberries and rhubarb. Make your own strawberry jam to avoid a possible product with synthetic color and flavor plus real timothy seed. Canned by usual methods, strawberries lose texture and color, but sliced and frozen they are excellent. Fresh or frozen strawberries rank high in Vitamin C. Their appetizing appearance makes them effective decoration for salads and baked foods.

Both cultivars and wild species are attractive ground covers and edging plants. They are excellent in containers; strawberry barrels and ceramic jars being perhaps the most popular. Novelty cultivars may be grown from seed and become climbing or hanging plants, whose runner plants bloom and bear fruit but form no roots.

Grapes

Description

The grape is the only food-bearing member of the large Vitaceae family. At present there is great interest in viticulture, especially in areas near the Great Lakes, in the Pacific Northwest, and in the American Southwest, in addition to California, by far the leading grape-growing area in North America. Four groups offer grapes for almost any home garden: (1) historic cultivars of the European grape, *Vitis vinifera*, introduced into California by the Spanish; (2) native North American species, especially those of the fox grape, *V. labrusca;* (3) hybrids of the two, the so-called "French hybrids" of European origin and modern cultivars bred in the United States and Canada; and (4) the very vigorous cultivars of the muscadine grape, *V. rotundifolia*, of the cotton belt of southern United States.

The grape plant is a woody vine (Fig. 12-9). Like other woody plants it produces leaves with axillary buds on shoots of the current season. Leaves are palmate, three or five lobed, except in the muscadines with nearly oval, toothed leaves. At certain nodes tendrils develop and panicles of small, greenish, fragrant flowers may appear, from which develop small to very large bunches of grapes. Each grape is a berry typically with up to four seeds. However, both old and new seedless cultivars are available. Muscadine flower clusters are small and loose, yielding three to fifteen berries that ripen unevenly, so are picked one by one. Other grapes are picked by the bunch.

Fruit of the grape is described as black, red, or white, although it may actually be pink,

FIGURE 12-9 Grapevine with a central trunk and four side canes trained on two wires. Note the abundant crop of fruit. (Courtesy of U.S.D.A.)

brown, green, yellow, blue, or purple. European grapes have thin, adhering skin, while North American cultivars are slip skin, having easily removed coats. Modern hybrids are often thin skinned.

Culture

Too many bunches of grapes per vine result in poor quality; so we prune, leaving a number of branches suitable to the particular cultivar and its pattern of training. After leaf fall, the growth of the current season, a flexible shoot with leaves and axillary buds, becomes a cane for one year, after which it is called "older wood." Cutting off part of a cane lets remaining buds develop shoots the next growing season. Typically cultivars of European habit bear fruit on shoots from basal nodes of canes, while North American types produce from middle nodes. This difference in bearing dictates differences in pruning, American vines often being trained to a cane system and European types to a spur system.

To carry out a pattern, tie strong canes to supports where needed and cut off to desired length. For each cane saved cut off two other canes to spurs of two buds each for the next year's renewal. Remove all other canes. The next pruning season remove the old cane and replace with a new one from a spur. Again leave spurs for the next renewal. Where older wood is necessary, as for the trunk of a vine, or for a **cordon** (a permanent horizontal branch) prune a cane to desired length and tie to a support; instead of removing it the next year, train canes from it and also leave renewal spurs.

In the cane system of pruning, customary with American cultivars, first develop an upright trunk. Then tie bearing canes to horizontal supports. One or two wires may be used. With two wires four renewal canes are maintained, the pattern being an upright trunk with four horizontal arms (Fig. 12-10). If using a spur system, in which a branch needs only two or a few bearing nodes, only the upright stem with its renewal area may need support. Tie it to a stake until it is strong enough to stand alone. One or more horizontal canes from an upright stem may be left to become cordons, each with one or more renewal areas. Cordons are often needed for covering a wall or an arbor and in high pruning of muscadines.

As with other fruits, grape culture varies with locality and cultivar. A well-drained location on a south slope with good air drainage and rich, light to medium soil at least three to four feet deep is ideal. A soil pH of 6 to 7.5 is desirable. The need for nitrogen and water varies. Too much of either delays ripening. A long growing season 150 to 180 days,

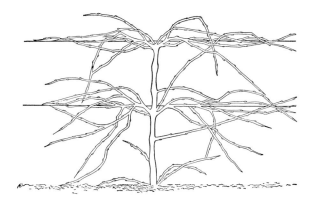

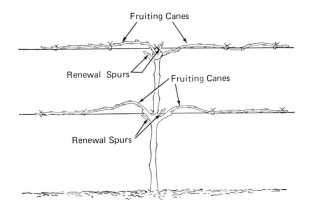

FIGURE 12-10 Pruning and training grapes. (Courtesy of U.S.D.A.)

is necessary for development of best quality in most cultivars.

Weed control is accomplished by hand, mulching, shallow cultivation, mowing, or use of herbicides. Vines are very sensitive to injury by 2,4-D.

Propagation

Grape vines are easily propagated by layering or by cuttings made from winter prunings. Pieces of cane about a foot long with at least three nodes are allowed to callus in moist peat or sawdust and planted out in spring for rooting. Plants are ready for the vineyards in one or two years. Such plants are on their own roots and are suitable only where root pests are not a problem.

Diseases and Insects

Grapes are susceptible to serious pests, a most important one being the root louse, *Phylloxera*, which attacks European and many American cultivars. Where this insect occurs plants must be grafted on resistant, virus-free roots. Japanese beetles are also destructive to grapevines causing defoliation. Serious fungus diseases are powdery mildew, controlled by dusting with sulfur, and downy mildew, by a sulfur preparation or by Bordeaux mixture. Gray mold (botrytis blight) may attack stems, leaves, or fruit, its severity varying with area, cultivar, and weather. Other pests may be injurious locally.

In France in 1882 to 1885 history was made in plant pathology when a preparation of copper sulfate and lime, applied to grape vines to discourage pilferers, was seen to control downy mildew. Bordeaux mixture, as it was called, became very important not only in vineyards but also in control of late blight of potatoes, where its use led to discovery of copper as an essential element.

For up-to-date control practices in your area, consult your Agricultural Extension Service.

Harvest and Use

Fruit should be harvested when it is vine-ripe—in best eating condition. Grapes do not improve in quality after harvest as do apples and pears. Some kinds ripen unevenly, so they may be picked over several weeks. It is well to harvest before fall rains cause skin-cracking or migrating birds take all. A covering of nylon net may solve the latter problem.

Bunches of high-quality grapes laid one layer deep on newspaper in a lug box in a cool place will keep a surprising number of weeks. Bunches cut with stalks about eight inches long and each placed in its own bottle of water

may keep in storage at 40 to 50°F for use into winter.

While most of the world's commercial crop is used in wine making, many home gardeners grow grapes for fresh dessert and salad fruit, juice, jellies, and conserves. Grape pie is an old favorite, its preparation being easier with newer seedless kinds. With modern dryers it is easy to prepare raisins. Grape leaves are used for wrapping small quantities of various meat-vegetable mixes for baking.

Grape vines trained along a fence or against a wall add interesting form and pattern to a landscape. Their distinctive leaves glow bright yellow in autumn, with bunches of colorful grapes adding to the picture.

GROWING TREE FRUITS

In this section we examine a number of our important temperate region tree fruits. In contrast to the tropical and subtropical tree fruits such as citrus, papaya, and avocado, those of the temperate zone are primarily deciduous. All are long-lived perennials, and, while they may not come into production for several years, they may continue to produce for a century or more. Thus, growing the tree fruits is definitely an investment in the future, an investment that—if properly nurtured—can give abundant yields.

Apple

Description

Of all deciduous tree fruits grown in North America, the apple is doubtless the best loved and most widespread. It grows in almost every state of the United States, across southern Canada to the Rocky Mountains, and in many parts of British Columbia. A few hardy sorts survive in Alaska; most of Florida and Hawaii are too warm.

Because apples are grown so widely they will receive more attention than other tree fruits. Much of what is said, however, applies to the others as well.

Apples are members of the rose family, Rosaceae, as are pears, peaches and nectarines, plums, cherries, strawberries, and the brambles. *Hortus Third* uses *Malus pumila* to designate the common apple. Although our cultivars are mixed with other species, especially *M. sylvestris, M. pumila* shows the garden apple's wild origin, probably in southwest Asia. However, certain pomologists prefer and use *M. domestica* instead of *M. pumila.* For home gardeners the genus name *Malus* and a cultivar name such as 'McIntosh' distinguish one kind of apple from all other plants. More simply so does the 'McIntosh' apple. A wide selection of apple cultivars is available to the home gardener (Fig. 12-11).

The cultivated crabapples are descendants of *Malus baccata*, the Russian crab, or are mixtures of *M. baccata* and other species or varieties. They are identified by *Malus* and the cultivar name, as *Malus* 'Hyslop' or the 'Hyslop' crabapple.

For best development most apple trees need a sunny summer, dependable water supply, at least 100 frost-free days, and, to break the rest period, six or more weeks of temperature not over 45°F. A minimum winter temperature of −20°F is desirable, but not essential for the hardiest cultivars.

Early settlers in North America brought named varieties from Europe, where apples had been grown and selected since ancient

FIGURE 12-11 A wide selection of apple cultivars is available to the home gardener.

I first noticed this problem, Doctor, when I realized how many cultivars are available.

times. They planted seed and grew fruit, chiefly for cider and vinegar. High-quality kinds came to be recognized and named. As the art of grafting spread, desirable cultivars, both native and European, were carried about the continent.

Flowers and Fruits

Apple bears flowers in spring and a crop of fruit in late summer or fall. Each flower has five green sepals, five showy pink and white petals, about 20 stamens, and five carpels making up the inferior ovary. Each carpel is indicated by a separate style and stigma. The branched cluster of three to seven flowers, whose terminal flower opens first, is a cyme (Fig. 3-15). The fruit of apple is a pome (Fig. 3-23). The five, tough, papery carpels are enclosed by edible pericarp surrounded by the edible floral cup, the fused bases of sepals, petals, and stamens. Each carpel usually contains two seed, but some cultivars may have up to four. These seed germinate readily with afterripening but do not breed true. Each seedling is a unique individual and must be propagated vegetatively.

Fruit Production

Most fruit trees have a natural fruit drop after petal fall and another more noticeable one later, usually in June. Seedless and abortive fruits and others unable to compete are eliminated. After the "June drop," which usually lasts over a period of days, remove small, blemished, or excess fruits, leaving one per cluster and these spaced six to eight inches apart. For exhibition apples, leave the one best fruit in its cluster that has developed from a flower other than the terminal one, as apples from terminal flowers are often atypical in form.

When a tree bears, branches heavy with fruit may need to be propped or supported to keep them from breaking. Removing excess fruits helps prevent breaking of limbs by reducing their load. It also improves quality by saving only the best and by reducing competition for food and other nutrients.

Standard and Dwarf Tress

For many years scions have been grafted on seedling roots, producing what we call **standard trees.** Common practice has been to grow plants from seed collected at processing plants and bud the seedlings in summer or store them to be bench grafted in winter.

A typical mature tree on seedling roots may grow 40 or more feet tall with a spread of up to 50 feet. It may live 100 or more years and produce half a ton of fruit in one season.

Present trends are away from the use of the standard cultivars in order to reduce labor, make the best use of space, and avoid the variability of seedling roots. Smaller trees now available are easier to care for and harvest, behave predictably, and produce more fruit per unit of space. Although the trees are smaller, their fruits are not. Indeed, their fruits are often larger than those of standard trees. The smaller, dwarf and semidwarf trees normally bear early in life, sometimes in the second season. These many positive attributes make the smaller trees ideal for home gardens.

Naturally dwarf apple trees have been known since ancient times, and grafting scions on roots of such dwarfs to produce small trees is an old practice. Modern horticulturists are able to produce trees of predictable size because the roots exert control on the growth of the scion. Dwarf trees are produced in one of two ways (Fig. 12-12). The older practice is to graft standard tree scions on roots of dwarfs to produce small trees. Alternatively, a stem piece of a dwarf, known as an **interstem,** can be grafted between a seedling of clonal root and a desired cultivar giving a dwarfing effect on the top.

Final tree size depends not only on the dwarfing effect of the rootstock, but also on

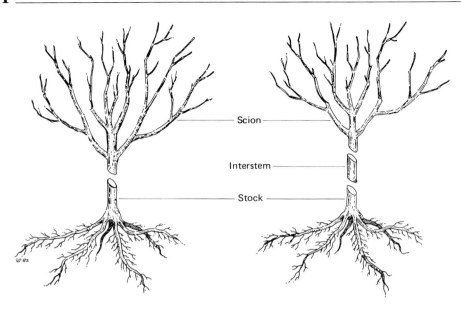

FIGURE 12-12
Creating dwarf trees. (Courtesy of Guerry Dean.)

Scion

Interstem

Stock

the natural size of the cultivar used as a scion. A cultivar such as 'Tydeman's Red', which is naturally larger than 'Golden Delicious,' will give a larger tree when the two are grown on the same rootstock. Conversely, cultivars of different natural size can be made approximately equal in size by choosing the appropriate rootstock for each.

A number of clonal rootstocks, both dwarfing and standard, have been used widely in Europe. Over many years they had become so mixed up that identities were doubtful. In this century at East Malling Experiment Station, Kent, England, stocks have been sorted out and assigned numbers. Those in the first series, containing the old natural dwarfs, are called **Malling** (M) (formerly East Malling—EM) and each has a number under 100. A later series, **Malling-Merton** (MM) have numbers over 100. Each designation, as M 7 or MM 111, names a clone with predictable behavior propagated by **stooling,** a kind of mound layering used in the nursery industry. The approximate relative size of apple trees on selected rootstocks is shown in Fig. 12-13.

Malling stocks produce trees from very dwarf through standard size. Malling 9 (M 9), one of the dwarfing stocks, is widely adaptable and useful in the home garden. The 6- to 12-foot tree occupies about one-tenth as much area as a standard tree and yields from one to several bushels of fruit a year. Pruning and harvesting can be done from the ground or a short ladder. A small hand-sprayer or duster is adequate for control of diseases and insects.

Dwarf trees on the market are often on M 9 rootstock and semidwarf, about half as large as a standard tree, on M 7, M 26, or MM 111. A recent release, M 27, produces a four-foot tree likely to be useful in containers and greenhouses as well as in the garden. It may become useful as a stock for pollinizers.

Problems with Clonal Roots

All trees on clonal stocks must be planted with the graft union a few inches or centimeters above ground. If the scion is permitted to root, the effect of the rootstock is lost. Dwarf trees may not thrive on poor, light soils. They need

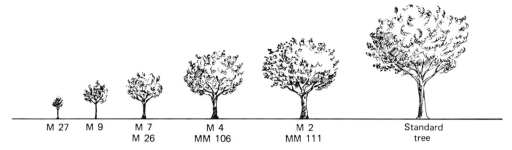

| M 27 | M 9 | M 7 | M 4 | M 2 | Standard |
| | | M 26 | MM 106 | MM 111 | tree |

FIGURE 12-13 The approximate relative size of apple trees on selected rootstocks. (Courtesy of Guerry Dean.)

more attention to nutrients and water supply than trees with more vigorous roots. Smaller root systems may provide poor anchorage. Trees heavy with fruit may blow over or break off. Malling 9, the most popular with home gardeners, has brittle wood and needs staking from time of planting. Other trees may need support as soon as planted or when bearing age is reached. Malling 9, 7, and 26 are only medium in hardiness, about the same as most peaches. In cold climates they suffer winter injury. Hardy stocks are needed in much of Canada and adjacent United States. Canada has developed the very hardy but nondwarfing Robusta 5, and a dwarf from Sweden, Alnarp 2 (A 2), is proving hardy and well anchored. Research continues at experiment stations in Canada, the United States, and several European countries. Doubtless, valuable new stocks will be bred or found.

Spur-type Dwarfs

Apple fruits are produced on short, long-lived branches termed **spurs** and terminally on one-year-old wood. The spur-type dwarfs are cultivars that have arisen as mutations (sports) and are natural dwarfs forming small, slow-growing trees with shorter internodes and spurs closer together than usual. They often produce larger crops than normal types. With proper pruning they form strong framework but may need staking when bearing age is reached. With extra care a spur-type on dwarfing roots pruned to a central leader may be a distinctive, dual-purpose, food plant. However, for best bearing with moderate care they are often budded on nondwarfing roots and trained to multiple leader form.

Choosing Cultivars

Of the thousands of apple cultivars, a few have become dominant in each growing area since World War II. The 'Red Delicious' with its mutations is the leading cultivar in North America. At its best it is good for eating fresh and for sweetening cider. It may now be outstripped in the world by the all-purpose 'Golden Delicious,' second in importance in the United States and third in Canada, which has been planted in great numbers in the Eastern Hemisphere, many on M 9 roots. Second in Canada and third in the United States has been 'McIntosh,' an old Canadian seedling. Because of its excellent all-purpose quality and resistance to winter injury, it retains a high position in northeastern United States, eastern Canada, and British Columbia. 'Granny Smith,' a green apple from Australia is rivaling 'McIntosh' for third place. It requires a longer growing season and more heat units than does 'McIntosh.'

Present trends—back to the land, home and organic gardening, emphasis on natural foods—are causing backward looks to cultivars popular in the nineteenth and early twentieth centuries. Cultivars remembered or legendary for special dessert, cooking, drying, and squeezing qualities are sought for home use and for display gardens established by museums and orchard societies. Nurseries are advertising "antique" apples. The old 'Yellow Transparent' and crabapples such as 'Transcendent' and 'Siberian Crab,' being very hardy, are useful in the colder parts of the apple's range.

Certain newcomers (also a few oldies) are resistant to apple scab fungus and some to scab and mildew. Others are more or less resistant to fire blight, a very destructive bacterial disease, and to root aphids. Using such cultivars reduces the need for disease control, but insect pests must still be dealt with.

While choosing favorites, old or new, that do well in your area, you must also consider fruitfulness, the ability of a cultivar to set fruit with its own pollen or that of other cultivars. A few common apples are more or less self-fruitful, but even these usually set a better crop when other cultivars with compatible pollen are in bloom nearby. Most apple cultivars are self-unfruitful, but cross-fruitful with most others. A cultivar whose pollen is compatible with that of another is a pollinizer of it; the two are usually cross-compatible or cross-fruitful.

Since orchard bloom may continue a month or more, but cultivars bloom early, midseason, or late in the period, it is necessary to plan for overlapping bloom as well as compatibility, since with insect pollination the pollinizer must be in bloom at the same time as the pollinized. In the absence of pollinizers, bouquets of compatible flowers in pails of water may be hung high in trees.

While sales catalogs are useful for study and inspiration, a distant dealer may not be the best source of plants, apples or others. An established nurseryman is likely to know cultivars suited to the area and stock them at planting time. Try your local dealer first. Order early, perhaps a year ahead, while a choice may be had. Always buy the best grade and one-year whips, if available. These have unbranched scions, one growing season from budding or grafting, and usually make better growth than older transplants. Second choice is two-year-old trees, with the main trunk and one-year-old branches ready for training (cutting back and thinning out) as soon as planted.

Planting

Fall planting often gives a slight advantage in growth and is practiced in mild areas. Where winters are severe with soil freezing and heaving, spring planting is best, as soon as soil can be worked. Roots must not be exposed to drying, freezing, sunshine, or rough handling. Carefully cut off broken, injured, or extra-long roots. Store or heel in trees until planted. The ideal situation is to dig the hole before the tree arrives, a hole large enough for the roots without bending or crowding. A hole about two feet deep and wide is usually adequate. Fill in with topsoil from the hole mixed with a pailful or more of moist peat, being careful to avoid air pockets. The topmost roots should be no more than two inches below the soil surface. One sage nurseryman advises, "Always plant a little high; the soil will settle." Most horticulturists give the same advice. Use no fertilizer in the hole or around the tree the first year. With soil prepared the season before planting, potassium and phosphorus will be present in the root zone and the topsoil around the roots will supply nitrogen. Also do not use any weed killers the first year.

Staking and Protecting

Certain dwarf trees should be staked at planting. A stout wood or metal post driven two and one-half feet into the ground on the windward side and held with one or two soft ties or wires inside a piece of hose is recommended. To protect against mouse or rabbit damage, a circle of ground at least two feet out from the trunk should be kept bare. Place a loose cylinder of quarter-inch mesh hardware cloth about 10 inches high around the base of the trunk, or wrap the trunk with strips sold especially for bark protection.

Fertilizing

Young apple trees, dwarf or standard, need to reach bearing size as early as possible. Usually a moderate amount of nitrogen fertilizer is applied in early spring beginning the year after planting. Two ounces of ammonium sulfate for each year of tree age is applied in a ring under and a little beyond the spread of the branches. If ammonium sulfate is not available, ammonium nitrate or urea is suitable. A complete fertilizer may be used, but surface applications of phosphorus are seldom beneficial and results with potassium are variable. After bearing begins, standard trees may not need fertilizer. If the growth length of most of last season's bearing branches was six inches to a foot or more and the leaves had healthy green color, no fertilizer is needed. Small trees may continue to need nitrogen applications. Timed-release fertilizers and slow-acting organic materials including animal manures must be used with caution, especially in cold areas. They encourage late growth of twigs and late development of fruit. This growth competes for food with flower bud formation for the following year.

Mulching

Under most conditions apple trees should be grown in sod without mulch except the in-place grass clippings. A fertilized, watered lawn is a suitable environment. In very hot, dry weather or in cold areas without snow cover, a mulch may be helpful, but remember to maintain protection against mice and rabbits and try to prevent late growth.

Space Requirements

A dwarf specimen tree needs at least a 10-foot circle of ground, while a standard type may dominate several times as much. Dwarfs grown as a hedge are set five to six feet apart with side branches pruned, spread, and tied so as to form a living wall, or planted at a 45° angle and tied to a support. Dwarf trees have been grown under crowded conditions, but far more work is necessary to keep them healthy and attractive. Espaliered trees are pruned severely to form a two-dimensional pattern against a trellis or wall (Fig. 14-11).

Training and Pruning

These chores are necessary to develop and maintain a manageable, productive tree. Training must begin at planting, but until a crop is borne, wood should be removed only for shaping and grooming. Remember that all pruning is dwarfing. The bulk of an unpruned tree will be greater than that of a pruned tree under similar conditions. Pruning is one method of dwarfing a tree or a branch, because reducing the leaf area reduces the food supply. For this reason a young, nonbearing tree is pruned as little as possible. However, shoots induced to grow by **thinning out** (taking out entire branches or shoots) and **heading back** (cutting off part of branches or shoots) are larger and more vigorous than those from unpruned branches because competition has been reduced. The implementation of these basic pruning cuts is illustrated in Fig. 12-14.

In the home garden, dwarf and spur-type trees are best trained to conical or **central-leader** form; semidwarf to either central or **multiple-leader;** and standard trees to multiple leader (Fig. 12-15). A multiple-leader tree has four **scaffold** branches arranged around the trunk with their centers at least 8 inches apart vertically. This form distributes light, lets sprays penetrate, and provides ladder holes—items that are not problems with small

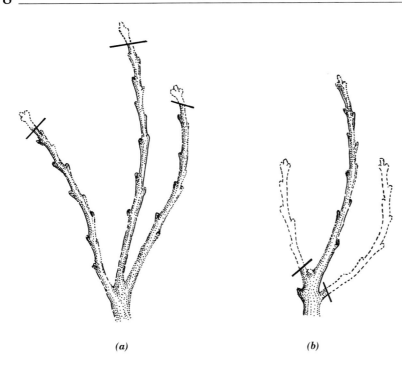

(a)　　　　　*(b)*

FIGURE 12-14 The two basic pruning cuts, heading back (*a*) and thinning out (*b*). Heading encourages lateral growth; thinning results in a more open structure. (Courtesy of Esther McEvoy.)

trees. For either type of training, a one-year whip is headed back to about 24 to 30 inches from the ground at planting.

Training a dwarf or spur-type tree to central-leader form takes careful pruning its first two to four years. During the first growing season branches will develop from the headed whip and tend to grow upward. A vigorous shoot near the top is chosen to be the leader and its top is cut back one-third to one-half to induce continued strong upright growth.

Other shoots are headed back one-quarter to one-third of their length at a bud on the upper side. The bud nearest a cut tends to make the strongest growth. Continue heading back each year until the desired height and framework are developed. While training to form, it may be helpful to remove young fruits from the leader to keep it from bending. Lateral branches should be spread while young to form an angle of about 60 degrees with the trunk. This is necessary because narrow-an-

gled branches are weak and tend to break under pressure, whereas wider angles promote structural strength (Fig. 12-16). Spread branches are held in place by homemade or commercial spreaders, tie-outs, or if trees are trained to a support of wood and wires, branches may be held in place with clothespins (Fig. 12-16). When they bear, they will be lowered further by weight of crop.

After form and size of tree have been established summer pruning will keep it in shape. Crowding branches may be thinned out and small, new, unwanted ones pinched out or encouraged to form flower buds by heading back. Where narrow crotches occur one branch should be removed or pruned severely to limit its growth. Dead branches and those growing in wrong directions should also be removed. In mild areas pruning may be done any time after leaf fall and before bloom. In severe climates wait until January or later. If possible, do not prune when the temperature is

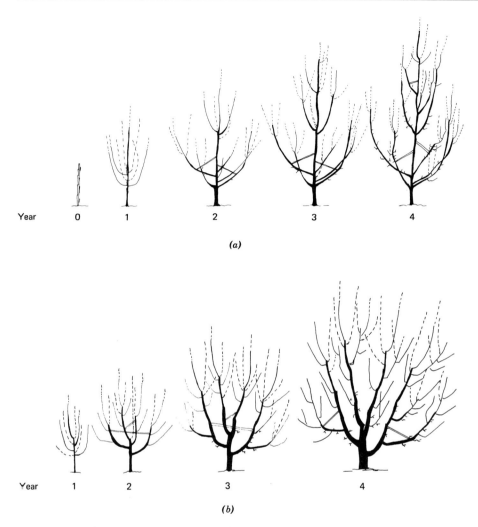

Year   0   1   2   3   4

(a)

Year   1   2   3   4

(b)

FIGURE 12-15 Training and pruning. Here trees are trained to a central leader (*a*) and multiple leaders or open center form (*b*). (Courtesy of Oregon State University Agricultural Extension Service.)

below freezing or likely to become so. Wounded (pruned) tissues are more susceptible to freezing injury than uninjured ones. Pruning in summer is not injurious and is needed with dwarf trees. Wounds heal faster in summer, but the pruning task is easier when leaves are absent. Advice has been to cut off branches flush with the branch from which they are removed. Recent observations, however, suggest that a short stub will heal faster. Suckers at the base of the trunk should be pulled off while young.

Pruning Tools

Only a few tools are necessary. Most important is a pair of strong, quality pruning shears,

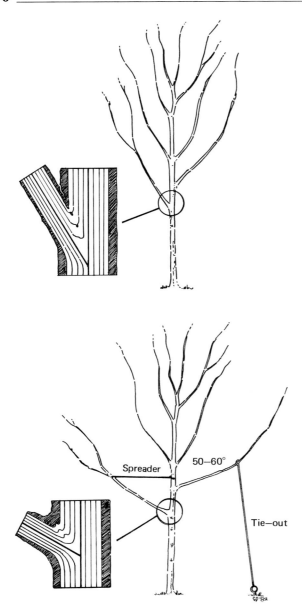

FIGURE 12-16 Proper crotch training is essential to develop strong branches. (Courtesy of Guerry Dean.)

comfortable in the hand. Much thinning out and heading back can be done with hand shears and surprisingly easy cuts may be made by bending the wood away from the shears at the point of cutting. Do not use power shears. Where reaching is necessary, a pair of lopping shears is used—strong but as light in weight as possible. Heavy lopping shears can take the joy out of pruning. The third essential is a pruning saw—a curved blade about a foot long, opening from a wooden handle, used for cuts too large for shears. For pruning taller trees use a pole saw or an orchard ladder.

Wound Treatment

In addition to the above tools you will need pruning compound and pruning paint to protect fresh cuts. Usually small cuts heal without treatment and in some areas neatly trimmed saw cuts seem not to heal faster or be protected against rots with wound dressing. In the author's garden a polyvinyl-acetate-base, plastic-type wound sealer applied with a paintbrush, is used on cuts an inch or more in diameter. Treated cuts heal without problems, but similar cuts left untreated may develop rot. If all our trees were dwarf, pruned in summer, the treatment problem would likely disappear. If sealing is practiced, use only preparations made for that purpose. Paints or sealers designed for other uses may be toxic to living tissues and delay or prevent healing.

Insects and Diseases

Apples are attacked by numerous disease organisms and insects. Consult your Agricultural Extension Service for more complete information for your locality. For many, an all-purpose spray or dust is adequate. Some of the ingredients may not be needed and therefore are wasted. Scab is the most widespread and serious fungus disease of apples in North America (Fig. 9-11). It causes dark or black, disfiguring spots on leaves and fruits. For control, in addition to sanitation, trees may need spraying or dusting with a fungicide from dormancy as long as the weather is moist. Spots may enlarge and new ones appear even on stored fruit. The "worm" in a "wormy" apple is its most serious insect pest. It is the larva of the small, bluish-gray codling moth, which lays eggs on young leaves and on or near young apples. A larva enters the fruit to live and eat there. Up to four broods of moths in a season may lay eggs, so a control program may be necessary from before bloom until harvest. Nearby vegetation—trees, shrubs, grass, and weeds—also need to be sprayed since they may be a protective winter home and present resting places for the insects.

In many areas the "railroad-worm," or apple maggot, is a serious pest. Small larvae make fine brown trails throughout the flesh of the fruit, leaving it unattractive and subject to rot. Larvae from fallen fruit enter the soil to pupate and overwinter. Most emerge the next spring to lay eggs, but some may remain in the soil to emerge two or three years later. Where practical, keeping all fruit picked up and destroyed helps in control, but is rarely enough. A spray program similar to that for codling moth is usually necessary and is included in the Agricultural Extension spray schedule for an infested area. Traps of various kinds, especially strips of sticky flypaper have been found useful.

Spider mites or red spiders are minute insects that are most troublesome during warm, dry weather, when they build up not only on apple trees, but on many other fruits, vegetables, ornamentals, and even weeds. Control is a cool rain, a vigorous hosing, or a miticide.

Harvesting and Uses

Proper handling and harvesting are important in producing quality fruit. Over 40 years ago the author was instructed, "Handle apples like eggs, only more so." That is good advice. Bruised fruit spoils quickly.

Apples picked immature can never attain best flavor. A mature apple is capable of ripening after it is picked. A ripe apple is one in prime eating condition. A fresh-picked or a stored apple often tastes better after a day or so at room temperature. When fruits are falling, they are usually ready to be picked. Apples hang with their tops (blossom ends) down. A mature one grasped and bent upright usually breaks off. Be careful not to pull the apple from its stalk, leaving a wound open to infection and spoilage. Also, the stalk is important in identification of cultivars and must be present on fruit for exhibition.

Apples stored for use from harvest to harvest keep well at 31 to 32°F. When this temperature cannot be maintained, strive for one not over 40°F but do not expect the fruit to keep as long as at lower temperature. Apples kept outdoors must be protected against cold. Freezing can start if the temperature of the fruit drops to 28°F. If you think apples have been frozen slightly, do not move or handle them for a few days. They may be usable.

Apple is one of the few fruits with an entire cookbook* devoted to it. There are dishes for any course in a dinner menu or for any meal of the day. The use of apples in sauce, pies, and other baked goods, jellies, juice, cider, vinegar, mincemeat, apple butter, and other dishes is well known. Home apple drying has become a popular activity. An instant applesauce can be made by putting home-dried apples through a blender before storing to save shelf space. When sauce is wanted, simply add boiling water and stir. To many, the highest use of apples is in pie. For people on low-fat diets fruit prepared as for pie but baked without crust in a deep covered dish are delicious and healthful. Choice cultivars eaten out of

*Demetria Taylor, *Apple Kitchen Cook Book*. Popular Library, New York, 1971.

hand provide the best of snacks and an apple at the end of a meal is often recommended as a substitute for tooth brushing.

Apples contain vitamin A and potassium and are low in sodium. Certain triploid cultivars contain more vitamin C than diploids. They are a source of pectin for both home and commercial jelly making. Scraped or blended fresh apple has been used in the treatment of digestive disorders, calcium pectate or pectin being considered the healing factor. Unharvested apples supply survival rations for birds and other wildlife.

From earliest times apple trees have added their special beauty with utility to dooryards and farms. Now we have a new assortment for changing life-styles. Those on growth-controlling stocks with their range of sizes and predictable behavior are tailormade for the home garden. All have ornamental value. Almost any garden has room for one or more small trees, at least one in a container, a conical dwarf or semidwarf as an accent plant, an espalier, or a bit of a living fence. Larger sizes also may provide bloom, shade, screening, framing for a building or a view, or other landscape function. Changes in form, color, texture, and pattern present year-round interest and often are hard to beat for downright good looks and attention getting.

Pears

Description

The pears, genus *Pyrus,* are woody perennials also in the rose family. Well-known cultivars with melting, "buttery" fruits, mostly pear-shaped, are descendants of *P. communis,* disseminated in Europe in ancient times from the region of the Caucasus Mountains. This group, sometimes called French pear, contains some of the world's choicest fruits. In North America a few grow in zone 5 and most in zones 6 and 7 (Fig. 6-11). There is a pear

cultivar for almost every home garden where apples may be grown.

The sand pear or pear apple, *P. pyrifolia* (formerly *P. serotina*), from eastern Asia, is distinguished by apple-shaped fruits with firm, gritty, sometimes crisp, and juicy texture. Needing little or no cold period to break dormancy, its cultivars are important south of the range of *P. communis.* Chinese and Japanese cultivars from *P. pyrifolia* are cherished in California by settlers from eastern Asia and their descendants. While not immune to fire blight, Asiatic pears are resistant and suffer little from this bacterial disease. The Callery pear, also from eastern Asia, is becoming more and more important as a rootstolk, especially for Asiatic cultivars.

Standard pear trees may reach a height of 25 feet or more and have a spread of about 20 feet; dwarf trees may be half this size. The deep-green, shining, simple leaves of many cultivars turn red in autumn. Masses of flowers, white or tinged with pink, appear in spring a little before bloom in apples, whose flowers they resemble in form, size, and number of parts. Pear cultivars may have pink anthers. Three to seven flowers form a branched cluster whose lowermost flower opens first, making the cluster a corymb (Fig. 3-15). The pear fruit is a pome, its flesh characterized by clusters of thick-walled cells that impart a gritty texture. Fruits are borne chiefly on long-lived spurs, much like those of apple.

Most pears are cross fruitful except 'Bartlett' and 'Seckel,' two somewhat self-fruitful cultivars. Some are self-fruitful in southern parts of their range, but most do better with a pollinizer. Bees sometimes pass over pear blossoms for others more to their liking.

Propagation

Pears, like apples, do not come true from seed so cultivars must be propagated vegetatively. Standard trees are budded or grafted on seedling roots, those of *P. communis* being used the most. The sand pear, *P. pyrifolia,* formerly an important stock because of its ability to tolerate both heat and cold, is no longer used, but mature trees on sand pear roots are still found in home gardens. Cultivars of common pear grafted on sand pear roots are susceptible to pear decline, a destructive mycoplasm disease. Callery pear seedlings are now a preferred stock, since they are resistant to pear decline. They are well anchored, tolerate both wet and dry soil, and are almost as hardy as peaches.

Dwarfing Stocks

Strains of quince roots, *Cydonia oblonga,* called, 'Angers' and 'Provence,' are used for dwarfing pears. Clones from East Malling Experiment Station, Quince A and Quince C, both English selections of 'Angers,' are good stocks for home gardens in mild climates where fire blight is not a problem. They are about as hardy as peaches. 'Bosc,' 'Duchess,' and a few lesser known cultivars may be grafted directly on quince, but many common cultivars, including 'Bartlett,' are graft-incompatible and need an interstock. A dwarf tree offered for sale has Quince A roots, an eight-inch stem piece of 'Old Home,' and the cultivar. Trained as a bush, with more than one trunk, it will become about 15 feet tall with about the same spread. Staking is needed. Rootstocks of 'Old Home x Farmingdale' may be more or less dwarfing than Quince A. Compatible with many cultivars, long-lived, blight resistant, and cold hardy, they deserve further trial.

Culture

Common pears, like apples, are temperate-zone fruits requiring cool, cloudy summer temperature for growth, and most needing winter cold to break rest. Freezing injury to pear wood limits their growth northward in North America and hot, moist summers favoring fire blight limit them southward.

Common pears grow best in rich, deep, heavy, but well-drained loam. They withstand considerable soil moisture, but size and quality of crop improve with adequate drainage. They require a dependable water supply with gradual withholding in fall.

The planting of pears is similar to that of apples and other trees. The graft union must be above ground to prevent root development on the interstock or the cultivar. Each tree or shrub needs a bare zone about its trunk for rodent control. It may be grown successfully in sod, a fertilized and watered lawn being suitable.

Pears thrive on a somewhat richer soil than apples, usually needing a moderate application of a nitrogen fertilizer in early spring, applied in a broad ring extending a little beyond the spread of the branches. However, if the bearing branches of a tree are making 6 to 12 inches of shoot growth each year, nitrogen is not needed. Trees in fertilized, watered lawn areas usually make sufficient growth without extra fertilizer or irrigation. Certain pear cultivars including 'Magness' and 'Seckel' seem able to use more nitrogen than others. The home gardener must work for moderate, not vigorous, growth in pears. The more vigorous the shoot, the more susceptible it is to fire blight.

The desirability of mulching depends on the location. Will the mulch improve moisture and nutrient supply? Protect against root freezing? Or will it promote late or soft growth, increasing susceptibility to fire blight or winter injury?

Prune newly set trees only enough to compensate for loss of roots in transplanting. After the first season pruning is necessary to maintain central leader form (Fig. 12-15), but prune young, nonbearing trees as little as possible, and in thinning out and heading back mature trees, avoid forcing vigorous shoot growth. Pruning only for shaping and grooming, while keeping the growth moderate, promotes pear tree health.

For a given pear cultivar, a large pear usually tastes better than a small one. As it matures, it stores sugar and develops best flavor. When fruit is heavy, thinning out after the June drop is needed to prevent limb breakage and improve quality. Imperfect, small, and excess pears should be removed, leaving fruits no closer than six inches apart. Fruit from dwarf trees is often larger than that from standard trees. At harvest time pick the largest fruits first. Those left a few days will continue to size up and sweeten.

Pests

The most important diseases of pears are fire blight, decline, and scab, and the worst insects, codling moth and psylla. Controlling these helps control others. Fire blight, a bacterial disease affecting apples and pears, but in most areas less destructive to apples, forms lesions in bark and causes leaves to dry and turn brown, looking as if they had been scorched. Tips of growing shoots bend over and hang down. Vigorous growth and warm, humid weather favor development of the disease. Pruning out lesions as soon as observed while disinfecting tools before each cut is recommended. Bacterial sprays may be necessary. Choosing a healthful area and planting resistant cultivars are most helpful. Decline mycoplasma, which defoliates and kills trees, is spread by the insect pear psylla. Rootstocks of sand pear, *P. pyrifolia,* are to be avoided. The insecticides recommended to control psylla and their application are regulated by law in some states. Biological control has given favorable results where tested. Scab, a fungus disease favored by moist weather, causes black, sometimes deforming spots on leaves and fruits. It is controlled by one to three or more sprays of a fungicide. Codling moth in pear is similar to that in apple. From one to seven sprays of insecticide may be needed to prevent "wormy" fruit. Obtain the spray schedule of the Agricultural Extension Service in your area for details.

Harvesting and Uses

Besides the nature of the cultivar and good culture of trees in a favorable climate, the most important factors in pear quality are proper harvest time and ripening. Most common pears must not be tree ripened. They are picked when the fruit is still green or a small change in color toward lighter or yellow-green appears. When they begin to fall, they are usually mature and will break off when the fruit is bent upright with a slight twisting motion. Handle even more carefully than eggs or apples.

To ripen 'Bartlett' and most other pears, keep at room temperature (68 to 70°F). Use when the flesh near the stalk yields slightly to pressure and they taste good. Do not wait for the skin to turn yellow, as they soon become soft and brown inside. Certain winter pears reach best quality when ripened after a few weeks in cold storage. 'Bosc' is one that should be tree ripened and does not need cold storage.

Before the days of sophisticated food processing and storage, pears were made into perry, a juice similar to apple cider. Perry was especially well known in France and colonial America. Now becoming popular again, it is good fresh, canned, frozen, or blended, and in recipes calling for cider or wine.

Pears peeled, cored, halved, dipped in brine, and dried, are a choice confection and snack, light in weight for carrying and stored without refrigeration. Like apples, dried pears may be made into instant sauce.

Pear trees have various forms, masses of flowers, autumn color, interesting winter patterns, and fruits varying in size from tiny 'Old Home' to hefty 'Pound.' They may function like any other tree or large shrub in the home garden where serious diseases are absent or can be controlled. They make excellent specimen trees, hedges, or espaliers, 'Tyson' being one of the best for training to special forms. A standard or dwarf tree of 'Howell' is said to be exceptionally ornamental. In some cities the 'Bradford' pear and other selections of tiny-fruited *P. calleryana* are used in parks and as street trees.

Peaches

Description

The peach, yet another member of the rose family, is an ancient fruit, cultivated in eastern Asia before written records and spreading westward across Asia, Persia, and the Caucasus region into Europe. It was brought to North America early by both English and Spanish settlers. Next to the apple the peach is the most widespread temperate-zone tree fruit in North America and in the world. Its high status has given us the word "peachy." Peaches are grown in parts of Canada, in all 48 states of the continental United States, and in Mexico.

There is only one species of peach, *Prunus persica*, believed by ancient Europeans to have originated in Persia. Races described, but not clearly delimited, include North China or Chinese Cling, the South China or Honey, and the Peento race. All have been hybridized. Many of our best-known cultivars are descended from Chinese Cling, introduced in 1850 from China by way of England. Other oriental peaches were brought to Mexico by the Spanish and have been important in early breeding programs. The South China or Honey peach is a very sweet, beaked fruit, while the Peento race has fruit flattened end to end with a flat seed. Honey and Peento cultivars and hybrids are important in areas where there is insufficient cold weather to break the rest period of common cultivars.

Peach trees are small by nature, seldom reaching height and spread of more than 20 feet—not too large for many home gardens. By severe pruning, which they need, they may be kept even smaller. Standard trees are usually cultivars grafted on nursery-grown peach seedlings. Recommended selections for rootstocks are 'Siberian C' seedlings in the North

and 'Nemaguard' (nematode resistant) in the South.

Peaches dwarfed by grafting on roots of sand cherry, *Prunus besseyi;* Nanking cherry, *P. tomentosa;* and the 'St. Julien' plum are not usually satisfactory except in colder, peach-growing areas, in dry-soil regions.

Genetic dwarfs trained as shrubs grow four to eight feet tall under field conditions. A few are as hardy as common cultivars. Widely known bush types produced in California from cold-tender, naturally dwarf Chinese peaches have showy semidouble flowers and are favorite container and landscape plants in mild areas. When blooming in a greenhouse, they may need hand pollination for fruit production.

The peach lacks long-lived fruit spurs like those of the apple and pear, instead bearing from flower buds formed on the growing shoot the previous season (Fig. 12-17). In winter examine the youngest (terminal) growth of a peach tree. Find the leaf scars, each of which shows the position of a node. At many nodes you will find three buds just above the leaf scar, a slender one with a plump one on each side of it. In early spring each plump bud will open to form one or two pink flowers. Typically each has five green sepals, five (or more) petals, many stamens, and a pistil of one carpel with one style and stigma. Some cultivars have double or semidouble flowers. One or occasionally two seeds develop in a sculptured pit, which is the endocarp of the fruit, not part of the seed. The simple, fuzzy fruit is a drupe.

The flesh of a peach may be white or yellow, rarely red, and may separate readily from the pit (**freestone**) or adhere more or less firmly (**clingstone**). Most home gardeners prefer yellow freestones, but there are also excellent white and clingstone sorts.

Of the many cultivars of peach, some are adapted to local areas and others to a wide range. New ones appear often, as several experiment stations and private breeders have active breeding programs. To find out which

Age of wood
(years)

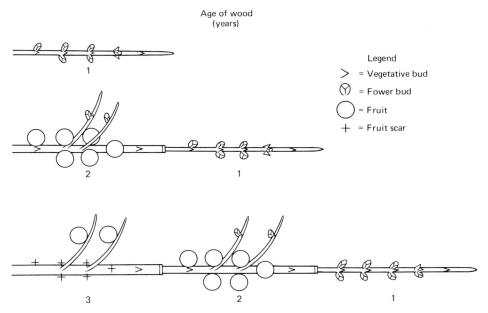

Legend

> = Vegetative bud

ⓨ = Fower bud

◯ = Fruit

+ = Fruit scar

FIGURE 12-17 Fruiting habit of peach. Fruit is borne on the second season's wood, which is itself developed during the previous season. (Reprinted with permission of McGraw-Hill Book Company from J. B. Edmonds, T. L. Senn, F. S. Andrews, and R. G. Halfacre, *Fundamentals of Horticulture,* 4th ed. Copyright © 1975 by McGraw-Hill, Inc.)

ones are successful in your area consult the Agricultural Extension Service and local nurseries.

As most peach cultivars are self-fruitful, usually only one tree is necessary for production. Old cultivars 'J. H. Hale' and 'Early Crawford,' however, are self- and cross-unfruitful and need a pollinizer, any other in bloom at the same time. Even with self-fruitful cultivars, however, bees and weather favorable for flight are necessary for good pollination.

Culture

As trees of *P. persica* are not so hardy in wood and buds as apples, they are grown where the winter temperature is not too low, but where there is enough cold to break a rather short rest period. The ideal climate is one with mild winters, minimum temperature not below −15°F, a few weeks at or below 45°F, warm dry springs for successful pollination and fruit set, and hot sunny summers. Zones 5 to 7 and parts of zone 8 are most favorable (Fig. 6-11).

Peach buds are easily injured by a cold spell after unseasonable warmth in winter has allowed buds already out of the rest period to begin growth. Also, flower buds become progressively more sensitive to cold in spring as they swell, open, and set fruit. A frost just after bloom may be more harmful than the same degree of cold a few days earlier. However, it is said that a peach tree can lose up to 95 percent of its flowers and still produce a moderate crop. To reduce the chances of frost damage choose a slope with good air drainage. On a frosty night one or a few trees may be protected by covering or by misting. A plant in a container may be moved inside.

Light or medium sandy loam or stony soil that holds heat is desirable, provided it is deep, well drained, and has water available throughout the growing season.

Clean cultivation with or without cover crop in late summer is the rule in peach orchards, but in the home garden mulching with organic material or black plastic sheets may save labor, control weeds, conserve moisture, and add warmth. Extra fertilizer is needed under a mulch high in organic material such as straw. Soil should be prepared before planting with incorporation of organic matter and complete fertilizer, lime, and other nutrients as needed. Each year peach trees require nitrogen for vigorous growth.

Each slender bud on a peach shoot contains a growing point capable of becoming a leafy branch. A section of branch produces leaves only once—in its growing season—and fruit only once—its second season (Fig. 12-17). Yearly production of fruit demands a yearly supply of new vigorous shoots. Trees are trained with about four scaffold limbs around an open center, needed to let in light and to make room for new growth (Fig. 12-15). Branches are thinned out to reduce competition and headed back one-third to one-half their length to improve size of fruit. Weak and interfering branches are removed first. Severe pruning annually is necessary to stimulate vigorous new growth. When a heavy crop sets, thin fruits to improve size and prevent breakage of branches. When young fruits are a little over half an inch in diameter, remove the small, defective, and excess, leaving one fruit about every six or eight inches (Fig. 12-18).

Pests

Troublesome diseases of peaches are brown rot, coryneum blight, and peach leaf curl. Each disease has its own calendar of development,

FIGURE 12-18 Peach before and after thinning.

demanding careful timing of control measures. To control coryneum blight, apply a spray of copper-containing fungicide after harvest, but before the fall rains begin, but not later than Columbus Day. To prevent leaf curl, apply copper sprays before bud scales have started to separate—at Thanksgiving, on Christmas or New Year's Day, and possibly again on Groundhog Day. Even with winter treatment, brown rot sprays (captan or copper fungicide) are needed at the popcorn stage and at petal fall, also in summer when the weather is moist to protect the ripening fruit. The Agricultural Extension agents for your locality can provide more details on control of these pathogens.

Spray programs for peaches also include control of borers and curculio beetles. With one or a few trees, borer larvae may be dug out with a jack knife, if **frass** (sawdustlike material) is seen near the soil line (sometimes below), on the trunk, or at the base of large branches. Curculio beetles may be shaken onto tarps and destroyed. Home gardeners with spare time might try wrapping fruits, as do the Chinese and Japanese. A few types of cover are under test and some are marketed.

Harvesting and Uses

Many occasions are celebrated with peach shortcake or cobbler, but the ripening of a choice peach cultivar is itself sufficient reason for celebration. Prime-ripe peaches are one of the world's most delicious fruits. Canned and dried peaches are useful items for your emergency shelf and the frozen product serves as fresh fruit. Peach leather is a good snack or backpack item.

Peach leaves and wood must not be used for tea or for nibbling by man or beast, as these tissues contain the potentially danger-ous glucoside, amygdalin, one of whose break-down products is cyanide.

Any well-cared-for peach tree or shrub is an attractive ornamental. However, it must be placed where it can receive the care necessary to keep it attractive. A tree suffering leaf curl or defoliation from other causes can spoil a landscape picture. 'Early Elberta' has larger, more showy flowers than many others. The semidouble ones of certain dwarfs such as 'Bonanza' have also received attention.

Nectarines

Description

Nectarines are regarded simply as fuzzless peaches. Their close relationship is illustrated by the fact that a peach seed may produce a nectarine and vice versa. A peach tree may produce a nectarine bud sport, but a mutation back to peach has been observed only if the nectarine arose as a bud sport of a peach. Of the countless nectarine trees that have origi-nated from peach seeds, none has ever been seen to produce a peach bud sport.

Hortus Third lists nectarines as *Prunus persica* var. *nucipersica*, indicating its close relation-ship with the peach, *P. persica*. Nectarine fruits tend to be smaller and firmer then those of peach, with richer aroma and flavor, charac-teristics believed correlated with smooth skin.

Many cultivars are grown in California, the leading nectarine state, where the ripening season extends from early June into Septem-ber. Fewer kinds are found in other peach-growing areas. Genetic dwarfs, as well as stan-dard trees, are available. *Hortus Third* rec-ommends cultivars for each growing area.

Culture

The culture is the same as for peaches and even more care is needed. The two crops are subject to the same diseases, but most nectar-ine cultivars are more susceptible to brown rot. In climates with warm, moist springs and summers both blossoms and fruits must be sprayed for brown rot control.

Nectarines are currently increasing in pop-ularity as a seasonal fresh fruit. Essentially all of the commercial crop is sold on the fresh market. However, they have all the uses of the peach.

Cherries

Description

Cherries, too, belong in the genus *Prunus* of the rose family. They are trees and shrubs characterized by small round drupes without waxy bloom and with round plump pits. Flower parts are similar to those of peaches. The twigs and simple, toothed leaves should not be eaten, made into tea, or grazed; like peaches they are potentially poisonous, especially when bruised.

The two leading species, *P. avium*, the sweet cherry, and *P. cerasus*, the sour or pie cherry, probably originated in the Caspian-Black Sea region of southwestern Asia and southeast-ern Europe, where they still grow wild.

Cherry culture had spread throughout much of Asia, Europe, and North Africa before there were written records. North American settlers carried cherry stocks across the continent. In 1847, the Llewellyn brothers started a nursery in Milwaukie, Oregon, where they developed the 'Bing' and 'Lambert' cherries, two of America's leading cultivars. Their successes marked the beginning of commercial and scientific cherry growing in North America.

Pie cherries are home garden fruits in much of zones 4 to 6 and at higher elevations farther south (Fig. 6-11). Sweet cherries are more limited in distribution, being adapted to locations on the West Coast, near the Great Lakes, and in New York's Hudson Valley besides smaller areas elsewhere. The very large trees of *P. avium* are about as winter hardy as peaches, but do not withstand so much heat. They need both mild winters and mild summers. Rain toward harvest time may cause destructive cracking of fruits followed by brown rot infections. Where there is room for only one tree of *Prunus avium*, more than one cultivar may be budded into it, or 'Stella' or 'Compact Stella' may be planted.

Sweet cherries are firm-fleshed but juicy. They may be light—yellow or cream with or without red blush, or dark—red to very dark red. Most cherries are dark.

Most sweet cherries are self-unfruitful, and all are compatible with any other except in the case of three leading cultivars, 'Bing', 'Lambert', and 'Royal Anne'. These are not only self-unfruitful, but each is cross-unfruitful with the other two. Other old and new cultivars are compatible. The old 'Black Republican' and 'Yellow Spanish' are good pollinizers and are among the more hardy sorts. 'Stella' and 'Compact Stella,' are self-fruitful. 'Garden Bing' is smaller, but needs a pollinizer. Bees are necessary for pollination with both self- and cross-fruitful cultivars.

The climatic areas favorable to *P. avium* suit also the Duke cherries, *P. avium* var. *effusus*, hybrids between sweet and sour cherries. These are good home garden fruit, yielding juicy red cherries that ripen over several weeks beginning in June. They need little or no sugar and are delicious fresh, cooked, in pies, frozen, or dried. The tree in bloom looks like a huge popcorn ball. Self-fruitful in mild areas,

it needs an early blooming sweet cherry as pollinizer in colder regions. Being hybrid, they are usually poor pollinizers and form few good seed. Trees are medium large, ornamental, and the cherries are very attractive to birds. They may serve as a trap crop, to protect other fruits ripening at the same time.

Sour cherries, *P. cerasus*, grow wherever apples grow and withstand dryness better than other tree fruits. These cherries are tart, bright red, juicy, and resistant to cracking. Genetic dwarfs of pie cherry are available for small areas or containers. Like other genetic dwarfs these may be kept smaller in containers. Cultivars of *P. cerasus* are self-fruitful, but bees are needed as pollinators.

The Nanking or Chinese bush cherry, *P. tomentosa*, often grown as an ornamental, yields edible fruit resembling pie cherries. These are useful, dual-purpose, medium to tall shrubs hardy in severe areas. Hansen bush cherries are selections of *P. tomentosa*.

The North American native western sand cherry, *P. besseyi*, also is useful in difficult areas. The four-foot bush is very hardy and yields small, dark, sometimes good-flavored fruits.

Cultivars of sweet cherry are usually budded on seedlings of *P. avium* called mazzard, a rootstock that does not sucker. Sour cherries may be budded on mazzard or *P. mahaleb*, the perfumed cherry. Those on mazzard are larger and longer-lived where they will grow. Special strains of mazzard or other stocks may be found in local areas. Genetic dwarf cherries are also budded, the easiest way to propagate them. Dwarfing roots are not needed since the scions are already dwarfs.

Culture

For sour cherries choose a site suitable for apples; for sweet cherries, one for peaches in a mild-summer area. Any good garden soil is suitable if it is deep and well drained. Sweet cherries prefer light loam. Preparation before planting should be made as for other fruit trees.

Planting is done in spring in cold climates, elsewhere in spring or fall. Handling and planting is the same as for other fruits. Sunny, uncrowded space must be provided, sweet cherries needing a 30-foot circle to dominate, while the smaller sour cherries need 20 feet.

Trees are trained to multiple leader form (Fig. 12-15). Because wood of sour cherries is brittle, young trees need their branches spread to develop wide angles with strong crotches (Fig. 12-16). Pie cherry trees are not for climbing; large branches may break off. Certain sweet cherries also tend to form narrow crotches and need spreading, for example, 'Lambert.'

In the home garden cherry trees are usually pruned very little. Since fruit is borne on long-lived spurs, severe pruning each spring to induce vigorous new growth is not necessary. Dead and interfering branches are removed and some thinning out and heading back may be needed to control growth and admit light.

In an orchard situation cherries may be treated the same as peaches, but are far less demanding. They may also be grown in sod with a bare area maintained around each trunk or along the tree row to control weeds, make mowing easier, and protect against rodents. A fertilized, irrigated lawn is suitable and extra fertilizer and water may not be necessary.

Like other fruit trees, cherries respond chiefly to nitrogen and a moderate application may be made each spring unless shoot growth the previous season was overvigorous, averaging more than 18 inches. Large, old, sweet cherry trees may need nitrogen to improve set and size of fruit.

Ordinarily thinning of cherries is not worthwhile. However, to produce fruit for exhibition it might help to reduce an extra heavy set as early as possible.

Pests

Cherries may have few or many problems, depending somewhat on geographical area. Birds may be the worst pest, sometimes taking the whole crop of a small planting. Covering trees with nylon net made for the purpose is the best protection. Because cherries are susceptible to certain virus diseases, it is important to start with trees certified virus free. A proprietary spray for stone fruits is often adequate for control of pests of cherries in the home garden. It may contain captan, malathion, and diazinon or methoxychlor. With cherries it is important to time spray applications in relation to harvest and at the same time control late broods of the cherry fruit fly, the most frequent insect pest, with curculio a possible runner-up east of the Rocky Mountains. In some localities the Agricultural Extension Service sends out dates of the emergence from the ground of the fly. Ask to receive notices for cherry growers and obtain the spray schedule for your area, which will also include control of diseases. Cherries, like peaches, are susceptible to brown rot, and in moist areas to bacterial canker with gummosis and dead bud. A control for the latter destructive disease is a spray of Bordeaux mixture 12-12-100 around Columbus Day, as soon after New Year's Day as possible, and in spring as soon as the flower buds begin to show color, plus pruning out dead branches using disinfected shears. Cultivars vary in susceptibility; 'Meteor' is said to be resistant. Brown rot sprays, as for peaches, will also control cherry leaf spot disease. Black knot of plum and cherry, which forms black, crusty growths along branches, is controlled by a dormant spray of a copper fungicide after the pruning out of knots from garden trees and wild cherries in the neighborhood. The most important single spray for cherries is the one just before bloom when the buds begin to show color. A cover spray after petal fall may also be needed. San Jose and other scale insects may need special attention in some areas.

Harvesting and Uses

Cherries are tree ripened and for cooking and canning are picked firm-ripe. For fresh use they may hang on the tree longer. They do not improve in flavor after harvest and soon lose best flavor even in cold storage.

Tree-ripened cherries are choice dessert, salad, and decorative fruit. Cherry pie rivals apple pie in popularity. Cherries baked without crust are also delicious, and a dish containing both sweet and sour cherries is a real treat. All kinds are useful for cooking, freezing, drying, and making fruit leather, also for juice, syrups, and toppings. Most are excellent for canning. Maraschino cherries may be made at home, but at present the question of a suitable red dye is controversial. Sweet cherries with stalks left on can be washed, put in plastic bags, and frozen without sugar or sirup to be eaten as after-school snacks.

Cherries have ornamental value, some more than others. In severe climates, native species and the Nanking cherry are especially valuable; medium to tall shrubs or small trees are often used in hedges and as windbreaks in the Central and Great Plains. However, in small gardens thought must be given to falling fruits, dusts or sprays, and visiting birds in or near living areas.

In the author's garden a big, old Mazzard seedling with small dark cherries is helpful in keeping birds away from other cherries and especially from nearby blueberries.

Plums

Description

The plums constitute yet another group of the genus *Prunus* in the rose family. They are distinguished by drupe fruits with **bloom,** a waxlike coating over the skin, and elongate, flattened, unsculptured pits. Flowers are similar to those of peaches and cherries. Leaves are simple.

Our most important species, *P. domestica,* was cultivated for plums and prunes in the region of the Caucasus Mountains and the Caspian Sea at least 2000 years ago, and it still grows wild there. Spread in Europe by invaders from Asia, European plums were brought to eastern Canada by the French and to southern United States by the Spanish. The Pilgrims and other early comers grew them in the American colonies.

Plums are the most variable of the tree fruits. Distinct groups within *P. domestica* plus other species and hybrids provide plants suitable for almost every home garden in north temperate America. *P. domestica* include the prunes, the Lombard types, the Green Gages, the Yellow Egg group, and others. Also important are the Japanese plums, *P. salicina,* the Damsons, *P. insititia,* the cherry or Myrobalan plums, *P. cerasifera,* and several native American species.

Prunes. Prunes are cultivars with blue or purple fruit, firm flesh, and high sugar content, suitable for drying whole, or cut in halves, but also an excellent all-purpose fruit. Canned Oregon purple plums are fresh prunes. Large quantities of prunes are dried in the West, California producing French or sweet prunes, many of which are sun dried, and Oregon, mostly tart Italian prunes, dryer dried. Familiar cultivars are 'Italian Prune,' 'Brooks,' 'German Prune,' 'Stanley,' and the sweet 'Agen.'

Lombard plums. The Lombard group has the largest trees (up to 20 to 25 feet) with some of the largest fruits, most of them reddish-purple in color muted with bloom. A few garden favorites are 'Bradshaw,' 'Lombard,' 'Pond,' and 'President.'

Green Gages. The Green Gage group includes cultivars with mostly small green or yellow plums of excellent flavor—good for dessert, cooking, and canning. A well-known cultivar called 'Green Gage' or 'Reine Claude' is considered the standard of quality among plums.

Yellow Egg plums. The Yellow Egg group is typified by a cultivar of the same name. Besides 'Yellow Egg,' 'Golden Drop' and 'Red

Magnum Bonum' belong here. These large, attractive cooking and canning plums are very hardy and thrive where better cultivars do not perform as well.

Japanese plums. Next in popularity to cultivars of *P. domestica* are those of the Japanese or Oriental plum, *P. salicina*, which includes red and yellow-skinned and some red-fleshed kinds that are attractive, good-flavored, all-purpose fruits. Cultivars include 'Abundance,' 'Burbank,' 'Elephant Heart,' 'Santa Rosa,' 'Satsuma,' and 'Shiro.' Most are about as hardy as peaches or a little less. They withstand more heat than European plums, so may be grown farther south if diseases are controlled. Trees are small, requiring a space about 15 to 18 feet in diameter.

Damsons. Closely related to *P. domestica* is the Damson group, *P. insititia*, with small trees and small, dark blue, tart fruits—old favorites for sauce and jam. More hardy than most *P. domestica* cultivars and largely self-fruitful, they grow farther north than most cultivars of *P. domestica* and only one tree is required. Coming almost true from seed, they have formed thickets in many places. Cultivars are 'Blue Damson', 'Shropshire' (best known), and the larger 'French Damson.' 'Mirabelle' is a yellow-fruited cultivar.

Native species. Native North American plum species offer selections for home gardens in difficult areas: *P. nigra* in the North, *P. americana* from Canada to Florida and from Alaska into Mexico, *P. subcordata* in the West, and the wild goose plums, *P. hortulana* and *P. munsoniana*, in much of the Central and Great Plains. The beach plum, *P. maritima*, a medium shrub, is used as an ornamental and sand binder along the Northeast coast and adjacent islands. Plum Island, off Massachusetts, is a namesake.

Fruits of native species tend to be small and variable, but excellent for jam. A few cultivars are listed in *Hortus Third*, and some may be found in nursery catalogs. Hybrids of North American natives, especially *P. besseyi* and Japanese cultivars, have the hardiness and disease resistance of the natives and approach the quality of the Orientals. These increase the range of dessert plums both northward and southward.

Propagation of most plums is by budding cultivars on stock of the Myrobalan or cherry plum, *P. cerasifera*, of Eurasian origin, which is widely adaptable to cultivars and soils. Selected clonal and hybrid stocks are resistant to root-knot nematode, bacterial canker, and the oak root fungus. Myrobalan stock is not dwarfing, but most plum cultivars are naturally small trees or shrubs, needing an area only 15 to 18 feet in diameter. Cherry plums, *P. cerasifera*, are not to be confused with the cherry-plum cultivars that are hybrids of the Japanese plum, *P. salicina*, and the western sand cherry, *P. besseyi*. Plums may also be budded on seedlings of peach. These are suited to areas where peaches succeed. Native species are grown from root cuttings, divisions, or seed.

Domestica plums need moderate climate—free of extremes of heat and cold, wetness and dryness. They are a little more hardy than the peach. Somewhat more cold resistant are the Damsons and growing even farther north is *P. nigra* (to zone 2) and *P. americana*. Other natives resist both cold and dryness. Somewhat less hardy than the domesticas are the Japanese plums, most kinds adapted to zones 5 and 6 (Fig. 6-11). Many cultivars thrive in California and parts of the Southwest. Hybrids of these and *P. besseyi* are resistant to both cold and heat, growing both north and south of the range of *P. salicina*. The range of the very hardy *P. americana* extends into the Gulf states and Mexico.

Most European plums are both self- and cross-fruitful. Apparent unfruitfulness is sometimes lack of bees. Cultivars of *P. domestica* and of *P. insititia* are hexaploids and are compatible with each other, but neither is compatible with the diploid *P. salicina* and other diploids. Most Japanese plums are self-unfruitful and all set better when another *P. salicina* cultivar is in bloom nearby.

Culture

Culture of plums is similar to that of peaches and cherries. Train trees to multiple leader form (Fig. 12-15). Domesticas and Damsons produce fruit on spurs, so need little pruning, but Japanese plums bear on spurs and also on one-year wood, like peaches. Vigorous shoots need constant renewal with thinning out and heading back. To save space, it is sometimes suggested that three compatible cultivars of plum be planted in the same hole. Some cultivars may set heavily; depending on possible size of fruit and its use, thinning to 4 to 6 inches apart very early may be worthwhile. Plums have about the same pests and diseases as peaches and cherries and control is similar.

Harvesting and Uses

Plums, like cherries, do not improve in quality after harvest and soon lose best flavor in storage. Besides the uses already mentioned, plums (or almost any other fruit) are delicious in graham cracker pudding. To prepare, pit, cut up, and sugar a quantity of fruit. Added cinnamon is good with fresh or dried, cooked prunes. Place a layer of buttered graham crackers (crumbled or not) in a baking dish. Add a layer of plums and a layer of crackers until all fruit is used. Bake 25 to 35 minutes in a moderate oven. Serve hot or cold with ice milk and a spoonful of honey or other topping.

Plums have ornamental value where disease can be controlled and falling fruit is not a problem. Common ornamentals with usable fruit and valuable rootstocks are the purple- and red-leaved forms of *P. cerasifera*. The Japanese plum, 'Burbank,' pruned judiciously forms a small, attractive accent or specimen tree. American natives have many landscape uses in severe climates. Plum thickets provide food and protection for wildlife. Plums for exhibit at shows or fairs should have the coating of bloom present. Do not wash or wipe it off. Handle fruit very carefully.

GROWING NUT CROPS

Possibilities for home nut-tree cultivation include the almond, walnuts, hickories and pecan, hazelnuts and filberts, and chestnuts. Most of these can be grown throughout much of the temperate United States, although almond has a very restricted adaptability to moderate, wet winters with warm, fairly dry summer areas, where blossom-damaging frosts can be avoided in the late winter bloom period. Pecans will not mature its fruit without a sufficiently long growing season, about 160 to 180 days for trees from a northern origin to 190 to 220 days for those from a southern provenance.

In the home garden, nut trees usually can be treated in much the same manner as typical shade trees. Training and pruning need be far less rigorous than with fruit such as peach or apple, although almond does need attention to proper framework branching during the first few years. Fertilization and irrigation usually can be part of the general lawn and garden program. Disease and insect control may be necessary but not on the regular schedule necessary for most tree fruit. The husks of many nuts can cause a mess under the trees, especially the spiny chestnuts and staining black walnut. Compared to other tree fruit, it takes a long time for most nut trees to bear their first good crops, particularly hickories and pecans. Bars to self-pollination in various nut trees make it necessary to plant more than one tree.

Besides providing a supply of otherwise expensive or commercially unavailable nuts, many of these trees are excellent ornamentals. Almonds, about the stature of closely related peaches, are attractive in bloom, which occurs earlier than any other showy flowered tree. Hickories, especially shagbark and shellbark, and pecan have long been used as tall shade trees, even though they are rather slow growing. The medium-to-large chestnuts form rounded crowns of attractive large, dark green, glossy foliage. Relatively small hazelnuts and large walnuts have little ornamental value. Except where almonds can be grown, nut trees that are to be grown for their ornamental qualities as well as fruit are definitely trees for large properties.

Almond

The almond, *Prunus amygdalus*, of the Rosaceae is an ancient food plant resembling the peach in tree characters and fruit structure. The "nut" like a peach pit, comprises one or two seeds within the endocarp of the drupe fruit. Outer layers of the pericarp become a dry husk. Almonds require mild, wet winters and warm, rather dry summers. Frost-free springs are important, since bloom occurs very early and insect pollination occurs. While they are said to occupy more acres in North America than any other nut crop, almonds are grown chiefly in parts of California. Cultivars for other sections must be chosen carefully. Hardy seedlings often have bitter nuts, but selections with sweet nuts are available. Most almonds are self-unfruitful and **dichogamy** (differential maturation of male and female flowers) may occur; so more than one cultivar is necessary. Culture is similar to that of peaches. Almond cultivars are grafted on bitter almond seedlings, especially for growing on high-lime soils. For acid-soil areas, peach or 'Nemaguard' stocks are used.

Walnuts

The Persian or English walnut, *Juglans regia* of the Juglandaceae, leading nut crop in North America, finds its most favorable habitat in parts of California and Oregon's Willamette Valley (Fig. 12-19). Even in these regions, however, trees have in recent years suffered severely from winter injury, walnut husk fly, bacterial blight, and a disease called black line.

Two strains of walnut somewhat more cold resistant than common California and Northwest cultivars, are available for use both within and outside commercial areas. Carpathian walnuts, brought to North America by settlers from eastern Europe, are more hardy and bear over a wider range than standard West Coast cultivars. Meats of seedlings may be bitter, but good-flavored selections have been propagated. Carpathians bloom early, succeeding best in areas free of spring frosts. The Manregian strain proved more hardy than most old strains in the test winter of 1972. Trees tend to bloom late and mature their crop early. Nuts are large, well-filled, and good flavored but darker-skinned than other kinds. There are several excellent named cultivars awaiting further trial, especially in areas other than the Pacific Coast states.

Until recently walnut cultivars have been propagated on rootstocks of the California black walnut, *J. hindsii*. After 20 or more years

FIGURE 12-19 Walnuts are the leading nut crop in North America. (Courtesy of U.S.D.A.)

trees often develop black line, a virus disease appearing near the graft union, from which the tree slowly dies. Trees should be grafted on stock of standard *J. regia*, Carpathian, or Manregian, whichever is recommended for a given area.

Walnuts are monoecious and wind pollinated. Most seem more or less self-fruitful, but not all. It is better to have more than one cultivar and some are better pollinizers than others, often because of timing. Male flowers shed pollen when it is mature, whether or not stigmas are receptive. As in almonds, chestnuts, and filberts, dichogamy occurs. A different cultivar shedding at a different date may be necessary for pollination.

Trees bear fruit when they are six or more years old. They need little pruning while young, but older trees may need to be kept at spraying height.

An important native species of *Juglans* is the butternut or white walnut, *J. cinerea*. The hardiest, and considered by some the best flavored of nuts, it is grown in home gardens across the northern United States, in adjacent Canada, and as far south as Arkansas.

Another native hardy into New England and Canada, but ranging widely to the West and South, is the black walnut, *Juglans nigra*, so abundant in earlier times that its wood was used in home building as well as in furniture. It also provided tannin and dye. It is now grown commercially as a timber crop. Many wild nuts are harvested. Named cultivars with nuts superior in flavor, size, and ease of cracking are available. However, the cracking is still difficult—it is best accomplished with a hammer and an iron block, or a metal vise. A chemical, juglone, found in its roots inhibits growth of tomatoes and certain annual and perennial plants. Nuts fall with their husks on. These are broken off by hand or tramped off. Dye in the husks stains hands brown by combining with protein in the skin. The color must wear off. Black walnut shells are made into buttons and other hand-crafted items.

Somewhat less hardy is the Japanese walnut or heartnut, *J. ailantifolia* var. *cordiformis*, whose range merges with that of *J. regia* and *J. cinerea*. The heartnut hybridizes with the butternut to form buartnuts, some of which are hardy. While the various walnut species are dichogamous, all are cross-fruitful, pollinizing one another. Butternuts, heartnuts, and their hybrids are grafted with difficulty on black walnut stock. Better rootstocks are needed. Walnut broom, a mycoplasmlike disease, may attack butternuts, heartnuts, and their hybrids.

Hickories and Pecans

Also in the walnut family two wide-ranging native species, shagbark hickory, *Carya ovata*, and shellbark hickory, *C. laciniosa*, bear delicious nuts besides being important timber trees. Many wild nuts are harvested and cultivars are available. Other species of *Carya* may be important in local areas. Hickories tend to come into bearing late. Mature ones are excellent large shade trees.

The pecan, *Carya illinoensis*, closely related to the hickories, bears one of our choicest nuts and also is a cabinet wood. Its natural range includes the valleys of the Mississippi and other southern rivers, extending northward into Iowa and southward into Mexico, but is cultivated over a much wider area. It withstands poor drainage while dormant, but not during the growing season. While pecans for the North are advertised, their success is doubtful in most seasons as they need more heat units and a longer growing season than most northern summers can supply. More adaptable cultivars are sought. However, within its range it is most desirable. There are several strains, each adapted to a different area of southern United States or Mexico.

Mature pecan trees tend toward alternate bearing, because nuts ripening over a long period use large amounts of photosynthate, thus competing with flower bud formation.

Like other family members they become large shade trees.

Filberts and Hazelnuts

Filberts and hazelnuts, species of the genus *Corylus* of the birch family, are among the smaller nut trees, being spaced about 20 feet apart. Thickets of native species, called hazelnuts, occur in many parts of North America and have been brought into cultivation and hybridized to a limited extent. They furnish food and cover for wildlife and genes for development of cultivars for difficult areas.

Filberts are mostly hybrids of *C. avellana*, grown from ancient times in southwestern Asia and in Europe, and of *C. maxima* of southeastern Europe. They were introduced into California in the nineteenth century. Rather exacting in soil requirements, they succeed best on the rich, well-drained river bottom lands of Oregon's Willamette Valley where they are an important commercial crop and a common home garden plant. They need mild summers and winters, with moisture while developing root systems. Bearing trees often do not need irrigation. They may respond to nitrogen, potassium, or boron. Plants are propagated by layering or stooling and grow on their own roots because no satisfactory rootstock has been found.

Since filberts are self-unfruitful, dichogamous, and wind pollinated, a pollinizer is needed nearby. Nuts are picked up when they fall beginning in September and are dried in a warm place or a dryer.

Filberts are used raw or toasted as snacks and in baked dishes, salads, and confections such as toasted hazelnut ice cream. Leaves are simple, moderately coarse, deep green and dense, making a good screen or shade. There are yellow- and red-leaved forms useful as accents and also one with contorted branches grown as a conversation piece and to supply material for artistic arrangements. A few hybrids of Old World and native species are available, as well as established and new cultivars of filbert.

Chestnuts

The American chestnut, *Castanea dentata*, of the beech family, Fagaceae, was the most important deciduous tree in North America until it was nearly wiped out early in this century by chestnut blight, a fungus disease from Asia. It furnished easily worked but very durable wood, bark high in tannin, and nutritious nuts supplying food for humans and animals. Today growth from old roots occurs, but is eventually killed back by the blight. Among several species, *C. dentata* is the hardiest and produces the best nuts, but survives blight only in a few small areas.

For the home garden the most successful chestnut species is the Chinese, *C. mollissima*, or hybrids of it. It is about as hardy as peach trees, growing in zones 5 to 8 (Fig. 6-11). It may be grown from seed, but for nut quality named cultivars should be planted. Available ones come into bearing young, have large, sweet nuts, and may produce up to a bushel or more at 10 years of age. As chestnuts are usually self-unfruitful and dichogamous, two or more cultivars are required. Trees are not immune to blight, but under most conditions are not injured seriously. They become large shade trees with shining leaves, having few cultural problems on well-drained soils.

The Japanese chestnut, *C. crenata*, also grown in North America, is less hardy (from zone 6) and shorter-lived than *C. molissima*.

At present there is considerable demand for chestnut trees for home gardens as a source of food and feed. Nuts contain 50 percent or more carbohydrates, 5 percent protein, and 5 percent fat. Chestnuts must be picked up promptly and in a lawn or yard the prickly husks must be kept cleaned up. For eating fresh nuts may be dried outdoors for several days allowing starch to change to sugar. In drying, texture changes from firm to spongy to firm again. They taste best in the spongy stage. Dried nuts may be stored at 32 to 40°F in plastic bags or glass jars.

There are other fruit and nut plants important in local areas and in individual gardens. Examples are medlar, paw paw, persimmon, pistachio, and saskatoon. These are good subjects for student and home gardener research.

Questions for Discussion

1. Define the following terms:
   (a) Fruitfulness   (d) Heading back
   (b) Self-fruitful   (e) Prune
   (c) June drop   (f) Juglone

2. Indicate the values of having perennial food plants in the home garden.

3. Discuss the importance of climate as a factor affecting the success of the perennial garden. Be sure to include in your discussion the effects of temperature, light, and moisture.

4. Your neighbors have a single large apple tree in their front yard. The production of fruit from this tree is generally quite poor. Aware that you are taking a gardening course at a nearby college they seek your help. What would you do to determine why fruit production is poor, and what steps would you suggest to remedy this situation?

5. Comment on the soil requirements for blueberry, including aeration, drainage, pH range, and organic matter content.

6. Present guidelines for the timing of harvest of blueberries, pears, and cherries.

7. Identify three important bramble fruits. What botanical features do the three have in common? How do they differ?

8. Discuss the culture of strawberries, including soil preferences, planting depth, and training systems.

9. What is Bordeaux mixture, and how was it important in the history of grape production?

10. Why are apples not an important crop in Florida and Hawaii?

11. How are dwarf apple trees produced?

12. Outline planting procedures for apple or other tree fruit of your choice.

13. Starting with a one-year-old apple whip, use a series of diagrams to illustrate how training and pruning should be accomplished.

14. What are the symptoms of apple scab disease, and how can its incidence be reduced in the home garden?

15. Describe where flower buds and flowers are formed in pear, peach, and cherry.

16. Outline a fertilization program applicable to all tree fruits.

17. Plums are the most variable of the tree fruits. What is meant by this statement?

18. List the major North American nut crops and identify those that are best suited to your area.

19. Compare the nut trees to the tree fruits as relates to their culture management.

Selected References

Childers, N. 1976. *Modern Fruit Science*. Horticultural Publications, New Brunswick, N. J.

Halfacre, R. and J. Barden. 1970. *Horticulture*. McGraw-Hill, New York.

Hortus Third. 1976. Macmillan, New York.

Jaynes, R. (Ed.). 1975. *Handbook of North American Nut Trees*. W. F. Humphrey Press, Geneva, N. Y.

Ortho Books. Staff. 1976. *All About Growing Fruits and Berries*, Chevron Chemical Company, San Francisco.

Reiley, H. and C. Shry. 1979. *Introductory Horticulture*. Delmar Publications, Albany, N. Y.

Shoemaker, J. 1975. *Small Fruit Culture*. AVI Publications, Westport, Conn.

Southwick, L. 1975. *Dwarf Fruit Trees for the Home Gardener*, 3rd Printing. Garden Way, Charlotte, Vermont.

Teskey, B. and J. Shoemaker. 1975. *Tree Fruit Production*. AVI Publications, Westport, Conn.

Weaver, R. 1976. *Grape Growing*. Wiley, New York.

Westwood, M. N. 1978. *Temperate-Zone Pomology*. W. H. Freeman, San Francisco.

LEARNING OBJECTIVES

By the time you have finished this chapter you should be able to:

1   Discuss briefly the major developments in the history of garden flower use that led to modern cultivars of annuals and biennials.

2   Discuss the criteria to be considered in deciding what flowers should be planted in the garden.

3   Decide when flower seed of selected cultivars should be sown, and whether they should be started indoors or out.

4   List environmental factors that can be modified to produce optimum seed germination indoors and describe specific modifications to fit needs of several species with different requirements.

5   List environmental factors that can be modified to produce optimum growth of flower seedlings, indoors and out, and describe specific modifications to fit needs of several species with different requirements.

6   Discuss necessary care of several flower species of different cultural requirements as they are grown to maturity in the garden.

7   Explain the essentials of effectively using flowers in the landscape.

8   Describe the basics of propagating perennials by four different methods.

9   Discuss the All-America Selections system and its importance to the grower of flowers.

10   Distinguish between inbred and hybrid flower cultivars and discuss their importance to the gardener.

11   Discuss the value of collecting seed from garden flowers and explain what results should be expected from hybrids and from inbreds.

13

GROWING GARDEN FLOWERS

by George Whiting, Department of Horticulture, Temple University

"Flowers seem intended for the solace of ordinary humanity;
children love them; quiet, tender, contented, ordinary people
love them as they grow; luxurious and disorderly people rejoice
in them gathered."

John Ruskin

HISTORY OF GARDEN FLOWER USE

When plants were first used as ornamentals, perhaps as much as 5000 years ago in Mesopotamia, Egypt, and China, nearly all garden flowers were perennials. Probably most of these were originally used for food, medicinal, cosmetic, or religious purposes, as were saffron crocus, foxglove, bluebells, Madonna lily, and some chrysanthemum species. Most of the annual flowers we know today would have been entirely unfamiliar to gardeners even several hundred years ago. Our most commonly used garden flower, the South American native petunia (Fig. 13-1), was not cultivated until after 1800.

The first annual flowers in garden settings were surely wildflowers growing there by chance, perhaps having simply moved into gardens from nearby meadows. It was not until the fifteenth and sixteenth centuries that annuals were cultivated in Europe and not until considerably later that there was any noticeable difference between cultivated annuals and their wild relatives.

Annuals used in early American colonial gardens were imported primarily from England, France, and Holland. Some were New World natives introduced to cultivation in Europe, then brought back. Marigolds, *Tagetes patula* and *erecta*, are Mexican natives that were introduced to colonial gardens from Europe before 1700. Four o'clocks, *Mirabilis jalapa*, native to the American tropics, were widely

FIGURE 13-1 'Appleblossom' petunia, bronze medal winner, 1965 All-America Selections.

grown in early colonial gardens. The European and Asiatic annuals most commonly grown in colonial gardens prior to 1700 were calendula, balsam, bachelor's button, globe amaranth, and two *Amaranthus* species—*A. tricolor*, Joseph's coat, and *A. caudatus*, love-lies-bleeding. Also grown in early colonial times were the snapdragon, morning glory, larkspur, pansy, nasturtium, globe candytuft, sunflower, strawflower, and about ten other species of little present-day importance. The two most popular biennials were hollyhocks and sweet William. The foxglove, Canterbury bell, and English daisy were also grown, although the latter was often treated as a perennial. During the 1700s the China aster, sweet pea, and crested cockscomb became widely used. The zinnia, introduced in the late 1700s, quickly became a favorite. Nearly all our other commonly grown annuals were nineteenth-century introductions, including portulaca, sweet alyssum, salvia, verbena, impatiens, coleus, cosmos, wax begonias, Drummond phlox, ageratum, plume cockscomb, and Chinese pinks. Other pinks and carnations were common perennials in colonial gardens.

The vast majority of presently grown annual and biennial cultivars are twentieth-century introductions. Tender perennial geraniums and dahlias, and hardy chrysanthemums, are now widely grown from seed. Another interesting modern development has been production of annual and perennial cultivars of such biennials as hollyhock and foxglove, considerably improving their usefulness.

Breeding programs during the 1900s, particularly since the 1940s, have improved the color range, range of flower and plant form, vigor, uniformity, and earliness of almost all commonly grown annuals and biennials. Of particular importance to gardeners in recent years has been development of cultivars resistant to important diseases. Improvements in disease resistance and development of resistance to insects, which is still in its infancy, are areas in which significant accomplishments can be expected in the future.

POPULAR FLOWER SPECIES

Since the 1940s, at least, the petunia (Fig. 13-1) has been the most commonly grown annual flower. According to a 1975 National Garden Bureau survey petunia was third behind zinnia and marigold in packet seed sales, but very large sales of young plants moves them to number one on the popularity list. For the same reason, the marigold is somewhat ahead of zinnia in total popularity. The top dozen in seed packet sales in 1975 were:

1. Zinnia.
2. Marigold.
3. Petunia.
4. Nasturtium.
5. Sweet alyssum.
6. Aster.
7. Morning glory.
8. Portulaca.
9. Snapdragon.
10. Sweet pea.
11. Pansy and violets.
12. Dianthus.

Relatively few biennials are widely grown and only sweet William and hollyhock even ap-

proach the popularity of the above annuals. There are no completely accurate data to indicate the order of popularity if sales of young plants were taken into account, but there would undoubtedly be some restructuring of the list and possibly a few additions and subtractions.

The many species and cultivars available can make selection a bewildering task. The intent of this chapter is to explore growth requirements of our important annual (Table 13-1) and biennial (Table 13-2) flowers, and discuss how cultural practices fit plants to particular garden conditions. We emphasize (1) sowing and germination of seed, (2) growing seedlings indoors and out, (3) transplanting and maintaining flowers in the garden, and (4) a variety of special topics concerning characteristics and culture of annual, biennial, and perennial flowers. However, first we review the naming of these flowering plants, particularly the scientific names.

NAMING GARDEN FLOWERS

All garden flowers have genus and species names as assigned by the first person to properly describe the species, as explained for vegetables in Chapter 2. Also, most flowers have widely accepted common names. For instance, *Antirrhinum majus* is the common snapdragon, *Ageratum houstonianum* is simply known as ageratum, *Cleome spinosa* is spider flower, and *Ipomoea purpurea* is morning glory. Some scientific names include an x before the species name, as in *Petunia* x *hybrida*, indicating that the species is a hybrid and does not exist as a species entity in the wild.

In some cases characteristics of cultivars are sufficiently different from the usual features of wild species to cause controversy. Larkspur, *Consolida orientalis*, has been assigned by various horticultural taxonomists to several different species, usually *Delphinium ajacis*, but also occasionally *Consolida ajacis* and *C. ambigua*. The authors of *Hortus Third*, the author-

TABLE 13-1 70 Annual Flowers: Culture and Characteristics

Scientific and common name	Sowing time, germination conditions — In garden	Sowing time, germination conditions — Indoors[a]	Days to germination	Indoor growing conditions for seedlings[b]	Time to move to garden	Bloom period	Flower color[c]	Height, inches	Plant form[d]	Hardiness[e]	Breeding type[f]	Comments
Ageratum houstonianum, Ageratum	After soil has warmed in spring press into soil surface	L 70° 8–10 weeks	5	70° for 1 week, then 50° until transplanting	After last frost date	All summer	W,P B,V	6–36	LB MB	T HH	I F₁	Tolerant of partial shade, easily propagated by cuttings, good pot plant
Amaranthus tricolor, Joseph's Coat	After soil has warmed in spring	70° 6–8 weeks	10	70° for 1 week, then 60°, needs sun	After last frost date	Summer	Y,R G	36–72	TB	T	I M	Grown for leaf color, usually sown in place, best color in relatively poor soil
Antirrhinum majus, Snapdragon	After soil has warmed in spring	L 65° 6–10 weeks	10	50°	After last frost date	Spring until frost	W,Y O,P R	4–48	LB MB S	H	I F₁ F₂ M P	Perennial in moderate climate, good cut flower
Arctotis stoechadifolia, African Daisy	Early spring	55° 8–10 weeks	10–21	55° needs some sun	Last frost date	Summer until frost	W,Y R,B	12–24	MB	HH	M	Good in sandy soil, good cut flower
Begonia x semperflorens-cultorum, Wax Begonia	After soil has warmed in spring	L 70° 12–15 weeks	15+	70° for 2 weeks, then 60°	After last frost date	Spring, summer	W,P R	6–12	LB MB	T	I F₁ M	Perennial in frost-free areas, shade tolerant
Browallia americana and *speciosa,* Browallia	After soil has warmed in spring	L 70° 8–12 weeks	15	60°	After last frost date	Spring until frost	W,B	12–36	LB MB	T	I	Tolerant of partial shade, good in hanging baskets
Calendula officinalis, Pot Marigold	Early spring as soon as soil can be worked, late fall	D 70° 6–8 weeks	10	50–55°	Last frost date	Late spring till fall	Y,O	12–30	MB	H	I M	Tolerates poor soil, requires good drainage, good pot plant, fine cut flower
Callistephus chinensis, China Aster	After soil has warmed in spring	70° 6–8 weeks	8	60°	After last frost date	Late summer until frost	W,P R,V	8–36	MB TB	HH	I M	If frost-free season is less than 5 months, start indoors, long lasting cut flower

Plant	When to Start	Indoor Temp / Weeks	Days to Germinate	Germination Temp	When to Transplant	Bloom Time	Color	Height	Light	Hardiness	Type	Remarks
Capsicum annuum, Ornamental Pepper	After soil has warmed in spring	70–75° 10–12 weeks	14–20+	65°	After soil has warmed	Fruit, summer until frost	G,Y O,V R	9–18	MB	T	I M	Good winter pot plant, tolerant of partial shade but best in full sun, fruit becomes bright red when ripe
Catharanthus roseus, Madagascar Periwinkle	After soil has warmed in spring	D 70° 8–10 weeks	7–15	70–75°	After last frost date	Late spring until frost	W,P R	6–24	LB MB	HH	I M	Perennial in frost-free areas
Celosia cristata, Cockscomb	After soil has warmed in spring	70° 4–6 weeks	10	65° grows rapidly	After all frost danger	Summer until early autumn	Y,O R	10–36	MB TB	T	I M	Heat and drought tolerant, retains true color when dried
Centaurea cyanus, Bachelor's Button, Cornflower	Early spring	D 65° normally sown in place	10	55°	Last frost date	Late spring to early summer	W,R B,V	12–36	MB TB	H	I M	Does best in light soil with neutral pH, good cut flower
Chrysanthemum x morifolium, Chrysanthemum	Early spring or summer for sturdy plants next year	70° 8–10 weeks	8	60°	Last frost date	Late summer until frost	W,Y O,P R,V	6–36	LB MB TB	H	F₁ M	Hardy perennial, fine cut flower, mainstay of fall garden
Cleome spinosa, Spider Flower	Early spring	L 55–60° 5–6 weeks	10	55° needs sun	Last frost date	Summer until mid-autumn	W,P V	36–60	TB	HH	I M	Heat and drought tolerant, grows best in sandy soils
Coleus x hybridus, Coleus	After soil has warmed in spring	L 65°	10	60° sow thinly	After last frost date	Grown for leaf color	Leaf Y,R G,O	6–24	LB MB	T	I M	Perennial in frost-free areas, good pot plant, pinch out flower clusters
Consolida orientalis, Larkspur	As early in spring as soil can be worked	D 55° 8–10 weeks	20	Grow cool, no higher than 55°	Average last frost date	All summer and early fall	W,P B,V	18–48	S	H	M	Two basic types exist, one with a single non-branching flower stalk, another with basal branching
Coreopsis tinctoria, Calliopsis	As soil is warming in spring	55–60° 6–8 weeks	8	Not above 60° needs sun	Average last frost date	Summer until hard frost	Y,O	8–36	MB TB	HH	I M	Needs sunny spot to flower

TABLE 13-1 (*continued*)

Scientific and common name	Sowing time, germination conditions		Days to germination	Indoor growing conditions for seedlings[b]	Time to move to garden	Bloom period	Flower color[c]	Height, inches	Plant form[d]	Hardiness[e]	Breeding type	Comments
	In garden	Indoors[a]										
Cosmos bipinnatus, Cosmos	After soil has warmed in spring	70° 6–8 weeks	5–14	65° well-drained medium	After all frost danger	Late spring, early summer, late summer	W,Y O,R P	24–48	MB TB	T	I M	Does well in sandy soils
Dahlia hybrids, Dahlia	After soil has warmed in spring	70° 6–8 weeks	5	55° until transplant, then 65°	After all frost danger	Summer until frost	W,Y O,R P,B V	15–96	MB TB	T	M	Perennial in frost-free areas, tuberous roots can be dug after tops killed by frost and stored until spring, long-lasting cut flowers
Dianthus caryophyllus, Carnation	After soil has warmed in spring	70° 8–10 weeks	20	55–60° pinch at 3 inches to induce branching	Average last frost date	Midsummer to fall	W,P R,Y O	12–48	MB TB	H	I F₁ M	Perennial except in areas with very cold winters, fine cut flower
Dianthus chinensis, Chinese Pinks	After soil has warmed in spring	70° 6–8 weeks	5–10	50–60°	Average last frost date	Late spring until late summer	W,P R	8–15	LB MB	H	I F₁ M	Perennial except in northern areas
Dimorphotheca sinuata, Cape Marigold, Star of the Veldt	After soil has warmed in spring	70° 6–8 weeks	10	60–65° needs sun	Average last frost date	Summer	W,Y P,O	12	MB	HH	I M	Best in hot, dry weather, blossoms open only in full sun
Eschscholtzia californica, California Poppy	As early in spring as soil can be worked	55° 6–8 weeks	10–14	50–55°	Average last frost date	Spring until frost	W,Y O,P R	12–24	MB	H	M	Does well in poor, dry soils, best in light, well-drained soils, reseeds readily
Gaillardia pulchella, Painted Gaillardia	After soil has warmed in spring	70° 6–8 weeks	20	60° plant out before buds form	Average last frost date	Summer until hard frost	Y,O R	10–18	MB	H	I M	Heat and drought tolerant, flowers can be air dried

Plant	When to sow	Germination temp/time	Days	Seedling temp/notes	When to transplant	Bloom time	Color	Height	Type	Hardiness		Remarks
Gazania rngens, Gazania	After soil has warmed in spring	D 60° 6–8 weeks	8	60° needs sun	Average last frost date	Summer until hard frost	W,Y O,R P	6–18	P LB	H	M	Perennial in mild climates, tolerant of poor, dry soils
Gerbera jamesonii, Transvaal Daisy	As soil warms in spring	L 65–70° 10–16 weeks	10–15	55–60° needs sun and well-fertilized medium	Average last frost date	Summer	W,Y R,P O,V	12–18	LB	HH	M	Perennial where temperatures do not go under 30°, excellent cut flowers
Gomphrena globosa, Globe Amaranth	After soil has warmed in spring	D 65° 8–10 weeks	15	65–70°	1–2 weeks after average last frost date	Summer until late autumn	W,P V	8–36	MB TB	T	I M	Heat tolerant, good cut flower and excellent quality air dried
Gypsophila elegans, Baby's Breath	After soil has warmed in spring	70° 6–8 weeks	10	65°	Average last frost date	Summer	W,P	12–24	MB	HH	I M	Best in neutral to slightly alkaline soil, drought resistant
Helianthus annuus, Sunflower	After soil has warmed in spring	D 70° 4–5 weeks	5	Best about 65°	After frost danger	Midsummer, early autumn	W,Y O,R	24–120	Tall	T	I M	Tolerates relatively dry soil, very tall, edible seed
Helichrysum bracteatum, Strawflower	After soil has warmed in spring	L 70° 6–8 weeks	5	60–65° do not overwater	1–2 weeks after average last frost date	Summer until midfall	W,Y O,R P	18–36	MB TB	HH	I M	Needs full sun, flowers easily air dried
Iberis amara, Rocket Candytuft	In spring after soil has warmed	70° 5–6 weeks	8	55–60°	Average last frost date	Spring only	W	10–15	MB	H	I	Not heat tolerant, fragrant
Iberis umbellata, Globe Candytuft	In spring as soil warms	70° 5–6 weeks	20	About 55–60°	Average last frost date	Late spring, early summer	W,P R,V	8–15	LB MB	H	I M	Best in relatively cool summer climate, good cut flower
Impatiens walleriana, Impatiens	In spring after soil was warmed	L 70° 8–10 weeks	15	60–65° grow in filtered sun	After last frost date	Spring until first frost	W,P R,V O	6–24	LB MB TB	T	I F₁ M	Perennial in frost-free areas, very shade tolerant
Ipomoea x multifida, Cardinal Climber	After soil has warmed in spring	65°	10–20	65–70°	After last frost date	Summer	R	to 240	Vine	T	I	Attractive fern-like foliage, fast growing
Ipomoea purpurea, Morning Glory	After soil has warmed in spring	65° 3–5 weeks	5 or 10–21	65–70° needs direct sun	After last frost date	Late spring until first frost	W,P R,B	12, 90–120	LB Vine	T	I M	Best to soak seed 24 hours before planting, needs full sun

TABLE 13-1 (continued)

Scientific and common name	Sowing time, germination conditions		Days to germination	Indoor growing conditions for seedlings[b]	Time to move to garden	Bloom period	Flower color[c]	Height, inches	Plant form[d]	Hardiness[e]	Breeding type[f]	Comments
	In garden	Indoors[a]										
Kochia scoparia, Summer Cypress	After soil has warmed in spring	L 70° 5–6 weeks	15	70° until transplant, then 65°	Average last frost date	Leaf color effective until frost	G,Y R leaf	24–36	MB TB	HH	I	Heat and drought tolerant, low, fine-textured hedge
Lathyrus odoratus, Sweet Pea	As early as soil can be worked in spring	D 55° 5–7 weeks	15	55°	2 weeks before average last frost date	Spring, fall if late planting made	W,P R,B V	12–72	MB Vine	H	I M	Few cultivars heat tolerant, excellent cut flower
Limonium bonduellii and sinuatum; Psylliostachys suworowii, Statice, Sea Lavender	As soil warms in spring	70° 6–8 weeks	15	55–60°	About a week after average last frost date	Spring, summer	See comments.	See comments.	TB	H	I M	L. bonduellii is sometimes biennial; L. sinuatum may also be biennial or perennial; L. bonduellii is 24 to 36 inches tall, yellow; L. sinuatum is about 18 inches tall, pink, white and lavender; P. suworowii is about 18 inches tall, white, red, violet and blue
Lobelia erinus, Edging Lobelia	Rather slow starting, best started inside in late-spring areas	L 70° 8–12 weeks	20	50°	Average last frost date	Summer until frost	W,R B	4–8	P LB	HH	I M	May be perennial in frost-free areas, tolerates some shade
Lobularia maritima, Sweet Alyssum	As early as soil can be worked in spring	70° 5–7 weeks	5	50–55°	Week before average last frost date	Spring, Summer	W,P V	4–10	P LB	H	I	Benefits from cutting back in midsummer, reseeds readily
Lupinus hybrids, Lupine	Early spring	55° 6–8 weeks, soak seeds before sowing	20	55°	Average last frost date	Summer	W,Y P,B V	12–36	S	H	I M	Tolerant of partial shade, short bloom period

Name	Sow outdoors	Indoor sowing	Days	Germination temperature	Transplant/sow outdoors	Flowering	Color	Height (in.)	Location	Hardiness	Type	Remarks
Malcomia maritima, Virginia Stock	Early spring	55° usually sown directly outdoors	5–10	55° needs sun, cool temperatures	Average last frost date	Spring, fall, from separate sowings	W,P V,R	4–8	LB	HH	M	Will begin to bloom 6–8 weeks after sowing in cool weather
Mirabilis jalapa, Four-O'Clock, Marvel-of-Peru	After soil has warmed in spring	70° 6–8 weeks	5	65–70°	After last frost date	Late summer until frost	W,P V,R Y	36–48	TB	T	I M	Perennial in frost-free areas, tuberous roots can be dug and stored, flowers open in afternoon
Moluccella laevis, Bells of Ireland	After soil has warmed in spring	70° 6–8 weeks	25+	60°	Average last frost date	Late summer and fall	G	18–36	S	HH	I	Small fragrant white corolla in center of large green calyx, cut flowers easily air dried, re-seeds readily
Nemesia strumosa, Nemesia	After soil has warmed in spring	D 65° 6–8 weeks	5	55°	After last frost date	Spring until end of summer	W,Y O,R P,B V	10–18	LB MB	T	M	Burns out in hot weather, pinch tallest branches to encourage bushiness
Nicotiana alata, Flowering Tobacco	After soil has warmed in spring	L 70° 6–8 weeks	20	60°	After last frost date	Summer until frost	W,G R,V P	10–36	MB TB	T	I M	Perennial in frost-free areas, fast growing
Nigella damascena, Love-in-a-Mist	Early spring as soon as soil can be worked	60° 6–8 weeks	8	50–55°	Week before average last frost date	Summer	W,P R,V	12–15	MB	HH	I M	Attractive lacy foliage, dry fruit attractive in arrangements
Papaver nudicaule, Iceland Poppy	In early spring as soon as soil can be worked	D 70° 6–8 weeks, difficult to transplant	10	60°	Average last frost date	Summer	Y,W O,R P	12–24	MB	H	I F$_1$ M	Actually a hardy perennial easily treated as an annual
Papaver rhoeas, Shirley Poppy	Early spring as soon as soil can be worked	D 70° 6–8 weeks, difficult to transplant	10	60°	Average last frost date	Late spring early summer	W,P R	24	MB	H	I M	Seed germinates well at temperatures well below 70°

TABLE 13-1 (continued)

Scientific and common name	Sowing time, germination conditions		Days to germination	Indoor growing conditions for seedlings[b]	Time to move to garden	Bloom period	Flower color[c]	Height, inches	Plant form[d]	Hardiness[e]	Breeding type[f]	Comments
	In garden	Indoors[a]										
Pelargonium x hortorum, Common or Zonal Geranium	After soil has warmed in spring	70° 10–12 weeks	15	60°	After last frost date	Late spring until frost	W,P R,O	18–24	MB	T	I F₁ F₂ M	Perennial in frost-free areas where they can become quite tall and woody, cuttings easily rooted, good indoor plant for year-round bloom
Petunia x hybrida, Petunia	After soil has warmed in spring	L 70° 8–12 weeks	10	60°	Week after average last frost date	Spring until hard frost	W,Y P,R V	8–15	LB	HH	I F₁ F₂	Perennial in frost-free areas, pinch at about 6 inches to induce good branching
Phaseolus coccineus, Scarlet Runner Bean	After soil has warmed in spring	70–75° 4–6 weeks	5–10	70°	After last frost date	Summer	R	72–120	Vine	T	I	There are white and pink flowering cultivars, but the red ones are much more showy, beans edible, pick for eating when pods are 4 inches long
Phlox drummondii, Drummond Phlox	As early in spring as soil can be worked	D 65° 6–8 weeks	10	55°	Average last frost date	Late spring until early fall	W,P R,Y V	6–18	LB	HH	I M P	Does not flower well in very hot weather
Portulaca grandiflora, Portulaca	After soil has warmed in spring	D 70° 6–8 weeks	10	60° grow slightly dry to avoid damping off	After last frost date	Summer until late autumn	W,Y O,R V	4–6	P	T	I F₁ M	Tolerant of dry soil, heat, drought, good ground cover on poor soil
Rudbeckia hirta, Gloriosa Daisy	As soil is beginning to warm in spring	70° 6–10 weeks	14–21	60°	Average last frost date	Summer until hard frost	Y,O R	24–30	TB	H	I M	May be biennial or short-lived perennial, drought tolerant

Botanical/Common Name	Sow Outdoors	Sow Indoors	Days to Germinate	Germination Temp.	Set Out	Blooming Period	Color	Height (in.)	Type	Hardiness	Notes
Salpiglossis sinuata, Painted Tongue	After soil has warmed in spring	D 70° 6–8 weeks	15	60°	After last frost date	Summer until mid-autumn	Y,R B,V	24–36	TB	T	M	Pinch to encourage branching, may need staking
Sanvitalia procumbens, Creeping Zinnia	After soil has warmed in spring	L 70° 6–8 weeks	10	60°	Week after last frost date	Summer until frost	Y	6	P	HH	I	Needs light, well-drained soil
Scabiosa atropurpurea, Sweet Scabius	After soil has warmed in spring	70° 6–8 weeks	10	55°	Average last frost date	Summer until frost	W,P R,B V	18–48	TB	HH	I M	Drought tolerant, pinch to encourage bushy growth
Schizanthus pinnatus, Butterfly Flower	After soil has warmed in spring	60° 6–8 weeks	20+	50–55°	After last frost date	Summer	W,P R,V	12–24	MB	T	M	Tolerates some shade, heat sensitive, good cut flower
Senecio cineraria, Dusty Miller	After soil has warmed in spring	L 75° 6–8 weeks	10	60° sow thinly, susceptible to damping-off	Average last frost date	Grown for foliage	Gray leaf	8–15	MB	H	I	Perennial in moderate climate, shade tolerant, stands frost exceptionally well
Tagetes erecta, African Marigold and *patula,* French Marigold	After soil has warmed in spring	70° 6–8 weeks	5	60° until transplant, then 65°	After last frost date	Spring until first frost	Y,O R	6–42	LB MB TB	T	I F_1 M P	Plant deep when transplanting, some cultivars have pungent odor
Thunbergia alata, Black-Eyed-Susan Vine, Clockvine	After soil has warmed in spring	70° 6–8 weeks	10	60°	After last frost date	Summer	W,Y O	60–72	Vine	T	I M	Perennial in frost-free areas, good in hanging baskets
Tithonia rotundifolia, Mexican Sunflower	After soil has warmed in spring	D 70° 8–10 weeks	10	60°	After last frost date	Summer until mid-autumn	O,R	48–60	TB	HH	M	Tolerant of dry soil, flowers best in hot weather
Torenia fournieri, Wishbone Flower	After soil has warmed in spring	70° 6–8 weeks	15	60°	After last frost date	Spring, until first frost	W,B V	12	MB	T	I M	Best in light shade except in cool summer areas, good in pots or hanging baskets
Tropaeolum majus, Nasturtium	After soil has warmed in spring	D 65° 6–8 weeks	8–10	50–55°	After last frost date	Late spring until frost	W,Y O,R	8–96	MB Vine	T	I M	Best in loose, well-drained, relatively poor soil, both bush and trailing types, young leaves are good in salads

409

TABLE 13-1 (*continued*)

Scientific and common name	Sowing time, germination conditions		Days to germination	Indoor growing conditions for seedlings[b]	Time to move to garden	Bloom period	Flower color[c]	Height, inches	Plant form[d]	Hardiness[e]	Breeding type[f]	Comments
	In garden	Indoors[a]										
Verbena x hybrida, Verbena	After soil has warmed in spring	D 65° about 8 weeks	20	60° until transplant, then 65°, needs sun	After last frost date	Late spring until frost	W,P R,B V	6–12	P LB	HH	I M	Perennial in frost-free areas, drought resistant
Viola tricolor, Pansy	As early in spring as soil can be worked	D 65° 10–12 weeks	10	50° transplant at same level as seedlings were growing	2 weeks before average last frost date	Late spring until hard freeze	W,Y R,B V	6–8	LB	H	I F_1 F_2 M	May be short-lived perennial in mild areas, will bloom all summer in areas with cool summer temperatures, cut plants back to about 3 inches to promote late season flowering
Zinnia elegans, Zinnia	After soil has warmed in spring	70° 6–7 weeks	4–6	60° until transplant, then 65°	After last frost date	Late spring, summer	W,Y O,R V,G	8–36	LB MB TB	T	I F_1 M P	Pinch first buds to promote lateral growth, bloom best in hot weather, susceptible to powdery mildew, very good cut flower

[a]D = dark, L = light; these conditions apply equally to germination in the garden. Temperatures listed are controlled, constant temperatures in degrees Fahrenheit. Weeks given are those before planting out time.

[b]Temperatures listed (°F) are night temperatures.

[c]Flower color: B = blue, P = pink, W = white, R = red, O = orange, Y = yellow, V = violet, G = green.

[d]Plant form: LB = low bushy, MB = medium bushy, TB = tall bushy, S = spire, P = prostrate.

[e]Hardiness: T = tender, HH = half hardy, H = hardy.

[f]Breeding type: I = inbred, F_1 = F_1 hybrid, F_2 = F_2 hybrid strain, M = mixture, P = polyploid.

TABLE 13-2 10 Biennial Flowers: Culture and Characteristics

Scientific and common name	Sowing time, germination conditions — In garden	Sowing time, germination conditions — Indoors[a]	Days to germination	Indoor growing conditions for seedlings[b]	Time to move to garden	Bloom period	Flower color[c]	Height, inches	Plant form[d]	Hardiness[e]	Breeding type[f]	Comments
Alcea rosea, Hollyhock	As early in spring as soil can be worked	60° 8–10 weeks	10	60°	Average last frost date	Late spring until mid-autumn	W,Y P,R	24–72	S	H	I M	Annual and perennial cultivars also available
Anchusa capensis, Cape Forget-Me-Not, Summer Forget-Me-Not	After soil has warmed in spring	70° 6–8 weeks	10	60°	Average last frost date	Late spring until midfall	W,P V	8–24	MB	H	I	Annual and perennial cultivars also available
Bellis perennis, English Daisy	In spring as early as soil can be worked	L 70° 6–8 weeks	5–8	50–55°	2 weeks before average last frost date	Spring	W,P V	3–8	LB	H	I F₁ M	Actually perennial, but probably best grown as a biennial
Campanula medium, Canterbury Bells	Early spring, summer for bloom the following year	70° 6–8 weeks	10–20	60°	Summer until early fall	Summer	W,B P,V	24–48	MB TB	H	I M	Best in rich soil, good cut flower, an annual cultivar also exists
Dianthus barbatus, Sweet William	After soil has warmed in spring	70° 6–8 weeks	5	60°	Average last frost date	Late spring, early summer	W,P R	3–12	LB MB	H	I M	Annual and perennial cultivars also available
Digitalis purpurea, Foxglove	Best in late spring–early summer	L 70° 10–12 weeks	20	55°	Summer to early fall	Late spring until late summer	W,Y P,V R	30–96	S	H	M	Annual and perennial cultivars as well as biennial, shade tolerant
Erysimum hieraciifolium, Siberian Wallflower	Late spring–early summer best	55–65° 6–8 weeks	5–14	50–55°	Summer, average last frost date for annual types	Spring, or summer	O	15–36	MB TB	H	I	Annual cultivars as well as biennial; some plants may tend to be short-lived perennials

TABLE 13-2 (continued)

Scientific and common name	Sowing time, germination conditions		Days to germination	Indoor growing conditions for seedlings[b]	Time to move to garden	Bloom period	Flower color[c]	Height, inches	Plant form[d]	Hardiness[e]	Breeding type[f]	Comments
	In garden	Indoors[a]										
Lunaria annua, Honesty, Money Plant	Early spring as soon as soil can be worked	55–60° 6–8 weeks	10	50–55°	Average last frost date	Summer	W,V R	30–36	TB	H	I M	May behave as annual if sown early enough, flowers fragrant
Matthiola incana, Stock, Gilliflower	Late summer– early fall	L 70° 8–10 weeks	10	50°	After last frost for early seeded types, fall for others	Late spring until end of summer	W,P R,V	15–36	S	HH	I M	Annual and perennial cultivars as well as biennial, good quality only in cool summer areas
Myosotis sylvatica, Forget-Me-Not	Late summer– early fall	D 55° 6–8 weeks	8	50°	Average last frost date	Spring	B,P W	6–20	LB MB	HH	I M	Annual cultivars are available; self seeds readily

[a]Sowing time, germination conditions, Indoors: D = dark, L = light. Temperatures listed are controlled, constant temperatures in degrees Fahrenheit. Weeks given are those before planting out time.
[b]Temperatures listed (°F) are night temperatures.
[c]Flower color: B = blue, P = pink, W = white, R = red, O = orange, Y = yellow, V = violet, G = green.
[d]Plant form: LB = low bushy, MB = medium bushy, TB = tall bushy, S = spire, P = prostrate.
[e]Hardiness: T = tender, HH = half hardy, H = hardy.
[f]Breeding type: I = inbred, F₁ = F₁ hybrid, F₂ = F₂ hybrid strain, M = mixture, P = polyploid.

ity used for determination of correct scientific names, suggest that most cultivars of larkspur probably can be assigned to *Consolida orientalis*. Another interesting case is that of *Dianthus chinensis*, the Chinese or common garden pink, so named because the flowers appear to have been edged by pinking shears. Most cultivars of garden pinks grown as annuals are regarded as originating from the old cultivar 'Heddewigii.' However, a careful evaluation of characteristics, particularly vegetative, indicates that some cultivars have a small proportion of traits that originated from hybridization with one or more other *Dianthus* species. These cultivars are still identifiable with *D. chinensis* 'Heddewigii,' or at least with *D. chinensis,* and placing them there avoids unnecessary taxonomic confusion.

For various taxonomic reasons several flowers have been moved to different genera or species from ones used for many years. Madagascar periwinkle was long classified as *Vinca rosea,* but is now called *Catharanthus roseus.* Cockscomb, previously *Celosia argentea,* is now classified as *C. cristata.* Impatiens has gone through two name changes, from *Impatiens sultanii* to *I. holstii,* and finally *I. wallerana.*

Wax begonia and coleus have long been known as *Begonia semperflorens* and *Coleus blumei.* It will probably be many years before these names fade from use in favor of the recently applied eipthets, *Begonia semperflorens-culto-*rum and *Coleus* x *hybridus,* which acknowledge the hybrid origin of these species. Our cultivars of lupine and dahlia also are apparently of hybrid origin, but these cultivars do not seem to constitute unified species, consequently they are now called simply *Lupine* hybrids and *Dahlia* hybrids. Name changes for other popular flowers are given in Table 13-3.

Plants often have numerous and confusing common names. For example, *Impatiens wallerana* is called impatiens, sultana, patience plant, patient Lucy, and busy Lizzie. Dusty miller has been applied to *Senecio cineraria, Centaurea cineraria, Artemesia stelleriana, A. albula* 'Silver King' and *A. schmidtiana* 'Silver Mount' because of their grayish foliage. The common name nasturtium for *Tropaeolum majus* might seem a strange choice for this brightly colored flower because there is a genus *Nasturtium* that is taxonomically quite different. The name originated from the likeness in taste of the leaves of *Tropaeolum majus* and watercress, which is *Nasturtium officinale.*

SOWING AND GERMINATION OF SEED

Some annuals are relatively difficult to grow from seed, because of small seed size, slow development, and/or erratic germination rates.

TABLE 13-3 Some Examples of Popular Flowers with New Scientific Names

| Common name | Scientific name | |
	Old name	New name
Cardinal climber	*Quamoclit sloteri*	*Ipomoea* x *multifida*
Gazania	*Gazania splendens*	*Gazania ringens*
Hollyhock	*Althaea rosea*	*Alcea rosea*
Purple cup flower	*Nierembergia caerulea*	*Nierembergia hippomannica*
Rat tail statice	*Limonium suworowii*	*Psylliostachys suworowii*
Siberian wallflower	*Cheiranthus allioni* and *Erysimum asperum*	*Erysimum hieraciifolium*
Spider flower	*Cleome spinosa*	*Cleome hasslerana*
Sweet alyssum	*Alyssum maritima*	*Lobularia maritima*

Such annuals as begonia, coleus, impatiens, and petunia are usually started indoors or in greenhouses, and are most commonly raised by commercial growers to be sold in market packs by garden centers. Others, particularly easy to grow when direct seeded in the garden, or difficult to transplant, such as nasturtium, morning glory, Shirley poppy, California poppy, bachelor's button, and scarlet runner bean, are seldom sold as young plants. Many others can be started indoors or out, depending on when flowering plants are desired.

Because of various cold tolerance of garden flower species, a knowledge of the average date of the last spring frost is vital for determining indoor and outdoor sowing and planting dates. These data are presented in Fig. 6-12, and for selected localities in Appendix D. In addition, many provincial and state agricultural services (Appendix C) and other agencies publish bulletins detailing frost information.

Outdoor Sowing Times

Outside sowing times can be conveniently divided into three categories: (1) early spring, as soon as soil can be worked, (2) midspring, as soil is beginning to warm, and (3) late spring, after soil has warmed. Flowers such as sweet alyssum, bachelor's button, and larkspur are usually sown early; snapdragon, Chinese pinks, and rudbeckia should be sown in midspring; while zinnia, Madagascar periwinkle, and ornamental pepper must be sown after conditions are settled and warm.

Unusually late or hard frosts may injure or even kill some plants, but gardeners should be prepared to replant, particularly if chances have been taken by sowing before recommended times. Species that germinate readily and initially grow best at relatively low temperatures, such as calliopsis, rudbeckia, and forget-me-not, are reasonable choices for early planting. Numerous others, including cocks-comb, zinnia, and scarlet runner bean need warm temperatures and definitely should be planted later.

Indoor Sowing Times

Many species need 6 to 8 weeks from sowing to the right maturity to be placed in the garden, but column 3 of Tables 13-1 and 13-2 indicates that the range is from as little as 3 to 5 weeks for morning glory and Virginia stock to 12 to 15 weeks for begonia and browallia and 10 to 16 weeks for gerbera. About the midpoint of the timetables should be attained if the seedlings are grown at the listed night temperatures and in the vicinity of 10 degrees warmer during the day. Many species can be grown as much as 10°F cooler or warmer than the listed temperatures, with about a week increase or decrease in time needed to produce seedlings ready for the garden.

Temperature and Seed Germination

Seed will often germinate at temperatures well above and below the listed temperatures, although the time to germination may be longer and/or percent of germination may be reduced. The temperatures presented in Tables 13-1 and 13-2 are for constant temperature germination in a controlled indoor situation. Some species may germinate better with fluctuating temperatures, as provided by nature, but there is little scientific information on this subject. Many species are known to germinate relatively poorly at constant temperatures lower than those listed (Table 13-4). Others, as well as some of the same species, do not germinate well at constant temperatures above those listed (Table 13-5). Species in both these lists will germinate poorly if not near the optimum temperature a reasonable proportion of the time. The amount of time necessary at or near optimum temperatures is open to conjecture.

TABLE 13-4 Flowers Whose Seed is Known to Germinate Poorly at Constant Temperatures Lower than Those Listed in Tables 13-1 and 13-2

Ageratum	Larkspur	Lupine	Sweet scabius
Joseph's coat	Carnation	Nemesia	Dusty miller (*Senecio*)
Snapdragon	Gaillardia	Flowering tobacco	Black-eyed Susan vine
Wax begonia	Gazania	Petunia	Wishbone flower
Browallia	Globe amaranth	Scarlet runner bean	Nasturtium
Calendula	Strawflower	Phlox	Pansy
Ornamental pepper	Impatiens	Portulaca	Zinnia
Madagascar periwinkle	Summer cypress	Gloriosa daisy	Hollyhock
Cockscomb	Sweet pea	Painted tongue	Canterbury bells
Bachelor's button	Statice	Mealy-cup sage	Forget-me-not
Coleus	Lobelia	Scarlet sage	

Effect of Light on Seed Germination

Some seed require either light, indicated by an L in column 3 of Tables 13-1 and 13-2, or dark, indicated by a D, for germination. Species with no L or D designation will germinate equally well in light or dark.

Seed requiring light for germination should remain uncovered, only pressed into the surface of the germination medium, or covered lightly. Those requiring dark must be adequately covered to exclude light. For others it makes no difference except that large seed ordinarily should be planted considerably deeper than small seed. Most light-requiring seed are small, while species needing dark for germination have relatively large seed. A conspicuous exception is gerbera, which has large seed but requires light for germination. Although it has been said that best results with

gerbera are attained by placing the seed on end and covering all but the tip, the usual successful practice is simply to cover lightly. Plant most seed an eighth to a quarter inch deep, except such large seeded species as scarlet runner bean, which should be sown at a depth of about a half inch.

Selecting Growing Media

Germination media should be pasteurized. Pasteurization of moist soil can be done as discussed in Chapter 14. Artificial media, which are excellent for germination, need not be pasteurized because they are already free of harmful organisms. Whatever medium is selected, it should be loose and well drained while also having good water-holding capacity. A near neutral pH is important, particularly for marigold and ornamental pepper, which germinate poorly in acid soil, and pe-

TABLE 13-5 Flowers Whose Seed is Known to Germinate Poorly at Constant Temperatures Above Those Listed in Tables 13-1 and 13-2

Snapdragon	Cape marigold	Flowering	Nasturtium
Browallia	Gazania	tobacco	Verbena
Bachelor's button	Sweet pea	Petunia	Pansy
Coleus	Statice	Phlox	Hollyhock
Cosmos	Lupine	Mealy-cup sage	Butterfly
Carnation	Nemesia	Scarlet sage	flower

tunia and pansy, which do not germinate well in alkaline soil. A pH about 6.5 is good for most common annuals or biennials. Minimum amounts of fertilizer are usually used, particularly for Madagascar periwinkle and wax begonia. Soluble salts should be low, especially for snapdragon and scarlet sage. For very small seed the surface of the medium should be relatively tight, either lightly pressed down or very finely screened. The surface of the medium should never become dry because this may cause crusting. Also avoid excess water at the surface to prevent damping-off, a seedling disease discussed in Chapter 7. Damping-off can also be prevented by covering seeds with fine sphagnum moss, which contains a natural fungus inhibitor. Although a pasteurized medium may be all that is necessary to prevent damping-off, the sphagnum will eliminate recontamination. Whether sown in rows or broadcast, seeds should be sown thinly to promote strong seedling growth by avoiding excess competition and to avoid tangling of roots, lessening transplant shock. If a sphagnum cover is not used, seeding in rows will prevent spread of damping-off should it occur.

Primary considerations for outdoor sowing are (1) to finely and evenly pulverize the seed bed and covering soil, (2) to sow thinly, and (3) to prevent drying of the soil surface. The last can be accomplished by periodic sprinkling, covering with a light shade cloth, or covering with a light mulch. Well-spaced sowing is difficult outdoors, particularly with small seeded species. These might be mixed with sand or some similar sterile material to ensure more uniform spacing.

The day a good stand of seedlings has developed from indoor sowing, they should be moved from germination conditions to the initial growing environment. When a good stand has developed outdoors, it is time to remove their protective covering or discontinue surface spraying.

GROWING SEEDLINGS OUTDOORS

Growth of outdoor-started seedlings is influenced mainly by moisture control, fertilization practices, pest control, and spacing. Some light and temperature modification is possible, but under most outdoor conditions plants must exist with what nature provides.

Moisture Control

Moisture control can begin by removing any light mulch from seed beds, pulling it between rows to retain soil moisture. This will reduce soil temperature on sunny days as well as provide weed control if enough mulch is used to exclude light. Watering an inch a week is a good rule of thumb throughout the growing season. Deep watering from the start will encourage deep root penetration. Overhead watering of young seedlings should be done with a fine spray to avoid knocking over small plants. On very hot days, occasional sprays will elevate humidity, reducing transpiration and avoiding wilting.

Fertilization

Fertilization can be done with watering or by incorporating manure or dry inorganic fertilizers into the soil. Unless the soil is already high in organic matter, at least some manure or other organic material should be added, for good growth throughout the year as well as during the seedling stage. The best time for adding organic matter is fall or very early spring because time is needed for decomposition. However, organic fertilizers alone are seldom enough unless soil fertility is high initially or an adequate organic fertilization program has been carried through several years.

Inorganic fertilizers can be applied either before or after seeding has been done, although they are simpler to apply before seeding so they can be easily raked in. Application of two to three pounds of 10-10-10 or equivalent per hundred square feet when first working the soil should be sufficient until about a month after growth has started, when the same amount can be repeated. If fertilization has not been done early, one application of 10-10-10 or equivalent at about two to three pounds per hundred feet of row about two weeks after initial growth should be enough during early development. Another excellent procedure is to use a liquid feed about every two weeks with a soluble fertilizer of about a 1-1-1 ratio. If 20-20-20 were used there should be no more than two level tablespoons mixed per gallon water. Slow-release fertilizers, applied when preparing soil in spring, or along rows later, are also effective. If the soil is low in phosphorus a fertilizer high in this nutrient, such as 10-50-10, should be used at the first feeding. A high-phosphorus liquid fertilizer used when watering in transplants will stimulate quick root growth. As soon as flower buds begin to form a shift should be made to a low-nitrogen fertilizer, whether using a dry or soluble form.

Pest Control

Early pest control is usually concerned with preventing damping-off, controlling slugs, and beginning the fight against weeds. Damping-off of succulent seedlings can be minimized by thin seedsowing. Fungicide treatment can totally eliminate the problem. Several fungicides will work with all kinds of damping-off fungi, others should be used in combination. For instance, Dexon and Terraclor need to be applied together while Banrot is effective alone. Damping-off is not always troublesome outdoors and chemical treatments should only be used in problem areas.

Slugs often feed on young seedlings, particularly in damp weather. They are difficult to eliminate, but some control is possible using meal or pellet slug bait or a "home style" bait such as a shallow dish of beer. Most commercial meal and pellet baits are highly poisonous and care should be taken to keep pets and children away from them, possibly putting them out only at night. Chewing insect larvae can be controlled by several inorganic and organic insecticides.

Without adequate control from the beginning, weeds are likely to reduce quality and quantity of garden flowers more than any other pest. Much of the weed battle can be won by mulching between rows. Nearly anything that will exclude light can be effective, such as grass clippings, chopped weeds, newspapers, old rug strips, or black plastic. Organic materials have the advantage of contributing to the development of good soil structure. Some weeds will still have to be pulled by hand, but the job will be much easier. Without mulches, the weeds will yield to hand and hoe, but the time necessary to do a good job will be considerable.

Birds, rodents, or rabbits often pluck out or chew off young seedlings. The only practical home remedy is a screen covering that will keep animals from reaching the plants.

Spacing

Thinning of rows can be accomplished any time after seedling identification is possible, which is usually after about two pairs of true leaves have developed above the cotyledons. Seedlings may be removed carefully from rows and reset elsewhere or pulled up and discarded, but to avoid root damage to plants to be left it is better to snip off excess seedlings. The best procedure if seedlings are to be transplanted elsewhere is to clip off very closely spaced plants, leaving plants about an inch apart, then transplanting after slightly more growth when the row is thinned to its final spacing. The distance between plants will depend upon size of the mature plant and whether the gardener desires a solid stand or a series of individual specimens. Use of the spacing recommendations on seed packets typically will result in plants being just together at maturity.

GROWING SEEDLINGS INDOORS

Environmental modification for growing seedlings indoors begins with temperature and light control. Moisture control, fertilization, pest control, spacing, transplanting, and pinching are also discussed.

Temperature Control

Night temperature should be optimum, as listed for particular species in column 5 of Tables 13-1 and 13-2. Day control, usually more difficult but not as critical, should be about 10°F higher. A common regime is a short initial period at the germination temperature followed by reduction to a growing temperature that should be maintained until transplant into market packs or small pots, after which the plants are grown somewhat warmer. Some species do not need different temperatures before and after transplant, but many develop most vigorously if temperature

is reduced immediately after transplant. Temperatures 10°F lower than optimum will result in unacceptably poor growth of begonia, celosia, coleus, and zinnia. A temperature of 10°F higher is too warm for calendula, celosia, bachelor's button, dahlia, dianthus, impatiens, phlox, salvia, Madagascar periwinkle, zinnia, and ornamental pepper.

Light Control

Lack of adequate sunlight is usually the major factor limiting the production of strong seedlings on windowsills. Most seedlings, except those that tolerate shade, need at least six hours of direct sun daily to avoid etiolation and weak growth. Flowers noted in the culture tables (Tables 13-1 and 13-2, column 5) as needing sun may not develop as strongly as possible even with six hours of sunlight and definitely need all they can get. This six hours of light are seldom available on a windowsill, so alternatives are either to supplement natural light or grow entirely under artificial light. Details of indoor lighting are discussed in Chapter 14. Cool white fluorescent tubes operating over a 16-hour period are recommended for most species. Some species definitely need long days because they form flowers at a small size and never develop normal stature under short days, whereas others need

TABLE 13-6 Flowers Requiring Long Days and Short Days for Best Flowering

Long days		
Celosia	Sunflower	Scarlet sage
Chrysanthemum	Cardinal climber	Marigold
Cosmos	Morning glory	Zinnia
Dahlia		

Short days			
Snapdragon	Gaillardia	Shirley poppy	Gloriosa daisy
China aster	Globe amaranth	Petunia	Salpiglosis
Bachelor's button	California poppy	Drummond phlox	Scabiosa
Calliopsis	Iceland poppy	Portulaca	Verbena

short (10- to 12-hour) days because of poor branching or flowering when too small if grown under long days. Examples of flowers in each of these categories are given in Table 13-6.

Most species grow well in open sunlight, but some shade-loving plants, such as begonia, coleus, and impatiens, require filtered sun. Filtered sun for a few days after germination can be helpful for most species, particularly on very bright days, but only the Drummond phlox seems to require this treatment.

Moisture Control

The medium for indoor-grown seedlings should be neither too soggy nor too dry. Overwatering can decrease absorption of water and nutrients as well as initiating damping-off. Also, toxic compounds are more readily formed in poorly aerated soil. Strawflower growth is seriously reduced by waterlogged soil. Cockscomb, portulaca, and zinnia should be kept slightly on the dry side. Cockscomb and scarlet sage are particularly sensitive to any growth check; the correct amount of water for them is essential.

Fertilization

Fertilization programs can either incorporate quick- or slow-release fertilizer into the soil or apply soluble nutrients with watering. A simple and effective method is to mix soluble fertilizer into the water, beginning with the first watering after transplanting. A first fertilization with a solution such as 10-40-10 or 10-52-17 will stimulate root production. After that, growth will be best using a balanced fertilizer such as 15-15-15 at one tablespoon per gallon, or the equivalent, at about 7- to 14-day intervals. Because they will remain in seedling flats so long, slow-growing species should be fertilized before transplanting, at about half the above rate. Remember that visible deficiency symptoms mean reductions in both quality and quantity of growth have al-

ready taken place. Overfertilization can cause equally serious growth reduction.

Fast-growing plants such as sunflower, zinnia, morning glory, and celosia seem to suffer more from inadequate fertilization than such slow developers as begonia, browallia, gerbera, and impatiens, but this is a deception. Deficiency symptoms and decreases in growth take longer to become obvious in slow-growing species, but as much reduction in vigor can occur. Gerbera is particularly sensitive to low nutrient levels and needs a well-fertilized medium from the start. Mature plants of species such as gaillardia, portulaca, and California poppy do well in poor soils; some others, notably nasturtium, produce too much vegetation and not enough flowers in rich soils. Young seedlings of all species, however, grow best in well-fertilized media.

Pest Control

With indoor-grown seedlings, preventing damping-off is critical. A Dexon-Terraclor or Banrot or similar drench should be used as a precautionary measure on all seedling populations as they are germinating, unless a sphagnum cover has been used, and at first transplant. If damping-off does appear, treat again with a fungicide after discarding affected plants and surrounding soil. The most common indoor insect pests are white flies and aphids, which can be controlled by several organic and inorganic insecticides. Sometimes aphids can simply be washed off. Spider mites can also be very destructive; these are usually controlled with a miticide such as Kelthane or the organic materials pyrethrum or rotenone.

Spacing

Unless sowing has been done in individual pots or seed have been spaced well apart, transplanting is necessary to avoid overcrowding. Vigor of crowded seedlings of most species, particularly geranium, will decrease quickly as development proceeds. Usually seedlings should have two to four true leaves at transplant time. Most flowers are best transplanted as individual plants, handling them by the leaves rather than stems, although lobelia, sweet alyssum, and portulaca can be moved in small clumps. Roots and stems of these three will intertwine and a small group will appear to develop as a single plant. For minimum disturbance of roots when plants are moved to their final locations, it is best to transplant seedlings into individual pots or sections of a divided container. Most commonly used are inexpensive plastic cell packs with 4 to 12 compartments, peat pots or bands, and special containers such as Jiffy-7's and Kys-Kubes (Figs. 7-23, 7-24). Another possibility for seedlings such as sweet alyssum, portulaca, and nemesia that are to be in rows at close spacing is a flat with lengthwise divisions only and a removable end so rows can be easily slipped out intact. This sort of flat will also work well for seeding of difficult-to-transplant flowers such as sweet pea and larkspur. Also, young plants can be spaced several inches apart in open flats. With this system, growth to the point of obvious crowding must be carefully avoided to prevent root injury when transferring plants to permanent locations.

Transplanting

Most species transplant well if given normal care, but several tap-rooted species are notoriously difficult, including larkspur, lupine, the poppies, statice, sweet pea, and morning glory. Lupine is easy to handle when it is only an inch or two tall, but large plants are difficult to relocate without serious setback in growth. Statice can be transplanted successfully, but only if great care is taken not to injure the tap root. California poppy is very difficult to transplant successfully as a small seedling, but if several seed are sown in spots a few inches apart in open flats and the excess

later snipped out, the plants can be divided with a sharp knife when they are ready to be moved to the garden. Most species will germinate fairly evenly and can be transplanted all at once. A few others, such as begonia, germinate over a longer period, thus prolonging the transplant period.

Media used for seedlings after transplant should be pasteurized, well drained, and have good water-holding capacity. A good mix has one part soil, one part peat or leaf mold, and a third sand, vermiculite, or perlite (Table 14-2) to improve drainage. Good drainage is particularly important for amaranthus, cosmos, Cape marigold, and California poppy. A medium with the above characteristics will be loose, so when plants are ready to be put into permanent locations they can be removed from flats with a minimum of root damage if they are not too large.

Flower Bud Formation, Branching, and Pinching

Smaller species and dwarf cultivars can be expected to set flower buds and sometimes flower before being planted in permanent locations, as the dwarf marigolds in Fig. 13-2. This is the best stage for moving these plants from flats to garden. The compact branching habit sometimes needs to be induced by **pinching,** removal of the terminal bud or a slightly longer portion of the stem tip (Fig. 13-3), particularly in carnation, and sometimes in petunia, zinnia, snapdragon, and others. The formation of long internodes or thin stems are indications of too much or undesirably weak growth. Because budded plants would be too crowded in flats, there should be no floral bud formation on transplants of (1) tall cultivars of snapdragon, China aster, dahlia, strawflower, marigold, and zinnia, (2) any cultivar of calendula, bachelor's button, chrysanthemum, carnation, and gaillardia, (3) such naturally

FIGURE 13-2 Dwarf French marigolds, ideal size for transplanting to permanent locations. Grown by vocational horticulture students, Hermitage High School, Henrico County, Virginia.

tall growing flowers as cleome, sunflower, and Mexican sunflower, or on (4) vining species.

Hardening

The last step before transplanting flowers to the garden is to prepare them for their new environment. Time-honored methods of hardening for this experience involve reducing temperature, water, and fertilizer, and exposing plants to some wind. Proper harden-

ing, as discussed in Chapter 7, is necessary so that plants will have adapted anatomically and physiologically to begin growth as soon as they are set into the garden. To make this possible, expose indoor-grown plants to the minimum of hardening necessary to prevent wilting and damage when setting outdoors. Excessive hardening will result in slow growth. Probably most effective is to set seedlings in a cold frame (Fig. 7-26) and gradually expose to outdoor conditions. If cold frames are not available,

flats can be set outdoors on favorable days, then brought in whenever night temperature or other conditions are not conducive to good growth.

CARE OF FLOWERS IN THE GARDEN

Flower production can only be successful if plants have the right temperature, water, nutrients, pest control, and such cultural aids as pinching and staking.

Temperature

Buildings and walls usually raise temperatures of nearby areas. Light colors and reflective surfaces can cause overheating if plants receive the reflected light. Many building materials, particularly stone, brick, and other types of masonry, absorb and retain heat, stabilizing temperatures at higher levels than open areas. Thus, flowers may be planted near buildings

FIGURE 13-3 Pinching of zinnia 'Gold Medal' to induce branching, (a) Unpinched plant, (b) Plant with terminal half inch removed, (c) Left, plant two weeks after pinch; right, plant developing without pinch.

earlier than in unprotected locations. Structures and plants can also reduce and alter wind flow, thereby reducing transpiration.

Some plants thrive in heat, whereas others develop poorly in hot locations, blooming best in cool situations (Table 13-7). Heat resistant cultivars are available for some of the more sensitive species, including calendula, carnation, sweet pea, and bachelor's button. A **trysomic** (possessing one or more extra chromosomes) cultivar of stock, 'Trysomic Seven Week,' will bloom all summer.

Moisture

Throughout the growing season the rule of thumb calls for about an inch of water a week for good growth of garden plants. There are no common annuals or biennials that will not grow well with this amount of moisture, although some are particularly tolerant of less, especially plants in this list:

African daisy	Summer cypress
Cockscomb	Portulaca
Cleome	Rudbeckia
Cape marigold	Scabiosa
California poppy	Mexican sunflower
Gaillardia	Verbena
Gazania	Sunflower
Baby's breath	

Overwatering, as previously mentioned, will decrease growth. Most drought-resistant plants are susceptible to overwatering particularly cockscomb, African daisy, and Cape marigold. Also sensitive to overwatering, and needing particularly good drainage, are snapdragon, calendula, creeping zinnia, and nasturtium.

Fertilization

Fertilization of flowers in the garden can be carried out in several ways. After transplant-

ing watering can be done with a starter solution such as 10-40-10 or a similar high-phosphorus material, mixed at one level tablespoon per gallon of water and applied inside a dike formed around the plant. This can be followed by liquid or granular application along rows. A soil analysis (Chapter 7) will indicate the exact fertilization program that should be followed. Lacking this, one possibility is about two cups of 5-10-5 or equivalent banded along each 100 feet of row, two or three times during the growing season. The general rule is relatively light fertilization and definitely lower nitrogen after floral buds begin to form. This is particularly important for nasturtium and some other species with an indeterminate growth habit, including morning glory, petunia, and scarlet runner bean. Lots of vegetation and few flowers is the result of too much nitrogen. Pansy is one annual that does flower best in rich soil. Conversely, amaranthus, calendula, California poppy, gazania, portulaca, and especially nasturtium have well-deserved reputations for tolerance of poor soil.

Pests

Weeds are likely to reduce growth and quality of flowers more than all diseases and insects combined. Fortunately, weeds can be suppressed by a variety of mulches (Table 8-2), and obviously can be hand pulled or hoed. Herbicides that might be used in a commercial operation are not practical on a home garden scale, as mentioned in Chapter 9, because of the proximity of differing crops, some of which would suffer the same fate as weeds. Organic mulches such as grass clippings, partially rotted hay, or sawdust are effective and aesthetic in flower beds. Nitrogen deficiency may result, particularly with sawdust, but nitrogen fertilization will compensate.

Most common garden flowers are not damaged by serious insect and disease pests. However, some are susceptible to specific prob-

TABLE 13-7 Flowers That Thrive in Heat and Flowers That Thrive in Cool Situations

Flowers that thrive in heat		
African daisy	Gazania	Globe amaranth
Cockscomb	Portulaca	Salvia
Cape marigold	Mexican sunflower	Zinnia

Flowers that thrive in cool situations			
Snapdragon	Globe candytuft	Nemesia	Butterfly flower
Calendula	Rocket candytuft	Iceland poppy	Pansy
Carnation	Sweet pea	Shirley poppy	Stock
Pinks	Virginia stock	Drummond phlox	Bachelor's button

lems. **Powdery mildew,** caused by a fungus, is the most common disease on a wide variety of flowers, especially zinnia, gloriosa daisy, and larkspur. The disease begins as white or gray powdery spots, usually on lower leaves. Later the spots enlarge and cause disfiguring or even leaf death. Several common fungicides are effective in prevention and early suppression of powdery mildew, but once the condition is in an advanced stage, elimination is very difficult. Another common fungal disease is **rust,** afflicting roses, snapdragon, and hollyhock. This disease begins as small brown or orange spots, usually on undersurfaces of older leaves and lower stems. It is controlled by fungicides or sulphur, the latter dusted on plants at about two-week intervals. Hollyhocks will get rust anywhere they are grown and treatment must be begun early to prevent disfigurement. If rust spots appear, infected leaves should be removed. Hollyhocks grown as biennials should have all leaves removed as growth is beginning in spring, even if no evidence of the disease is apparent.

Aster yellows, causes yellow, deformed leaves, and stunted growth of China aster and marigold, ruining both. It is carried by the six-spotted leaf hopper, which must be con-trolled to eliminate the disease. There is no treatment for aster yellows. Infected plants should be pulled up and buried or burned.

Wilt diseases of several sorts can affect various annuals and biennials. These soil-borne diseases are prevalent under wet conditions, particularly in China aster and carnation. The most common causal agents are *Verticillium* and *Fusarium* fungi, and bacteria, which affect carnations particularly severely. Resistant cultivars should be grown. Wilted plants should be promptly destroyed. In general, most diseases are best prevented by using resistant cultivars whenever these are available.

A number of insects may also cause problems (Table 13-8). Inasmuch as insect resistance is not yet an effective tool, dependence on insecticides will be necessary if insects threaten to get out of hand. Most insecticides are not particularly specific, so considerable care should be exercised in their use to avoid killing pollinators, predators, and other beneficial types. Even organic insecticides, although low in toxicity to humans, should be directed at target populations only. Rotenone, which is toxic to fish, and pyrethrum, a common ingredient of household pressurized insecticides, are the most readily available com-

TABLE 13-8 Some of the More Common Insect Pests and Flower Species Commonly Attacked

Insect	Species
Aphids of various species	Nasturtiums, sweet peas, chrysanthemum, ornamental pepper, lupine, Iceland and Shirley poppies, roses
Stem borers, several species	Dahlia, ornamental pepper, sunflower
Leaf miners	Chrysanthemum, ornamental pepper (in hot areas), verbena
Leaf-chewing larvae of various species, such as cabbage looper and several cutworms	A wide variety of garden flowers
Mexican bean beetle	Scarlet runner bean

mercial organics. Home preparations, such as ground garlic, are also sometimes useful. A soap solution will wash off some insects, including aphids. Control of these and other insects can be achieved by a variety of methods, including those mentioned above. Almost all insects can be controlled by insecticides; specific recommendations should be acquired from the appropriate agricultural information agencies listed in Appendix C. These sources will keep gardeners abreast of the latest trends in pest management.

Several mite species are commonly destructive to many flowers, including carnations and pinks, phlox, marigolds, and thunbergia. Mites are difficult to see, being very tiny and living mainly on the undersides of leaves. Their presence should be suspected if upper surfaces of leaves are finely stippled with yellow spots. Webs of spider mites are easier to spot than the minute arachnids themselves. Mites are sometimes controlled by broad spectrum insecticides, but there are specific miticides such as the previously mentioned Kelthane.

Slugs will eat numerous flower crops, but are probably worst on calendula during moist fall weather. These can be handled in the same manner as they were with seedlings. When using any pesticide, apply only as directed on the label.

Pinching

Several annuals will benefit from pinching of the terminal shoot and sometimes lateral branches to induce fullness (Fig. 13-3). Even if this was done when plants were small, a repeat might be necessary, particularly with salpiglosis, petunia, and many coleus cultivars. If no pinching was done before transplanting to the garden, it will be beneficial for most cultivars of calendula, carnation, nemesia, zinnia, and the above three. The best stage for pinching is usually before, but no later than shortly after, formation of the first floral bud. The pinch should be made just above a lateral bud. While it may be effective later, the pinch will have to be made in older tissue, which usually means a longer wait for formation of good branches.

Often the first flush of flowering seems the most beautiful, but later bloom can be just as good with removal of old flowers and cutting back excess growth. When the central spike is left on such flowers as snapdragon, foxglove, and larkspur, seed will be formed and the plant's strength will be channeled into development of that seed to the detriment of further vegetative and floral growth. The same problem exists with petunia, pansy, and many others, although from these, individual flowers should be removed as they fade. It is apparent that removing old flowers from some species, for instance, lobelia, sweet alyssum, and ageratum, would be impractical even though it might be beneficial. Also, many of these very floriferous types do continue to produce flowers well even though they are also forming seed. Eventually, however, most plants except sterile ones will have relatively few flowers and many fruit if normal development is allowed to proceed. Some species can be rejuvenated when this happens by cutting them back severely, often to within a few inches of the ground. Lobelia, sweet alyssum, petunia, pansy, and numerous others will respond well to shearing in midsummer, although they will appear quite bare for a brief period. Chinese pinks can even be cut down with a lawn mower at the high setting.

Reseeding

Some species that ordinarily form seed will reseed themselves in areas that are not completely cultivated the next spring. Some efficient reseeders are sweet alyssum, California poppy, ageratum, English daisy, bells of Ireland, flowering tobacco, and, in mild areas, marigold and wax begonia. Foxglove and sometimes snapdragon will reseed readily if the spikes are not all removed and many petunia and pansy cultivars will also, if faded flowers are not pinched off. In frost-free areas many flowers will reseed themselves if the ground is not kept mulched or cultivated.

Staking

A few flowers may benefit from staking, including salpiglosis, larkspur, foxglove, and tall cultivars of snapdragon, carnation, and dahlia. Snapdragon might also be kept upright by hilling soil around bases of plants as they develop toward mature size. Staking is not always necessary, but should be used as a precautionary measure in windy areas or with topheavy cultivars such as 'Americana', bachelor's button, 'Sensation' cosmos, and 'Giant Imperial' scabiosa. Stakes should be placed next to plants when they are young to avoid root injury, with ties being made as growth occurs.

Flowers in the Landscape

The appearance of mature plants is important as it relates to their landscape use. Except for a few tall species with open growth habit, such as sunflower (Fig. 13-4), the desired vegetative form is compact and bushy. This is true whether the plant is low and spreading (Fig. 13-5), upright but of moderate height (Fig. 13-6), or tall with flowers in spike form, as the scarlet sage in Fig. 13-7. Even vines such as morning glory or thunbergia are at their best when relatively compact, with internodes short enough to give the plants an appearance of being entirely clothed with leaves. Cultivars with unbroken flower cover usually give the most satisfying display, as shown in the scarlet sage (Fig. 13-7), begonia (Fig. 13-8), and impatiens (Fig. 13-9) beds. The begonias were placed so their branches slightly intertwine, while the upright scarlet sage plants just touch. The large impatiens plants also just touch, giving an undulating appearance to the bed, which also illustrates pleasing use of a color mix. Various species likewise can be effective together in random or pattern designs (Fig. 13-10). Depending on the situation, impressive creations are equally possible from various styles of formal (Figs. 13-7, and 13-8),

FIGURE 13-4 Tall, unbranched sunflowers behind fence of Apothecary Shop Garden, Colonial Williamsburg, Virginia.

flowing free form (Fig. 13-9), or carefully patterned informal (Fig. 13-10) designs. Within any basic design, consideration must be given to color, texture, and growth form in deciding which species and cultivars to include. Colors may be harmonized by blending tones and basic colors that are near each other in the spectrum, as was done with the red, pink, lavender, and white impatiens of Fig. 13-9. Colors that contrast, lying in opposite portions of the spectrum, are complimentary and when juxtaposed will produce dramatic effects, as with the blue-violet ageratum and yellow marigolds of Fig. 13-10. This same design also illustrates an integration of elements brought about by the curving line of the subordinate blue-gray, light, fine-textured lavender cotton, a woody perennial. Subordination of some colors, emphasizing dominance of others, will help concentrate attention and prevent designs from appearing discordant and confused. Combining various plant forms, upright with spreading, for instance, can add to the interest of designs. Otherwise, form and size of a cultivar will dictate relative placement, tall cleome as a background for salvia, for example. These comments on utilization of color, form, and texture as developed in

FIGURE 13-5 Prostrate sweet alyssum, All-America Trial Gardens, Virginia Polytechnic Institute and State University, Blacksburg, Virginia.

mature flowers should provide a partial basis for understanding the complexities of flower garden design.

A CAPSULE REVIEW OF FLOWERING HERBACEOUS AND WOODY PERENNIALS

Flowering perennials are as important to gardeners as annuals and biennials, perhaps even more so. Each perennial has its individual growth requirements as do all other plants, and a few representative, widely grown examples will be considered. Unlike annuals and biennials, asexual reproduction is very important in propagating cultivars of most perennials. Our examples will illustrate a broad range of vegetative structures and propagation methods described in Chapter 3, and will include seed-grown material.

Flowering perennials have the same basic needs as annuals and biennials, although some cultural practices may be emphasized differently. For instance, various pruning practices are often important in maintaining perennials, particularly woody ones, compared to little or no necessity for pruning most annuals and biennials. Watering and fertilization are often considered unimportant for

FIGURE 13-6 Fifteen-inch upright, well-branched China aster, All-America Trail Gardens, Virginia Polytechnic Institute and State University, Blacksburg, Virginia.

many perennials, which seem to grow well on what nature provides. However, growth and flowering of perennials can usually be improved with a regular watering and fertilization program similar to that for other garden plants. Insect, disease, and weed control are also important with perennials. Care is necessary to avoid leaving overwintering stages, such as spores of fungi, insect larvae or pupae, or weed seeds, on or near plants.

Propagation

All perennials except sterile ones can be seed propagated, but many are commonly reproduced asexually, either because seed grown plants are inferior to asexually propagated cultivars or seed reproduction is difficult or slow. Table 13-9 lists some widely grown perennials and methods commonly used for their propagation. Often one method is used commercially while another is more likely employed by home gardeners. In general, grafting and budding are most commonly used by commerical growers. This is certainly true of roses, for instance, which are often reproduced from stem cuttings by home gardeners.

Seed Propagation

Perennials such as oriental poppy, chrysanthemum, crape myrtle, and rhododendron can be grown from seed in the same manner as annuals and biennials. However, another group, including phlox, dogwood, hawthorn, and magnolia must be artificially stratified or sown in fall for spring germination. **Stratification** involves thoroughly cleaned seeds being subjected to a cold (34 to 50°F) moist period of one to four months, although in some cases a warm (55 to 75°F constant or 68 to 86°F alternating night-day) period of as much as seven months is necessary prior to the cold. Exact temperatures and times needed depend on species. The most common treatment is 40°F for about three months, and if a warm pretreatment is necessary usually three to five months at room temperature is sufficient.

Crown Division

Propagation by division is a simple procedure. Almost all gardeners who grow perennials increase plants by digging and separating crowns or some underground structure. Division of day lilies (Fig. 3-7) is typical of the operation with most crown-forming perennials. About every five years there tends to be much foliage and few flowers, so in fall plants can be dug, pried apart into several sections about six inches across, then promptly replanted in well-worked soil. Bloom may be a bit sparse the next year, but by the second year they will be producing a beautiful summer display, lasting about a month. A dividend from day lilies is that they will bloom nearly as well in part shade as in full sun. They can also be used as a large ground cover, and while flowering will not be

TABLE 13.9 Propagation Methods for Selected Herbaceous and Woody Perennials

Scientific and common name[a]	Seed		Cuttings		Division							Layering	Grafting		Budding
	Nonstratified	Stratified	Stem	Root	Crown	Suckers	Bulbs	Corms	Rhizomes	Tubers	Tuberous roots		Stem	Root	
Ajuga reptans, Ajuga	X				X										
Allium species, Ornamental Onions	X		X				X								
Aster species, Aster	X				X										
Begonia species, Begonia (tuberous)	X										X				
Caladium species, Caladium	X									X					
Camellia japonica and *C. sasanqua*, Camellia			X										X		
Canna generalis, Canna	X								X						
Chrysanthemum morifolium, Chrysanthemum			X		X										
Clematis species, Clematis			X									X		X	
Cornus species, Dogwood	X	X	X												
Crataegus species, Hawthorn	X	X												X	X
Crocus species, Crocus								X							X
Dahlia hybrids, Dahlia	X				X						X				
Delphinium elatum, Delphinium	X		X												
Forsythia x intermedia, Forsythia			X												
Gladiolus species, Gladiola								X							
Hemerocallis species, Day Lily					X										
Hibiscus rosa-sinensis and *H. syriacus*, Hibiscus			X												

430

Plant										
Hosta species, Hosta	x									
Hyacinthus orientalis, Hyacinth		x								
Hydrangea species, Hydrangea	x		x		x					
Hydrangea quercifolia, Oak Leaf Hydrangea		x								
Iris species, Iris (rhizomatous)							x			
Lagerstroemia indica, Crape Myrtle	x	x								
Lilium species, Lily	x				x					
Lonicera species, Honeysuckle	x	x							x	
Magnolia species, Magnolia	x	x	x						x	x
Malus species, Crabapple					x				x	x
Narcissus species, Daffodil		x								
Nerium oleander, Oleander	x		x		x					
Paeonia species, Peony	x	x	x							
Papaver orientale, Oriental Poppy		x		x						
Philadelphus species, Mockorange	x		x	x						
Phlox paniculata, Phlox	x	x	x							
Prunus species, Ornamental Cherries	x		x						x	x
Rhododendron species, Rhododendrons and Azaleas	x									
Rosa species, Rose		x							x	x
Spiraea species, Spirea		x	x						x	x
Syringa species, Lilac		x	x	x					x	x
Viburnum species, Viburnum		x	x							
Vinca minor and *V. major*, Periwinkle		x							x	
Wisteria sinensis and *W. floribunda*, Wisteria	x		x						x	x

*Perennials listed as "species" commonly include hybrids not distinctly assignable to a particular species, as well as individual species.

FIGURE 13-7 Round, formal bed of scarlet sage, Apothecaries Shop, Yorktown, Virginia.

as prolific as with periodically divided plants, there will be an acceptable display. In general, division is best done when plants are dormant or when they are just beginning to grow in spring.

Many other perennials can also be propagated by crown division. Summer-blooming phlox and fall-blooming asters can be divided in either spring or fall, leaving several old stalks in each portion. Summer-flowering delphiniums and hosta as well as late summer-fall blooming chrysanthemums are usually divided in spring, with the latter best reduced annually to single stems and their roots. Oriental poppies can be separated in summer after the leaves have died down. Peonies are usually divided to three to four buds in early fall, when more plants are desired, because, unlike most other herbaceous perennials, they can

continue to produce a first-rate display almost indefinitely. Peonies produce quite fleshy crowns with comparatively large, obvious buds. The root mass should be washed off, then sliced apart with a sharp knife. Division of woody plants such as mockorange or spirea is usually a rough operation with a spade or even an ax, ordinarily done in spring just before onset of growth. A partial division is sometimes utilized with such plants as lilac, in which suckers and their roots are removed from the side of the main plant.

Division of Bulbs and Other Fleshy Underground Structures

An entirely separate group of perennials is propagated by division of fleshy underground structures: bulbs, corms, rhizomes,

FIGURE 13-8 Tightly planted formal bed of wax begonias, Meadowbrook Hall, Rochester, Michigan.

tubers, and tuberous roots (discussed in Chapter 3). Tulips and daffodils produce **offsets,** large lateral bulbs that are easily separated from the base of the main bulb (Fig. 3-5*b*). Hyacinths and lilies also form offsets, but slowly, thus they are usually propagated by **bulblets,** small bulbs formed on underground structures. Lilies also often form usable **bulbils,** small bulbs produced on aerial parts. Bulblets produced at the base of individually planted bulb scales can be used to propagate most lily species. Hyacinths can be encouraged to form bulblets by scooping out or coring the center of the bulb base, or by cutting several scores across the base. Orna-

mental onions are usually propagated by bulblets, or by bulbils formed in the flower heads. Crocus and gladiolus will sometimes produce two or rarely more new corms after flowering and ordinarily form many small **cormels,** new miniature corms, between the old and new corms (Fig. 3-5*a*). These will usually flower in two years, larger ones sometimes in one. Bulbs and corms are divided at the end of their growth period, after the leaves have died. Rhizomes of iris and canna can be cut into sections with at least one vegetative bud, either at the end of the growing period or just as the next begins. The tuber of caladium can be cut into sections, like a potato, each usually

FIGURE 13-9 Flowing free-form bed of mixed shades of red, pink, lavender, and white impatiens, National Botanic Garden, Washington, D.C.

with two beginning shoots. Tuberous roots of dahlias (Fig. 3-6) and tuberous begonias are very different in appearance, but division of both must include one or more buds that are located at the base of the crown, or depending on how one looks at them, at the top of the tuberous root. Most of these plants flower much more strongly with periodic division. Tulip, hyacinth, and gladiolus benefit particularly from division every year or two.

Propagation by Cuttings

Hardwood-cutting propagation (fully woody, mature stems, usually taken during dor-

mancy) of such plants as forsythia, honeysuckle, spirea, some dogwood species, Rose-of-Sharon, crape myrtle, mockorange, and roses is usually easier for the home gardener than use of leafy cuttings. However, hardwood cuttings of species such as hydrangea, several magnolias, viburnum, deciduous azaleas, and some dogwoods, including *Cornus florida*, flowering dogwood, do not root well. For these **softwood** cuttings (immature stems, not yet woody at base) taken from new growth are best. **Semihardwood** cuttings (partially mature stems, beginning to become woody at base) are usually employed with evergreen rhododendrons and azaleas, Chinese hibiscus, ca-

FIGURE 13-10 Carefully blended informal pattern integrated by rows of blue-gray lavender cotton curving through the design; yellow marigolds (foreground, curving to right); blue-violet ageratum (left, below sword-leaved yucca), deep red verbena (center and curving right under small tree); Horticultural Research Institute of Ontario, Vineland Station, Canada.

mellia, clematis, and the very easily rooted oleander. Softwood cuttings taken in spring are usually used for herbaceous material, including chrysanthemum and delphinium, but some species, such as phlox, can be rooted most any time from spring through fall. Layering in spring is an easy method of propagating such vines as clematis, honeysuckle, periwinkle, and wisteria, and can be used with some shrubs that have branches near the ground.

Grafting

Most grafting, as with camellia, ornamental cherries, and crabapples, is done in spring just as growth is ready to start, although magnolia is grafted in mid to late summer when the rootstock is in active growth. Budding of crabapples, ornamental cherries, hawthorn, roses, and various other species, using dormant bud material, is usually accomplished in spring or fall when bark is slipping best (in most active growth, with maximum water supply to the cambium).

Special Cultural Requirements for Selected Perennials

Roses have a reputation for cultural difficulty. Compared to a nearly problem-free flower such as peony they do require more care, with attention to pruning and pest control being critical. Otherwise, they respond well to much the same treatment as do most other flowering perennials. Some perennials do have special cultural requirements, or preferences, generally not difficult to supply. Clematis needs slightly alkaline soil (pH 7.0 to 7.5) and cool roots (shaded), and care should be taken not to injure the rather brittle stems to prevent entrance of disease. Lilac is strongest on neutral soil, while camellias, rhododendrons, and azaleas require acid (pH 4.5 to 6.0) soil. Camellias, rhododendrons, and azaleas, as well as tuberous begonias and most hosta need some shade. Those excellent ground covers ajuga, periwinkle, and pachysandra, and most dogwood, viburnum, and aster species also grow well in varying amounts of shade. Although not requiring it, oleander and some aster species do very well in dry situations. On the other hand, New England aster and rose mallow do well in wet soils.

Climatic conditions have a distinct effect on choice of flowering perennial; for instance, delphiniums and most rhododendrons and azaleas are at their best under cool summer conditions while crape myrtle flowers well only in hot summer areas. Cold hardiness limits the use of crape myrtle, oleander, camellia, Chinese hibiscus, and many other flowering perennials. Lack of sufficient cold restricts the southern range of such winter-chill-requiring plants as crabapple, peony, and most of the northern spring flowering bulbs. There are also numerous diseases and insects that can afflict particular perennials, but none that should prevent any from being grown. Despite potential problems or specific cultural requirements, there are flowering perennials suitable for any location.

A POTPOURRI OF FLOWER-GROWING TOPICS

All-America Selections

Familiarity and experience with species and their cultivars will greatly aid gardeners in choosing plants. Another help is to consider All-America Selections, or selections from similar organizations in other countries, such as Fleuroselect in Europe. All-America Selections, as discussed in Chapter 7, is a cooperative program sponsored by seed-testing associations throughout North America. Each year seed packets are distributed to a series of trial gardens (Fig. 13-11) where they are grown and judged in comparison to the closest equivalent cultivar on the market. Usually few cultivars are awarded All-America Selections medals; there may even be years with no selections. Cultivars may be honored with bronze or silver medals or the more rarely bestowed gold medal, an indication of exceptional excellence. Flowers awarded All-America Selections medals, even ones from many years past, should be considered to be among the best cultivars of their species.

The greatest number of All-America medal winners over the years have been from the most popular annuals, petunias, marigolds, and zinnias. This is natural because there has been more breeding work on these and consequently more entrants. For instance, in 1965, petunia 'Appleblossom' (Fig. 13-1) received a bronze medal and in 1978 zinnias 'Cherry Ruffles' and 'Yellow Ruffles' were awarded silver medals and zinnia 'Red Sun' was a bronze winner. There have also been many medals awarded within other species, including such popular cultivars as phlox 'Twinkle,' pansy 'Majestic Giant,' verbena 'Amythest' and 'Blaze,'

FIGURE 13-11 All-America Trail Gardens, Lexington, Kentucky; flower judge Hugo Davis looking at 'Redskin' dahlia, All-America bronze medal winner, 1975.

sweet alyssum 'Rosie O'Day' and 'Royal Carpet', cosmos 'Diablo' and 'Sunset,' hollyhock 'Silver Puffs' and 'Majorette,' foxglove 'Foxy,' several geraniums, snapdragons, pinks, and cockscombs.

Growing Biennials

A biennial grows vegetatively the first growing season, then in the second it flowers and completes its life cycle. An annual goes through the full process in one growing season. The advantage of growing annuals would seem obvious, and all the biennials listed in Table 13-2 have annual cultivars or will behave as annuals if started early enough. Why, then, should one bother to grow plants as biennials? There is satisfaction in successfully growing such a plant, but there are practical reasons as well. Usually a biennial grown as such will flower earlier than an annual cultivar of the species. The biennial is usually a stronger plant during flowering. The vegetative phase may be quite attractive and well used as a subordinate element in a bed or for the textural effect of the foliage. In addition, there are very attractive cultivars that can only be grown as biennials.

It is often recommended that biennials should be sown in late summer or early autumn. This may work well in many cases, but the natural cycle of flowering and seed production, as well as speed of vegetative growth, should be considered in deciding on actual sowing dates. Biennials that flower in spring will produce strongest vegetative growth when sown in late spring or early summer because they will have plenty of time to grow yet will not have enough time or the right environmental conditions for floral initiation. Summer-flowering Canterbury bells, stock, and forget-me-not are at their vegetative best from the late summer-early fall sowing. Biennials can be protected over winter in a cold frame, a common practice in very cold areas or if their planned locations are occupied at autumn planting time. If grown in open locations anywhere except very mild areas, mulching the plants will avoid frost heaving. Earliest blossoming will ordinarily be produced by well-protected plants grown without a spring transplant, although this is a common practice for all biennials.

Woody Annuals

A woody plant that will produce an effective flower display the first year from seed is a very recent development. The only one now available is a crape myrtle, *Lagerstroemia indica,* called 'Crape Myrtlettes.' Bloom in shades of red, rose, pink, lavender, and white will develop in four and a half to five months on plants about a foot tall. A perennial where temperatures do not fall below about 10°F, 'Crape Myrtlettes' bloom most profusely where summers are hot and sunny.

A few other woody plants display attractive foliage color and texture the first year from seed if started early enough, particularly thyme, *Thymus vulgaris,* or possibly other species, and the santolinas, green lavender cotton, *S. virens,* and lavender cotton, *S. chamaecyparissus,* that winds gracefully through the design of

Fig. 13-10. Flowers should not be expected on any of these.

Annual Grasses

Annual grasses, both for foliage and flower, deserve much more use in our gardens. They are easily grown for display of their fine-textured foliage or graceful feathery flower heads. Some annual grasses are short lived and are usually grown only for cutting. These and some others will often self-sow if allowed to reach full maturity and can become weedy, particularly hair grass, *Aira capillaris* var. *pulchella,* loose silky bent grass, *Apera spica-venti,* and brome grasses, various *Bromus* species. Long-lived grasses are better suited for use in flower borders, including ruby grass, *Rhynchelytrum roseum;* quaking grass, *Briza maxima;* and hare's tail grass, *Lagurus ovatus.* Whether grown for display or cutting, all annual grasses produce floral heads that are easy to dry, needing only to be hung upside down with good air circulation.

Questions for Discussion

1. Define the following terms:
   (a) Damping-off   (d) Pinching
   (b) *Hortus Third*   (e) Powdery mildew
   (c) 10-52-17   (f) Bulbil

2. List several garden annuals that were popular in colonial gardens prior to 1700. Are any of these flowers popular today?

3. Identify the five most popular annual flowers and offer suggestions as to the reasons for this popularity.

4. Some scientific names include an "x" before the specific epithet. Why?

5. Describe techniques and procedures to be used in starting flowers indoors from seed.

6. Identify the important characteristics of a good seed-starting medium.

7. What fertilization schedule would you recommend for plants started from seed up to the time they are transplanted to the garden?

8. What are the reasons for hardening seedlings for transplant to the garden? How is hardening accomplished?

9. While fertilization is as important for garden flowers as for vegetables, more fertilizer does not necessarily mean more and better flowers. Explain.

10. Identify five flowers suited to hot locations and five that develop and bloom best in cool situations.

11. Discuss the statement "for best flowering it is important that individual flowers be removed as they fade."

12. Identify several flowers that are efficient reseeders.

13. Suggest guidelines for the effective utilization of color, form, and texture in flower garden design.

14. What roles do organizations such as All-America Selections play in the gardening industry?

15. Most biennial flowers have annual cultivars or will behave like annuals if started early enough. Why, then, should one bother to grow plants as true biennials?

16. The author states that ornamental grasses deserve much more use in our gardens. In this regard, go to a local garden center or nursery and determine which grasses are available for use in your area.

Selected References

Bailey, L. H., E. Z. Bailey, and Staff of the Liberty Hyde Bailey Hortorium. 1976. *Hortus Third,* Macmillan, New York.

Ball, V. (Ed.). 1972. *The Ball Red Book,* Twelfth Edition. Geo. J. Ball, Inc., West Chicago, Ill.

Brooklyn Botanic Garden Editors. 1974. A handbook on annuals. *Plants and Gardens,* 30(2), Brooklyn Botanic Garden, Brooklyn, N.Y.

Cathey, H. M. 1970. Growing flowering annuals, slightly revised. *Home and Garden Bulletin No. 91,* U. S. Department of Agriculture, Washington, D. C.

Crockett, J. U. 1973. *Annuals,* Revised Edition. Time-Life Books, New York.

Crockett, J. U. 1977. *Crockett's Victory Garden.* Little, Brown, Boston.

Favretti, R. F. and G. P. DeWolf. 1972. *Colonial Gardens,* Barre Publishers, Barre, Mass.

Haring, E. 1967. *The Complete Book of Growing Plants from Seed.* Hawthorn, New York.

Hyams, E. 1971. *A History of Gardens and Gardening.* Praeger, New York.

Mastalerz, J. W. (Ed.). 1976. *Bedding Plants: A Manual on the Culture of Bedding Plants as a Greenhouse Crop.* Second Edition, Pennsylvania Flower Growers, University Park, Pa.

Meyer, M. H. 1976. *Ornamental Grasses.* Scribner's, New York.

Seymour, E. L. D. 1970. *The Wise Garden Encyclopedia.* Grosset and Dunlap, New York.

Sunset Editors. 1974. *How to Grow Annuals,* Second Edition. Lane Books, Menlo Park, Cal.

Wyman, D. 1977. *Wyman's Gardening Encyclopedia.,* Revised and Expanded Edition. Macmillan, New York.

LEARNING OBJECTIVES

By the time you have finished this chapter you should be able to:

 1 Present guidelines to be used by a gardener in planning and organizing a garden of less than 100 square feet in area.

 2 Explain why vegetables can be regarded as multipurpose plants.

 3 Identify three vegetables that have important ornamental qualities.

 4 Explain what is meant by the phrase "vertical gardening" and name three vegetables that lend themselves to vertical gardening efforts.

 5 Discuss what constitutes a good container for growing vegetables.

 6 Identify some of the common ingredients used in potting mixes and tell of their importance.

 7 Describe the natural lighting and temperature conditions at different window exposures in the home.

 8 Discuss some of the factors to be considered in choosing and using artificial light sources for indoor gardening.

 9 Explain why accumulations of stagnant air present unfavorable conditions for plant growth.

10 Describe the exchange of air, water, and mineral salts in a porous clay pot as compared to a plastic pot.

11 Tell why an effective fertilization program is especially important for growing quality vegetables in containers.

12 Identify measures that can be taken to enhance pollination in vegetables grown indoors.

14

GARDENING WITH LIMITED SPACE

"Small Is Beautiful."

   E. F. Schumacher

In limited-space gardening, it is essential to first conduct a space inventory and to locate gardening space where there often seems to be none. Perhaps a piece of lawn or landscaping can be converted into a garden, a bank can be terraced, or vegetables can be used in border plantings. Patios, balconies, roof tops, windowsills, walls, ceilings, bedrooms, closets, and even attics can be used for garden crops (Fig. 14-1).

Once having found a space you will want to use it most effectively. If you plan to use a typical but small outdoor area an understanding of practices such as intercropping, succession, wide row, and staggered plantings will become doubly important because space is at a premium. For container gardening to be

effective, containers, planting media, and locations must be carefully selected.

Finally, limited space gardeners will need a special understanding of the plants they work with. Vertical space utilization may require the use of climbing vegetables, or vegetables that do well in hanging baskets. Container gardening may call for the use of minivegetables (Fig. 14-2), or vegetables that do well in soil of limited depth. Many gardeners will, of necessity or choice, do all of their gardening indoors; an environment much different than that of the backyard garden.

In spite of the requirements outlined above limited space gardening is not difficult. One does not have to battle the elements to such an extent, disease and pest problems are re-

FIGURE 14-1 Most homes have areas suitable for gardening. (Courtesy of National Garden Bureau, Sycamore, Illinois.)

crops but over a longer period of time. Much of the crop is eaten as harvested, providing a supply of fresh vegetables without the need for preservation. Vegetables and herbs adorning yards, patios, and living rooms add aesthetic appeal and serve an educational role in the home (Fig. 14-3). The minigarden converts solar energy into delicious food.

Our discussion of limited-space gardening is divided into two sections: (1) small-space yard gardening and (2) container gardening. The first section deals with small-scale gardens, landscape gardens, and vertical gardens—the kinds of gardens that can be grown if small spaces can be found anywhere in the yard. The second section deals with container gardens, both outdoors and indoors.

duced, and once a gardening system is established it is practically foolproof. With small space and container gardening expenses are much less. Cash layouts for rototillers, sprinklers, and sophisticated tools are eliminated. Traditional backyard gardeners grow a large crop that they then preserve for the long winter ahead. Small-scale gardeners grow smaller

SMALL-SCALE GARDENS

Site selection is as important for the small garden as for any other, and frequently it is more important. Small yards are more likely to present problems of shading by houses or fences, interference by tree roots, and poorly drained or low quality soils. Where space is limited, gardeners may be required to make

FIGURE 14-2 Gardeners can use plants that have been bred especially for containers. (Courtesy of Petoseed Co., Inc., Saticoy, California.

FIGURE 14-3 Vegetables are truly multipurpose plants. (Courtesy of National Garden Bureau, Sycamore, Illinois.)

compromises. For example, it may be impossible to find a garden site that enjoys full sun for most of the day. Remember that some vegetables need more sun than others. As a general rule: (1) Leafy vegetables (cabbage, chard, lettuce, spinach) can tolerate the most shade; (2) root vegetables (beets, carrots, radishes, turnips) can tolerate some shade; (3) vegetables grown for their fruits (corn, squashes, tomatoes) need full sun.

In general, leafy vegetables require at least four hours of sunshine daily, root crops six hours, and vegetables grown for their fruits need at least nine hours for good yields.

In the minigarden space utilization becomes critical. Fortunately, some vegetables make better use of space than others. The following standard vegetables are especially suited to the small garden:

Beans	Eggplant	Spinach
Beets	Leaf lettuce	Swiss chard
Broccoli	Onions	Tomatoes
Cabbage	Peas	Turnips
Carrots	Radishes	

Space consumers such as corn, melons, and some of the squashes should generally be omitted. For vegetables such as bean and tomato both determinate and indeterminate growth forms are available. The determinate forms take up less space, but this may be offset by the efficient utilization of vertical space possible with the indeterminate forms. Recently, seed producers have introduced a number of "midget" or "minivegetables" (Fig. 14-2). These midget cultivars make nearly every vegetable a candidate for the small-space garden.

There are other considerations when choosing vegetables for the small garden. The crops should be the ones that your family most enjoys. You may wish to delete from your list those vegetables where freshness is not a consideration. While squash, celery, and potatoes purchased at the supermarket are usually of

excellent quality, nothing compares with the taste of freshly harvested tomatoes, peas, and asparagus. Other favorites for the small garden are expensive or hard-to-get crops. You may want to choose plants that produce over a long period of time. Leaf lettuce can be harvested only one time, whereas Swiss chard, endive, carrots, and beets are vegetables with a long harvest period. For vegetables where both pole and bush types are available, the pole types will bear over a longer period, and will generally outproduce bush types.

The problem of how much to grow is accentuated for the small-space gardener because space is scarce. Generally, one wants a continuous supply of fresh vegetables—not too much or too little. The yield figures presented in Table 7-1 will aid in making approximations, but only experience will dictate correct amounts. Using vegetables with long harvest periods can reduce somewhat the problem of overproduction.

How much garden space is needed? The bigger the better, of course, but a space as small as two feet by three feet—or even smaller—can be used effectively (Fig. 14-4). A great many vegetables can be grown in a plot of this size, and an almost endless number of combinations is possible. The space can be used either at ground level or as a raised bed. If raised beds are employed they should be framed, using 2-inch by 10-inch boards or comparable material for framing, in order not to diminish the already small growing area. If a larger space is available, say four by six feet or four by nine feet, it can be treated as a number of two- by three-foot modules (Fig. 14-5). Each module can be planted to a specific group of crops, for example, salad crops, root crops, perennials, herbs, and so on. The modular approach also facilitates record keeping, rotation of crops, and numerous cultural operations. Individual modules can also represent the open sections in an area paved by bricks or other material. The open areas, in addition to providing garden space, have

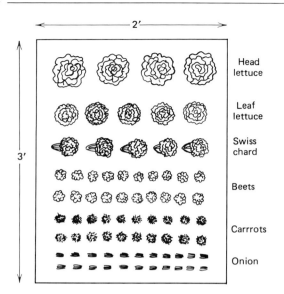

FIGURE 14-4 An area as small as 6 square feet can supply an abundance of vegetables. (Courtesy of Guerry Dean.)

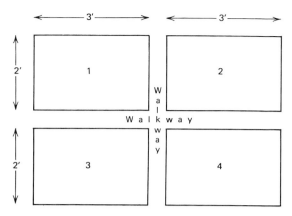

FIGURE 14-5 A modular approach to garden layout. Each module can be planted to specific crops. Cultural operations are also facilitated by this arrangement.

aesthetic appeal. The paved areas help to hold moisture, prohibit growth of weeds, provide working space, and absorb solar insolation. The modular approach can be extended to a garden of any size. As in other plans care should be taken to see that tall vegetables do not shade those of lesser stature.

A small garden (90 square feet) with a more traditional layout and containing nine different vegetables is shown in Fig. 14-6. A garden similar to this at Burpee Seeds' Fordhook Farms in Pennsylvania produced the following amounts of vegetables over a single season:

Beans	10 lb
Beets	3½ lb
Carrots	3½ lb
Cucumbers	147 individual
Lettuce	12 heads
Onions	8 lb
Radishes	X
Squash (zucchini)	63 individual
Tomatoes (large)	50 individual
Tomatoes (small)	54 individual

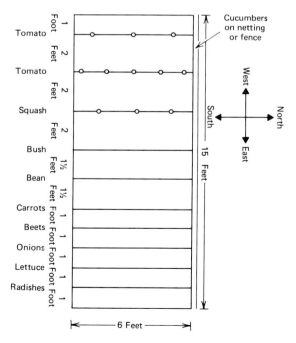

FIGURE 14-6 A 6 by 15 foot minigarden with vegetables and spacings.

Clearly, a small space can produce a lot of food!

If you can find a somewhat larger space than mentioned above you will be able to grow a greater variety of vegetables and probably produce some surplus for canning and freezing. Northrup King provides an excellent plan for a small (12 by 15 foot) basic vegetable garden (Fig. 14-7). Note that the fence or trellis is located at the north end of the garden so as not to shade the rest of the garden.

The Northrup King garden can be either flat or in the form of raised beds. Three beds each 12 feet long, and just slightly narrower than that shown in Fig. 7-12 could be fitted into this space. Regardless of planting scheme this basic garden offers excellent opportunities for a large yield from a small space, particularly if succession planting or other space-intensive techniques are used. Using the succession planting cross-reference chart for southwestern Illinois (Table 7-7) we find that two successive crops of radishes can be grown and following this the space will be available for beans, okra, Swiss chard, or some other crop. Or, in the carrot row an early planting of carrots can be followed by Swiss chard, zucchini, or beans, and then another crop of carrots. The possibilities are many and the po-

tential yields great. In theory, any "in" crop can follow an "out" crop as long as there is no time overlap. In practice, however, some choices are better than others. For example, a specific vegetable, or a closely related crop, replanted in a given space may encourage a build-up of harmful organisms, and at the same time create an excessive drain on the soil nutrients required by that crop.

LANDSCAPE GARDENS

In home landscaping, vegetables would once have seemed out of place and without desirable aesthetic qualities. However, the growth of our population, the high cost of food, the chemical contamination of commercial food products, and the desire of people to return to the land has brought about a dramatic shift in cultural attitudes toward the ornamental attributes of vegetables. The backyard garden is becoming the frontyard garden; vegetables are becoming a part of the landscape.

The increased use of garden plants in landscaping has been facilitated by several forces. First, through educational efforts the public has come to recognize that vegetables are multipurpose plants; that is, in addition to being

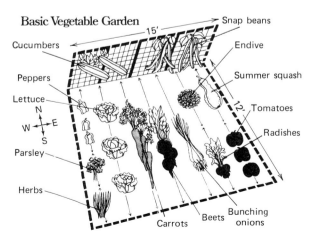

Basic Vegetable Garden — Snap beans
Cucumbers, Peppers, Lettuce, Parsley, Herbs, Endive, Summer squash, Tomatoes, Radishes, Carrots, Beets, Bunching onions
15'
12'
N W E S

FIGURE 14-7 A basic vegetable garden in which you can grow 13 popular vegetables in a space measuring only 12 by 15 feet. You can vary this plan according to your own preference. (Courtesy of Northrup King Seed Co., Minneapolis, Minnesota.)

a food source they are also seen to have outstanding ornamental and decorative qualities. 'Rhubarb Chard' (also commonly known as 'Ruby Red' chard), for example, is an exceptional ornamental plant with its translucent, reddish stalks and crinkly green leaves. Second, plant breeders and seed distributors have become increasingly aware of the demand for vegetables that have both ornamental and utilitarian value. The recently introduced midget vegetables can be as handsome in a container as any foliage plant, and at the same time yield fresh produce (Fig. 14-2).

Vegetables and herbs such as lettuce, cabbage, endive, and parsley have leaves of striking color and texture and serve well in border plantings or mixed with flowers. Other vegetables have beautiful flowers and fruits—okra, eggplant, and pepper are prime examples—and should have a place in the flower border. Corn, when planted in clumps, can look as tropical as bamboo or banana, and can be used to give balance or emphasis in the landscape. Perennial herbs such as rosemary and small fruits such as blueberry are handsome shrubs and are worthy of a place in the shrub border for their beauty of form, foliage, and flower. It is important to remember that in landscape plantings, vegetable plants will assume a more individualistic character if disassociated from their typical row style of planting. Site selection should be based on the same principles as employed for other ornamentals, thus creating plantings of great beauty as well as utility.

A listing of some common garden vegetables and their uses in the landscape is presented in Table 14-1. The list is not intended to be exhaustive, and the creative gardener can readily find other useful vegetables. Note that Table 14-1 excludes the herbs, because their use as ornamentals was discussed in Chapter 11. The reader is directed especially to the parts of Chapter 11 dealing with the uses of basil, chives, dill, marjoram, mints, parsley, rosemary, sage, savory, and thyme.

One word of caution, however; Chapter 11 deals almost entirely with herbs of culinary value. Many others, including southernwood, sweet woodruff, calendula, lavender, the scented geraniums, germander, nasturtium, and sweet violet are herbs known chiefly for their ornamental uses. Many of the prostrate, spreading herbs go well along walkways where their foliage and flowers are clearly displayed to passers-by. In such locations wayward foliage is frequently crushed afoot, resulting in the release of fragrant, volatile oils into the air.

Certain small fruits (Chapter 12) are becoming important landscape plants. Among these, strawberries, blueberries, currants, and gooseberries offer the most possibilities. Strawberry can be fitted into borders, grown in beds, or placed in containers (Fig. 14-8). Blueberry, currant, and gooseberry are shrubs with handsome foliage, flowers, and fruits, deserving of a place in the shrub border (Fig. 12-2). In small-fruit plantings space may be a limiting factor and consequently one may only want to plant enough of each for a reasonable supply of fruit. The number of plants estimated as adequate for a family of four would be as follows:

Small Fruit	Number of Plants	Yield (qt)
Strawberry	20	16–24
Blueberry	4	12–16
Currant	2	6–12
Gooseberry	1	3–6

The tree fruits, especially the dwarf forms, also provide excellent landscaping material for those with limited space. Apple and pear are most popular, but peach and plum are useful too. Specimens can be grown as lawn trees, used as a hedge, or made a part of the shrub border. The cultivars available at your local nursery will be adapted to your area. A word of warning, however; be sure to determine

TABLE 14-1 Some Vegetables of Ornamental Value and Suggested Uses

Plant	Form(s) or cultivar(s)	Part(s) displayed	Characteristics	Use in landscape
Artichoke	'Green Globe'	Foliage Flowers	Silvery-green leaves Purplish, thistlelike flowers to 6 inches in diameter	Up to 4 feet tall and spreading; background plant, accent plant
Asparagus	All cultivars	Foliage	Feathery appearance of leaves	Tall, background plant; accent plant
Beet	All cultivars	Foliage	The simple leaves are colorful	Border plant
Cabbage	Red forms	Foliage	The red leaves have an unusual bluish cast	Border plant; with flowers; raised beds
Carrot	All cultivars	Foliage	Finely divided, green leaves	Border plant
Corn	All cultivars	Foliage	When grown in clumps gives a lush, tropical appearance	A delightful patio plant; serves also as a background and accent plant
Cucumber (pumpkin and squash too)	All cultivars	Foliage Flowers Fruit	Attractive tropical-appearing leaves and large yellow flowers; colorful and shapely fruits	Goes well in pots, tubs and containers; grow atop retaining walls or climbing on a trellis
Eggplant (and pepper)	All cultivars	Foliage Flowers Fruit	Handsome plants with beautiful foliage, small flowers, and fruits	Moderate height; grow in full sun with flowers or alone
Kale	Blue curled forms Flowering form	Foliage	The curled form appears fernlike; the flowering form has an unusual reddish-pink color	The curled form goes well in a bed or as a border plant. The flowering form (also edible) is a good accent plant; does well in pots too.
Lettuce	Bibb and leaf types	Foliage	A wide range of forms and colors (light green to reds); some varieties wrinkled	Excellent for borders, raised beds, and containers. Color variation allows pattern plantings.
Okra	All cultivars	Foliage Flowers	Handsome foliage and showy, yellow hibiscus-type flowers; varieties from 2 to 5 feet high	Goes well in the flower border
Rhubarb	All cultivars	Foliage	Lush, tropical foliage; perennial leaves poisonous	An excellent tub plant; also goes well along borders, but choose a permanent location
Swiss chard	'Rhubarb Chard'	Foliage	An attractive edible ornamental	Almost everywhere; excellent as a border plant or mixed with flowers

FIGURE 14-8 Strawberries thrive in containers. (Courtesy of Guerry Dean.)

the pollination requirements of the cultivars chosen because many require a cross-compatible cultivar for pollination (see Table 3-3).

VERTICAL GARDENS

Nearly all of us have employed some form of vertical gardening. Whenever we grow tall peas as opposed to the short forms or pole beans as opposed to the bush forms we are taking advantage of vertical space. **Vertical gardening** is simply gardening in vertical space; extending plants vertically instead of horizontally. The chief advantage of vertical gardening is that plant growth can be extended vertically *without* adding to the soil area required for growth. There are other advantages or reasons for vertical gardening too. Vertical gardens lend themselves to restricted (horizontal) areas such as borders. The vertical growth of plants can be used to hide undesirable features of the landscape, to create areas

of privacy, and to take advantage of climatic influences.

Some of our vegetables, most notably peas and beans, have a natural climbing habit. Others, such as the cucurbits and tomato, use vertical space well if they are tied or trellised (Fig. 14-9). With cucumbers vertical growth is reported to result in improved pollination with greater fruit set, increased production of clean, easy-to-pick fruits, and less disease and insect problems.

Trellises can be designed in many different ways. They can either be vertical, or an A-frame that provides its own support (Fig. 14-10). The framework of either type can be left open or covered with a coarse mesh wire, depending on need. The A-frame type is more versatile if it is hinged and adjustable. It allows harvest of vegetables from the inside, can be folded flat for storage, and when covered with plastic can become a portable greenhouse to provide frost protection or shade.

FIGURE 14-9 Cucumbers on a trellis. (Courtesy of National Garden Bureau, Sycamore, Illinois.)

FIGURE14-10 A versatile A-frame trellis. (Courtesy of Guerry Dean.)

Some of the small fruits—blackberries, raspberries, and grapes, for example—can, and should, be trained so as to make maximum use of vertical space. The tree fruits, particularly dwarf apple and pear, can be trained to grow along a wall, railing, or trellis, known as an **espalier.** Although an espaliered plant (also called an espalier) requires much care and labor, the end result can be an attractive and useful ornamental (Fig. 14-11). The techniques involved in creating an espalier involve rigorous pruning and also the bending of young, flexible shoots. The later growth of curved portions results in retention of created shapes. Trees espaliered on the south side of a building will bloom and mature fruit earlier than normal and may require protection from frost.

Pyramid gardens—vertical planting areas of successively lesser diameter—also present an excellent way for taking advantage of vertical space. A garden only 6 feet across can, because of the successive layers, provides nearly 50 feet of row for planting. Pyramid gardens can be circular, square, or rectangular depending on space availability and preference. The diameter and height of successive tiers

can vary too, but shading effects must be considered. Normally it is suggested that rows be at least 1 foot wide. Construction can be of wood or metal. Pyramid gardens work well with flowers, vegetables, and, of course, the versatile strawberry.

WHY CONTAINER GARDENING?

Interest in container gardening is growing as the number of apartment and condominium dwellers increases. For many of these people container gardening is their only way to grow vegetables. Container gardening allows the use of rooftops, patios, and windowsills. Containers can be moved so as to take advantage of daily and seasonal weather fluctuations. With containers there is less area to weed, water, and maintain, and since containers can be elevated above ground level these operations can be managed more easily. Disease and pest problems are also lessened. Container gardening can, of course, be carried out the year around, and well-chosen containers full of choice vegetables are a source of interest and beauty.

With these thoughts in mind let us explore the topic of container gardening. For convenience we will divide our subject matter into three areas, as follows: (1) Containers and planting mixes; (2) plants and planting; and (3) care of container-grown plants.

CONTAINERS AND PLANTING MIXES

Containers

Containers may be of metal, plastic, clay, wood, paper, pulp, or styrofoam; they may be square, rectangular, round, shallow, or deep; they may be readymade or built at home. Any combination of characteristics will suffice so long as

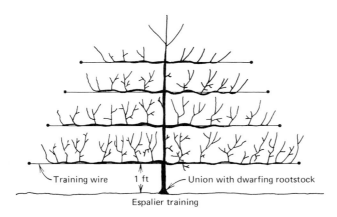

FIGURE 14-11 Creating an espalier. (Modified from Oregon State University Cooperative Extension Service.)

Training wire   1 ft   Union with dwarfing rootstock

Espalier training

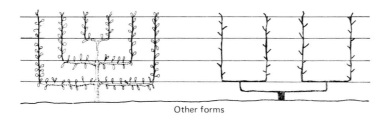

Other forms

the container has certain basic features. Foremost among these features is that the container must provide *good drainage*. Many aspects of container culture, including selection of media, aeration, and watering center around the subject of drainage.

As a general rule, containers less than 10 inches in diameter need two holes ⅜ inch in diameter to provide good drainage. Containers over 10 inches need four or more holes. Holes spaced evenly around the sides of the container at the very bottom are more effective in promoting drainage than holes in the bottom. If insufficient drainage is provided, the planting medium usually becomes waterlogged resulting in an imbalance in the uptake of oxygen and water by roots and an increased frequency of root rot. Overwatering, stemming from improper drainage, is the greatest cause of failure in container growing.

The manner in which containers are filled with soil mix is also important in relation to drainage. If the bottom hole of the container is large enough to present a soil-leakage problem, then a small piece of broken crockery or a piece of screening can be placed over the hole to remedy the problem and at the same time allow proper drainage. Numerous authors also recommend the practice of putting a layer of stones or coarse, broken pottery in the bottom of containers. This promotes drainage and holds in the potting soil. Most professionals advise against this practice, however. The layer of stones forms a collection place for water and organisms, creating conditions favorable for root diseases. The layer also takes up available space that could be used by growing roots. Remember, a container needs space at the top of the container for watering. It is not necessary or wise to give up space (volume) at the bottom of the container too.

Two other areas for caution: first, when using containers that have been used previously be sure that they are clean. Cleaning can be accomplished by scrubbing with a brush or steel wool in soapy water and rinsing in hot water. This kills disease organisms and keeps your pots in a better state of repair. Second, *always* use pot saucers with your pots. Saucers retain excess water, preventing damage to floors, tabletops, and windowsills. The above does not apply to porous clay saucers, which, because of their porosity, allow passage of some water and salts. It is a good idea to elevate the bottom of the pot above the surface of the saucer by use of some small pebbles or other objects. In this manner water that runs into the saucer will not touch the bottom of the pot and will not be taken up into the soil again. The drainage water collecting in the saucer can also help to create a higher relative humidity in the vicinity of the plant.

Planting Mixes

Successful container gardening requires the use of a good soil mix in the container, and, as indicated above, the soil mix chosen also has a significant bearing on proper drainage. Some container gardeners simply use common garden soil in their containers. While convenient, soil becomes compacted easily in containers. This creates drainage and aeration problems. Also, when garden soil dries it tends to shrink away from the sides of the container. With subsequent watering you may find that the water runs over the surface and down the sides between the soil and the container and out the bottom, without thoroughly soaking the soil. In addition, garden soil weighs a lot, often contains disease organisms, and for some people is hard to obtain.

An alternative is to use a manufactured potting mix. A **mix**, as the name implies, is a combination of compounds, each of which is added to give certain favorable properties to the final product. Some of the more popular

are presented in Table 14-2. Note that certain of the mixes contain soil, whereas others are soilless. The soilless—**synthetic soil**—mixes, whether made at home or purchased at the store, offer several advantages: (1) they are free of plant disease organisms and weed seeds; (2) they provide good drainage; (3) they hold moisture and plant nutrients well; and (4) they are lightweight. Two of the most popular of the soilless mixes were perfected at Cornell University and the University of California. The mixes are of similar composition, with the exception that the Cornell mix substitutes vermiculite for the fine sand used in the UC formula. Both mixes are available commercially—the Cornell mix under such trade names as Jiffy Mix, Pro Mix, and Redi-Earth, and the UC mix under the names First Step and Super Soil.

A word of caution about packaged potting mixes: the demand for these mixes was created by the widespread interest of gardeners of all types in growing plants in containers. Many of the mixes on the market were formulated by commercial growers or university personnel. Others, however, have little or no growing experience behind them. There are no current guidelines or standards for manufacturers to follow, either in the production of mixes or in their packaging and labeling. Furthermore, although some manufacturers employ elaborate quality control checks others do not. Thus, while the vast majority of packaged mixes are of high quality, problems have been encountered with some. Chief among these problems are: (1) excessive content of heavy metals such as cadmium, (2) deficiency of nutrients, particularly phosphorus, (3) unfavorable (usually low) pH, and (4) excess salt content.

There are several reasons why the reader should not be overly alarmed by the above comments. First, most mixes are of outstanding quality and will serve the gardener well. Second, when manufacturers have been apprised of deficiencies in their mixes they have

TABLE 14-2 Some Common Potting Mixtures

Mix	Ingredients
Peter Chan's potting mix	1 part soil 1 part sand (or perlite) 1 part peat moss 1 part manure
A Brooklyn Botanic Garden mix	4 parts soil 2 parts sand 2 parts leaf mold or peat moss 1 part dried cow manure ½ cup bone meal/2½ gallons of the mixture
A Cornell University mix	8 quarts vermiculite 8 quarts shredded peat moss 2 tablespoons superphosphate 2 tablespoons ground limestone 8 tablespoons dried cow manure or steamed bone meal

A modified Univ. of Calif. mix	To make one cubic yard of mix
	  1 part peat moss
	  1 part fine sand
	  1 part perlite
	  2 pounds single superphosphate
	  1 pound sulfate of potash
	  ½ pound iron sulfate

A USDA potting mix	1 bushel vermiculite 1 bushel shredded peat moss 1¼ cups ground limestone ½ cup 20% superphosphate 1 cup 5-10-5 fertilizer

remedied these problems. Third, manufacturers are improving quality-control techniques.

The kind of mix chosen by any individual will be influenced by cost, availability, and preference. If you choose to use a mix that incorporates garden soil you will probably want to **sterilize** (kill all organisms) or **pasteurize** (kill harmful organisms) the soil first. This can be done in the home oven using one of the following schedules:

Temperature*	Time	Results
140°F	30 min	Kills most pathogenic fungi and bacteria
160°F	30 min	Kills soil insects
180°F	30 min	Kills most weed seeds

*Actual soil temperatures.

Soil to be heated should be moistened and placed in a shallow layer in a suitable container. Care should be taken not to overcook, since this can release toxic substances in the soil. Compost should also be treated to kill harmful organisms. If sand is used in a potting mix avoid ocean beach sand because of its high salt content. When peat moss is used in mixtures it should first be broken apart and wetted, then mixed.

The gardener can more intelligently choose or modify potting mixtures if he or she understands the properties of the basic components of the mix. Certain of these properties are shown in Table 14-3. Both sand and soil are heavy. A chief function of the other ingredients is simply to bring about an overall weight reduction. Sand, an excellent material for drainage, has little ability to hold water and nutrients. Compost, perlite, and vermiculite provide good drainage while paradoxically having good qualities of moisture retention too. Peat moss retains high levels of moisture. Soil, although quite variable in composition, ranks high in fertility. The other ingredients, with the exception of compost and some forms of peat moss, are low in nutrients. Consequently a supplementary fertilization (feeding) program is usually necessary. Fertilization and other maintenance operations are discussed later in the chapter.

PLANTS AND PLANTING

Much of the information presented in earlier chapters applies directly to container-grown plants, and need not be repeated here. For example, information on germination processes, photosynthesis and respiration activities, hardiness characteristics, and cropping patterns finds application in both growing situations. On the other hand, certain aspects of plants and planting are specific to container growing.

Vegetables for Containers

Almost any vegetable can be grown in a container, but some are better choices than others. Selection will depend on whether the containers are to be placed outdoors or indoors, plant-growth characteristics, sizes of containers available, and personal preferences.

Some common cool-season and warm-season garden vegetables with their ratings for outdoor and indoor container growth are presented in Table 14-4. Over twice as many vegetables receive a good or better rating for outdoor growth, as compared to indoor growth. This is to be expected, since outdoor conditions approach more closely those of the gar-

TABLE 14-3 Potting Mix Ingredients and Selected Properties

Ingredient	Properties			
	Weight	Water-holding capacity	Drainage	Natural fertility
Sand	Very heavy	Poor	Very good	Very low
Soil	Heavy	Variable	Variable	High
Compost	Light	Good	Good	Moderate to high
Peat moss	Light	Very good	Moderate	Low to moderate
Perlite	Very light	Good	Very good	Very low
Vermiculite	Very light	Good	Good	Low

TABLE 14-4 Some Warm-season and Cool-season Vegetables with Ratings as Outdoor and Indoor Container Crops

Crop	Warm season or cool season	Outdoor containers	Indoor containers
Bean	WS	Fair	Poor
Beet	CS	Excellent	Excellent
Broccoli	CS	Fair	Poor
Brussels sprouts	CS	Good	Good
Cabbage	CS	Good	Poor
Carrot	CS	Excellent	Excellent
Cauliflower	CS	Good	Fair
Corn	WS	Good	Poor
Cucumber	WS	Excellent	Fair-good
Eggplant	WS	Good	Fair
Kale	CS	Good	Fair
Lettuce	CS	Excellent	Excellent
Muskmelon	WS	Good	Poor
Mustard greens	CS	Excellent	Excellent
Okra	WS	Fair	Poor
Onion	CS	Good	Good
Pea	CS	Fair	Poor
Pepper	WS	Excellent	Fair
Radish	CS	Excellent	Excellent
Rhubarb	CS	Good	Poor
Rutabaga	CS	Good	Poor
Spinach	CS	Good	Good
Squash, summer	WS	Good	Poor
Swiss chard	CS	Good	Good
Tomato	WS	Excellent	Excellent
Turnip	CS	Good	Fair
Watermelon	WS	Good	Poor

den. The beginner should start out with those plants that experience has shown do well in containers.

In recent years plant breeders and seed producers have developed small-sized vegetables especially suited to growing in pots and tubs. These "midget varieties" or "minivegetables" are available through most mail-order houses (see Appendix E) and at some garden stores. These plants are usually one-fourth to one-half the size of standard cultivars. They grow well in small containers, and most mature earlier—generally 10 to 15 days—than their standard-sized counterparts. Some of these "container varieties" are given in Table 14-5. Obviously, the vegetables in this group are good candidates for the container gardener, and all have received high ratings as container crops (Table 14-4).

Several standard vegetables also mature rapidly and are of interest to the container gardener.

Vegetables that mature rapidly[a]	
Radishes	22 days
Mustard greens	35
Loose-leaf lettuce	40
Green onions	40
Spinach	42
Turnips	45
Bush snap beans	48
Summer squash	50
Early peas	55
Kale	55

[a]Modified from Stokes Seeds Inc., Buffalo, N. Y.

These can be used for a quick harvest, intercropping, and, since many are harvested for their vegetative parts, in locations that enjoy less than full sun.

Choosing Containers

Sizes and shapes of containers available have much to do with the kinds of vegetables grown. To illustrate, if carrots are container grown and the carrot tip touches the bottom of the container, the carrot will not develop to maturity in a normal manner. Cultivars such as 'Imperator' develop a harvestable root nine inches long or more and require a deep container. Conversely, some of the midget forms such as 'Little Finger' and 'Short 'n Sweet' have storage roots less than four inches long and are ideal for shallow containers. These same cultivars would thrive in shallow or heavy soils where root development of longer cultivars would be inhibited.

While vegetables can be grown in almost any container, many lend themselves well to particular container situations. Some examples are as follows.

Container	Vegetables
Trough or window box	Beets, broccoli, Brussels sprouts, bush beans, cabbage, carrots, celery, collards, eggplant, garlic, herbs, lettuce, mustard, onion, peas, spinach, Swiss chard, turnip
Trough with trellis	Cucumbers, melons, pole beans, tall peas, tomatoes
Hanging baskets	Cherry tomatoes, cucumbers, herbs, peas, sweet potatoes
Pots and tubs	Beet, cabbage, carrot, eggplant, garlic, lettuce, okra, onions, peppers, Swiss chard, summer squash, tomatoes

The containers chosen must be large enough to comfortably accommodate the vegetables you wish to grow. They must also be strong enough to hold the planting mix and withstand intermittent wetting and drying.

Yields of many container-grown vegetables are tied not only to container depth but also to total container volume. Work done in Massachusetts has provided some guidelines on the proper sizes of containers for different crops and given some indication of expected yields (Table 14-6). Other sources suggest different container sizes, and yields will vary a great deal from season to season. Container experience to date suggests that peppers, potatoes, tomatoes, lettuce, summer squash, carrots, beets, radishes, and onions give the largest yields. In one study a single cherry tomato plant growing in a five-gallon container produced 488 fruits. Overall, much more technical data are needed, however, on the subjects of container size and yields from container-grown plants.

Planting in Containers

Planting vegetables in containers is similar to planting them in the ground and many of the same rules apply. Plants such as bush beans, beets, carrots, peas, radishes, spinach, and turnips should be planted directly since they

TABLE 14-5 Some Midget Vegetable Cultivars with Days to Maturity

Vegetables	Cultivars	Days to maturity
Beet	Baby Canning	48
Cabbage	Baby Head	65
	Dwarf Morden	55
	Little Leaguer	72
	Short 'n Sweet	60
	Midget Muskmelon	60
Carrot	Baby Finger Nantes	65
	Gold Nugget	71
	Little Finger	65
	Midget	65
	Planet	55
	Short 'n Sweet	68
	Tiny	62
Corn	Golden Midget	55
	Midget Sweet	65
	White Midget	74
Cucumber	Baby	55
	Cherokee 7	60
	Little Minnie	52
	Mini	55
	Patio Pik Hybrid	51
	Pot Luck Hybrid	55
	Tiny Dill Cuke	
Eggplant	Golden Yellow	65
	Morden Midget	
Lettuce	Tom Thumb	65
Muskmelon	Minnesota Midget	65
Pea	Little Gem	60
	Mighty Midget	60
Tomato	Patio Hybrid	50
	Pixie	52
	Presto	55
	Small Fry	70
	Tiny Tim	55
	Tumblin' Tom Hybrid	48
Watermelon	Golden Midget	65
	Lollipop	70
	Market Midget	69
	Petite Sweet	65

TABLE 14-6 Some Common Vegetables with Recommended Container Size and Reported Yields

Vegetable	Container volume (gal)	Yield (lb) (6/8 to 10/1/76)
Beet	2	4.1
Broccoli	5	2.2
Cabbage	1	4.5
Chinese cabbage	2	4.7
Cucumber	5	6.9
Eggplant	1	2.2
Kohlrabi	2	1.6
Lettuce, 'Buttercrunch'	1	2.3
Lima bean	2	1.8
Muskmelon	5	21.1
Okra	2	.8
Onion	1	1.2
Pea, 'Little Marvel'	1	.3
Pepper	2	5.6
Potato, 'Katahdin'	5	3.7
Summer squash 'Straightneck Prolific'	5	9.1
Swiss chard	1	2.3
Tomato, large fruited	5	8.8
Tomato, 'Jet Star'	2	6.8

do not transplant well. Others, such as members of the cabbage, cucumber, and tomato families can be started from seed indoors and transplanted to their permanent containers later, much as outlined in Chapter 7. Spacing is more critical with container plants inasmuch as space is at a premium. In this regard the intensive planting guidelines presented in Table 7-6 will be useful. Regardless of techniques employed, one ultimately (or optimistically) ends up with containers full of young seedlings "eager to face the world"—a world that may be quite different from that of the traditional backyard garden. In the next section we discuss the care and maintenance of these container-grown seedlings.

CARE OF CONTAINER-GROWN PLANTS

A sage botanist once remarked that "Mother Nature never made a house plant" and how true this is. As a matter of fact, to this statement could be added an important corollary: Mother Nature never made a container plant either. Plants grown in containers are in environments foreign to their being and therefore require special understanding and care. It is important that the gardener appreciate these environmental parameters—light, temperature, and air—as they affect growing plants. In addition, one needs to be familiar with cultural practices such as proper watering and fertilization, and in some cases biological phenomena, particularly pollination. Let us briefly examine these topics.

Light

Light is essential for plant growth. As stated earlier, plants make themselves out of carbon dioxide and water in the presence of sunlight by a process termed photosynthesis. The water provided by the soil solution also contains nutrients essential for growth. Vegetables grow best in full sunlight, but some can thrive with less light than others. For outdoor containers, light per se is usually not a problem, but for indoor container gardening the available light can be critical. The amount of natural light received indoors is a function of exposure (Table 14-7). The data presented represent averages; light intensity for a given exposure is influenced by shading from eaves and other adjacent structures, latitude, and by season of the year. Normally, light intensities are greater in the summer than the winter. In summer it may even be necessary to screen plants from a too-hot sun by use of a light curtain or shade. In winter, and for poor exposures, artificial light may be required to supplement the sunlight.

Not too long ago indoor gardening was limited to the growing season, when the avail-

ability of natural light was greatest. Now, a great variety of light sources and lighting fixtures are on the market. This ready availability of inexpensive light sources has greatly increased the popularity and efficiency of indoor gardening. Artificial light can be used to supplement sunlight, or plants can be grown by artificial light alone. Artificial light offers certain advantages in that its amount remains constant the year around and the daily exposure to light can be closely regulated. Furthermore, since artificial light sources are often portable, they can be used in any room or moved from room to room. This is not to say that artificial light is identical to natural light in either quality or quantity. As a matter of fact, light sources vary greatly in spectral quality (Table 14-8). A major goal in artificial lighting is to approach as closely as possible the conditions provided by natural light.

Choosing the Best Lighting Sources

When visible light from the sun is passed through a prism it is broken into its component colors. Plants do not use all parts of the visible spectrum with the same relative efficiency. Photosynthesis uses primarily blue and red light and transmits or reflects the green wavelengths. (The green color we see in plants is, of course, derived from the transmission of these green wavelengths.) Other plant pro-

TABLE 14-7 Light Intensity, Supplemental Light Requirements, and Types of Vegetables to Grow Based on Window Exposure During the Outdoor Growing Season

Window exposure	Light intensity	Supplemental light required	Vegetables to grow
South	High	No	All types
West	Moderate–high	Possibly	Most types
East	Moderate	Possibly	Leaf and root types
North	Low	Yes	Few without supplemental light

TABLE 14-8 Relative Energy Output of Various Fluorescent and Incandescent Lamps and Combinations in Different Spectral Regions

Lamp	Quality of light			
	Blue	Yellow-green	Red	Far-red
Incandescent	L[a]	H	VH	M
Fluorescent				
Cool white	H	VH	L	VL
Daylight	VH	VH	VL	VL
Warm white	M	VH	M	VL
Natural white	L	VH	VH	VL
Gro-Lux	VH	L	H	VL
Plant-Light	VH	L	H	VL
Agro-Lite	M	H	H	L
Wide-Spectrum Gro-Lux	M	H	H	L
Combinations				
Cool white/warm white[b]	M-H	VH	L-M	VL
Cool white/warm white/inc.[c]	M-H	VH	M	L
Gro-Lux/Wide-Spectrum Gro-Lux[b]	H	M	H	L

From Raymond P. Poincelot, Gardening Indoors with House Plants, *Rodale Press, Inc., Emmaus, Pa. 18049. Reprinted by permission of Raymond P. Poincelot. Copyright 1974.*

[a]Abbreviations: Very low, VL; Low, L; Medium. M; High, H; Very high, VH.

[b]1:1 ratio.

[c]1:1:2 ratio, using 25-watt incandescent bulbs.

cesses require light of different wavelengths. Red and far-red energy, for example, is necessary to promote flowering. A primary goal with many indoor vegetables is to achieve high rates of photosynthesis—and concomitant vegetative growth—and therefore we must select light sources that emit strongly in the regions where photosynthetic activity is greatest. For other vegetables we want flowering— and subsequent fruit formation—as well so we need light sources that emit energy in the red and far-red regions of the spectrum.

Two basic types of light sources are available to us—incandescent and fluorescent. In general, incandescent sources emit strongly in the red and far-red regions and weakly in the blue region (Fig. 14-12). Thus, they provide favorable light for flowering but are not well balanced for photosynthesis. Ordinary fluorescent lights emit strongly in the blue region, but not so strongly in the red and far-red regions. Thus they are suitable for vegetative growth through photosynthesis but not for flowering. Several specialized forms of fluorescent lamps are designed especially to promote plant processes. These lamps, and particularly the "wide spectrum" types, emit strongly in the red and far-red regions as well as in the blue region (Fig. 14-12). Thus, these lamps are suitable for promoting both vegetative and reproductive development. Characteristics of these light sources may be summarized in Table 14-9.

What light source or sources should the indoor gardener use? A time-tried combination is to use one cool-white and one warm-white fluorescent tube (each 40 watts) in combination with two 25-watt incandescent bulbs. Fixtures are available that accommodate both types of bulbs in a single unit. A second possibility

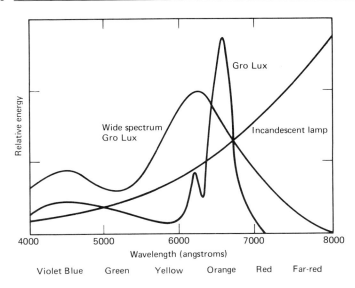

FIGURE 14-12 Distribution curves of spectral energy from incandescent and selected fluorescent lamps.

is to use selected fluorescent lamps only; either of one type or in combinations (Table 14-8). This method is easier, and still provides light of the desired quality. The use of fluorescent lamps alone offers other advantages too. They are approximately three times more effective than incandescent lamps in converting electrical energy into light and thus produce less waste energy in the form of heat. Fluorescent lamps are also less expensive as their lifetime is 15 to 20 times that of incandescent lamps.

In summary, each gardener must choose artificial light sources that: (1) yield the quality and quantity of illumination needed, (2) are compatible with the home environment, and (3) are available and not too expensive. To illustrate, a gardener growing leafy vegetables, where flowering is not a factor, may choose different light sources than a gardener growing crops that produce fruits. Growth of many indoor vegetables may require light intensities unsuitable for living areas. Consequently the gardener can either alter the kinds of plants grown, the light sources used, or the location of the growing area.

Putting Light to Work

Once light sources and fixtures have been selected they must be properly installed and used. A popular unit is the standard fluorescent tube reflector fixture that accommodates two 48-

TABLE 14-9 Energy Output and Plant Processes Favored by Incandescent and Fluorescent Lamps

Lamp type	Energy output		Plant process favored
	Blue	Red and far-red	
Incandescent	Low	High	Flowering
Fluorescent			
Ordinary	High	Low	Photosynthesis
"Plant growth and wide-spectrum" types	High	High	Both of above

inch lamps. These can be suspended from the ceiling by adjustable chains or made part of an adjustable stand. Either way, the lights can be raised as the plants grow. For example, during germination of seeds, the lights can be placed close to the pots and left on continuously (Fig. 14-13). The heat generated will hasten germination and emergence. Once seedlings have emerged the duration of lighting should be decreased to 12 to 16 hours a day (say 7 A.M. to 11 P.M.). An inexpensive automatic timer will aid greatly in regulating light/dark periods. The dark period provided not only saves electricity; the day/night cycle established is also essential to the normal development of the plant.

After seedlings have emerged, the height of the light source should be adjusted so as to make maximum use of light in promoting plant growth. Due to their design, fluorescent

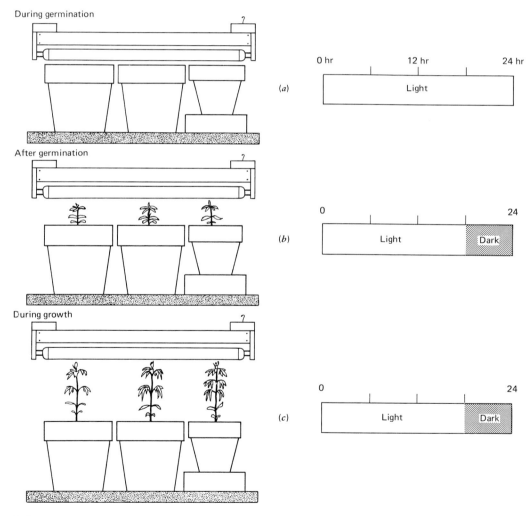

FIGURE 14-13 The height of lights and the duration of lighting should be manipulated to maximize plant growth and development.

lamps generate much less heat than do incandescent sources. The smaller amount of heat generated by the fluorescent lamps allows their placement closer to plants. Normally, fluorescent lamps should be two to eight inches above the tops of plants, whereas incandescent light sources should be kept at least a foot away from the tops of plants. As a rule of thumb, if your hand feels warm when held at the level of the plant tops then the light is too close.

For the purposes of home gardeners light intensity is often expressed in terms of footcandles, abbreviated f.c. Many inexpensive light meters are available commercially and some of these read directly in footcandles. Light intensity readings can also be made using a camera with a built-in exposure meter by one of the following procedures:

1. Procedure 1
   (a) Set the film-speed dial to ASA 100.
   (b) Read the light in the proposed plant location by aiming the camera at a white piece of paper placed in the position leaf surface will occupy.
   (c) Adjust camera to correct exposure.
   (d) The shutter speed opposite f/4 will give the approximate footcandles of illumination (e.g., 1/250 sec = 250 f.c.)

2. Procedure 2
   (a) Set film-speed dial to ASA 200.
   (b) Set shutter speed at 1/125 second.
   (c) Read light as in 1(b) above and adjust the f-stop until a correct exposure is shown in the light meter.
   (d) The f-stop reading can be converted to footcandles using the following table:

F-stop	Footcandles
2.8	32
4	64
5.6	125
8	250
11	500
16	1000
22	2000

For good growth of vegetables USDA scientists recommend 1000 to 2000 f.c. of light that includes light from the red and far-red regions of the spectrum. But what about existing light intensities? Outdoors, light readings may range from about 10,000 f.c. at noon on a clear summer day to 500 f.c. on an overcast winter day. Indoors, light readings are much lower, as illustrated by the following examples:

Direct sun entering a window	4000–8000 f.c.
Shady window on bright side of house	400–600 f.c.
Shady window on shady side of house	200–300 f.c.

Light intensity will, of course, vary for each location but it is clear that supplemental light is frequently essential for indoor gardening.

Reflector-type fluorescent light banks with two 40-watt fluorescent lamps provide about 850 f.c. of light to plants six inches beneath the tubes. This intensity is none too high for good vegetable growth. The gardener, therefore, must make every effort to maximize upon the light energy available. This can be done in several ways. First, plants can be kept in close proximity to the light source. Growing tips need be only two or three inches from the light. Second, the light intensity is greatest below the center of the tube and diminishes in either direction. Plants with the highest light requirement should be given preferential placement. Third, it frequently happens that plants are of variable height. In this case they should be lined up by height and the light bank adjusted accordingly (Fig. 14-14). Fourth, because light output from fluorescent lamps decreases as the lamp ages they should be replaced when they have reached 70 to 75 percent of their stated service life; generally after about two years of use. Fifth, lights should be positioned so that they do not reduce the intensity of incoming natural light. Finally, surfaces of high reflectivity such as white walls

FIGURE 14-14 Plants should be lined up by height and the light bank adjusted accordingly.

and mirrors should be used to reflect maximum light, natural and artificial, to growing plants.

A number of specialized lamps are available that give greater light intensities. In the fluorescent line High Output (HO), Very High Output (VHO), and Super High Output (SHO) lamps are available. Additionally, a relatively new family of lamps, termed High-Intensity Discharge (HID) lamps, are available. The mercury vapor lamps used as street lights are an example of this lamp type. The HID lamps have been used successfully in commercial plant production, but their cost is very high. There is little doubt, however, that these innovative light sources will be more widely used by indoor gardeners in the future.

Temperature

It is difficult to separate the effects of temperature on plants from those of light. For example, plants grow rapidly at high light intensities. If night temperatures are too high, however, plant respiration will also be high. The net result will be smaller plants (see Fig. 6-14). As a matter of fact, any time that light strikes an object, part of its energy is dissipated as heat. Within limits, this heat is beneficial to plants, but too high, or too low, temperatures can be detrimental to growth. Let us look at some of the influences of temperature as they relate to container-grown plants.

Fortunately, the range of temperatures that is comfortable for us—65 to 88°F—is suitable for our plants too. True, plants may show vigorous growth over a broader range of temperatures, but photosynthesis, for example, begins to decline above about 86°F, and few plants show appreciable growth below 40°F. Every plant has its own optimum temperature range, and this range varies with stage of development and between day and night. A good example is tomato, with day temperatures of 75 to 80°F and night temperatures of 59 to 64°F. Fruit set has been found to be closely correlated with nighttime temperatures. This helps explain why fruit set is greater and more reliable in some parts of the country—such as California—than others, and why the extent of fruit set varies from year to year.

The sensitivity to temperature applies equally to plants in and out of containers. Container plants grown outdoors are subject to the normal fluctuations of temperature unless they can be moved so as to avoid extremes. Indoor plants are usually exposed to a narrower range of temperatures, but this is not necessarily good. For vegetables generally, a daytime temperature of 65 to 75°F and a nighttime temperature 10 to 15° lower is quite satisfactory. The lower night temperatures not only reduce respiratory breakdown of sugars, but often are required for some formative response such as flowering.

The cool-season vegetables are grown primarily for their vegetative parts and do best at daytime temperatures near 70°F. If temperatures are much higher the plants become weak and spindly, flower prematurely, or show other unfavorable responses. Warm-season vegetables are usually grown for their fruits and actually do better at temperatures above 70°F. It is a major challenge to provide for both of these types of plants within the home. One should carefully select vegetables for specific microclimates, taking into consideration daily and seasonal temperature (and light) variations at different window exposures, circulation patterns in the home, and so on. The near-window temperatures, for example, correspond closely to the intensity and duration of incoming solar radiation. Thus, temperatures at eastern exposures are greater than those at northern exposures but less than those at west and south facing windows.

Unfavorable temperatures can have a profound effect on both the shoot and root systems of container plants. Outdoors, patios, decks, and other surfaces can absorb heat from the sun, and if containers are placed on these surfaces their bases can become too hot. To prevent such overheating, aeration should be provided under containers by use of bricks or boards. This elevation also promotes drainage, but care must be taken to see that the drainage water does not damage surfaces. The containers themselves also absorb heat, which causes the growing medium to dry out faster and many also injure sensitive roots. To prevent overheating containers should either be made of material that is not excessively heat absorbent or shaded from intense sun.

Heat injury can also occur indoors. Direct sun can bring the temperature a short distance behind a window to well over 100°F. To prevent your vegetables from literally being "cooked alive" they should be screened from the direct sun, moved further away from the window, or put in some more favorable location. During the winter when temperatures are low plants placed close to windows may freeze. To prevent this simply move the plants five or six inches away from the window, or place a piece of cardboard between the plants and the window.

Air

In this section we wish to discuss certain other properties of air—moisture content, quality, circulation—as they relate to container plants. Outdoors these properties are either mitigated or largely beyond our control. Thus, our discussion will center mainly on the indoor environment.

Relative Humidity

The amount of moisture in the air is measured in terms of its relative humidity. **Relative humidity** refers to the amount of water vapor in the air at a given temperature, expressed as a percentage of the maximum amount the air could hold at that temperature. (As air temperature increases the amount of water it can hold also increases.) A relative humidity of around 50 percent is considered optimal for human comfort and our plants do well at this percentage also. However, in many homes the relative humidity is well below the 50 percent value. Low humidity values can be caused by the type of heating used (forced air, for example), use of air conditioning, and the humidity of the locale in question. Vast areas of the continent have summer humidity values of under 40 percent (Fig. 14-15).

The relative humidity of the air is important from a gardening standpoint because as the humidity decreases the loss of water from the soil (evaporation) and plant surfaces (transpiration) increases. The relative humidity of the intercellular spaces within stems and leaves is nearly 100 percent. As the humidity of the external air decreases the diffusion gradient between internal and external spaces increases and loss of water vapor to the external air is greater. To make matters worse, if a leaf or stem is at a higher temperature than the surrounding air—as can happen when the sun shines directly on a plant—the leaf air would hold even more water vapor and consequently the gradient of diffusion would be still greater.

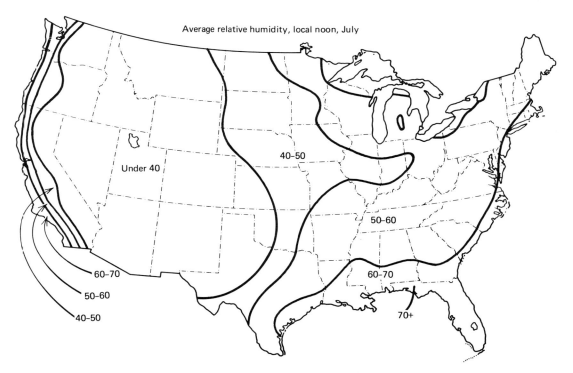

FIGURE 14-15 Average relative humidity across the continental United States at local noon in July. (Reprinted with permission of Samual R. Aldrich, Walter O. Scott, and Earl R. Leng, from *Modern Corn Production*, Copyright © 1975 by A. & L. Publications.)

Low humidity can be elevated in several ways. Plants can be grouped together in trays filled with pebbles to which water is added (care must be taken to see that the water level remains lower than the bottoms of the pots). Or, containers of water can be placed among vegetables. In either case the relative humidity increases around the plants. Inexpensive plastic sprayers are also available and periodic sprayings can be used to raise the humidity around plants. Finally, more expensive, but usually more effective, ways to raise humidity include installing a humidifier as part of the central heating system, or purchasing any of the room humidifiers available commercially.

Air Quality

Plants need carbon dioxide (CO_2) for growth through photosynthesis. The normal level of atmospheric CO_2 is about 0.03 percent, but reaches higher levels (up to 0.05 percent) in cities. Plants, can, however, respond through greater photosynthetic rates to increased levels of CO_2 up to about 0.5 percent. While it is generally not feasible for us to artificially elevate CO_2 levels it is important that we maintain natural levels as high as possible. As it turns out plants can bring about a localized depletion of CO_2 in the vicinity of leaf surfaces if they are photosynthesizing rapidly in calm air. Out-of-the-way windowsills, little used rooms, and closed, stuffy apartments offer ideal settings for CO_2 depletion. Proper air circulation can prevent this problem. Plants can be placed in areas of human movement or windows can be opened enough to create a natural flow of air through the home.

Air can contain gases harmful to plants. Fumes from manufactured gas contain ethylene and carbon monoxide, both of which can cause injury to plants. Cooking or heating with this form of gas does not present problems; it is only the unused, or leaked gas, that harms plants. A young tomato plant placed in a room where a gas leak is present will respond by bending its leaves sharply downward within 24 hours after exposure. Natural gas, which is used more widely nowadays, does not contain these harmful compounds.

Fumes from manufactured gas are not the only air quality problem, of course. Unclean air characterizes large urban areas everywhere, and neither animals nor plants are immune to its harmful contaminants. Some, such as sulfur dioxide, do serious injury when they react with plant surfaces. Other contaminants, such as dust, dirt, and grime, cover plant surfaces and thereby reduce the effectiveness of sunlight and slow the exchange of gases by clogging stomates. In areas of high pollution it may be necessary to wash foliage of exposed plants to prevent this kind of damage. Special care is also required in harvesting vegetables in areas of high pollution, a subject discussed in Chapter 15.

Watering

In container gardening the roots of vegetables are *confined to the containers* in which they are growing. The plants cannot explore a large volume of soil for water and minerals; instead, they depend on you, the gardener, for a constant supply of these life-giving substances. Recognition of this fact is central to successful container gardening. That is, proper watering and fertilization are essential if one is to grow abundant crops of vegetables in containers. In principle, watering and fertilization are distinct acts and we treat them as such. In the soil, however, their interaction is so complete as to make this division unrealistic.

Frequency of Watering

The frequency of watering container plants depends on many factors, not the least of which is the kind of container used. Unglazed clay pots **breathe;** that is, their porous nature allows for the movement of air, water, and mineral salts through the container walls (Fig. 14-16). The movement of mineral salts is evidenced by the whitish accumulations that build up on the outsides of these pots. Glazed clay pots and plastic pots serve as examples of pots that do not breathe. Pots of this type lose water from the soil surface through evaporation and from the bottom through drainage, but *not* through the walls of the container proper. With porous clay pots it is estimated that about half of the water added is lost through the sides by evaporation. Thus, to supply adequate water to plants in a clay pot would require about twice as much water as for plants in a plastic pot. The characteristics of the pot itself must be understood for effective watering. Failure in this regard can result in either overwatering or underwatering.

Frequency of watering depends on other factors as well. One must consider the kinds of plants, size of plants in relation to pot size, water-holding capacity of the soil mixture, temperature, and relative humidity. The idea is to provide your vegetables with ample moisture without keeping the soil too wet or too dry. Overwatering creates a waterlogged soil low in oxygen and results in poor growth and root rots, whereas underwatering results in stunted growth and is often expressed by wilting of plants. Generally it is best to provide your plants with a wet-dry cycle by watering thoroughly on a regular basis rather than by sprinkling frequently. Excess water must be allowed to drain from pots and then removed from the saucers. As the growing plants use the available moisture the amount of air in the soil increases. The key is to let this trend continue as long as the plants have available water, and yet to water before the plants show signs of moisture stress.

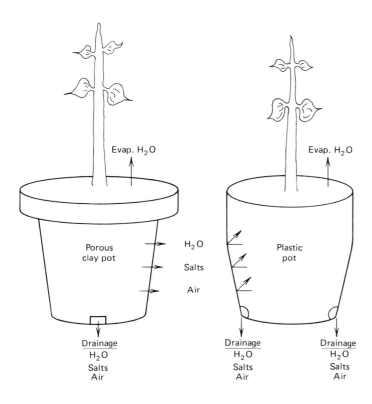

FIGURE 14-16 Exchange of air, water, and mineral salts in a porous clay pot and a plastic pot.

When to Water

How does one know when to water? One technique is to water each time the soil becomes dry down to a depth of about one inch. This method works well for clay pots because they tend to dry out rather uniformly. It does not work so well with plastic pots, where the surface soil may be quite dry while the soil two or three inches down is still wet. A good method for the beginner is to use one of the inexpensive probe-type water meters available on the market. These meters give a good indication of moisture content at different soil levels within a pot, and they can be used in the outdoor garden as well. Alternatively, some gardeners thrust a toothpick or a lollypop stick into the soil and then pull it out. If soil particles cling to the wood no water is needed, but if the toothpick or stick comes out clean, water should be added. All of the methods suggested give approximations at best. Only through experience can the container gardener learn the many variables that influence watering in a given situation.

How to Water

Should pots be watered from the top or from the bottom? Authorities differ in their answer to this question, but for container vegetables the majority recommend watering from the top. Generally, fertilizers are added at the soil surface, and watering from the top drives the fertilizer salts down into the root zone, whereas watering from the bottom drives these salts to the top where they can form harmful deposits. Furthermore, the plants can be watered in place; they do not need to be moved to a large tray or tub. Top watering is easier and faster. When watering from the top, avoid watering one side of the plant only. Water should be added all around the pot about midway between the stems and the sides of the pot (Fig. 14-17). The water should be added carefully, using a long-necked watering can which allows control of the rate and direction

FIGURE 14-17 Water should be added all the way around the pot, about midway between the stems and the sides of the pot.

of flow. Watering can be done any time of day. The important thing is to water thoroughly. The accumulation of water in the pot saucer will indicate that enough water has been added, but remember, the drainage water in the saucer carries nutrients with it. The greater the amount of drainage water the more nutrients are lost. Be sure to promptly remove the excess water which accumulates in the saucer.

Water Quality

Water is water, right? Wrong! There is cold, warm, and hot water, soft and hard water, chlorinated water, fluoridated water, and more. All of these do not serve equally as water for container plants, and some kinds of water are actually harmful to plants. Water that comes directly from the tap is commonly several degrees below room temperature. If added directly to the soil this cold water lowers the temperature in the root zone and slows growth. For this reason it is best to collect water ahead of time and let it stand in containers until it is at room temperature. Letting water stand in uncapped containers for about 24 hours offers an additional advantage in that the chlorine, which is present in most water sources, diffuses into the air. The chlorine does not harm vegetables—it is actually essential for certain plants—but is believed to be detrimental to some important soil organisms.

Virtually all water sources contain **fluorides**—compounds of fluorine—either naturally or as commercial additions. Normal levels of fluorides are not toxic to plants, but at high levels can cause tip and marginal burn on leaves of some plants. The high levels of fluoride emission in the vicinity of certain industrial plants can be harmful to plants. Most tap water used by gardeners should present no problems with regards to fluorides. If a problem should exist it can be remedied easily by addition of a small amount of limestone to the potting mix. The calcium present in limestone renders the fluoride insoluble and therefore unavailable to plants.

Hard water contains an overabundance of calcium and magnesium salts. In general these salts are not harmful to vegetables, but excesses can be bad for acid-loving plants such as azaleas. Some water softeners exchange the calcium and magnesium salts in the water for sodium. Excess sodium imparts unfavorable properties to soil, and is harmful to plants. Thus, if your water is softened by means of ion exchangers (e.g., Zeolite) it is better to take water for your vegetables before it passes through the water softener.

Fertilization

It is important to emphasize that the principles of soil fertility and mineral nutrition outlined in Chapter 5 apply to all vegetables, including those grown in containers. However, the fertilization of container-grown plants presents certain problems, because (1) the reservoir of nutrients in the planting mix may be low initially, (2) container plants cannot exploit a large volume of soil for nutrients (they depend on you for regular additions), and (3) it is easier to create deficiencies or excesses in the nutrient supply. For these reasons it is important that the container gardener have a clear understanding of kinds of fertilizers available, fertilization techniques, and the fate of fertilizers after application.

Kinds of Fertilizers Available

Plants need at least 16 different elements to complete their life cycle (Fig. 14-18). These elements are obtained either from the air, water, or the soil solids (Table 5-2). Our discussion of fertilization centers around the three nutrients taken from the soil in greatest amounts—nitrogen (N), phosphorus (P), and potassium (K). These elements are called the "fertilizer elements" and any fertilizer that contains all three is said to be a complete fertilizer. The amounts of these elements in a complete fertilizer is expressed as a combination of figures such as 10-6-8. These figures list in order the percentage of nitrogen (10 percent), phosphorus (6 percent), and potassium (8 percent).

Each element needed by plants induces specific responses. Nitrogen promotes a lush growth of dark green foliage. Phosphorus promotes root growth, flowers, and fruits. Potassium has a less clear role, but appears to help regulate nutrient utilization, strengthen stems, and aid resistance to insects and diseases. The indoor gardener can greatly influence plant development by choice of fertilizers. For example, a tomato plant with an abundance of deep green foliage but with few fruits probably has too much nitrogen. All vegetables grown for their fruits need a more favorable balance of phosphorus to promote flowering and fruit development.

Complete fertilizers come in many forms. They may be from organic or inorganic

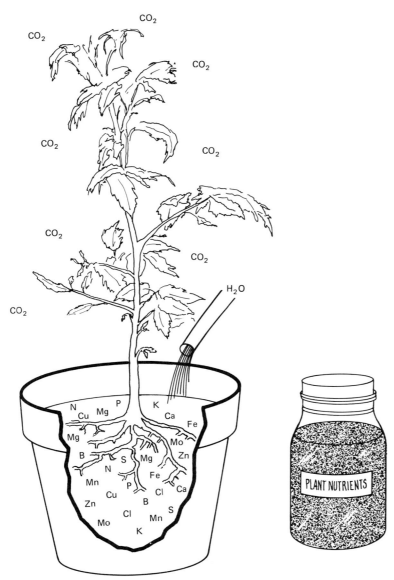

FIGURE 14-18 Plants need at least 16 elements to complete their life cycles. These elements are contributed by soil, water, and air. The trace elements are needed in minute amounts. Each element has specific roles in the plant.

sources, liquids or solids (powders, pellets, tablets, sticks), quick-release or timed-release, and with or without a specified content of trace or other essential elements. Choice of type used will depend on several factors including availability, price, potting mixture, plants grown, and preference.

Fertilization Techniques

Potting mixtures that contain soil, manure, compost, or similar organic ingredients will have a relatively high nutrient content. With mixes of this sort addition of the three fertilizer elements should be adequate. Some of the soilless mixes, on the other hand, have a relatively low nutrient level. With these you will need to add all of the nutrients commonly obtained by the plant from the soil, just as if you were growing vegetables using hydroponics. Fish emulsion fertilizer, seaweed fertilizer, and other organic fertilizers contain all of these needed elements. Also a number of chemical fertilizers are formulated with trace elements as part of their guaranteed analysis, and many others contain most or all of the needed elements as "impurities."

In any container program it is well to remember that vigorously growing vegetables need regular and systematic fertilization. Some of the pelleted time-release fertilizers can be added to the soil before or at the time of planting. These slowly release nutrients over a three- to four-month period; thus, with most vegetables just one fertilization is adequate. Many of the so-called "houseplant" fertilizers, on the other hand, are formulated for application biweekly or monthly. In general, most container gardeners feel that, with fertilizers of this type, small applications at frequent intervals are better than larger doses at infrequent intervals. Always moisten the soil well before adding any solid fertilizer to prevent root burn. Finally, a number of liquid organic and chemical fertilizers are formulated for use with each watering. Use of fertilizers of this type assures a constant and uniform supply of nutrients and at the same time eliminates the need for remembering when to fertilize your plants. A number of these fertilizers, as well as some of those mentioned earlier, can be applied either to the soil, or to the foliage. The nutrients applied to foliage are absorbed through leaf surfaces and through the stomates.

What Are the Fates of Applied Fertilizers?

Most fertilizer salts are taken up by plants and used in growth. Top watering causes a continuous downward movement of the salts, however, and if watering is excessive much fertilizer is leaked from the pot. Bottom watering causes an upward movement of salts, which can eventually accumulate on the soil surface as harmful deposits. Alternate top-bottom watering would cause a downward-upward movement of fertilizers, and for this reason is advocated by some container growers. Porous clay pots lose moisture through their walls and this moisture carries fertilizer salts with it (Fig. 14-16), as evidenced by whitish deposits on the outsides of used pots. Consequently, the fertilization rate for vegetable in porous clay pots must be greater—as much as doubled—than for vegetables grown in plastic containers. Overfertilization can cause an uptake of excess salts by plants and result in injury. Symptoms of overfertilization include wilting, premature loss of lower leaves, and, perhaps most commonly, browning of the margins and/or tips of leaves. Lower leaves do die eventually, however, so not all mortality is caused by fertilizer.

Pollination

When plants are grown indoors there is a decreased likelihood of pollination in many plants because the pollinating agents are scarce or absent. The gardener must therefore step in and help in pollen transfer. This can be done in different ways. The solanaceous plants—eggplant, tomato, pepper—are commonly self-pollinated (Table 3-2), but outdoors this process is facilitated by wind shaking the plant. Indoors, the gardener must do the shaking; by gently shaking the plant by hand or by vibrating it with an electric toothbrush or similar instrument. The best time to shake or vibrate plants seems to be in midday. Cucumbers and other cucurbits are normally pollinated by bees. Indoors the gardener generally needs to transfer the pollen. This can be done using a camel's hair brush to move the pollen from the male flowers to the female flowers of the same or an adjacent plant. We have omitted discussing topics relating to the long-term maintenance of potted plants because, while they apply to house plants in general, they are not relevant to vegetables in particular. Many of the topics covered would have required more extensive treatment if we were dealing with the whole range of house plants. Finally, we have omitted discussion of hydroponics. While hydroponics is unquestionably a challenging and fascinating area of interest, it ranks low as a means of productive container gardening. Certain of the references at the end of the chapter expand on this interesting topic.

Questions for Discussion

1. Define the following terms:
   (a) Pyramid garden
   (b) Synthetic soil
   (c) Peat moss
   (d) Wide-spectrum lamp
   (e) HID lamp
   (f) Relative humidity

2. What approaches can the gardener with limited space employ for effective space utilization?

3. List several vegetables that are well suited to small spaces. Also indicate two or three vegetables that in general should not be grown if space is at a premium.

4. Draw up plans for a garden measuring 8 × 10 feet. Indicate crops to be grown, planting patterns to be used, and for each vegetable specify whether seed or transplants are utilized.

5. The authors state that vegetables are multipurpose plants, providing both utility and beauty. Offer strong evidence in support of this statement.

6. Cite at least three instances where you have either observed or utilized some form of vertical gardening.

7. What basic features do all good plant containers have in common?

8. List three advantages and three disadvantages of clay containers as they relate to indoor gardening.

9. Discuss the roles of sand, compost, and vermiculite in potting mixes.

10. Identify five vegetables that you feel are especially suited to container gardening.

11. Discuss light intensity as related to window exposure.

12. Should the same planting mix be used year after year in container gardening? What is the basis for your answer?

13. Compare the spectral emission patterns of fluorescent and incandescent lamps. How do these patterns relate to plant growth and development?

14. How is it possible to take light-intensity readings using a camera with a built-in exposure meter?

15. Discuss the effects of temperature as relates to the growth of plants indoors. Include important aspects of both daytime and nighttime temperatures.

16. What aspects of air quality need to be taken into consideration in indoor gardening?

17. Discuss at length the watering of container-grown plants.

18. Identify some of the hazards associated with the fertilization of plants in containers.

Selected References

Baker, K.F. (ed.). 1957. *The U.C. System of Producing Healthy Container-Grown Plants.* Manual 23, California Agricultural Experiment Station.

Bickford, E. and S. Dunn. 1972. *Lighting for Plant Growth.* Kent State University Press, Kent, Ohio.

Douglas, J. 1975. *Hydroponics: The Bengal System with Notes on Other Methods of Soil Cultivation.* Oxford University Press, New York.

Frank, M. and E. Rosenthal. 1974. *Indoor/Outdoor Highest Quality Marijuana Grower's Guide.* And/Or Press, San Francisco.

Keen, J. 1975. What's watt in lighting. *American Orchid Society Bulletin*, 44: 292–297.

Kranz, F. and J. Kranz. 1971. *Gardening Indoors Under Lights.* Viking Press, New York.

Newcomb, D. 1976. *The Apartment Farmer*. J.P. Tarcher, Inc., Los Angeles.

Olkowski, H. and W. Olkowski. 1975. *The City People's Book of Raising Food*. Rodale Press Inc., Emmaus, Pa.

Poincelot, R. 1976. *Gardening Indoors with House Plants*. Rodale Press, Emmaus, Pa.

Ray, R. (ed.). 1974. *House Plants Indoors/Outdoors*. Ortho Books, San Francisco.

Ray, R. (ed.). 1975. *The Facts of Light About Indoor Gardening*. Ortho Books, San Francisco.

Sherman, C. and H. Brenizer. 1975. *Hydroponic Gardening at Home*. Nolo Press, Occidental, Cal.

Teuscher, H. (ed.). 1958. *Gardening in Containers*. Brooklyn Botanic Garden, Brooklyn, N.Y.

U.S. Department of Agriculture. 1974. *Minigardens for Vegetables*. Home and Garden Bulletin No. 163. U.S. Government Printing Office, Washington, D.C.

University of Illinois. 1962. *Hydroponics as a Hobby*. Circular 844, Urbana.

LEARNING OBJECTIVES

By the time you have finished this chapter you should be able to:

1 Name the major foodstuffs consumed by humans.
2 State the importance of crude fiber in the diet.
3 Identify four fat-soluble vitamins.
4 Explain the importance of ruminants in agricultural systems.
5 Discuss the importance of regulating respiration and transpiration in stored fruits and vegetables.
6 Define "coasting."
7 Identify two "heavy metals" and tell why they must often be considered when harvesting vegetables.
8 State the "three R's" of vegetable cooking.
9 Given a specific vegetable, describe its most favorable storage conditions in the home.
10 Describe the steam-pressure and water-bath methods of canning.
11 Identify the organism that causes botulism and explain how this organism can be dealt with in home canning.
12 Define "blanching" and tell of its place in freezing vegetables.
13 Present a flowchart for the drying of vegetables and fruits.
14 State the advantages and disadvantages of each of the major methods of food preservation.

15

HARVESTING, STORING, AND PRESERVING GARDEN CROPS

"Eat your vegetables."

Guess who!

Every chain has its weakest link and so too every gardener has strengths and weaknesses. Planning, preparing, planting, growing (those innumerable cultural manipulations), harvesting, storing, and preserving are the links in the chain of gardening events. For many gardeners the growing of a fine crop of vegetables is not a major problem, and they may even set aside nice-looking products for later use. But looks are not everything! The fruits of many vegetables have a critical harvest period, beyond which their nutritional value and eating qualities decline rapidly. Vegetables grown for their vegetative parts must also be harvested properly. And, once harvested, how are vegetables best handled for cooking, storing, and preserving? If vegetables are canned, for example, are we sure they are safe to eat?

What vegetables freeze well? How does drying affect the nutritional value of vegetables? What vegetables are good sources of vitamin A? Vitamin C?

The answers to the above questions are, all too often, not known by the gardener. This chapter is designed to help remedy these weaknesses and to provide critical information on harvesting, storing, and preserving garden crops.

HARVESTING VEGETABLES

Harvesting is the physical separation of a given plant part from the parent plant. The portion harvested may be small relative to the size of the parent plant, as when we remove a few tomatoes, peas, or lettuce leaves. In other cases the portion harvested is large, as when we harvest head lettuce, cabbage, or carrot. Regardless of the size of the harvested portion, harvesting has profound effects on both the portion removed and the parent plant. The more important of these effects are discussed below.

The Effects of Harvesting

Food, water, hormones, vitamins, and minerals are supplied to developing parts by the parent plant. At harvest this supply is abruptly interrupted; the harvested portion becomes a separate living entity (Fig. 15-1). Since the harvested part is living it still needs a source of energy. This energy comes from its own stored food reserves; starch is converted into simple sugars and these are utilized in respiration, as follows:

$$\text{simple sugars} + \text{O}_2 \text{ (oxygen)} \rightarrow$$

$$\text{useful energy} + (\text{carbon dioxide}) \text{CO}_2 + \text{H}_2\text{O} + \text{heat}$$

Note that respiration also results in the production of carbon dioxide, water, and heat—which must be dealt with if the harvested portions are to be stored.

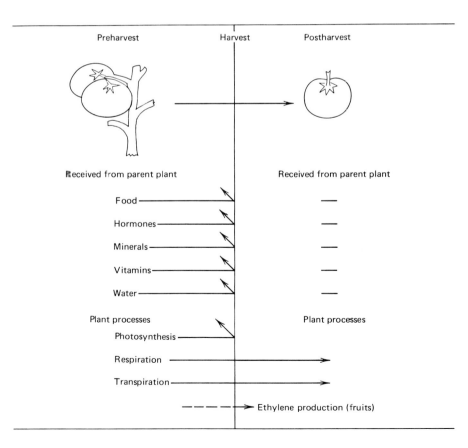

Preharvest   Harvest   Postharvest

Received from parent plant   Received from parent plant

Food —
Hormones —
Minerals —
Vitamins —
Water —

Plant processes   Plant processes
Photosynthesis —
Respiration
Transpiration

Ethylene production (fruits)

FIGURE 15-1
Conditions relating to a tomato, or other vegetable, before and after harvest.

Transpiration, the loss of water vapor from plant surfaces, also continues in harvested fruits and vegetables. Water loss can occur quite rapidly with succulent plant parts such as asparagus spears and leafy vegetables, less rapidly with more mature plant parts such as cabbage heads, celery, and potatoes. Keeping plant parts cool and moist will reduce transpiration losses in vegetables stored for any length of time.

In leafy vegetables and certain other products photosynthesis is occurring just prior to harvest. This is not an important consideration generally, for the harvested parts are either quickly consumed or else stored in the dark. Certain other cellular activities are, however, of great importance. In corn, for example, enzymes within the kernel begin converting sugars into starch within moments after picking. Because of this, corn is best if placed immediately into boiling water. Many other vegetables, including asparagus, beans, and peas, also lose sweetness and flavor within a few hours of harvest. These qualities are best maintained if vegetables are used—raw or cooked—soon after harvest.

Harvesting also has important consequences for the parent plant. Often these effects are beneficial. For example, once a plant part is removed the parent plant will channel its resources into younger, developing parts. Fruit-yielding vegetables such as zucchini produce more if harvested frequently. In hot weather its fruit will develop to harvestable size just a few days after flowering. If left on the plant they will become oversize, overripe, and of poor quality. Furthermore, if the fruits are not harvested the production of new fruits will slow greatly or stop altogether.

Leafy vegetables respond to proper harvesting too. Leaf lettuce, endive, and Swiss chard, like other plants, produce leaves outwardly from a central growing point. That is, the oldest leaves are found at the outside of the plant. Removal of outer leaves as they reach harvestable size encourages the production and growth of more leaves from the center of the crown. If harvest is delayed, the growth of new leaves slows, a condition known as **coasting**. Proper harvest prevents this "productive laziness."

When to Harvest

Most vegetables should be harvested just as they reach their peak of quality (it is safe to assume that the quality will not improve significantly after harvest). Manifestations of peak quality will, of course, vary from vegetable to vegetable. Maturity dates, as presented on seed packets and in seed catalogs, can be used as a guide to time of harvest for many vegetables, especially if adjustments are made for growing conditions. Specific guidelines for harvesting individual vegetables are presented in Chapter 10.

Harvesting cannot always be done at the correct time. An unexpected frost, for example, can drastically alter harvesting plans. Many vegetables can be covered to protect against damage in case of early frost. Portable lath houses, inverted V-shaped troughs, plastic-covered frames, or plastic sheets all protect plants. Another practice employed by many gardeners is to surround individual plants with inner tubes full of water (Fig. 8-20b). Water is very effective in storing heat and, as temperatures lower, cools only gradually as compared to most liquids and air. Thus, the inner tube with water reduces temperature extremes in the microclimate of the tube center, much as a lake modifies climate of adjacent land areas. Vegetables such as beets, carrots, chard, parsnips, and peppers will usually survive frosts given adequate protection.

Frost-tender vegetables such as most solanaceous and cucurbitaceous crops (Appendix A) are more difficult to protect from frost, and even if protected the fruits remaining on the plant may not ripen satisfactorily. Thus, while some gardeners report success in covering these crops, many feel the best approach is to harvest fruits, unripe and ripe, at the first signs of a frost. As an alternative, some gardeners pull entire plants and hang them in a protected place. In either case, the rescued fruits of these warm-season crops will ripen and the harvest can be eaten fresh or preserved.

Nutrient Content of Harvested Vegetables

Three classes of organic compounds—carbohydrates, fats, and proteins—provide the bulk of human nutrition. In the typical American diet the contribution of each of these foodstuffs to the daily energy requirements is as follows:

Major Foodstuff	Percent Contribution
Carbohydrates	45–55
Fats	40–50
Proteins	12–15

In addition, small amounts of vitamins and minerals are present in virtually all foodstuffs. While vitamins and minerals are needed in much smaller quantities than the three major foodstuffs, they are nonetheless absolutely essential to existence. Vegetables are an excellent source of all of these food materials (Appendix F).

Carbohydrates

Starches, sugars, and cellulose are the most common forms of carbohydrate in plants. Of these, only starches and sugars can be used directly by humans to meet our energy requirements. These compounds, which make the major contribution to our energy needs, provide the body with a relatively inexpensive source of energy as compared to other products. Cellulose is the main structural component of plant cell walls, and, as we shall see, is of indirect nutritional value.

Vegetables harvested for their stem, leaf,

flower, and immature fruits are generally low in usable carbohydrates in proportion to their cellulose content. Thus, a full serving of many of these vegetables provides fewer than 30 **calories**.[1]

Some examples in this category are:

Asparagus	Chard	Peppers
Beet greens	Cucumber	Radish
Cabbage	Eggplant	Spinach
Cauliflower	Endive	Tomatoes
Celery	Lettuce	Turnip greens

Servings of several other vegetables provide an intake of 50 calories or less (Appendix F). These low-calorie vegetables are also excellent sources of vitamins and minerals. Small wonder they are so popular among weight-conscious adults.

In contrast to the low-calorie crops mentioned above, vegetables harvested for their storage roots or mature fruits usually have a high proportion of usable carbohydrate. The cereals, legumes, and potatoes are rich sources of starch, and are mainstays of human diets throughout the world. In our bodies the large starch molecules are **digested** (broken down) into sugars, and these provide a readily accessible form of energy.

The Importance of Cellulose in the Diet

Cellulose is an important plant carbohydrate, forming the framework of plant cell walls. Cellulose is the most abundant material in plants, but unfortunately, perhaps, it is not a universal food source. Mammals do not possess the enzymes capable of breaking cellulose

---

[1]A calorie is the amount of heat required to raise the temperature of 1 gram of water 1 centigrade degree at 1 atmosphere of pressure. A kilocalorie is 1000 times as large, that is, the heat required to raise 1 kilogram of water 1 centigrade degree. Caloric values of specific foods or diets always refer to kilocalories (also correctly written as calories).

down into simple sugars, but a number of bacteria do secrete these enzymes. Some mammals can utilize cellulose, however, by virtue of the fact that their digestive tract is modified and contains the appropriate bacteria. In animals with simple stomachs (Fig. 15-2a), such as the dog, pig, and human, the suitable bacteria are found in the **cecum** and **colon** (parts of the large intestine), and the food passes through these structures too quickly for cellulose digestion to be effective. In these animals nutrients are absorbed in the small intestine that precedes the cecum and colon. In the **ruminants** (cud-chewing animals), however, cellulose is utilized. Animals such as sheep, goat, and cow, have a more

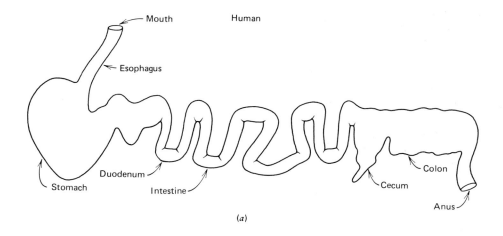

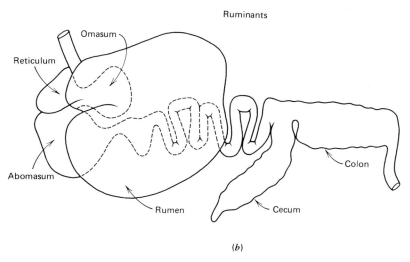

FIGURE 15-2 The anatomy of the digestive tract determines in large part whether or not an animal can derive sustenance from cellulose. The simple stomach of humans and certain other animals (a) cannot utilize cellulose, whereas the complex stomach of ruminants (b) can degrade this important plant product. (From Jules Janick, Carl Noller, and Charles Rhykerd, "The Cycles of Plant and Animal Nutrition." Copyright © 1976 by Scientific American Inc. All rights reserved.)

elaborate digestive tract consisting of a stomach with four separate compartments (Fig. 15-2b). The first compartment, the **rumen**, acts as a large fermentation vat in which cellulose and other materials are broken down into smaller units. Food materials then pass through the other compartments where they are acted on by gastric juices, and from there into the intestine. As a result, the full length of the intestine is available for absorption of nutrients. Because of their unique capabilities, the ruminants occupy a special place in agricultural systems.

The fact that cellulose is not digested by humans does not mean that it is unimportant in human nutrition. To the contrary, mounting evidence supports the view that cellulose and related wall constituents are a vital part of the human diet. Together these wall materials comprise what we call **crude fiber**, or simply **fiber**. Fiber, which is regarded as the "humus of the digestive system," has the following important properties. It absorbs up to six times its own weight in water, it adds bulk to the intestinal tract, and it favorably influences the character of the intestinal flora. Through these properties, fiber increases the motility of food through the digestive tract and makes the stool softer, thus allowing for easier elimination. There is much evidence that the effects of inadequate fiber in the diet may have serious consequences. For example, many medical researchers find a positive correlation between diseases of the intestine, such as diverticular disease, hemorrhoids, and colonic cancer, and inadequate dietary fiber.

It is important to recognize that our *only* source of dietary fiber is from plant products; there is *no* cellulose fiber in animal products. We occasionally make reference to the "fibrous" nature of a steak we may be eating, but this simply refers to its physical condition, not its chemical makeup. We must get our fiber from plant materials. All vegetables contain fiber. The vegetables referred to earlier as being low in usable carbohydrate in pro-

portion to cellulose are all good sources of fiber, as are fruits such as apple and pear (especially if they are eaten with the skins left on). While the above are good fiber sources, bran is unquestionably the outstanding source. **Bran** is technically the outside covering of the kernel of cereals such as wheat (Fig. 15-3). The bran layer covers the endosperm and embryo, or **germ** as it is called in the milling profession. Proteins, iron, and the B-vitamins are more highly concentrated in the bran layer and germ than in the inner portions of the kernel. Unfortunately, in Western culture, the bran layer and germ are removed from cereals in the milling process. As a result bread made from white flour has only about one-tenth the fiber of bread made from whole-grain wheat. (The lost vitamins and minerals are readded to create our **enriched flour**.)

Fats

Fats are an important part of the diet of Americans, contributing anywhere from 40 to 50 percent of our daily energy requirements. Fats perform several useful functions in the diet. For one, they are a concentrated source of energy, providing more than twice as many calories per gram as do carbohydrates or proteins. Fats are also important as carriers for vitamins A, D, E, and K—the **fat-soluble vitamins**. The presence of fats in foods contributes greatly to their palatability. Finally, because fats move through the stomach relatively slowly, they give the diner a satisfied feeling for a relatively long period of time.

There are many good sources of fats, both animal and plant. Certain fruits and seeds are particularly good sources as illustrated by the avocado, peanut, and soybean (Appendix F). Other familiar plant sources include cottonseed, corn, sunflower, and safflower. These plant sources provide the edible oil products of commerce, marketed as vegetable oil, salad oil, and margarine. While the above-mentioned plants represent forms in which large

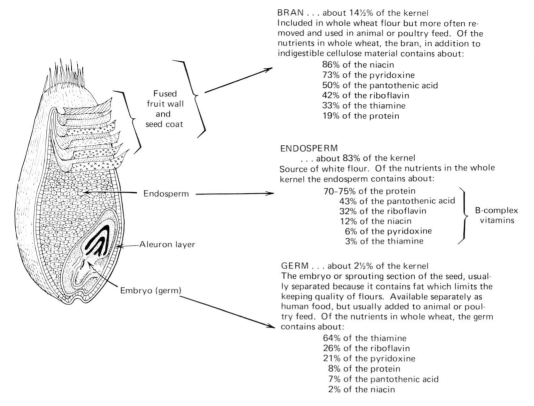

BRAN . . . about 14½% of the kernel
Included in whole wheat flour but more often removed and used in animal or poultry feed. Of the nutrients in whole wheat, the bran, in addition to indigestible cellulose material contains about:

    86% of the niacin
    73% of the pyridoxine
    50% of the pantothenic acid
    42% of the riboflavin
    33% of the thiamine
    19% of the protein

ENDOSPERM
    . . . about 83% of the kernel
Source of white flour. Of the nutrients in the whole kernel the endosperm contains about:

    70–75% of the protein
    43% of the pantothenic acid
    32% of the riboflavin   B-complex
    12% of the niacin   vitamins
    6% of the pyridoxine
    3% of the thiamine

GERM . . . about 2½% of the kernel
The embryo or sprouting section of the seed, usually separated because it contains fat which limits the keeping quality of flours. Available separately as human food, but usually added to animal or poultry feed. Of the nutrients in whole wheat, the germ contains about:

    64% of the thiamine
    26% of the riboflavin
    21% of the pyridoxine
    8% of the protein
    7% of the pantothenic acid
    2% of the niacin

FIGURE 15-3 A kernel of wheat. (Modified from Oswald Tippo and William L. Stern, *Humanistic Botany* with permission of W. W. Norton and Company, Inc. Copyright © 1977 by Oswald Tippo and William L. Stern.)

amounts of fat are concentrated in fruits or seeds, note that all plants contain at least a limited amount of fats throughout their vegetative bodies.

Proteins

Each of us is 18 to 20 percent protein by weight. This high value indicates the importance of protein in our diet. There are many functions of protein. Proteins are required for the growth and maintenance of body tissues. Proteins are a major component of skin, hair, nails, cartilage, tendons, muscles, and even the organic component of bones. Protein is in especially high demand for growing children, but is also needed by adults to replace tissues that are continually breaking down and to build other tissues that are continually growing (Fig. 15-4). Proteins are also needed to form many hormones, **antibodies** (substances that combat disease-producing organisms and give immunity to disease), and enzymes—all critically important body proteins. **Hemoglobin**, the oxygen-carrying molecule in the blood, is also a protein. Proteins are also important in maintaining proper body pH and the fluid balance of cells. Finally, like carbohydrates and fats, proteins sooner or later serve as a source of energy for the body.

Many animal products, including meat, eggs, milk, and cheese are good protein sources.

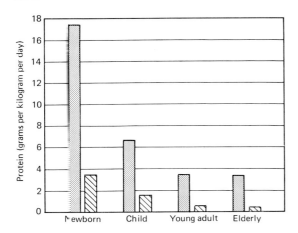

FIGURE 15-4 Protein requirements per unit of body weight decrease with age (cross-hatched colums). This decrease parallels the decline in protein synthesis by the body with advancing age (stippled columns). (From Nevin S. Scrimshaw and Vernon R. Young, "The requirements of Human Nutrition." Copyright © 1976 by *Scientific American, Inc.* All rights reserved.)

Plants are also good sources of protein. Two plant groups—the legumes and the cereals—stand out in this regard. Soybeans have a protein content approaching 35 percent of dry weight. Other beans, peas, and lentils average 20 percent of protein. The cereals are somewhat lower; rice averages 7 to 8 percent, corn 9 to 10 percent, and wheat 10 to 12 percent. Recently plant breeders have developed cereal cultivars of higher protein content, and although certain problems have been experienced with these cultivars, we are certain to see more of them in the future.

Minerals and Vitamins

In addition to the major organic foodstuffs, healthy humans require small but vital amounts of certain minerals and vitamins. The essential minerals are classified as either macronutrients or micronutrients (Table 15-1). The macronutrients are calcium, phosphorus, sulfur, potassium, chloride, sodium, and magnesium (carbon, hydrogen, oxygen, and ni-

trogen are here too). The micronutrients are iron, fluoride, zinc, copper, selenium, manganese, iodine, molybdenum, chromium, and cobalt. The vitamins are grouped on the basis of their solubility, as either water or fat soluble (Table 15-2). The **water-soluble vitamins** include the B-vitamin group and vitamin C. Fat-soluble vitamins include A, D, E, and K. A summary of characteristics of the minerals and vitamins is presented in Tables 15-1 and 15-2.

Vegetables are excellent sources of most of the minerals and vitamins (Appendix F). Plants are regarded as only a fair source of the mineral cobalt, and we rely on iodized salt and drinking water for much of our sodium, chloride, iodine, and fluoride. Aside from these possible exceptions, vegetables make a major contribution to our mineral needs. Vitamins needed to promote and maintain human life also occur abundantly in plants. The importance of citrus fruits in preventing scurvy among sailors is legend. Indeed, plants are the only good natural source of vitamin C, the vitamin that prevents scurvy. Of the numerous vitamins, only two—vitamins B-12 and D—are confined to animal products. Many nutritionists feel that the best way to get one's daily mineral and vitamin needs is to eat well-planned and carefully prepared meals. If a wide variety of foods is served, the diet should include suitable levels of all the vitamins and minerals. If supplemental vitamins are consumed it is important to realize that vitamins A and D are toxic in excessive amounts. None of the water-soluble vitamins is regarded as toxic, since they are readily excreted in the urine when taken in excessive amounts.

TRIMMING AND WASHING

Trimming

Harvested vegetables almost always need some trimming to remove damaged or diseased portions, and other inedible material. Trimming should be adequate but judicious, because any material removed reduces the amount of nutrients present. In broccoli, head lettuce, and cabbage the more nutritious parts can easily be trimmed away. In lettuce, the darker green, outer leaves are as much as 30 times higher in vitamin A than are the inner, bleached leaves. The dark leaves make up about 10 percent of head weight, but if they are discarded more than 75 percent of the vitamin A content is lost. Broccoli leaves, which are edible, have many times the vitamin A content of the stalks and flower buds. In other cases removing petioles, stems, and other fibrous parts may be beneficial. In collards, for example, the thin leaf parts have about 30 times more vitamin A than the midribs and petioles, and the leafy parts of turnip greens have about 20 times as much vitamin A. The pale color of stem and leaf axes of kale, and certain other leafy vegetables, in comparison to the dark-green color of the leaf blade proper indicates a vitamin A distribution similar to that of collards and turnip. The thin blade part of leaves also contains many times more vitamin C, and two to four times more iron than the stem and petiole parts. Thus, while the stems, petioles, and midribs of leaves account for more than half the weight, their nutrient contribution is much less.

Washing

Harvested vegetables should be washed thoroughly in preparation for eating, and in most cases, for storing. Washing is particularly important in urban areas and along freeways due to the presence of environmental contaminants, such as the **heavy metals** (cadmium, chromium, lead, mercury, nickel, zinc, and others), many of which are toxic to humans, even in very small quantities. Of these, lead has perhaps received the most attention. Lead is used in gasoline as an antiknock ingredient and is emitted in exhaust fumes. About 50

percent of automobile lead is deposited within the first 100 feet from a roadway, but significant amounts may travel up to $2\frac{1}{2}$ miles. This lead accumulates on foliage and in the soil. Studies show that 30 to 50 percent of foliage lead can be removed by thorough washing, but degree of removal is influenced by hairiness and roughness of plant parts. Soil-borne lead is taken up by plants, but usually not in excessive quantities. Interestingly, lead has been found in high concentrations in earthworms living among major highways, although without any visible damage symptoms.

The portion of the plant consumed is an important variable as relates to environmental contaminants. Many vegetables have a protective covering over their edible parts; corn has its husks, peas have their pods, and watermelons have their rinds. Root crops are relatively safe from airborne pollutants but susceptible to soil contaminants. Dietary patterns are also important. A mixed diet tends to minimize the danger related to contaminants in a single source. A properly placed barrier of evergreen trees or shrubs can help to screen out airborne pollutants. Finally, the garden site should be located as far away from pollution sources as possible.

USING FRESH VEGETABLES

Raw Versus Cooked Vegetables

Vegetables such as radish and lettuce are almost always eaten raw, while others such as beans and corn, are almost always cooked. Numerous others, however, are eaten both raw and cooked. Proponents of raw vegetable consumption stress the better flavor of the uncooked product and the loss of vitamins and minerals associated with cooking. These claims are well founded, but others seem less substantial. Among the latter are the claims that (1) raw vegetables provide humans with

TABLE 15-1 The Essential Mineral Elements and Their Characteristics in the Human Diet

Mineral	Amount in adult body (grams)	RDA for healthy adult male (milligrams)	Dietary sources	Major body functions	Deficiency	Excess
Macronutrients						
Calcium	1500	800	Milk, cheese, dark-green vegetables, dried legumes	Bone and tooth formation Blood clotting Nerve transmission	Stunted growth Rickets, osteoporosis Convulsions	Not reported in humans
Phosphorus	860	800	Milk, cheese, meat, poultry, grains	Bone and tooth formation Acid-base balance	Weakness, demineralization of bone Loss of calcium	Erosion of jaw (fossy jaw)
Sulfur	300	(Provided by sulfur amino acids)	Sulfur amino acids (methionine and cystine) in dietary proteins	Constituent of active tissue compounds, cartilage and tendon	Related to intake and deficiency of sulfur amino acids	Excess sulfur amino acid intake leads to poor growth
Potassium	180	2500	Meats, milk, many fruits	Acid-base balance Body water balance Nerve function	Muscular weakness Paralysis	Muscular weakness Death
Chloride	74	2000	Common salt	Formation of gastric juice Acid-base balance	Muscle cramps Mental apathy Reduced appetite	Vomiting
Sodium	64	2500	Common salt	Acid-base balance Body water balance Nerve function	Muscle cramps Mental apathy Reduced appetite	High blood pressure
Magnesium	25	350	Whole grains, green leafy vegetables	Activates enzymes Involved in protein synthesis	Growth failure Behavioral disturbances Weakness, spasms	Diarrhea
Micronutrients						
Iron	4.5	10	Eggs, lean meats, legumes, whole grains, green leafy vegetables	Constituent of hemoglobin and enzymes involved in energy metabolism	Iron-deficiency anemia (weakness, reduced resistance to infection)	Siderosis Cirrhosis of liver

Mineral	Amount in body	Recommended daily amount	Source	Function	Deficiency	Excess
Fluoride	2.6		Drinking water, tea, seafood	May be important in maintenance of bone structure	Higher frequency of tooth decay	Mottling of teeth, Increased bone density, Neurological disturbances
Zinc	2		Widely distributed in foods	Constituent of enzymes involved in digestion	Growth failure, Small sex glands	Fever, nausea, vomiting, diarrhea
Copper	0.1		Meats, drinking water	Constituent of enzymes associated with iron metabolism	Anemia, bone changes (rare in humans)	Rare metabolic condition (Wilson's disease)
Selenium	0.013	Not established (Diet provides .05-1 per day)	Seafood, meat, grains	Functions in close association with vitamin E	Anemia (rare)	Gastrointestinal disorders, lung irritation
Manganese	0.012	Not established (Diet provides 6-8 per day)	Widely distributed in foods	Constituent of enzymes involved in fat synthesis	In animals: poor growth, disturbances of nervous system, reproductive abnormalities	Poisoning in manganese mines: generalized disease of nervous system
Iodine	0.011	0.14	Marine fish and shellfish, dairy products, many vegetables, table salt	Constituent of thyroid hormones	Goiter (enlarged thyroid)	Very high intakes depress thyroid activity
Molybdenum	0.009	Not established (Diet provides .4 per day)	Legumes, cereals, organ meats	Constituent of some enzymes	Not reported in humans	Inhibition of enzymes
Chromium	0.006	Not established (Diet provides 0.05-0.12 per day)	Fats, vegetable oils, meats	Involved in glucose and energy metabolism	Impaired ability to metabolize glucose	Occupational exposures: skin and kidney damage
Cobalt	0.0015	(Required as vitamin B-12)	Organ and muscle meats, milk	Constituent of vitamin B-12	Not reported in humans	Industrial exposure: dermatitis and diseases of red blood cells

From Nevin S. Scrimshaw and Vernon R. Young, "The Requirements of Human Nutrition." Copyright © 1976 by Scientific American, Inc. All rights reserved.

TABLE 15-2 The Essential Vitamins and Their Characteristics in the Human Diet

Vitamin	RDA for healthy adult male (milligrams)	Dietary sources	Major body functions	Deficiency	Excess
Water soluble					
Vitamin B-1 (thiamin)	1.5	Pork, organ meats, whole grains, legumes	Coenzyme (thiamine pyrophosphate) in reactions involving the removal of carbon dioxide	Beriberi (peripheral nerve changes, edema, heart failure)	None reported
Vitamin B-2 (riboflavin)	1.8	Widely distributed in foods	Constituent of two flavin nucleotide coenzymes involved in energy metabolism (FAD and FMN)	Reddened lips, cracks at corner of mouth (cheilosis), lesions of eye	None reported
Niacin	20	Liver, lean meats, grains, legumes (can be formed from tryptophan)	Constituent of two coenzymes involved in oxidation-reduction reactions (NAD and NADP)	Pellagra (skin and gastrointestinal lesions, nervous, mental disorders)	Flushing, burning, and tingling around neck, face, and hands
Vitamin B-6 (pyridoxine)	2	Meats, vegetables, whole-grain cereals	Coenzyme (pyridoxal phosphate) involved in amino acid metabolism	Irritability, convulsions, muscular twitching, dermatitis near eyes, kidney stones	None reported
Pantothenic acid	5–10	Widely distributed in foods	Constituent of coenzyme A, which plays a central role in energy metabolism	Fatigue, sleep disturbances, impaired coordination, nausea (rare in humans)	None reported
Folacin	10.4	Legumes, green vegetables, whole-wheat products	Coenzyme (reduced form) involved in transfer of single-carbon units in nucleic acid and amino acid metabolism	Anemia, gastrointestinal disturbances, diarrhea, red tongue	None reported
Vitamin B-12	0.003	Muscle meats, eggs, dairy products, (not present in plant foods)	Coenzyme involved in transfer of single-carbon units in nucleic acid metabolism	Pernicious anemia, neurological disorders	None reported

Biotin	Not established (Diet provides 0.15–0.3 per day)	Legumes, vegetables, meats	Coenzyme required for fat synthesis, amino acid metabolism, and glycogen (animal-starch) formation	Fatigue, depression, nausea, dermatitis, muscular pains	Not reported
Vitamin C (ascorbic acid)	45	Citrus fruits, tomatoes, green peppers, salad greens	Maintains intercellular matrix of cartilage, bone, and dentine Important in collagen synthesis	Scurvy (degeneration of skin, teeth, blood vessels, epithelial hemorrhages)	Relatively nontoxic Possibility of kidney stones
Fat soluble					
Vitamin A (retinol)	1	Provitamin A (beta-carotene) widely distributed in green vegetables Retinol present in milk, butter, cheese, fortified margarine	Constituent of rhodopsin (visual pigment) Maintenance of epithelial tissues Role in mucopolysaccharide synthesis	Xerophthalmia (keratinization of ocular tissue), night blindness, permanent blindness	Headache, vomiting, peeling of skin, anorexia, swelling of long bones
Vitamin D	0.01	Cod-liver oil, eggs, dairy products, fortified milk and margarine	Promotes growth and mineralization of bones Increases absorption of calcium	Rickets (bone deformities) in children Osteomalacia in adults	Vomiting, diarrhea, loss of weight, kidney damage
Vitamin E (tocopherol)	15	Seeds, green leafy vegetables, margarines, shortenings	Functions as an antioxidant to prevent cell-membrane damage	Possibly anemia	Relatively nontoxic
Vitamin K (phylloquinone)	0.03	Green leafy vegetables. Small amount in cereals, fruits and meats	Important in blood clotting (involved in formation of active prothrombin)	Conditioned deficiencies associated with severe bleeding, internal hemorrhages	Relatively nontoxic Synthetic forms at high doses may cause jaundice

important enzymes; (2) uncooked fiber is more beneficial; and (3) cooking results in the formation of toxic or harmful products. Proponents of cooking stress the fact that cooking improves the digestibility of vegetables and allows our simple stomachs (Fig. 15-2a) to absorb many substances that would otherwise be lost. They also argue that cooking (1) destroys harmful organisms; (2) makes many vegetables more palatable; and (3) destroys otherwise harmful substances.

Guidelines for Cooking Vegetables

Nutritionists and home economists suggest that if vegetables are to be cooked we should follow the three R's of cooking that have an important bearing on the final quality of cooked vegetables:

1. Reduce the amount of water used.
2. Reduce the length of cooking period.
3. Reduce the amount of surface area exposed.

Of these three R's, reducing the volume of water used in cooking is most important. When vegetables are cooked in a large volume of water—more water than vegetable—the loss of soluble nutrients will be greater than when the volume is small. To illustrate, cabbage cooked quickly in about one-third as much water as cabbage retains nearly 90 percent of its vitamin C, whereas cabbage cooked in four times as much water as cabbage retains less than half. Some of the soluble nutrients can be retained if the cooking liquid is used along with the vegetables. Cooking techniques, such as steaming or use of the microwave oven, eliminate the need for immersing vegetables in water.

The second and third of the three R's are interrelated. The longer a vegetable is cooked, the greater the destruction of nutrients (unfortunately, most vegetables are commonly overcooked). Cooking time is shortened if the water is boiling before the vegetable is added. It has also been shown that if vegetables are placed in cold water fairly large nutrient losses occur even before the water begins to boil. Cooking time can be reduced by cutting vegetables into smaller pieces (the antithesis of the third R), but this permits a greater extraction of nutrients in the cooking process. It also allows for a greater oxidative destruction at the more extensive cut surfaces. Because smaller pieces cook faster the adverse effects related to the increased surface area may be offset, at least partially, by the shorter cooking time.

Methods for cooking vegetables to conserve their nutrients include use of a microwave oven, steaming, pressure cooking, and cooking rapidly in a tightly covered pan with only enough water to prevent scorching. The Chinese cut vegetables into relatively small pieces and cook them over high heat for a brief time. Cooking is done in a round-bottomed wok (an electric frypan or heavy skillet work equally well), and almost everything is stir fried quickly. In this method the high heat destroys disease organisms, while leaving the inner parts of cooked pieces crisp and only lightly cooked. A few vegetables, such as potato and sweet potato, require a long cooking or baking time.

STORING VEGETABLES

If fruits and vegetables are not to be consumed immediately, but are to be used within a relatively short period of time, they can be stored. Basically, **storing** is a method of placing the harvested materials in an environment where the life processes—most notably respiration and transpiration—are maintained at low levels. This preserves high-quality products. Proper storage is a difficult task, and requirements vary from vegetable to vegetable. The Irish potato illustrates the problems encountered in storage situations (Fig. 15-5).

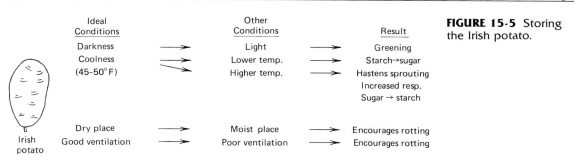

FIGURE 15-5 Storing the Irish potato.

Ideally, the Irish potato should be stored in a dark, cool (45 to 50°F), dry place having adequate ventilation. Varying any of these conditions has unfavorable consequences: light causes greening; lower temperatures result in conversion of starch to sugars, whereas higher temperatures hasten sprouting and can increase respiration rates; moist conditions encourage rots; and the absence of good air circulation causes diseases such as blackheart to develop.

Respiration During Storage

During respiration, sugars and other compounds are broken down within cells with the release of energy, carbon dioxide, water, and heat. The energy released is needed by the living cells of the stored product, and the metabolic water is used in transpiration. Carbon dioxide and heat are by-products that can be removed by adequate circulation of air. In storage, respiration should be kept at a minimum. Several factors regulate respiration rates. In general, within normal temperature limits, the higher the temperature the greater the rate of respiration (as indicated by the van't Hoff-Arrhenius law, Chapter 6), a feature illustrated in Fig. 15-6. The presence of soluble sugars in cells also influences respiration rate. In potato higher temperatures favor the conversion of sugars to starch (Fig. 15-5) and thus respiration rates increase only slowly with increase in temperature (Fig. 15-6).

The rate of respiration also varies directly with water content. Thus, at a given temperature succulent plant parts such as head lettuce respire more rapidly than nonsucculent products such as sweet or Irish potatoes, and immature fruits more rapidly than mature fruits. Finally, respiration rate is influenced by oxygen level. In respiration oxygen is absorbed and carbon dioxide is given off. Consequently, if products are stored in an airtight

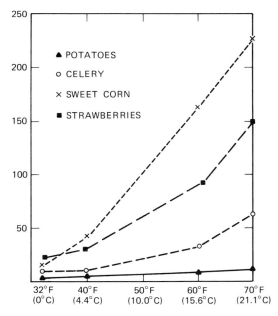

FIGURE 15-6 Rates of respiration of certain fruits and vegetables at various temperatures. (Courtesy of U.S. Department of Agriculture.)

container the oxygen level will decrease and carbon dioxide will gradually increase (Fig. 15-7). As a result, the rate of respiration gradually decreases. This principle is put to use in the commercial storage of certain horticultural products, particularly apples. Such storage methods are not for the home gardener, however; for if oxygen levels fall too low for the complete combustion of sugars, compounds form that are harmful to living tissues. Remember, maintenance of low respiratory levels is important to conserve stored food reserves.

Transpiration During Storage

Transpiration rates are also influenced by several factors. In general, highly succulent tissues lose water vapor more rapidly than do less succulent tissues. For example, young leaves of Swiss chard would be expected to lose water more rapidly than the leaves of a mature cabbage head. The greater water loss in succulent tissues is related in large part to

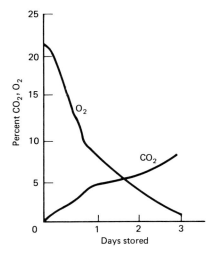

FIGURE 15-7 Changes in percentages of oxygen (O_2) and carbon dioxide (CO_2) sealed polyethylene bags containing green snap beans stored at 40°F. (Courtesy of U.S. Department of Agriculture.)

their less highly developed protective covering. A weakly cutinized epidermal layer provides less protection against drying than a strongly cutinized layer, and a strongly cutinized epidermis is less effective than a periderm. Asparagus and the leafy vegetables are examples of crops with an epidermis, whereas apple, pear, and the root crops form a periderm. Environmental factors influence transpiration rates too. Higher temperatures and lower humidities both promote transpirational losses. Thus, regulation of these parameters becomes an important aspect of storage. In some cases very high relative humidity encourages spoilage. Remember, maintaining low transpiration levels is necessary to eliminate shrinkage of stored crops.

The Refrigerator as a Storage Facility

Storage facilities available to the home gardener generally include the refrigerator, and space found in garages, basements, attics, and pantries. The refrigerator is most commonly used for short-term storage. Most refrigerators maintain a temperature near 40°F, but temperatures vary within the storage compartment (Fig. 15-8). In single-door models with a frozen-food storage unit, temperatures are generally lowest just beneath the storage unit. This cold air settles, forcing warm air near the vegetable tray upward along the sides. The circulation air is of low humidity and will dry out uncovered vegetables. The humidity in the vegetable tray can be maintained at a higher level by use of moist towels or an abundance of produce.

While many vegetables store well in the refrigerator for up to a week or more, certain storage precautions should be observed. For one, ripening fruits should not be stored together with vegetables. These fruits give off ethylene gas, causing yellowing of green vegetables, russet spotting on lettuce, toughening of asparagus spears, sprouting of potatoes,

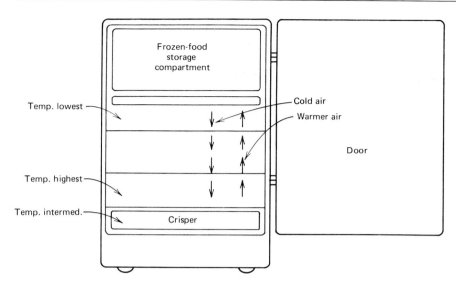

Frozen-food
storage
compartment

Temp. lowest —

Cold air
Warmer air

Door

Temp. highest—

Temp. intermed.—

Crisper

FIGURE 15-8 Single-door refrigerator with frozen-food storage compartment.

and a bitter taste in carrots. Most vegetables (probably all) produce volatile compounds that can cause off flavors in fruits and leafy vegetables. The abundance of these volatiles is illustrated by the large number that have been isolated from certain vegetables, as follows:

50 from cabbage

50 from onion

63 from carrot

79 from pea

119 from celery

174 from tomato

Many of these vegetables should be stored separately in the crisper or in the main compartment of the refrigerator.

Storage Requirements of Specific Crops

We have outlined above only some of the more general aspects of storage. Specific details on the storage requirements of a number of garden crops are presented in Table 15-3. Note from the table that many crops do not require cold storage. Some of these (Groups 2 and 3)

require cool, relatively moist conditions for best storage. Information on building simple storage facilities for storing these fruits and vegetables is available from virtually every federal, provincial, and state agricultural information agency (Appendix C).

Several fruit crops are excluded from Table 15-3. Berries should be sorted to remove injured fruits, and kept dry, whole, and refrigerated until ready for use. Ripe apples, peaches, pears, apricots, cherries, grapes, and plums all have a longer storage life under refrigeration. Other fruits are highly sensitive to chilling. This group includes avocado, banana (which suffers chill injury at temperatures below 56°F), grapefruit, lemon, and pineapple. These fruits are best stored at a cool room temperature. In general, the citrus fruits keep well for many days without refrigeration. When they are stored at temperatures lower than 50 to 55°F for a prolonged period of time, pitting of the skin and discoloration of the flesh may result. Unripe fruits of any kind should not be refrigerated. Many slightly underripe fruits that are sound will ripen in open air at room temperature, but they should not be placed in the sun.

TABLE 15-3 Fresh Vegetables Are Grouped in Four Groups According to Storage Requirements.[a]

Group 1: Keep under cold, moist conditions (32 to 41°F and 85 to 95% relative humidity). Store in the refrigerator crisper and maintain high humidity by keeping the crisper more than half full. Wash and drain vegetables well before storage.

Beet greens	Leeks
Chard	Lettuce
Collards	Mustard greens
Endive	Spinach
Escarole	Turnip greens
Green onions	Watercress
Kale	

Store the following vegetables in a crisper separate from the above vegetables or in plastic bags or containers in the main compartment of the refrigerator.

Artichokes	Mushrooms
Asparagus	Parsnips
Beets	Peas
Broccoli	Radishes
Brussels sprouts	Rhubarb
Cabbage	Sweet corn
Carrots	(unhusked; keep
Cauliflower	close to freezer
Celery	compartment)
Lima beans	Turnips

Group 2: Ideally, it is best to store these vegetables at 45 to 55°F and 85 to 90% relative humidity because of sensitivity to chilling injury. Since this is not possible in most homes, store in the refrigerator for no longer than 5 days. Use soon after removing from the refrigerator.

Bell peppers	Ripe melons
Chili peppers	Snap beans
Cucumbers	Summer squash

Group 3: Store in a cool place (50 to 60°F); lower temperatures cause chilling injury. Pantries, basements, or garages can provide a cool place during most of the year. However, noninsulated garages may be too warm in summer and too cold in winter. If you do not have such a space available, store eggplant and okra as described for the vegetables in Group 2. Store ripe tomatoes, hard-rind squashes and pumpkins, sweet potatoes, and potatoes as recommended for the vegetables in Group 4.

Eggplant	Potatoes
Okra	(protect from
Tomatoes (ripe)	light to
Hard-rind	prevent greening)
squashes and	Sweet potatoes
pumpkins	

Group 4: Store these vegetables at room temperature (65 to 70°F). Store them so they are away from direct sunlight.

Garlic, dry	Onions, dry
Melons (unripe	(in open-mesh
or partly ripe)	container)
until ripe	Tomatoes (mature
	green, partly ripe,
	and ripe)

Courtesy of University of California, Division of Agricultural Science, Leaflet 2989, 1977.
[a]Since it is not always possible to provide all these different conditions, compromise if the storage time is short (a few days)

PRESERVING GARDEN CROPS

The preservation of vegetables allows the gardener to enjoy the yield from the garden year round. It is not likely that there will be much overproduction from small gardens, but larger gardens commonly produce more vegetables than can be consumed fresh, stored, or even given away to neighbors. People who garden on a large scale often plan on large yields with the goal of preserving the excess. It is wise to plan the garden so that preserving can be done in short sessions rather than all at once. To get the most from your garden, preserving should be a season-long program.

Several methods of preservation are available to the gardener. Canning is perhaps the most familiar method of food preservation and has been practiced for nearly 200 years. In recent years freezing has gained in popularity and is now the most popular method of preserving home-grown vegetables. Still more recently the ancient art of drying—probably the oldest method known for preserving food—has come back into vogue. Gardeners by the thousands are trying their hand at preserving vegetables by this age-old method.

Canning

Canning is the method whereby food materials are packed into containers, which are heated to destroy spoilage organisms and sealed to keep out other organisms and air. The heat employed inactivates spoilage organisms—molds, yeasts, and bacteria—and enzymes present in the food. Most of the air in the container and in the food is driven out during heating and kept out by the tight seal that forms. If some air remains the food near the top of the container may darken gradually. This discoloration is unsightly, but not dangerous. Glass jars are the most commonly used containers. These jars must be clean, free of cracks, and have lids capable of forming tight seals.

Among spoilage organisms, the anaerobic bacterium, *Clostridium botulinum*, stands out from all others in that it produces an extremely toxic poison in foods. Furthermore, **botulism** cannot be detected by taste or by sight. Its absence *must* be insured by proper canning procedures. Generally, the bacterium that causes botulism is not found in foods, such as jams and jellies, that have been canned with a great deal of sugar. Nor is it common in high-acid fruits and vegetables, such as tomatoes. The major problem centers around the processing of **low-acid vegetables;** those low in organic acids. Fortunately, these vegetables can be processed safely in a steam-pressure cooker. The more acid products can be processed safely by boiling in a water bath. Let us explore briefly each of these canning methods.

The Steam-pressure Method

Foods canned by the steam-pressure method are processed by steam and under pressure. Most vegetables should be processed by this method to prevent botulism (Table 15-4). At

TABLE 15-4 Vegetables That Should Be Processed by the Steam-pressure Method, with Cooking Times and Pressure

| Vegetables | Minutes | | Pounds |
	Pints	Quarts	
Asparagus	25	30	10
Beans (green-wax)	20	25	10
Beans, lima	40	50	10
Beets	30	40	10
Broccoli	25	40	10
Brussels sprouts	45	55	10
Cabbage	45	55	10
Carrots	25	30	10
Cauliflower	25	40	10
Corn (cream style) pts only	85		10
Corn (whole grain)	55	85	10
Eggplant	30	40	10
Greens (all kinds)	70	90	10
Hominy	60	70	10
Okra	25	40	10
Peas (all shelled peas)	40	40	10
Peppers, bell	35	35	10
Potatoes, Irish	40	40	10
Pumpkin	60	80	10
Rutabagas	35	35	10
Soybeans	80	80	10
Squash (summer)	25	30	10
Squash (winter)	60	80	10
Sweet potatoes (wet pack)	55	90	10
Turnips	20	25	10

Courtesy of the Kerr Glass Manufacturing Corporation, 1981.

greater pressure water boils at a higher temperature. This principle is put into effect in pressure canning to kill the highly resistant spores of *C. botulinum*. At lower elevations 10 pounds of pressure is adequate to bring the boiling point of water to the desired 240°F. At high altitudes, however the pressure under which vegetables are cooked must be increased in order to reach the desired temperature, as illustrated in Table 15-5. Adjusting for altitude is of considerable importance in the western half of the continent where elevations generally exceed 2000 feet and in many areas exceed 5000 feet.

Vegetables to be canned by the steam-pressure method should be put into jars, and the jars placed onto a rack in the pressure canner. Enough hot water (usually one to two inches) should be present in the bottom of the canner to reach the rack on which the jars are placed. The lid should then be fastened tightly so that steam can escape only through the vent. Allow steam to pour from the vent for 7 to 10 minutes before closure, to exhaust the canner. Then close the petcock (or weighted gauge). When the required amount of pressure is shown on the gauge, start counting the processing time (Table 15-4). When the processing time is up remove canner from heat and allow to sit till pressure gauge returns to zero. Then open petcock gradually. If no steam escapes the lid can be safely removed. Remove jars from canner and allow them to sit two to three inches apart on several thickness of cloth or on a rack to cool. When the jars are cold test for seal by (1) pressing the flat metal lid; if it does not move the seal is good, (2) tapping the lid with a spoon; a clear ring indicates a good seal, and (3) turning the jar part way over to see if it leaks. The contents of unsealed jars should be used at once, or reprocessed.

The Water-bath Method

A number of garden products can be canned quite safely by the water-bath method (Table 15-6). In this method jars are filled, placed on racks, and put into a container nearly full of hot water. The water should cover the jars by about two inches, and at least two inches of space should remain between the water surface and the container top to allow the water

TABLE 15-5 Pressure Adjustment to Obtain 240°F at Various Altitudes

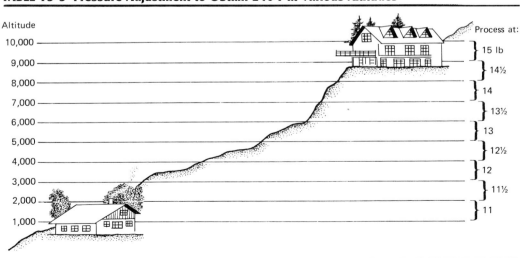

Courtesy of the Kerr Glass Manufacturing Corporation, 1981.

TABLE 15-6 Garden Products That Can Be Safely Processed by the Water-bath Method, with Processing Time

Product	Pints	Quarts
	Minutes	
Apples	20	25
Applesauce	25	25
Apricots	20	25
Berries (except strawberries)	15	20
Cherries	20	20
Cranberries	10	10
Currants	20	20
Dried fruits	15	15
Figs	30	30
Fruit juices	10	10
Grapefruit	20	20
Grapes	20	20
Nectarines	20	25
Peaches	20	25
Pears	25	30
Pineapple	30	30
Plums	20	25
Rhubarb	10	10
Strawberries	15	15
Tomatoes	35	45
Tomatoes (low acid)[a]	45	55
Tomato juice	10	10
Tomatoes, stewed	55	55

Pickles and relishes		Minutes
Bread and butter	qt	10
Bread and butter	pt	5
Chutney	pt	5
Cross cut slices	pt	5
Dill green beans	pt	5
Gherkins, sweet	pt	5
Piccalilli	pt	5
Pepper-onion relish	pt	5
Relish, corn	pt	15
Watermelon	pt	5
Fruit pickles		
Peaches	qt or pt	20
Pears	qt or pt	20

Start counting processing time as soon as water returns to boil.

Pickles—dill		
Fermented (whole)	qt	15

TABLE 15-6 (continued)

Product		Minutes
Unfermented (whole) (fresh-pack dills)	qt	20
Sauerkraut	qt	15

Courtesy of the Kerr Glass Manufacturing Corporation, 1981.
[a] Recent USDA research shows that low-acid tomatoes of good quality present no greater home-canning risk than other cultivars.

to boil rapidly. Cover the canner when the water comes to boil, and process for the times indicated in Table 15-6. At higher altitudes additional boiling time is needed to compensate for the lower boiling temperature of water (Table 15-7). Handle processed foods the same as for the steam-pressure method.

Additional Canning Guidelines

Several additional comments are helpful. Both cooked (**hot pack**) and uncooked (**raw pack**) vegetables can be canned. The raw pack is simple to do, and the final product is as good as or better than the precooked pack. On the

TABLE 15-7 Boiling Times for Garden Products Processed by the Water-bath Method with Adjustments for Altitude

Altitude	20 minutes or less	More than 20 minutes
	Increase processing time if the time called for is	
1,000 feet	1 minute	2 minutes
2,000 feet	2 minutes	4 minutes
3,000 feet	3 minutes	6 minutes
4,000 feet	4 minutes	8 minutes
5,000 feet	5 minutes	10 minutes
6,000 feet	6 minutes	12 minutes
7,000 feet	7 minutes	14 minutes
8,000 feet	8 minutes	16 minutes
9,000 feet	9 minutes	18 minutes
10,000 feet	10 minutes	20 minutes

Courtesy of Kerr Glass Manufacturing Corporation, 1981.

other hand, more precooked than raw food can be packed into a jar, a feature of importance if space is limiting. Most vegetables can be packed to within one-half inch of the jar top, but starchy vegetables such as bean and corn need about twice that space because they swell during processing. It is often difficult to visualize the yield of canned goods from a harvest or to know how much to harvest for a specific size processing, but estimates can be made using Table 15-8.

All canned produce should be stored in a cool, dry place away from light. For best stor-age, the temperature should be 70°F or lower. Higher temperatures tend to reduce eating quality. To illustrate, canned vegetables stored at 65° lose up to 15 percent of their thiamin in a year, but stored at 80° the losses increase to about 25 percent. Vitamin C comparisons are comparable: at 65° losses are about 10 percent in one year, as opposed to 25 percent at 80°F. Upon removal from storage for use, low-acid vegetables (Table 15-4) should be boiled in an open vessel for 10 minutes before tasting or using. This boiling is done as a pre-caution to denature any of the highly lethal

TABLE 15-8 Approximate Yield of Canned Goods from Freshly Harvested Garden Crops

Food	Fresh	Canned
Fruits		
Apples	1 bu (48 lb)	16 to 20 qt
Apricots	1 bu (50 lb)	20 to 24 qt
Berries, except strawberries	24-qt crate	12 to 18 qt
Cherries, as picked	1 bu (56 lb)	22 to 32 qt
Peaches	1 bu (48 lb)	18 to 24 qt
Pears	1 bu (50 lb)	20 to 25 qt
Pineapple	2 average	1 qt
Plums	1 bu (56 lb)	24 to 30 qt
Strawberries	24-qt crate	12 to 16 qt
Tomatoes	1 bu (53 lb)	15 to 20 qt
Vegetables		
Asparagus	1 bu (45 lb)	11 qt
Beans, lima, in pods	1 bu (32 lb)	6 to 8 qt
Beans, snap	1 bu (30 lb)	15 to 20 qt
Beets, without tops	1 bu (52 lb)	17 to 20 qt
Brussels sprouts	1 lb	1 pt
Carrots, without tops	1 bu (50 lb)	16 to 20 qt
Cauliflower	2 medium	3 pt
Corn, sweet, in husks	1 bu (35 lb)	8 to 9 qt
Eggplant	1 average	1 pt
Kale	1 bu	12 to 18 pt
Okra	1 bu (26 lb)	17 qt
Peas, green, in pods	1 bu (30 lb)	12 to 15 qt
Pumpkin	50 lb	15 qt
Spinach	1 bu (18 lb)	6 to 9 qt
Squash, summer	1 bu (40 lb)	16 to 20 qt
Sweet potatoes	1 bu (55 lb)	18 to 22 qt

Courtesy of Kerr Glass Manufacturing Corporation, 1981.

proteinaceous toxins of *C. botulinum* that may be present. Thus, we take precautions against botulism at each end of the processing route. The botulism organism is common in soils and as a part of soil particles adhering to fruits and vegetables. We eat the organism constantly without any ill effects. It is simply that jars of low-acid food devoid of air provide an ideal environment for growth of the organism and release of toxins, whereas the intestinal tract does not.

In our discussion of canning we have not mentioned the open-kettle method. This method should not be used for normal vegetable canning, but serves well for making preserves, jellies, and the like. Oven canning is also not advised for fruits and vegetables. The method is dangerous because the temperature in jars is frequently not high enough to assure destruction of spoilage organisms, and the jars may explode during processing. We also recommend that you do not use preservatives that are supposed to kill harmful organisms without high temperatures. More detailed advice on proper canning procedures is available from manufacturers of canning products and from governmental agencies.

Freezing

Freezing is the method whereby food materials are stored under very cold conditions—0°F or lower—to slow down the changes that affect quality and cause spoilage. Chief among the factors that cause deterioration of frozen foods are microorganisms, enzymes, and oxygen. Preservation by freezing is based on the principle that extreme cold retards the growth of these microorganisms, and slows down enzyme activity and oxidation.

A majority of garden crops can be frozen successfully, but some crops freeze better than others. As a general rule, crops that we cook are the ones most successfully frozen. This group includes:

Asparagus	Corn	Rhubarb
Beans	Eggplant	Spinach
Beets	Kohlrabi	Squash
Broccoli	Okra	Tomatoes
Brussels	Peas	(puree,
sprouts	Peppers	sauce,
Carrots	Pumpkin	juice)
Cauliflower	(puree)	

Vegetables that are commonly eaten raw, such as cabbage, celery, cucumber, lettuce, onions, and radish generally do not freeze well. Also, remember that for a given crop certain cultivars will freeze better than others. Information on the best cultivars to use can be obtained from agricultural information agencies or seed companies (see Appendixes C and E).

Maintaining Quality in Frozen Vegetables

For most vegetables, just simply freezing them is not enough; other steps must be taken to assure maintenance of high qualty. For example, many vegetables, and particularly the low-acid ones, require brief heating or scalding, known as **blanching,** to inactivate enzymes and reduce the number of microorganisms. Blanching also shrinks foods somewhat and helps to remove the skin from many fruits. Blanching should be done as soon as possible after vegetables are picked. The procedure can be carried out somewhat as follows: bring at least one gallon of water in a large kettle to a rolling boil. Then put about a quart-size portion of prepared vegetables into a perforated metal container or cheese-cloth bag, and immerse the contents in the water. Place a lid on the kettle and note the time. Blanching time varies with the vegetable (Table 15-9), and it is advisable to stick closely to the recommended times. Upon removing the blanched vegetables, plunge them quickly into cold water to stop the cooking, drain carefully, and pack immediately and put into the freezer. Blanching can also be accomplished using steam. The steam method is favored by

TABLE 15-9 Recommended Blanching Times for Selected Vegetables[a]

Vegetable	Minutes
Asparagus stalks	2 to 4
Beans, lima (or pods)	2 to 4
Beans, green or wax (1- or 2-inch pieces or frenched)	3
Beets	25 to 50 (until tender)
Broccoli stalks (split)	3
Brussels sprouts	3 to 5
Carrots (small, whole)	5
Carrots (diced, sliced, or lengthwise strips)	2
Cauliflower (1-inch flowerets)	3
Corn on the cob	7 to 11
Corn (whole-kernel and cream-style—cut corn from the cob after heating and cooling)	4
Peas, green	1.5
Spinach	2
Squash, summer	3

Courtesy of Kerr Glass Manufacturing Corporation, 1981.
[a]When a range of times is given, use the shortest time for small vegetables and the longest for large vegetables.

many because it reduces somewhat the loss of water-soluble vitamins and minerals.

Oxidation, as mentioned, also causes a deterioration in the quality of frozen foods. We are all familiar with the browning of fruits such as apple and pear when their fleshy tissues are exposed to air. The discoloration occurs when, through enzymatic action, oxygen is combined with substances in the fruit to produce dark-colored compounds. For years housewives have used lemon juice (or ascorbic acid) on freshly cut fruits for the dinner table to prevent browning. These substances work well as **antioxidants** because they lower the pH of the exposed tissue to a level where activity of the oxidase enzyme is inhibited. Lemon juice and ascorbic acid are also used occasionally to prevent oxidation in fruits being pre-

pared for freezing. For most garden crops, however, the best ways to prevent or reduce oxidation are to select packaging that will restrict the entry of air, and reduce to a minimum the amount of air left in the pack when packing and sealing. Acceptable packaging materials must not only prevent air from entering; they must prevent moisture from escaping. Glass, aluminum, plastic, and moisture-vaporproof paper all work well in this regard.

Freezing Procedures

Produce for freezing should be properly harvested, carefully trimmed, washed, usually blanched, quickly cooled, and put into the freezer in the shortest possible time. Generally it is best to pick vegetables early in the day when they are coolest. Washing—not soaking—should be done in cold water and the produce then drained thoroughly. Storage in the freezer should be at 0°F or lower to preserve quality, and even then most fruits and vegetables should be stored for no longer than 8 to 12 months. To illustrate, at 0°F broccoli, cauliflower, and spinach lose from one-third to three-fourths of their vitamin C in one year, but at −20°F losses are minimal. Other vitamins tend to show less rapid deterioration. Many gardeners feel that frozen vegetables retain better flavor, color, and nutritional content than do canned vegetables. On the other hand, the initial cost of a freezer and the continued cost of electricity make freezing a more costly form of preservation. Finally, it bears repeating that the frozen-food storage compartment in single-door refrigerators is not a freezing unit. Frozen foods should be stored in these units for no more than a few days.

Drying

Drying is the method whereby food materials are exposed to air and heat to reduce greatly their moisture content through dehydration.

The principle involved in preserving food through drying is simply that without moisture the microorganisms that cause food spoilage cannot grow. Drying can be easy and inexpensive. Dried foods are lightweight, easy to package, take little storage space, and do not require refrigeration. Furthermore, dried foods retain much of their natural color, flavor, and nutrients.

A flowchart of the drying process is presented in Fig. 15-9. Garden crops to be dried should be fresh, prime for eating, and of high quality. Wilted or otherwise inferior fruits and vegetables, many immature crops, and overmature produce will not yield a satisfactory dried product. Harvested materials should be trimmed of unusable and damaged parts, washed to remove soil and contaminants, drained to remove excess water, and sorted for quality and size. Generally, slicing, peeling, pitting, or other manipulative procedures are necessary in preparation for pretreatment and drying. These preparative steps are summarized for a number of vegetables and fruits in Table 15-10.

Drying Procedures

Most garden products require some kind of pretreatment before drying to slow down or stop the activity of enzymes that cause their colors and flavors to deteriorate during drying and storage. Blanching with steam or boiling water is the pretreatment recommended for most vegetables, but some such as beet should be completely cooked, and others such as cucumbers, onions, and peppers, do not need blanching. The blanching process is described in an earlier section.

Many fruits have a tendency to darken during drying and storage. To retain better color, some home gardeners expose fruits to the fumes of burning sulfur before drying, a process known as **sulfuring.** The sulfur, which can be purchased in drug stores as flowers of sulfur, also decreases loss of vitamins A and

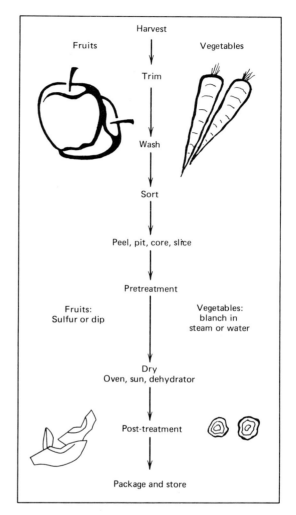

FIGURE 15-9 Flowchart for drying fruits and vegetables. (Modified from Oregon State University Cooperative Extension Service.)

C. Sulfuring should be done outdoors because the fumes are irritating to the eyes and nose and have an unpleasant odor. The process can be carried out in an apparatus like that illustrated in Fig. 15-10. Suggested sulfuring times are given in Table 15-10, but times will vary with texture and size of pieces, as well as method of drying to be employed. Sulfuring can also be done indoors, using a solution of

TABLE 15-10 Procedure for Drying Vegetables and Fruits

	Steps for drying vegetables		
Vegetable	Preparation	Pretreatment	Drying procedure
Beans, lima	Shell.	Choose one: Steam 10 to 15 minutes. Place in boiling water 5 minutes. Drain.	Spread on tray about ½ in. deep. Stir frequently at beginning. Drying time averages 6 to 10 hours in controlled heat. Dry until hard and brittle. Beans will break clean when broken.
Carrots	Select crisp, tender carrots, free from woodiness. Wash. Trim off the roots and tops. Cut into slices or strips about ⅛ in. thick.	Steam-blanch 8 to 10 minutes.	Arrange in a thin layer on trays. Dry until tough and leathery.
Corn *Any good table cultivar.*	Select tender, sweet corn. Husk. Sort ears on basis of maturity. Young corn requires longer blanching time. Cut from cob after blanching and cooling enough to handle.	Steam on the cob 10 to 15 minutes or until milk is set.	Spread kernels ½ to ¾ in. deep on trays. Stir frequently to prevent kernels from lumping. Drying time averages 6 to 10 hours in controlled heat. Dry until hard and brittle.
Onions *Select pungent cultivars.*	Peel. Remove outer discolored layers. Cut uniform slices ⅛ to ¼ in. thick.	No treatment necessary.	Spread thinly on trays. In controlled heat, dry 6 to 10 hours at 140°F. Dry until brittle and light colored. For onion powder, crush slices after drying.
Peas	Select young, tender peas of a sweet variety. Shell.	Choose one: Steam-blanch 8 to 10 minutes. Dip in boiling water 3 to 4 minutes.	Spread thinly on trays. Drying time averages 6 to 10 hours in controlled heat. Dry until hard and shriveled. Peas should shatter when hit with a hammer.

500

TABLE 15-10 (*continued*)

	Steps for drying vegetables		
Vegetable	Preparation	Pretreatment	Drying procedure
Pumpkin, deep orange Squash, hubbard	Cut into strips about 1 in. wide. Peel off rind. Scrape off fiber and seeds. Cut peeled strips crosswise into pieces about ⅛ in. thick.	Steam-blanch until tender, about 6 minutes.	Arrange in thin layer on trays. Dry until tough to brittle.
Squash, summer Crookneck Scallop Zucchini	Wash. Trim. Cut into ¼ in. slices.	Steam-blanch 6 minutes.	Spread in a thin layer on trays. Dry until brittle.
Herbs for seasoning	Gather when leaves are mature but before flowers develop. Wash thoroughly, remove excess water.	No treatment necessary.	Hang small bundles of stems in warm, dry, airy place. (May be enclosed in large brown paper bag.) Or Dry leaves on a cookie sheet in a warm oven. Dry until leaves become brittle and crumble easily.
Chili peppers, green	To loosen skin, rotate pepper over flame or scald in boiling water. Peel, split pods, and remove seeds and stem.	Choose one: No treatment necessary. Steam-blanch 10 minutes.	Spread in a thin layer on trays. Dry until crisp and brittle.
Chili peppers, red	Select mature pods. Wipe clean with damp cloth. String whole pods together with needle and cord or hang bunches, root side up, in airy place.	No treatment necessary.	Drying time in controlled heat, 6 to 10 hours at 150°F; at room temperature, several weeks. Dry until pods are shrunken, dark red, flexible.
	Steps for drying fruits		
Apples *Use cultivars of good dessert or cooking quality. Mature but not soft.*	Wash, peel, and core. Cut into slices or rings ⅛ to ¼ in. thick, or into quarters or eighths. Coat with ascorbic acid solution (2½ tsps. of	Choose one: Blanch in steam 10 minutes. Sulfur 30 to 60 minutes depending on size of pieces. Soak 10 to 15 minutes	Arrange slices on trays not more than two layers deep. Drying time approximately 6 hours in controlled heat.

TABLE 15-10 (*continued*)

	Steps for drying vegetables		
Vegetable	Preparation	Pretreatment	Drying procedure
	pure crystalline ascorbic acid to each cup cold water) to prevent darkening.	in sodium sulfite solution or saline solution.	Dry until leathery and suedelike. There should be no moisture when cut slices are squeezed.
Apricots *Use any cultivar. Fully ripened but not so soft they are easily mashed or lose shape during drying.*	Wash, cut in half, and pit. Do not peel. Coat with ascorbic acid solution (1 tsp. pure crystalline ascorbic acid in each cup of cold water) to prevent darkening.	Choose one: Sulfur 1 to 2 hours depending on size of pieces and ripeness of fruit. Riper fruit absorbs sulfur more slowly. Steam-blanch halves 5 to 10 minutes. Blanch in a hot syrup made of equal parts corn syrup and water or sugar and water. Soak 10 to 15 minutes in sodium sulfite solution or saline solution.	Arrange in single layer on drying trays, pit side up. Average drying time for halves up to 14 hours in controlled heat; 1 to 2 days in the sun. Dry until leathery and suedelike. There should be no moisture when cut slices are squeezed.
Berries *Firm type.*	Sort, wash, and leave whole.	Choose one: No treatment necessary. Steam ½ to 1 minute. Crack tougher skins by dipping 15 to 30 seconds in boiling water, then in cold water. Remove excess moisture.	Spread in layer not more than 2 berries deep. Cloth on tray may help keep berries from sticking. Drying time up to 4 hours in controlled heat. Dry until berries are hard and rattle when shaken on the tray. There should be no moisture when berries are crushed between the fingers.
Cherries	Sort, wash, leave whole or stem and remove pit.	Choose one: No further treatment necessary. Check skins of whole cherries in boiling water 15 to 30 seconds, cool immediately, and remove moisture. Syrup-blanch.	Dry in single layers on trays. Drying time up to 6 hours. Dry until leathery and slightly sticky.

TABLE 15-10 (*continued*)

	Steps for drying vegetables		
Vegetable	Preparation	Pretreatment	Drying procedure
Figs	Select fully ripe fruit. When not fully ripe, the sugar content is too low to produce a good dried product and fruit may sour. Wash, or clean whole fruit with damp terry cloth. If small or partly dried on tree, leave whole. Otherwise, cut in half lengthwise.	Choose one: Check skins of whole figs by dipping in boiling water for 30 to 45 seconds. Cool quickly, and remove excess moisture. Sulfur light colored cultivar (like Kadotas) for 1 hour or more. Sulfuring is optional with others. (Black figs will turn a mottled color when sulfured.) If cut, blanch 20 minutes. Syrup blanch.	Spread in single layers on trays. Stir or turn figs to keep from sticking. Drying time averages up to 5 hours for halves in controlled heat; about 3 days in the sun. Dry until leathery, with flesh pliable yet slightly sticky.
Grapes *Any seedless cultivar.*	Wash, and remove all defective fruit. Leave whole on stem; cut closely packed stems into small bunches; or remove from stems.	Choose one: No treatment necessary. Dip in boiling water 15 to 30 seconds to crack skins; then in cold water to cool immediately. Drain.	Spread in a single layer. Drying time averages up to 8 hours in controlled heat. Dry until pliable and leathery.
Nectarines	Peel if desired. Halve and stone. Cut into quarters or slices. To prevent browning during preparation, treat with ascorbic acid solution as for apricots.	Choose one: Steam-blanch halves 15 to 18 minutes, slices 5 minutes. Sulfur slices 1 hour, halves or quarters 2 hours. If blanched, sulfur 30 and 90 minutes, respectively. Soak 10 to 15 minutes in sodium sulfite solution or saline solution.	Dry like peaches.
Peaches *Any good table cultivar; freestones preferred. Ripe enough for eating but not fully ripe.*	Peel. Cut in half and pit. Leave in halves or cut in quarters or slices. To prevent browning during preparation, treat with ascorbic acid solution as for apricots.	Choose one: Steam-blanch halves 15 to 20 minutes, slices 5 to 7 minutes. Sulfur 1 to 2 hours, depending upon size of pieces. If steam-blanched before sulfuring, cut	Arrange in single layers on trays, pit side up to retain juices. Turn over halves when visible juice disappears. Drying time under controlled conditions averages up to 15

TABLE 15-10 (*continued*)

	Steps for drying vegetables		
Vegetable	Preparation	Pretreatment	Drying procedure
		sulfuring time about half. Soak 10 to 15 minutes in sodium sulfite solution or saline solution.	hours for halves and about 6 hours for slices. Dry until leathery and somewhat pliable.
Pears 'Bartlett' is best for drying.	Pare. Cut in half lengthwise and core. Cut in quarters or eighths or slices ⅛ to ¼ in. thick. To prevent browning during preparation, treat with ascorbic acid as for apricots.	Choose one: Steam-blanch 5 to 20 minutes, depending upon size of pieces. Sulfur as for peaches. Syrup-blanch. Soak 10 to 15 minutes in sodium sulfite solution or saline solution.	Spread in single layers on trays. Drying time under controlled conditions averages up to 15 hours for halves or 6 hours for slices. Dry until springy and suedelike. There should be no moisture when pieces are cut and squeezed.
Prunes Plums	Dry whole if small. Otherwise, cut into halves (pit removed) or slices.	Choose one: Dip whole fruit in boiling water 30 or more seconds to check skins. Cool and drain. Steam-blanch halves 15 minutes, slices 5 minutes. Sulfur whole fruit 2 hours, halves and slices 1 hour.	Spread on trays in a single layer. Drying time under controlled conditions averages 6 to 8 hours for slices and halves; up to 14 hours for whole fruit. Dry until pliable and leathery. A handful of pieces will spring apart after squeezing.

Courtesy of Oregon State University Cooperative Extension Service.

sodium sulfite, sodium bisulfite, or sodium metabisulfite. Be sure to use U.S.P. or reagent-grade sulfur materials to assure purity. Fruits treated this way should be soaked for 10 to 15 minutes in a gallon of water containing one to two tablespoons of sodium sulfite or its equivalent, and then drained well. Obviously, fruit treated by soaking will take longer to dry than fruit treated with sulfur fumes.

Fruits to be dried can also be pretreated by use of a saline solution (two to four tablespoons salt per gallon of water), ascorbic acid solution, or by blanching.

Some fruits, such as those of grape, plum, and cherry, have tough skins with a waxy covering. If these fruits are to dry properly the continuity of the skins must be broken. This can be accomplished by blanching. Following

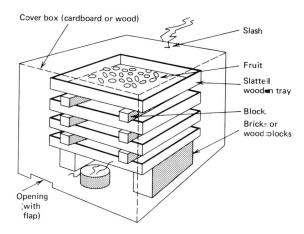

Cover box (cardboard or wood)

Slash

Fruit

Slatted wooden tray

Block

Bricks or wood blocks

Opening (with flap)

FIGURE 15-10 A simple apparatus for sulfuring. (Courtesy of Oregon State University Cooperative Extension Service.)

treatment the fruits should be thoroughly drained on absorbent toweling. The pretreatment procedures for several vegetables and fruits are summarized in Table 15-10.

Drying can be carried out in the sun, using a conventional oven, or with a dehydrator. Many of us have dried vegetables, such as onion, and herbs simply by hanging them in a warm, dry place. Drying can be done in the sun if there is an abundance of warm days with low humidity. For most, however the oven or dehydrator yields a superior product. Remember, when using these appliances that successful drying depends on (1) enough heat to draw out moisture but not enough to cook the material, (2) dry air to absorb the moisture released, and (3) circulating air to carry the moisture away. A properly constructed dehydrator is more effective than an oven, but requires an additional investment and space. A good approach for beginners is to try drying in an oven to determine if this method of preservation fits your needs. If so, move on to a dehydrator. Plans for constructing home dryers are available from governmental agencies, and many commercial models

are on the market. Drying procedures are summarized in Table 15-10.

Storing Dried Materials

Dried materials generally require special treatment before storage. At the least, they should be allowed to cool thoroughly before being put into containers. Materials dried in the sun may require hot air treatment to kill insects, and foods exposed to moist air may need similar treatment to remove the moisture. All dried foods should be packed tightly into clean, dry containers with tight-fitting lids. Glass jars, metal cans, and airtight plastic bags work well. Separate sulfured fruits from metal by use of plastic bags. Label each container and store in a cool, dry place away from light. As a general rule, plan to use dried products within a 10- to 12-month period.

This concludes our discussion of harvesting, storage, and preservation of garden crops. Discussion has been, of necessity, limited in each of these areas. To illustrate, our coverage of preservation was confined to the more popular methods; the subjects of brining, jelling, and juicing, are omitted even though these methods enjoy limited use. The reader is reminded that detailed information on the various subjects summarized in this chapter is available from governmental agencies and from companies engaged in the manufacture of products used in the preservation of garden crops.

Questions for Discussion

1. Define the following terms:
   (a) Coasting   (d) Hot pack
   (b) Ruminant   (e) Blanching
   (c) Bran   (f) Sulfuring

2. Discuss the physiological effects of harvesting from the standpoint of both the parent plant and the harvested portion.

3. List the major foodstuffs consumed by humans and discuss the importance of each in the diet.

4. According to the authors the cellulose present in plants is not digested by humans. If this is so, then why is cellulose regarded as a vital part of the diet?

5. Identify three fat-soluble vitamins and a like number of water-soluble vitamins. List good sources for each of these vitamins.

6. How do plants compare to animals as sources of the major foodstuffs, vitamins, and minerals?

7. In urban areas environmental contamination due to heavy metals is becoming increasingly common. Prepare a list of guidelines that would help the urban gardener to deal successfully with these toxic pollutants.

8. What are the three R's of vegetable cookery?

9. Give the most favorable storage conditions for freshly harvested lettuce, bell peppers, potatoes, and tomatoes. If these conditions cannot be met in your home suggest acceptable alternatives.

10. What is botulism?

11. How do the canning procedures for low-acid and high-acid garden produce differ?

12. Is a pressure cooker required for canning? Explain your answer.

13. Why do the authors recommend that many vegetables be blanched before freezing?

14. Outline the general procedures for drying fruits such as apple and pear.

15. Compare and contrast canning, freezing, and drying as methods for preserving garden crops. Which method provides the longest storage? The least nutrient loss? Is least expensive?

Selected References

Arthey, V. 1975. *Quality of Horticultural Products.* Halsted Press, New York.

Beyer, B. 1976. *Food Drying at Home the Natural Way.* J.P. Tarcher Inc., Los Angeles.

Chapman, H. (Ed.). 1966. *Diagnostic Criteria for Plants and Soils.* University of California, Berkeley.

Edmund, J., T. Senn, F. Andrews, and R. Halfacre. 1975. *Fundamentals of Horticulture.* McGraw-Hill, New York.

Fitch, K. and P. Johnson. 1977. *Human Life Sciences.* Holt, Rinehart and Winston, New York.

Hanson, G. 1976. Vegetables, freeways, and lead. *Lasca Leaves,* 26:8–10.

Janick, J., C. Noller, and C. Rhykerd. 1976. The cycles of plant and animal nutrition. *Scientific American,* 235:75–86.

Johnson, A., H. Nursten, and A. Williams. 1971. Vegetable volatiles: a survey of components identified. *Chemistry and Industry,* pp. 1212–1224.

Lappe, F. 1975. *Diet for a Small Planet.* Ballantine Books, New York.

Reuben, D. 1975. *The Save Your Life Diet.* Ballantine Books, New York.

Scrimshaw, N. and V. Young. 1976. The requirements of human nutrition. *Scientific American,* 235:51–64.

Stare, F. and M. Williams. 1977. *Living Nutrition.* John Wiley & Sons, New York.

United States Department of Agriculture. 1959. *Food. Yearbook of Agriculture,* U.S. Government Printing Office, Washington, D.C.

Wilson, E., K. Fisher, and M. Fuqua. 1975. *Principles of Nutrition.* Wiley, New York.

The Class, Family, Genus, Species and Botanical Variety to Which Several Vegetables Belong and Their Common Names. The Genera within Each Family Have a Close Natural Relationship.

Botanical groupings		Common names
Monocotyledoneae (the monocots)		
Dioscoreaceae (yam family)		
Dioscorea alata		Yam
Liliaceae (lily family)		
Allium ascalonicum		Shallot
Allium cepa		Onion
Allium cepa	var. *aggregatum*	Multiplier onion
Allium cepa	var. *viviparum*	Top onion
Allium fistulosum		Welsh onion
Allium porrum		Leek
Allium sativum		Garlic
Asparagus officinalis		Asparagus
Poaceae (grass family)		
Zea mays	var. *everta*	Popcorn
Zea mays	var. *rugosa*	Sweet corn
Dicotyledoneae (the dicots)		
Asteraceae (composite family)		
Cichorium endivia		Endive
Cynara cardunculus		Cardoon
Cynara scolymus		Artichoke
Helianthus tuberosus		Jerusalem artichoke
Lactuca sativa		Lettuce
Lactuca sativa	var. *capitata*	Head lettuce
Lactuca sativa	var. *crispa*	Curled lettuce
Lactuca sativa	var. *longifolia*	Romaine lettuce
Tragopogon porrifolius		Salsify
Brassicaceae (mustard family)		
Armoracia lapathifolia		Horseradish
Brassica caulorapa		Kohlrabi

Botanical groupings		Common names
Brassica chinensis		Bok choy
Brassica juncea		Mustard
Brassica napobrassica		Rutabaga
Brassica oleracea	var. *acephala*	Collards (and kale)
Brassica oleracea	var. *botrytis*	Cauliflower
Brassica oleracea	var. *capitata*	Cabbage
Brassica oleracea	var. *gemmifera*	Brussels sprouts
Brassica oleracea	var. *italica*	Broccoli
Brassica pekinensis		Chinese cabbage
Brassica rapa		Turnip
Raphanus sativus		Radish
Raphanus sativus	var. *longipinnatus*	Chinese radish
Chenopodiaceae (goosefoot family)		
Beta vulgaris		Beet
Beta vulgaris	var. *cicla*	Swiss chard
Spinacia oleracea		Spinach
Convolvulaceae (morning-glory family)		
Ipomoea batatas		Sweet potato
Cucurbitaceae (gourd family)		
Citrullus vulgaris		Watermelon
Cucumis melo		Muskmelon
Cucumis melo	var. *cantalupensis*	Cantaloupe
Cucumis melo	var. *chito*	Lemon cucumber
Cucumis melo	var. *reticulatus*	Netted melon
Cucumis sativus		Cucumber
Cucurbita maxima		Squashes, winter
Cucurbita moschata		Winter crook-neck squashes
Cucurbita pepo		Pumpkin
Cucurbita pepo	var. *melopepo*	Squashes, summer
Fabaceae (pea or legume family)		
Glycine max		Soybean
Phaseolus acutifolius		Tepary bean
Phaseolus aureus		Mung bean
Phaseolus limensis		Lima bean
Phaseolus vulgaris		Kidney bean
Phaseolus vulgaris	var. *humilis*	Bush bean
Pisum sativum		Pea
Pisum sativum	var. *macrocarpon*	Edible-podded pea
Vicia faba		Broad bean
Vigna sinensis		Cowpea
Malvaceae (mallow family)		
Abelmoschus esculentus		Okra
Polygonaceae (buckwheat family)		
Rheum rhaponticum		Rhubarb
Rumex patientia		Patience (or dock)

Botanical groupings		Common names
Solanaceae (nightshade family)		
Capsicum frutescens	var. *grossum*	Bell pepper
Capsicum frutescens	var. *longum*	Cayenne
Lycopersicon esculentum		Tomato
Lycopersicon esculentum	var. *cerasiforme*	Cherry tomato
Lycopersicon esculentum	var. *commune*	Common tomato
Lycopersicon esculentum	var. *grandifolium*	Potato-leaved tomato
Lycopersicon esculentum	var. *pyriforme*	Pear tomato
Lycopersicon esculentum	var. *validum*	Upright tomato
Lycopersicon pimpinellifolium		Current tomato
Solanum melongena	var. *esculentum*	Common eggplant
Solanum tuberosum		Potato
Apiaceae (parsley family)		
Apium graveolens	var. *dulce*	Celery
Apium graveolens	var. *rapaceum*	Celeriac
Daucus carota	var. *sativa*	Carrot
Pastinaca sativa		Parsnip
Petroselinum crispum	var. *tuberosum*	Hamburg (or turnip-rooted) parsley

Conversion Factors

The metric system originated in France in 1790 and spread throughout Europe, Latin America, and the East during the nineteenth century. Today, approximately 95 percent of the world's inhabitants use this system. Only in the United States and a handful of smaller nations is the metric system not widely used.

In 1975, however, President Gerald R. Ford signed the Metric Conversion Act, calling for voluntary conversion to the metric system and establishing a U.S. Metric Board to coordinate this conversion. Although change to the metric system is not mandatory, gradual changeover has occurred, and the pace seems to be increasing. It is estimated that by 1990 the United States will be a predominately metric country.

Because the majority of our audience is more familiar with the English system of measurements it is used predominantly in the text. To aid our readers in conversion to the more progressive metric system we offer the following conversion factors.

Linear Measure

12 inches = 1 foot
36 inches = 3 feet; 1 yard
1 rod = 16.5 feet
1 mile = 5280 feet; 1760 yards; 160 rods; 80 chains; 1.6094 kilometers
1 chain = 66 feet; 22 yards; 4 rods; 100 links
1 inch = 2.54 centimeters
1 millimeter = 0.394 inch or 1/25 inch
1 meter = 39.37 inches; 10 decimeters; 3.28 feet
1 micron (μ) = 1/1000 millimeter
1 kilometer = 1000 meters; 3280 feet; 0.62 mile
1 mile per hour = 1.47 feet per second; 88 feet per minute

Weight

1 gamma = 0.001 milligram
1 grain = 64.799 milligrams
1 gram = 1000 milligrams; 15.432 grains; 0.0353 ounce
1 pound = 16 ounces; 7000 grains; 453.59 grams
1 ounce = 28.35 grams
1 short ton = 2000 pounds; 907.2 kilograms
1 long ton = 2240 pounds
1 kilogram = 2.2 pounds

Liquid

1 gallon (U.S. = 3785.4 milliliters; 256 table-spoons; 231 cubic inches; 128 fluid ounces; 16 cups; 8 pints; 4 quarts; 0.8333 imperial gallon; 0.1337 cubic foot

1 liter = 1000 milliliters; 1.0567 liquid quarts (U.S.)

1 gill = 118.29 milliliters

1 fluid ounce = 29.57 milliliters; 2 table-spoons

3 teaspoons = 1 tablespoon; 14.79 milliliters

1 gallon of water = 8.355 pounds

1 cubic foot of water = 62.43 pounds; 7.48 gallons

1 pint = 1.043 pounds

1 quart = 0.86 pound

Area

1 township = 36 sections; 23,040 acres

1 square mile = 1 section; 640 acres; 259 hectares

1 acre = 43,560 square feet; 160 square rods; 4840 square yards; 208.7 feet squared; an area one rod wide and ½ mile long; 0.4 hectare

1 hectare = 2.471 acres; 10,000 square meters

Capacity (Dry Measure)

1 bushel (U.S.) = 4 pecks; 32 quarts; 35.24 liters; 1.244 cubic feet; 2150.42 cubic inches

Pressure

1 foot lift of water = 0.433 pound pressure per square inch (psi)

1 pound pressure per square inch will lift water 2.31 feet

1 atmosphere = 760 millimeters of mercury; 14.7 pounds; 33.9 feet of water

Sources of Agricultural Information

Canada

In Canada agricultural information can be obtained by writing to the Department of Agriculture (refer to the Parliament Buildings as the address) for the province of interest, as follows:

British Columbia Department of Agriculture-Victoria

Alberta Department of Agriculture-Edmonton

Saskatchewan Department of Agriculture-Regina

Manitoba Department of Agriculture-Winnipeg

Ontario Ministry of Food and Agriculture-Ottawa

Quebec Department of Agriculture-Quebec City

New Brunswick Department of Agriculture-Fredericton

Nova Scotia Department of Agriculture-Halifax

Prince Edward Island Department of Agriculture-Charlottetown

Newfoundland Department of Agriculture-St. John's

The Canadian Department of Agriculture is as follows:

Agriculture Canada
Information Division
Parliament Buildings
Ottawa, Ontario

United States

The Cooperative Extension Service is one of the best sources of information on vegetable gardening, or any other subject relating to agriculture or home economics. This service is one of the divisions of the land-grant college or university which is found in each state. These colleges were established as a result of the Morrill Act passed by Congress in 1862. This act granted land to each state to set up colleges which emphasized instruction in agriculture and mechanics arts.

Land-grant colleges or universities now have three divisions, and offer three distinct types of service. *The teaching division* provides courses for students in all aspects of agriculture, home economics, and related engineering subjects. *The research division* conducts experiments and develops new and improved production practices for all types of crops and livestock. *The extension division* carries the useful agricultural knowledge beyond the land-grant college campus to all the residents of the state.

The services of the Extension division are rendered by two types of employees. Specialists, who are usually located at the land-grant college or university campus, are the top-level resource people. The Extension agents are located in the counties and some cities in the state, and supply information directly to the home gardener, homemaker, or farmer. The specialists at the state level supply the Extension agents with publications and information they need to render service to the residents in their county or city.

The following is a list of the land-grant institutions.

Alabama
School of Agriculture, Auburn University, Auburn, AL 36380
Alabama Agricultural and Mechanical University, Normal, AL 37562
Tuskegee Institute, Tuskegee, AL 36088

Alaska
Department of Agriculture, University of Alaska, Fairbanks, AK 99701

Arizona
College of Agriculture, University of Arizona, Tucson, AZ 85721

Arkansas
Division of Agriculture, University of Arkansas, Fayetteville, AR 72701
Pine Bluff, AR 71601

California
Division of Agricultural Sciences, University of California, Berkeley, CA 94720
Davis, CA 95616
Riverside, CA 92502

Colorado
College of Agricultural Sciences, Colorado State University, Fort Collins, CO 80521

Connecticut
College of Agriculture, University of Connecticut, Storrs, CT 06268

Delaware
College of Agricultural Sciences, University of Delaware, Newark, DE 19711
Department of Agriculture, Delaware State College, Dover, DE 19901

Florida
Institute of Food and Agricultural Sciences, University of Florida, Gainesville, FL 32611
Florida Agricultural and Mechanical University, Tallahassee, FL 32307

Georgia
College of Agriculture, University of Georgia, Athens, GA 30601
Fort Valley College, Fort Valley, GA 31030

Guam
Institute of Resources Development, University of Guam, Agana, GU 96910

Hawaii
College of Tropical Agriculture, University of Hawaii, Honolulu, HI 96822

Idaho
College of Agriculture, University of Idaho, Moscow, ID 83843

Illinois
College of Agriculture, University of Illinois, Urbana, IL 61801
School of Agriculture, Southern Illinois University, Carbondale, IL 62901

Indiana
School of Agriculture, Purdue University, Lafayette, IN 47907

Iowa
College of Agriculture, Iowa State University, Ames, IA 50010

Kansas
College of Agriculture, Kansas State University, Manhattan, KS 66506

Kentucky
College of Agriculture, University of Kentucky, Lexington, KY 40506
Kentucky State University, Frankfort, KY 40601

Louisiana
Louisiana State University and Agricultural & Mechanical College, University Station, Baton Rouge, LA 70802
Southern University and Agricultural & Mechanical College, Southern Branch P.O., Baton Rouge, LA 70813

Maine
College of Life Sciences and Agriculture, University of Maine, Orono, ME 04473

Maryland
College of Agriculture, University of Maryland, College Park, MD 20742
Department of Agriculture, University of Maryland, Eastern Shore, Princess Anne, MD 21853

Massachusetts
Agricultural Experiment Station, University of Massachusetts, Amherst, MA 01002

Michigan
College of Agriculture and Natural Resources, Michigan State University, East Lansing, MI 48823

Minnesota
Institute of Agriculture, University of Minnesota, St. Paul, MN 55101

Mississippi
College of Agriculture, Mississippi State University of Applied Arts and Sciences, State College, MS 39726
Alcorn Agricultural and Mechanical College, Lorman, MS 39096

Missouri
College of Agriculture, University of Missouri-Columbia, Columbia, MO 65201
Lincoln University, Jefferson City, MO 65101

Montana
College of Agriculture, Montana State University, Bozeman, MT 59715

Nebraska
College of Agriculture, University of Nebraska, Lincoln, NB 68503

Nevada
Max C. Fleischmann College of Agriculture, University of Nevada, Reno, NV 89507

New Hampshire
College of Life Sciences and Agriculture, University of New Hampshire, Durham, NH 03824

New Jersey
College of Agriculture and Environmental Science, State University of New Jersey, New Brunswick, NJ 08903

New Mexico
College of Agriculture and Home Economics, New Mexico State University, Las Cruces, NM 88003

New York
New York State College of Agriculture, Cornell University, Ithaca, NY 14850

North Carolina
Schools of Agriculture and Life Sciences and Forest Resources, North Carolina State University, Raleigh, NC 27607
School of Agriculture, North Carolina Agricultural & Technical State University, Greensboro, NC 27411

North Dakota
College of Agriculture, North Dakota State University, State University Sta., Fargo, ND 58102

Ohio
College of Agriculture and Home Economics, Ohio State University, Columbus, OH 43210

Oklahoma
Oklahoma State University of Agriculture & Applied Science, Stillwater, OK 74074
Langston University, Langston, OK 73050

Oregon
School of Agriculture, Oregon State University, Corvallis, OR 97331

Pennsylvania
College of Agriculture, Pennsylvania State University, University Park, PA 16802

Puerto Rico
College of Agricultural Sciences, University of Puerto Rico, Mayaguez, PR 00708

Rhode Island
Colleges of Resource Development and Home Economics, University of Rhode Island, Kingston, RI 02881

South Carolina
College of Agricultural Sciences, Clemson University, Clemson, SC 29631
South Carolina State College, Orangeburg, SC 29115

South Dakota
College of Agriculture and Biological Sciences, South Dakota State University, Brookings, SD 57006

Tennessee
Institute of Agriculture, University of Tennessee, P.O. Box 1071, Knoxville, TN 37901
School of Agriculture and Home Economics, Tennessee State University, Nashville, TN 37203

Texas
College of Agriculture, Texas A&M University, College Station, TX 77843
School of Agriculture, Prairie View A&M University, Prairie View, TX 77445

Utah
College of Agriculture, Utah State University, Logan, UT 84321

Vermont
College of Agriculture, University of Vermont, Burlington, VT 05401

Virginia
College of Agriculture and Life Sciences, Virginia Polytechnic Institute, Blacksburg, VA 24061
School of Agriculture, Virginia State College, Petersburg, VA 23803

Washington
College of Agriculture, Washington State University, Pullman, WA 99163

West Virginia
College of Agriculture and Forestry, West Virginia University, Morgantown, WV 26506

Wisconsin
College of Agriculture and Life Sciences, University of Wisconsin, Madison, WI 53706

Wyoming
College of Agriculture, University of Wyoming, University Sta., P.O. Box 3354, Laramie, WY 82070

Climate Data for Selected Locations within Canada and the United States

To use the climate tables simply turn to your province or state and find the city or town nearest you (in some cases it may be in an adjacent province or state). Then follow the rows across for the averages for (1) the last 32°F frost in the spring, (2) the first 32°F frost in the fall, (3) the length of the growing season, (4) the July temperature, (5) the growing season precipitation, and (6) the annual precipitation. A few words of caution; remember that the figures represent long-term averages and no one year is ever quite average. Also, even if you are near one of the locations listed the climate in your garden site may be quite different due to local influences.

CANADA

(Adapted from Wernstadt, F. L. 1972. *World Climate Data*. Climatic Data Press, Lemont, Pa., and G. M. Hemmerick and G. R. Kendall, 1972. *Frost Data 1941-1970*. Atmospheric Environment Service, Dept. of Environment-Canada, Downsview, Ontario.)

Place	Temperatures				Precipitation	
Province and location	Last 32°F frost in spring	First 32°F frost in fall	Growing season (days)	Mean July temp.	Growing season (May-Aug.)	Annual
Alberta						
Calgary	May 28	Sept. 12	106	62.0	9.70	11.59
Edmonton	May 27	Sept. 6	102	63.1	10.75	13.26
Lethbridge	May 21	Sept. 16	118	66.0	7.82	10.18
Peace River	June 3	Aug. 31	88	60.6	6.89	9.21

Place	Temperatures				Precipitation	
Province and location	Last 32°F frost in spring	First 32°F frost in fall	Growing season (days)	Mean July temp.	Growing season (May-Aug.)	Annual
British Columbia						
Kelowna	May 2	Oct. 7	157	68.2	3.98	9.20
Prince George	June 10	Aug. 28	78	58.9	9.20	16.71
Prince Rupert	Apr. 19	Nov. 5	199	56.2	19.87	91.14
Revelstoke	May 14	Oct. 2	140	66.1	8.74	26.37
Vancouver	Mar. 31	Oct. 30	212	63.8	7.11	48.23
Manitoba						
Brandon	May 29	Sept. 13	106	67.3	11.00	14.12
Dauphin	May 31	Sept. 15	106	67.1	10.51	13.99
Winnipeg	May 25	Sept. 21	118	68.3	10.53	15.22
New Brunswick						
Fredericton	May 21	Sept. 26	127	67.0	13.76	33.93
Moncton	May 23	Sept. 23	122	65.4	12.06	29.16
St. John	May 18	Oct. 2	136	62.2	14.48	41.84
Newfoundland						
Corner Brook	May 25	Oct. 3	130	62.6	12.03	28.60
Grand Falls	June 7	Sept. 26	110	62.9	10.99	26.31
Port Aux Basques	May 21	Oct. 21	152	56.2	16.59	43.77
St. John's	June 5	Oct. 7	123	61.1	13.64	41.56
Northwest Territory						
Aklavik	June 12	Aug 30	77	56.5	3.26	4.02
Fort Smith	June 15	Aug 19	64	61.1	5.58	7.76
Yellowknife	June 15	Aug 28	108	60.6	4.50	6.24
Nova Scotia						
Halifax	May 15	Oct. 15	152	65.3	15.41	47.30
Sydney	May 23	Oct. 16	145	64.9	13.93	41.82
Truro	June 4	Sept. 17	104	64.8	12.23	34.93
Ontario						
Lindsay	May 16	Sept. 26	132	68.0	11.63	25.70
London	May 9	Oct. 6	149	69.6	12.45	29.94
Ottawa	May 11	Oct. 1	142	68.9	12.30	26.38
Timmins	June 3	Sept. 5	93	63.4	10.50	17.71
Toronto	Apr. 20	Oct. 30	192	71.5	10.54	25.07
Prince Edward Island						
Charlottetown	May 17	Oct. 15	150	66.8	12.50	32.50
Quebec						
Drummondville	May 12	Sept. 29	139	67.9	14.12	30.17
Montreal	May 5	Oct 7	154	70.8	14.20	31.33

Place	Temperatures				Precipitation	
Province and location	Last 32°F frost in spring	First 32°F frost in fall	Growing season (days)	Mean July temp.	Growing season (May-Aug.)	Annual
Quebec	May 18	Sept. 28	132	67.9	17.54	33.38
Sherbrooke	June 1	Sept. 11	101	68.2	14.22	29.65
Saskatchewan						
Prince Albert	June 5	Sept. 7	93	64.9	8.32	10.83
Regina	May 27	Sept. 12	107	66.0	8.89	11.26
Saskatoon	May 28	Sept. 15	109	66.8	7.67	10.04
Swift Current	May 28	Sept. 19	113	66.9	8.48	10.83
Yukon Territory						
Dawson	May 26	Aug. 27	92	59.8	6.15	7.68
Watson Lake	May 30	Sept. 3	95	59.1	6.51	8.73
Whitehorse	June 5	Sept. 1	87	57.5	4.28	5.49

UNITED STATES

(Adapted from *Climate and Man*, Yearbook of Agriculture, 1941, United States Department of Agriculture.)

Place	Temperatures				Precipitation	
State and location	Last frost in spring	First frost in fall	Growing season (days)	Mean July temp. F°	Growing season (Apr.-Aug.)	Annual
Alabama						
Birmingham	Mar. 16	Nov. 11	240	80.0	23.29	54.08
Madison	Apr. 5	Oct. 31	209	80.1	21.90	52.47
Mobile	Feb. 17	Dec. 12	298	81.6	27.92	60.67
Montgomery	Mar. 3	Nov. 19	261	81.3	20.88	51.44
Alaska						
Anchorage	May 25	Sept. 12	110	57.0	5.74	14.32
Fairbanks	May 29	Aug. 26	89	60.0	6.20	11.87
Juneau	Apr. 28	Oct. 17	172	56.6	26.99	83.25
Ketchikan	May 5	Oct. 17	165	57.5	45.40	150.89
Arizona						
Flagstaff	June 3	Sept. 29	118	65.2	8.68	20.92
Phoenix	Feb. 5	Dec. 6	304	90.3	2.56	7.62
Prescott	May 17	Oct. 7	143	72.5	8.36	20.71
Tucson	Mar. 19	Nov. 19	245	85.1	5.09	11.16
Yuma	Jan. 12	Dec. 26	348	91.0	0.96	3.58
Arkansas						
El Dorado	Mar. 19	Nov. 15	241	82.4	20.87	51.38
Fayetteville	Apr. 6	Oct. 25	202	78.9	21.81	43.90

Place	Temperatures				Precipitation	
State and location	Last frost in spring	First frost in fall	Growing season (days)	Mean July temp. F°	Growing season (Apr.-Aug.)	Annual
Jonesboro	Apr. 1	Nov. 4	217	81.4	19.19	47.43
Little Rock	Mar. 17	Nov. 13	241	81.2	19.54	46.12
California						
Bakersfield	Feb. 21	Nov. 25	277	83.5	1.10	6.12
Los Angeles	Jan. 3	Dec. 28	359	70.5	1.45	14.76
Redding	Feb. 28	Dec. 3	278	82.0	5.18	37.54
Sacramento	Feb. 6	Dec. 10	307	73.9	1.74	15.88
San Diego	Frost free		365	67.5	1.16	10.11
San Francisco	Jan. 7	Dec. 29	356	58.9	1.89	20.23
Colorado						
Burlington	May 4	Oct. 7	156	74.6	12.18	16.88
Colorado Springs	May 8	Oct. 8	148	68.2	10.34	14.19
Denver	Apr. 26	Oct. 14	171	72.5	8.45	13.99
Durango	May 28	Sept. 26	121	67.0	7.91	19.54
Grand Junction	Apr. 16	Oct. 24	191	77.9	3.82	8.76
Connecticut						
Hartford	Apr. 19	Oct. 18	182	72.5	18.51	42.86
New Haven	Apr. 11	Oct. 23	195	72.6	18.72	44.96
Delaware						
Dover	Apr. 19	Oct. 24	188	76.7	21.24	44.70
Newark	Apr. 18	Oct. 15	180	74.8	20.80	43.70
Wilmington	Apr. 17	Oct. 28	194	76.0	20.97	44.58
Florida						
Jacksonville	Feb. 15	Dec. 11	299	81.4	25.38	48.21
Miami	Frost free		365	81.7	29.56	59.18
Tallahassee	Feb. 25	Dec. 4	282	80.8	27.77	54.89
Tampa	Jan. 13	Dec. 27	348	81.6	28.18	48.35
Georgia						
Atlanta	Mar. 23	Nov. 9	231	78.5	19.64	47.58
Columbus	Mar. 12	Nov. 16	249	82.1	20.95	48.67
Savannah	Feb. 28	Nov. 28	273	81.2	23.57	44.67
Valdosta	Mar. 14	Nov. 16	247	81.4	25.23	49.23
Hawaii						
Hilo	Frost free		365	74.7	51.37	137.12
Honolulu	Frost free		365	77.5	5.65	25.28
Idaho						
Boise	Apr. 23	Oct. 17	177	74.2	3.78	12.47
Coeur d'Alene	May 8	Oct. 6	151	69.1	6.36	22.80

Place	Temperatures				Precipitation	
State and location	Last frost in spring	First frost in fall	Growing season (days)	Mean July temp. F°	Growing season (Apr.-Aug.)	Annual
Idaho Falls	May 15	Sept. 19	127	69.0	4.73	11.27
Lewiston	Apr. 5	Oct. 26	204	75.9	4.71	13.27
Moscow	May 6	Oct. 6	153	67.1	5.92	21.81
Pocatelo	Apr. 28	Oct. 6	161	72.2	5.49	13.34
Illinois						
Carbondale	Apr. 13	Oct. 20	190	80.0	19.89	43.27
Chicago	Apr. 13	Oct. 26	196	73.9	15.82	31.85
Rockford	May 6	Oct. 10	157	74.1	17.73	33.90
Springfield	Apr. 11	Oct. 22	194	78.0	17.42	34.59
Indiana						
Evansville	Apr. 2	Oct. 30	211	79.6	17.11	40.19
Indianapolis	Apr. 15	Oct. 24	192	76.3	17.54	38.26
South Bend	May 7	Oct. 15	167	73.7	16.66	34.83
Iowa						
Davenport	Apr. 18	Oct. 17	182	76.6	17.47	32.29
Des Moines	Apr. 19	Oct. 11	175	76.3	17.88	30.69
Mason City	May 8	Oct. 3	148	72.8	18.48	30.38
Sioux City	Apr. 25	Oct. 7	165	75.9	15.97	25.65
Kansas						
Dodge City	Apr. 15	Oct. 25	193	78.7	12.65	19.52
Hays	Apr. 29	Oct. 14	168	79.1	15.60	22.74
Topeka	Apr. 8	Oct. 20	195	79.5	18.85	32.27
Wichita	Apr. 10	Oct. 27	200	80.4	17.69	29.64
Kentucky						
Bowling Green	Apr. 14	Oct. 22	191	79.5	20.53	49.31
Lexington	Apr. 16	Oct. 22	189	76.4	17.84	41.12
Louisville	Apr. 9	Oct. 24	198	78.7	17.56	40.58
Paducah	Apr. 7	Oct. 24	200	81.3	19.47	46.91
Louisiana						
Lake Charles	Mar. 3	Nov. 23	265	82.5	25.76	56.72
Monroe	Mar. 11	Nov. 13	247	82.4	21.88	53.25
New Orleans	Feb. 20	Dec. 9	292	80.1	28.60	59.72
Shreveport	Mar. 8	Nov. 15	252	83.2	17.92	41.64
Maine						
Bangor	May 9	Oct. 6	150	70.8	15.83	39.52
Houlton	May 25	Sept. 19	117	67.3	13.73	32.78
Portland	Apr. 27	Oct. 17	173	67.8	16.08	42.05

State and location	Last frost in spring	First frost in fall	Growing season (days)	Mean July temp. F°	Growing season (Apr.-Aug.)	Annual
					Precipitation	
Maryland						
Baltimore	Apr. 8	Nov. 2	208	77.9	19.71	41.94
Cumberland	Apr. 27	Oct. 14	170	75.9	17.33	35.22
Salisbury	Apr. 20	Oct. 21	184	76.8	20.60	44.34
Massachusetts						
Boston	Apr. 13	Oct. 29	199	72.4	16.31	38.94
Pittsfield	May 10	Oct. 1	144	70.8	18.26	40.83
Worchester	Apr. 28	Oct. 14	169	71.7	17.18	39.74
Michigan						
Detroit	Apr. 24	Oct. 18	177	73.1	14.85	31.04
Grand Rapids	May 1	Oct. 17	169	73.0	15.09	32.34
Marquette	May 9	Oct. 14	158	65.3	14.33	31.73
Saginaw	May 5	Oct. 10	158	72.0	14.33	29.13
Traverse City	May 15	Oct. 11	149	70.1	12.20	27.47
Minnesota						
Bemidji	May 25	Sept. 16	114	68.9	14.83	23.46
Duluth	May 10	Oct. 5	148	64.9	15.14	26.51
Minneapolis	Apr. 25	Oct. 13	171	73.2	16.32	27.31
Rochester	May 15	Sept. 29	137	72.1	17.58	28.50
Mississippi						
Greenville	Mar. 21	Nov. 5	229	82.2	20.30	51.65
Hattiesburg	Mar. 13	Nov. 1	233	82.1	26.44	58.52
Jackson	Mar. 19	Nov. 8	234	81.5	22.29	51.29
Tupelo	Mar. 30	Nov. 1	216	80.7	20.93	51.49
Missouri						
Kansas City	Apr. 7	Oct. 28	204	80.0	19.68	35.73
Kirksville	Apr. 24	Oct. 13	172	76.9	21.45	38.73
Springfield	Apr. 8	Oct. 28	203	77.6	21.41	40.19
St. Louis	Apr. 2	Oct. 29	210	80.2	17.73	36.67
Montana						
Billings	May 15	Sept. 25	133	71.2	7.95	13.41
Butte	May 29	Sept. 16	110	64.8	7.65	13.51
Great Falls	May 9	Sept. 25	139	69.0	9.46	15.21
Havre	May 11	Sept. 22	134	68.3	8.37	13.07
Missoula	May 18	Sept. 23	128	67.6	6.63	14.01
Nebraska						
Chadron	May 10	Oct. 3	146	75.6	11.63	17.97
Grand Island	Apr. 29	Oct. 6	160	78.0	16.91	26.00

| Place | Temperatures | | | | Precipitation | |
State and location	Last frost in spring	First frost in fall	Growing season (days)	Mean July temp. F°	Growing season (Apr.- Aug.)	Annual
Lincoln	Apr. 18	Oct. 15	180	78.0	17.13	27.31
Omaha	Apr. 14	Oct. 20	189	78.1	15.23	25.49
Nevada						
Carson City	May 25	Sept. 19	117	68.6	1.75	9.29
Elko	June 1	Sept. 12	103	69.8	3.07	9.46
Las Vegas	Mar. 16	Nov. 10	239	86.1	1.87	4.84
Winnemucca	May 11	Sept. 29	141	71.9	2.73	8.20
New Hampshire						
Berlin	May 30	Sept. 16	109	66.2	15.49	38.58
Manchester	May 13	Sept. 29	139	67.5	16.39	34.81
New Jersey						
Atlantic City	Apr. 6	Nov. 7	215	72.9	17.61	40.91
Newark	Apr. 14	Oct. 28	197	74.8	21.58	47.28
Trenton	Apr. 13	Oct. 27	197	75.0	20.00	42.50
New Mexico						
Albuquerque	Apr. 13	Oct. 28	198	76.7	4.75	8.40
Bloomfield	May 12	Oct. 6	147	74.7	4.33	9.14
Carlsbad	Mar. 29	Nov. 4	220	80.5	7.20	13.13
Santa Fe	Apr. 24	Oct. 19	178	68.9	7.90	14.19
Silver City	Apr. 26	Oct. 23	180	71.9	8.37	16.83
New York						
Albany	Apr. 23	Oct. 14	174	73.2	15.50	33.11
Binghamton	May 2	Oct. 10	161	70.5	16.51	34.01
Buffalo	Apr. 26	Oct. 23	180	70.1	12.90	32.77
New York	Apr. 9	Nov. 6	211	74.4	18.39	41.63
Rochester	Apr. 26	Oct. 21	178	71.6	13.40	31.29
Utica	May 13	Oct. 7	147	70.7	18.86	41.20
North Carolina						
Asheville	Apr. 11	Oct. 22	194	71.7	18.17	38.47
Raleigh	Mar. 23	Nov. 9	231	78.8	22.30	45.58
Wilmington	Mar. 16	Nov. 17	246	79.1	23.35	45.40
Winston-Salem	Apr. 11	Oct. 25	197	77.5	21.17	44.90
North Dakota						
Bismarck	May 10	Sept. 25	140	70.9	10.48	15.43
Grand Forks	May 16	Sept. 25	132	68.5	12.65	19.80
Minot	May 23	Sept. 16	116	68.6	10.39	15.47
Williston	May 15	Sept. 25	133	69.4	9.55	14.08

Place	Temperatures				Precipitation	
State and location	Last frost in spring	First frost in fall	Growing season (days)	Mean July temp. F°	Growing season (Apr.- Aug.)	Annual
Ohio						
Cincinnati	Apr. 12	Oct. 25	196	76.8	17.20	37.21
Cleveland	Apr. 16	Nov. 5	203	72.4	14.19	31.89
Columbus	Apr. 19	Oct. 23	187	75.4	15.87	34.10
Lima	May 4	Oct. 14	163	73.7	17.20	36.66
Youngstown	Apr. 25	Oct. 17	175	75.4	16.06	33.17
Oklahoma						
Lawton	Mar. 31	Nov. 6	220	82.5	17.05	31.98
Oklahoma City	Mar. 28	Nov. 7	224	81.6	16.94	31.15
Tulsa	Mar. 25	Nov. 1	221	82.8	20.24	38.38
Oregon						
Bend	June 8	Sept. 7	91	65.1	3.82	12.64
Eugene	Apr. 13	Nov. 4	205	65.9	7.31	37.88
Medford	May 6	Oct. 14	161	72.0	3.77	16.48
Pendleton	May 3	Oct. 5	155	72.1	4.04	13.68
Portland	Mar. 6	Nov. 24	263	66.7	7.21	39.43
Pennsylvania						
Harrisburg	Apr. 9	Oct. 30	204	75.3	17.15	37.24
Philadelphia	Apr. 5	Nov. 2	211	76.7	19.42	41.86
Pittsburgh	Apr. 20	Oct. 20	183	74.2	16.56	34.77
Scranton	Apr. 22	Oct. 15	176	72.1	17.49	37.16
State College	May 2	Oct. 5	156	71.0	18.91	38.93
Rhode Island						
Bristol	Apr. 16	Nov. 4	202	69.7	16.08	39.60
Providence	Apr. 17	Oct. 26	192	72.5	16.31	39.84
South Carolina						
Charleston	Feb. 23	Dec. 5	285	81.4	20.94	40.26
Columbia	Mar. 15	Nov. 18	248	80.9	20.96	41.95
Greenville	Mar. 27	Nov. 10	228	76.9	22.39	52.10
South Dakota						
Aberdeen	May 11	Sept. 27	139	72.9	15.80	23.96
Rapid City	May 1	Oct. 4	159	71.9	14.00	18.09
Sioux Falls	May 6	Oct 3	150	73.6	17.42	26.43
Tennessee						
Chattanooga	Mar. 21	Nov. 11	235	78.6	21.70	51.35
Knoxville	Mar. 30	Nov. 2	217	77.9	20.56	46.85
Memphis	Mar. 17	Nov. 10	238	80.9	17.70	45.29
Nashville	Mar. 30	Oct. 30	214	79.3	18.96	44.77

Place	Temperatures				Precipitation	
State and location	Last frost in spring	First frost in fall	Growing season (days)	Mean July temp. F°	Growing season (Apr.-Aug.)	Annual
Texas						
Amarillo	Apr. 11	Nov. 2	205	76.8	13.04	20.96
Corpus Christi	Jan. 26	Dec. 27	335	82.7	10.92	25.54
Dallas	Mar. 18	Nov. 17	244	83.7	16.03	33.60
Houston	Feb. 10	Dec. 8	301	83.1	19.90	44.84
San Angelo	Mar. 25	Nov. 9	229	83.7	10.85	21.57
San Antonio	Feb. 24	Dec. 3	282	83.7	13.27	26.79
Utah						
Logan	May 7	Oct. 11	157	73.1	5.99	16.48
Moab	Apr. 24	Oct. 10	169	78.4	3.81	9.58
Provo	May 24	Sept. 25	124	72.3	5.22	15.33
Salt Lake City	Apr. 13	Oct. 22	192	77.0	6.01	15.79
St. George	Apr. 10	Oct. 23	196	82.9	2.98	8.73
Vermont						
Burlington	May 4	Oct. 10	159	69.4	15.22	31.87
Newport	May 14	Sept. 26	135	67.5	17.54	35.61
Rutland	May 16	Oct. 4	141	69.0	17.86	37.43
Virginia						
Charlottesville	Apr. 6	Oct. 16	213	76.9	22.29	45.22
Norfolk	Mar. 19	Nov. 16	242	78.3	20.50	40.45
Richmond	Mar. 29	Nov. 2	218	78.0	20.77	41.93
Roanoke	Apr. 15	Oct. 22	190	75.7	18.65	40.80
Washington						
Seattle	Mar. 14	Nov. 24	255	63.1	6.39	31.80
Spokane	Apr. 12	Oct. 13	184	69.0	4.34	14.62
Walla Walla	Mar. 31	Nov. 5	219	74.0	4.54	15.71
Yakima	Apr. 15	Oct. 22	190	71.4	1.70	6.79
West Virginia						
Bluefield	Apr. 28	Oct. 11	166	71.6	19.34	40.12
Charleston	Apr. 20	Oct. 23	186	76.9	21.35	45.80
Morgantown	May 1	Oct. 13	165	73.5	19.58	40.61
Parkersburg	Apr. 18	Oct. 19	184	75.4	18.17	37.89
Wisconsin						
Eau Claire	May 4	Oct. 1	150	71.5	18.77	32.86
Madison	Apr. 29	Oct. 17	171	72.1	16.42	30.60
Milwaukee	Apr. 22	Oct. 23	184	70.1	14.34	29.64
Wausau	May 22	Sept. 29	130	68.4	17.78	31.06

Place	Temperatures				Precipitation	
State and location	Last frost in spring	First frost in fall	Growing season (days)	Mean July temp. F°	Growing season (Apr.- Aug.)	Annual
Wyoming						
Casper	May 19	Sept. 29	133	72.1	8.23	14.99
Cheyenne	May 14	Oct. 2	141	67.0	9.91	15.82
Green River	June 3	Sept. 11	100	69.3	3.65	7.42
Laramie	May 29	Sept. 19	113	63.8	6.89	11.19
Sheridan	May 15	Sept. 22	130	68.9	8.65	15.31

Commercial Seed Companies Specializing in Mail Order Sales

Following is a list of several commercial seed companies that specialize in mail order sales, and generally have available catalogs describing their offerings. This list was formulated from responses to a questionnaire sent to over 100 companies. (Only companies that responded are included here.) The companies are separated on the basis of geographic region (by time zone). If a company indicated that it specialized in a particular product, this is so designated. Please bear in mind that the list is by no means exhaustive, but rather is intended to give a wide geographic distribution of seed companies. No endorsements are implied in the listing, and companies excluded are not regarded as inferior in any way.

Readers should be aware that there are strict laws regulating the international movement of plant materials. Canadian residents who wish to order plants from the United States must first obtain a mailing label from the Canadian Department of Agriculture. This label is then included with your order when you mail to the United States. Americans who order plant materials from Canadian sources must obtain a Phytosanitary Certificate from the Canadian Government to insure that the materials were grown in Canada and are free of pests and diseases.

Name	Address	Plant offerings[a]				Other[b] notations
		H	V	P	F	
Atlantic Time Zone						
1. Vesey's Seed Ltd.	York, P.E., Canada COA IPO		X		X	
Central Time Zone						
2. Earl May Seed & Nursery Co.	Shenandoah, IA 51603	X	X	X	X	
3. Farmer Seed & Nursery Co.	818 W. 4th St., Faribault, MN 55021	X	X	X	X	D
4. Gurney Seed & Nursery Co.	Yankton, SD 57079	X	X	X	X	B

Name	Address	Plant offerings[a]				Other[b] notations
		H	V	P	F	
5. Henry Field Seed & Nursery Co.	407 Sycamore St., Shenandoah, IA 51602	X	X	X	X	B
6. Hilltop Herb Farm	P.O. Box 1734, Cleveland, TX 77327	X		X	X	
7. J. A. Demonchaux Co.	225 Jackson St., Topeka, KS 66603	X	X			C
8. J .W. Jung Seed Co.	335 S. High St., Randolph, WI 53956	X	X	X	X	B,C
9. L .L. Olds Seed Co.	2901 Packers Ave., P.O. Box 7790, Madison, WI 53707	X	X	X	X	
10. McFayden Seed Co., Ltd.	P.O. Box 1600, 30-9th St., Brandon, MB Canada R7A 6E1	X	X	X	X	D
11. Midwest Seed Growers	505 Walnut St., Kansas City, MO 64106	X	X	X	X	
12. Porter & Son, Seedsmen	P.O. Box 104, Stephenville, TX 76401	X	X	X	X	B
13. R. H. Shumway, Seedsman	628 Cedar St., Rockford, IL 61101	X	X	X	X	
14. Reuter Seed Co., Inc.	320 N. Carrollton Ave., New Orleans, LA 70119	X	X		X	B
15. Stark Brother's Nursery & Orchard Co.	Louisiana, MO 63353			X	X	
16. W. Atlee Burpee Co.	P.O. Box B-2001, Clinton, IA 52732	X	X	X	X	
17. Yankee Peddler Herb Farm	Dept. PSU, Hwy 36 N., Brenham, TX 77833	X	X	X	X	
	Eastern Time Zone					
18. Abbott & Cobb, Inc.	4744 Frankford Ave., Philadelphia, PA 19124		X		X	
19. Burgess Seed & Plant Co.	67 E. Battle Creek St., Galesburg, MI 49053	X	X	X	X	B,C
20. Cameron Nursery & Ashby's Garden Centre	RR2, Cameron, ON, Canada KOM 1GO	X				
21. Capriland's Herb Farm	Silver St., Coventry, CT 06238	X	X	X	X	
22. Chas. C. Hart Seed Co.	Main & Hart St., Wethersfield, CT 06109	X	X	X	X	
23. Comstock, Ferre, & Co.	263 Main St., Box 125, Wethersfield, CT 06109	X	X	X	X	B
24. D. Landreth Seed Co.	2700 Wilmarco Ave., Baltimore, MD 21223	X	X	X	X	B
25. Geo. Tait & Sons, Inc.	900 Tidewater Dr., Norfolk, VA 23504	X	X	X	X	B
26. Geo. W. Park Seed Co., Inc.	Box 31, Greenwood, SC 29646	X	X	X	X	
27. Glecklers Seedmen	Metamora, OH 43540	X	X		X	C
28. Grace's Gardens	22 Autumn Ln., Hackettstown, NJ 07840	X	X		X	C

Name	Address	Plant offerings[a]				Other[b] notations
		H	V	P	F	
29. H. G. Hastings Co.	Box 4274, Atlanta, GA 30302	X	X	X	X	
30. Herb Shop	P.O. Box 362, Fairfield, CT 06430	X				
31. Herbst Brothers Seedsmen	1000 N. Main St., Brewster, NY 10509	X	X	X	X	C
32. Johnny's Selected Seeds	Albion, ME 04910	X	X			D
33. Joseph Harris Co., Inc.	Moreton Farm, Rochester, NY 14624	X	X	X	X	
34. Lakeland Nurseries Sales	340 Poplar St., Hanover, PA 17331	X	X	X	X	C
35. Le Jardin du Gourmet	Box 454, W. Danville, VT 05873	X			X	
36. Meyer Seed Co.	600 S. Caroline St., Baltimore, MD 21231	X	X	X	X	
37. Natural Development Co.	Box 215, Bainbridge, PA 17502		X			B
38. Ottis S. Twilley Seed Co.	P.O. Box 1817, Salisbury, MD 21801	X	X	X	X	B
39. Otto Richter & Sons Ltd.	P.O. Box 26, Goodwood, ON, Canada L0C 1A0	X				
40. R. L. Holmes Seed Co.	2125 46th St., N.W., Canton, OH 44709		X			
41. Rehoboth Herb Farm	74 Winter St., Rehoboth, MA 02769	X		X		
42. Seedway Inc.	Hall, NY 14463	X	X	X	X	
43. Spring Hill Nurseries	110 Elm St., Tipp City, OH 45366	X	X	X	X	B
44. Stillridge Herb Farm	10370 Rt. 99, Woodstock, MD 21163	X				B
45. Stokes Seed, Inc.	737 Main St., Box 548, Buffalo, NY 14240	X	X	X	X	
46. Stokes Seed, Inc.	Box 10, St. Catharines, ON, Canada, L2R 6R6	X	X	X	X	
47. Thompson & Morgan Inc.	P.O. Box 100, Farmingdale, NJ 07727	X	X	X	X	
48. W. Atlee Burpee Co.	300 Park Ave., Warminster, PA 18974	X	X	X	X	
49. Well-Sweep Herb Farm	317 Mt. Bethel Rd., Port Murray, NJ 07865	X				
50. William Dan Seeds	Hwy. 8, West Flamboro, ON, Canada L0R 2K0	X	X		X	
51. Wyatt-Quarles Seed Co.	P.O. Box 2131, Raleigh, NC 27602	X	X	X	X	B
Mountain Time Zone						
52. Alberta Nurseries & Seeds Ltd.	Bowden, AB, Canada T0M 0K0	X	X	X	X	D

| Name | Address | Plant offerings[a] | | | | Other[b] notations |
		H	V	P	F	
53. D. V. Burrell Seed Growers Co.	P.O. Box 150, Rocky Ford, CO 81067	X	X		X	
54. Dominion Seed House	Georgetown, ON, Canada L7G 4A2	X	X	X	X	A,C
55. Roswell Seed Co.	Box 725, Roswell, NM 88201	X	X	X	X	
56. Wilton's Organic Potatoes	Box 28, Aspen, CO 81611		X			
	Pacific Time Zone					
57. A. L. Castle, Inc.	P.O. Box 877, Morgan Hill, CA 95037		X			
58. Dessert Seed Co., Inc.	P.O. Box 181, El Central, CA 92243	X	X	X	X	
59. J. L. Hudson, Seedsman	P.O. Box 1058, Redwood City, CA 94064	X	X	X	X	C
60. Jackson & Perkins Co.	2 Rose Ln., Medford, OR 97501	X	X	X	X	B
61. Kitazawa Seed Co.	356 W. Taylor St., San Jose, CA 95110		X			C
62. Tsang & Ma International	P.O. Box 294, Belmont, CA 94070	X	X	X		C
63. W. Atlee Burpee Co.	P.O. Box 748, Riverside, CA 92502	X	X	X	X	

[a]Herbs, H; vegetables, V; perennials, P; flowers, F.
[b]Sales in Canada only, A; sales in U.S. only, B; uncommon vegetables, C; vegetables for northern climates (or high altitudes), D.

Nutritional Guide to Some Common Garden Crops[a]

(Numbers in parentheses denote values imputed—usually from another form of the food or from a similar food. Zero in parentheses indicates that the amount of a constituent probably is none or is too small to measure. Dashes denote lack of reliable data for a constituent believed to be present in measurable amount.)

Raw vegetable or fruit 100 grams (3½ oz.)	Water (%)	Food energy (cal)	Protein (g)	Fat (g)	Carbohydrate Total (g)	Fiber (g)	Calcium (mg)	Phosphorus (mg)	Iron (mg)	Sodium (mg)	Potassium (mg)	Vitamin A (I.U.)	Thiamine (mg)	Riboflavin (mg)	Niacin (mg)	Ascorbic acid (mg)
Apples	84.4	58	0.2	0.6	14.5	1.0	7	10	0.3	1	110	90	0.03	0.02	0.1	4
Apricots	85.3	51	1.0	0.2	12.8	0.6	17	23	0.5	1	281	2,700	0.03	0.04	0.6	10
Artichoke	85.5	b	2.9	0.2	10.6	2.4	51	88	1.3	43	430	160	0.08	0.05	1.0	12
Asparagus—spears	91.5	26	2.5	0.2	5.0	0.7	22	62	1.0	2	278	900	0.18	0.20	1.5	33
Avocado	74.0	167	2.1	16.4	6.3	1.6	10	42	0.6	4	604	290	0.11	0.20	1.6	14
Banana	75.7	85	1.1	0.2	22.2	0.5	8	26	0.7	1	370	190	0.05	0.06	0.7	10
Beans—common white	10.9	340	22.3	1.6	61.3	4.3	144	425	7.8	19	1,196	0	0.65	0.22	2.4	—
Beans—common red	10.4	343	22.5	1.5	61.9	4.2	110	406	6.9	10	984	20	0.51	0.20	2.3	—
Beans—fava	72.3	105	8.4	0.4	17.8	2.2	27	157	2.2	4	471	220	0.28	0.17	1.6	30
Beans—snap green	90.1	32	1.9	0.2	7.1	1.0	56	44	0.8	7	243	600	0.08	0.11	0.5	19
Beans—wax yellow	91.4	27	1.7	0.2	6.0	1.0	56	43	0.8	7	243	250	0.08	0.11	0.5	20
Bean sprouts	88.8	35	3.8	0.2	6.6	0.7	19	64	1.3	5	223	20	0.13	0.13	0.8	19
Beets	87.3	43	1.6	0.1	9.9	0.8	16	33	0.7	60	335	20	0.03	0.05	0.4	10
Beet greens	90.9	24	2.2	0.3	4.6	1.3	119	40	3.3	130	570	6,100	0.10	0.22	0.4	30
Blackeye peas	66.8	127	9.0	0.8	21.8	1.8	27	172	2.3	2	541	370	0.43	0.13	1.6	29
Blueberries	83.2	62	0.7	0.5	15.3	1.5	15	13	1.0	1	81	100	(0.03)	(0.06)	(0.5)	14
Broccoli—spears	89.1	32	3.6	0.3	5.9	1.5	103	78	1.1	15	382	2,500	0.10	0.23	0.9	113
Brussels sprouts	85.2	45	4.9	0.4	8.3	1.6	36	80	1.5	14	390	550	0.10	0.16	0.9	102
Cabbage—Chinese	95.0	14	1.2	0.1	3.0	0.6	43	40	0.6	23	253	150	0.05	0.04	0.6	25
Cabbage—red	90.2	31	2.0	0.2	6.9	1.0	42	35	0.8	26	268	40	0.09	0.06	0.4	61
Cabbage—white	92.4	21	1.3	0.2	5.4	0.8	49	29	0.4	20	233	130	0.05	0.05	0.3	47
Carrots	88.2	42	1.1	0.2	9.7	1.0	37	36	0.7	47	341	11,000	0.06	0.05	0.6	8

Cauliflower	91.0	27	2.7	0.2	5.2	1.0	25	56	1.1	13	295	60	0.11	0.10	0.7	78
Celeriac—root	88.4	40	1.8	0.3	8.5	1.3	43	115	0.6	100	300	—	0.05	0.06	0.7	8
Celery	94.1	17	0.9	0.1	3.9	0.6	39	28	0.3	126	341	240d	0.03	0.03	0.3	9
Chard—Swiss	91.1	25	2.4	0.3	4.6	0.8	88	39	3.2	147	550	6,500	0.06	0.17	0.5	32
Chives	91.3	28	1.8	0.3	5.8	1.1	69	44	1.7	—	250	5,800	0.08	0.13	0.5	56
Collards—leaves and stems	86.9	40	3.6	0.7	7.2	0.9	203	63	1.0	43	401	6,500	0.20	(0.31)	(1.7)	92
Corn—white and yellow	72.7	96	3.5	1.0	22.1	0.7	3	111	0.7	Trace	280	400c	0.15	0.12	1.7	12
Cucumbers	95.1	15	0.9	0.1	3.4	0.6	25	27	1.1	6	160	250	0.03	0.04	0.2	11
Eggplant	92.4	25	1.2	0.2	5.6	0.9	12	26	0.7	2	214	10	0.05	0.05	0.6	5
Endive	93.1	20	1.7	0.1	4.1	0.9	81	54	1.7	14	294	3,300	0.07	0.14	0.5	10
Garlic	61.3	137	6.2	0.2	30.8	1.5	29	202	1.5	19	529	Trace	0.25	0.08	0.5	15
Jerusalem artichoke	79.8	f	2.3	0.1	16.7c	0.8	14	78	3.4	—	—	20	0.20	0.06	1.3	4
Kale—leaves and stems	87.5	38	4.2	0.8	6.0	1.3	179	73	2.2	75	378	8,900	—	—	—	125
Kohlrabi	90.3	29	2.0	0.1	6.6	1.0	41	51	0.5	8	372	20	0.06	0.04	0.3	66
Leeks	85.4	52	2.2	0.3	11.2	1.3	52	50	1.1	5	347	40	0.11	0.06	0.5	17
Lettuce—butter head	95.1	14	1.2	0.2	2.5	0.5	35	26	2.0	9	264	970	0.06	0.06	0.3	8
Lettuce—crisp head	95.5	13	0.9	0.1	2.9	0.5	20	22	0.5	9	175	330	0.06	0.06	0.3	6
Lettuce—loose leaf	94.0	18	1.3	0.3	3.5	0.7	68	25	1.4	9	264	1,900	0.05	0.08	0.4	18
Melons—cantaloupes	91.2	30	0.7	0.1	7.5	0.3	14	16	0.4	12	251	3,400g	0.04	0.03	0.6	33
Melons—casaba	91.5	27	1.2	Trace	6.5	0.5	(14)	(16)	(0.4)	(12)	(251)	30	(0.04)	(0.03)	(0.6)	13
Melons—honeydew	90.6	33	0.8	0.3	7.7	0.6	14	16	0.4	12	251	40	0.04	0.03	0.6	23
Melons—watermelon	92.6	26	0.5	0.2	6.4	0.3	7	10	0.5	1	100	590	0.03	0.03	0.2	7
Okra	88.9	36	2.4	0.3	7.6	1.0	0.8	92	51	0.6	3	520	(0.17)	(0.21)	(1.0)	31
Onions—Welsh	90.5	34	1.9	0.4	6.5	1.0	18	40	—	—	—	—	0.05	0.09	0.4	27
Onions—white globe	89.1	38	1.5	0.1	8.7	0.6	27	36	0.5	10	157	40h	0.03	0.04	0.2	10
Onions—young green	89.4	36	1.5	0.2	8.2	1.2	51	39	1.0	5	231	(2,000)	0.05	0.05	0.4	32
Oranges—navel	85.4	51	1.3	0.3	12.7	0.5	40	22	0.4	1	194	(200)	0.10	0.04	0.4	(61)
Oranges—Valencia	85.6	51	1.2	0.3	12.4	(0.5)	40	22	0.8	1	190	(200)	0.10	0.04	0.4	(49)
Parsley	85.1	44	3.6	0.6	8.5	1.5	203	63	6.2	45	727	8,500	0.12	0.26	1.2	172
Parsnips	79.1	76	1.7	0.5	17.5	2.0	50	77	0.7	12	541	30	0.08	0.09	0.2	16
Peaches	89.1	38	0.6	0.1	9.7	0.6	9	19	0.5	1	202	1,330i	0.02	0.05	1.0	7
Peanuts—with skin	5.6	564	26.0	47.5	18.6	2.4	69	401	2.1	5	674	—	1.14	0.13	17.2	0
Pears	83.2	61	0.7	0.4	15.3	1.4	8	11	0.3	2	130	20	0.02	0.04	0.1	4
Peas—edible podded	83.3	53	3.4	0.2	12.0	1.2	62	90	0.7	—	170	(680)	0.28	0.12	—	21
Peas—green	78.0	84	6.3	0.4	14.4	2.0	26	116	1.9	2	316	640	0.35	0.14	2.9	27
Peppers—sweet green	93.4	22	1.2	0.2	4.8	1.4	9	22	0.7	13	213	420	0.08	0.08	0.5	128
Plums—Japanese and hybrid	86.6	48	0.5	0.2	12.3	0.6	12	18	0.5	1	170	250	0.03	0.03	0.5	6
Potatoes	79.8	76	2.1	0.1	17.1	0.5	7	53	0.6	3	407	Trace	0.10	0.04	1.5	20
Pumpkin	91.6	26	1.0	0.1	6.5	1.1	21	44	0.8	1	340	1,600	0.05	0.11	0.6	9

Raw vegetable or fruit 100 grams (3½ oz.)	Water (%)	Food energy (cal)	Protein (g)	Fat (g)	Carbohydrate Total (g)	Carbohydrate Fiber (g)	Calcium (mg)	Phosphorus (mg)	Iron (mg)	Sodium (mg)	Potassium (mg)	Vitamin A (I.U.)	Thiamine (mg)	Riboflavin (mg)	Niacin (mg)	Ascorbic acid (mg)
Radishes—common	94.5	17	1.0	0.1	3.6	0.7	30	31	1.0	18	322	10	0.03	0.03	0.3	26
Raspberries—red	84.2	57	1.2	0.5	13.6	3.0	22	22	0.9	1	168	130	0.03	0.09	0.9	25
Rhubarb	94.8	16	0.6	0.1	3.7	0.7	96	18	0.8	2	251	100	(0.03)	(0.07)	(0.3)	9
Rutabagas	87.0	46	1.1	0.1	11.0	1.1	66	39	0.4	5	239	580	0.07	0.07	1.1	43
Salsify	77.6	—ʲ	2.9	0.6	18.0ᶜ	1.8	47	66	1.5	—	380	10	0.04	0.04	0.3	11
Shallots	79.8	72	2.5	0.1	16.8	0.7	37	60	1.2	12	334	Trace	0.06	0.02	0.2	8
Sorrel	90.9	28	2.1	0.3	5.6	0.8	66	41	1.6	5	338	12,900	0.09	0.22	0.5	119
Soybeans—mature seeds	10.0	403	34.1	17.7	33.5	4.9	226	554	8.4	5	1,677	80	1.10	0.31	2.2	—
Spinach	90.7	26	3.2	0.3	4.3	0.6	93	51	3.1	71	470	8,100	0.10	0.20	0.6	51
Squash—summer	94.0	19	1.1	0.1	4.2	0.6	28	29	0.4	1	202	410	0.05	0.09	1.0	22
Squash—winter	85.1	50	1.4	0.3	12.4	1.4	22	38	0.6	1	369	3,700	0.05	0.11	0.6	13
Strawberries	89.9	37	0.7	0.5	8.4	1.3	21	21	1.0	1	164	60	0.03	0.07	0.6	59
Sweet potatoes	70.6	114	1.7	0.4	26.3	0.7	32	47	0.7	10	243	8,800	0.10	0.06	0.6	21
Tomatoes—red	93.5	22	1.1	0.2	4.7	0.5	13	27	0.5	3	244	900	0.06	0.04	0.7	23
Turnips	91.5	30	1.0	0.2	6.6	0.9	39	30	0.5	49	268	Trace	0.04	0.07	0.6	36
Turnip greens	90.3	28	3.0	0.3	5.0	0.8	246	58	1.8	(3)	—	7,600	(0.21)	(0.39)	(0.8)	139
Wheat—whole grains, hard red spring	13.0	330	14.0	2.2	69.1	2.3	36	383	3.1	(3)	370	(0)	0.57	0.12	4.3	(0)
Yams	73.5	101	2.1	0.2	23.2	0.9	20	69	0.6	—	600	Trace	0.10	0.04	0.5	9

ᵃFrom *Agricultural Handbook No. 8*, 1963 U.S. Department of Agriculture.

ᵇValues may range from 9 calories per 100 grams for freshly harvested raw artichokes to as many as 47 for stored product.

ᶜA large proportion of the carbohydrate in the unstored product may be inulin, which is of doubtful availability.

ᵈAverage for all cultivars. For green cultivars, value is 270 I.U. per 100 grams; for yellow cultivars, 140 I.U.

ᵉBased on yellow cultivars; white cultivars contain only a trace of cryptoxanthin and carotenes, the pigments in corn that have biological activity.

ᶠValues range from 7 calories per 100 grams for freshly harvested Jerusalem artichokes to 75 calories for those stored for a long period.

ᵍValue based on cultivars with orange-colored flesh; for green-fleshed cultivars, value is about 280 I.U. per 100 grams.

ʰValue based on yellow-fleshed cultivars; white-fleshed cultivars contain only a trace.

ⁱBased on yellow-fleshed cultivars; for white-fleshed cultivars, value is about 50 I.U. per 100 grams.

ʲValues for raw salsify range from 13 calories per 100 grams for the freshly harvested vegetable to 82 calories for the product after storage.

INDEX

Throughout the index the following special notations are used: 1) **bold face** numbers identify pages where words or concepts are defined in the text proper; 2) *italicized* numbers identify pages having general discussion of individual crops; 3) numbers followed by "t" identify tables; those followed by "f", figures.

Acid, **121**. *See also* pH
Adhesion, **102**
Aerobic, **126**
African daisy (*Arctotis*), 402t, 423, 424t
African marigold (*Tagetes*), 409t
African violet (*Saintpaulia*), 16, 59
Ageratum (*Ageratum*), 400, 401, 402t, 415t, 426, 427, 435
Agriculture:
  origins of, 3
  sources of information on, 512–515
Ajuga (*Ajuga*), 430t, 436
Albedo, **144**
Albuminous seeds, **74**
Alfalfa (*Medicago*), 126, 233
Alkaline, **121**. *See also* pH
All-America Selections, 174–175, 293, 436–437
Almond (*Prunus*), 60, 107t, 150t, 151t *393–394*
Altitude, **154**
Amaranth (*Amaranthus*), 5, 420, 423
Anaerobic, **126**
  composting, 187
Angelica (*Angelica*), 331t
Animals, pest, 268, 280–284, 377, 417, 425
Anion, **117**
Anise (*Pimpinella*), 331t
Annuals, **17**
  as flowers, 402t–410t
  grasses, 438
  as landscape plants, 438
  as weeds, 278
  woody, 438
Antibiotic, **265**
Antibodies, **481**
Antioxidant, **498**

Apical dominance, **52**
Apical meristem, **26**
Apple (*Malus*), 4, 18, 34, 35f, 49, 50, 141, 355, 358, *371–382*, 530t
  characteristics of, 17t, 68f, 76, 77f, 107t, 151t, 156t, 371–375, 490
  culture of, 50t, 60, 70, 73t, 134, 357, 375–380
  pests of, 246, 260, 261, 262, 273
  pruning and training, 377–380, 449, 450f
  as specialty crops, 446, 449
Apricot (*Prunus*):
  culture of, 50t, 60, 73t, 107t, 140, 151t, 157t
  harvest and processing of, 491, 495, 496, 502t, 530t
Arachnids, **268**
  as garden pests, 381
Artichoke, globe (*Cynara*), 6, 289, *323–324*, 359, 447t
  characteristics of, 17t, 28t, 66t, 150t
  culture of, 50t, 447t
  harvest and processing, 492t, 530t
Asexual reproduction, **49**
  artificial methods of, 56–62
  using natural structures, 50–56
Asparagus (*Asparagus*), 16, 17, 28, 31, 77, 167, 170, 289, *324–325*, 359, 443, 477
  as specialty crop, 447t
  characteristics of 17t, 28t, 64, 66–67t, 102f, 107t, 124t
  culture of, 24, 50t, 56, 192t, 194t, 203
  harvest and processing of, 166t, 171t, 325, 478, 490, 492, 493, 496, 497, 498, 530t

pests of, 266, 268
Assimilates, **43**
  movement in plants, 44–45
Aster (*Aster*), 16, 401
  culture of, 56, 430t, 432, 436
  pests of, 260
Aster yellows, **424**
Atmospheric circulation, 144–146
Atom, **115**
Auxin, **45**
Avocado (*Persea*), 5, 114, 140
  characteristics of, 76, 107t, 151t
  harvest and processing of, 480, 491, 530t
Azalea (*Rhododendron*), 434, 436

Baby's breath (*Gypsophila*), 405t, 423
Bachelor's button (*Centaurea*), 400, 403t, 414, 415t
  characteristics of, 418t, 424t
  culture of, 420, 423, 426
Bacillus thuringiensis, as natural predators, 255, 256
Baked-apple berry (*Rubus*), 364
Balm (*Melissa*), 331t
Balsam (*Impatiens*), 400
Bamboo, 51, 446
Banana (*Musa*), 139, 156, 446
  characteristics of, 74, 76, 150t, 151t
  culture of, 50t, 51
  harvest and processing, 491, 530t
Bark, **37**
Barley (*Hordeum*), 4, 19, 98t
Basic, **121**. *See also* pH
Basil (*Ocimum*), 331t, *337–338*, 446
Bean (*Phaseolus*), 5, 17, 19, 25, 26f, 30, 33, 126, 140, 141, 154, 164, 165f, 167f, 168, 170, 196, 197,